ADVANCES IN NATURAL GAS: FORMATION, PROCESSING, AND APPLICATIONS

ADVANCES IN NATURAL GAS: FORMATION, PROCESSING, AND APPLICATIONS

Natural Gas Sweetening

Volume 2

Edited by

MOHAMMAD REZA RAHIMPOUR
Department of Chemical Engineering, Shiraz University, Shiraz, Iran

MOHAMMAD AMIN MAKAREM
Department of Chemical Engineering, Shiraz University, Shiraz, Iran

MARYAM MESHKSAR
Department of Chemical Engineering, Shiraz University, Shiraz, Iran

Elsevier
Radarweg 29, PO Box 211, 1000 AE Amsterdam, Netherlands
125 London Wall, London EC2Y 5AS, United Kingdom
50 Hampshire Street, 5th Floor, Cambridge, MA 02139, United States

Copyright © 2024 Elsevier Inc. All rights are reserved, including those for text and data mining, AI training, and similar technologies.

No part of this publication may be reproduced or transmitted in any form or by any means, electronic or mechanical, including photocopying, recording, or any information storage and retrieval system, without permission in writing from the publisher. Details on how to seek permission, further information about the Publisher's permissions policies and our arrangements with organizations such as the Copyright Clearance Center and the Copyright Licensing Agency, can be found at our website: www.elsevier.com/permissions.

This book and the individual contributions contained in it are protected under copyright by the Publisher (other than as may be noted herein).

Notices
Knowledge and best practice in this field are constantly changing. As new research and experience broaden our understanding, changes in research methods, professional practices, or medical treatment may become necessary.

Practitioners and researchers must always rely on their own experience and knowledge in evaluating and using any information, methods, compounds, or experiments described herein. In using such information or methods they should be mindful of their own safety and the safety of others, including parties for whom they have a professional responsibility.

To the fullest extent of the law, neither the Publisher nor the authors, contributors, or editors, assume any liability for any injury and/or damage to persons or property as a matter of products liability, negligence or otherwise, or from any use or operation of any methods, products, instructions, or ideas contained in the material herein.

ISBN: 978-0-443-19217-3

For information on all Elsevier publications visit our website at
https://www.elsevier.com/books-and-journals

Publisher: Candice Janco
Senior Acquisitions Editor: Anita Koch
Editorial Project Manager: Anthony Marvullo
Production Project Manager: Paul Prasad Chandramohan
Cover Designer: Christian Bilbow

Typeset by TNQ Technologies

Contents

Contributors xiii
About the editors xvii
Preface xix
Reviewer acknowledgments xxi

Section I Natural gas sweetening concepts

Chapter 1 Introduction to natural gas sweetening methods and technologies 3
Syed Ali Ammar Taqvi and Aisha Ellaf

   1. Introduction 3
   2. Sweetening techniques 4
   3. Selection of the most suitable sweetening technique 24
   4. Conclusion and future outlooks 27
      Abbreviations and symbols 28
      References 29

Chapter 2 Natural gas sweetening standards, policies, and regulations 33
Nadia Khan and Syed Ali Ammar Taqvi

   1. Introduction 33
   2. Environmental policies and regulations 35
   3. Regulations for controlling emissions from equipment 42
   4. Operational and design standards for various sweetening processes 44
   5. Conclusion and future outlooks 48
      Abbreviations and symbols 49
      References 49

Chapter 3 Economic assessments and environmental challenges of natural gas sweetening technologies 55
Nadia Khan and Syed Ali Ammar Taqvi
   1. Introduction 55
   2. Methods of economic analysis 57
   3. Techno-economic analysis of natural gas sweetening process 59
   4. Environmental challenges 67
   5. Conclusion and future outlooks 68
   Abbreviations and symbols 69
   References 70

Section II Absorption techniques for natural gas sweetening

Chapter 4 Acid gases properties and characteristics in companion with natural gas 75
Sina Mosallanezhad, Hamid Reza Rahimpour and Mohammad Reza Rahimpour
   1. Introduction 75
   2. Natural gas origins 77
   3. Natural gas composition 77
   4. Removing acid gases from natural gas 78
   5. Definition of physical and chemical properties 79
   6. Physical and chemical properties of CO_2 and H_2S 84
   7. Conclusion and future outlooks 85
   Abbreviations and symbols 86
   References 86

Chapter 5 Application of amines for natural gas sweetening 89
Abdul Rahim Nihmiya and Nayef Ghasem
   1. Introduction 89
   2. Amine-based techniques for acid gas removal 92
   3. Amine-based absorption process 97

4. Current applications and cases 102
   5. Conclusion and future outlooks 108
   Abbreviations and symbols 110
   References 110

Chapter 6 Physical and hybrid solvents for natural gas sweetening: Ethers, pyrrolidone, methanol and other sorbents 115
Samuel Eshorame Sanni, Babalola Aisosa Oni and Emeka Emmanuel Okoro
   1. Introduction 115
   2. Physical and hybrid sorbents for NG sweetening 116
   3. Mechanism of solute take-up by physical and hybrid sorbents ... 126
   4. Conclusion and future outlooks 127
   Abbreviations and symbols 127
   References 128

Chapter 7 Natural gas sweetening by solvents modified with nanoparticles 135
Moloud Rahimi, Maryam Meshksar and Mohammad Reza Rahimpour
   1. Introduction 135
   2. NG sweetening techniques 139
   3. Conclusion and future outlooks 148
   Abbreviations and symbols 149
   References 149

Chapter 8 Encapsulated liquid sorbents for sweetening of natural gas 153
Babak Emdadi and Rasoul Moradi
   1. Introduction 153
   2. Encapsulated liquid sorbents 157
   3. NG sweetening using encapsulated liquids 166
   4. Conclusion and future outlooks 174

Abbreviations and symbols 175
References 176

Chapter 9 Cryogenic fractionation for natural gas sweetening.........185
Juan Pablo Gutierrez, Fabiana Belén Torres and Eleonora Erdmann

   1. Introduction 185
   2. Thermodynamic principles of CO_2 and CH_4 separation 187
   3. Cryogenic processes for acid gas removal 189
   4. Current applications and improvements 199
   5. Conclusion and future outlooks 202
   Abbreviations and symbols 202
   References 203
   Further reading... 205

Chapter 10 Absorption processes for CO_2 removal from CO_2-rich natural gas...207
Ali Behrad Vakylabad

   1. Introduction 207
   2. Amine absorption 212
   3. CO_2 removal flowsheet... 223
   4. CO_2 absorption plant: Unit operations and parameters 224
   5. Current applications and cases................................. 227
   6. Specific characterizations and properties of CO_2 absorption .. 234
   7. Novel methods and solvents..................................... 235
   8. Scales up of the ionic liquid–based technologies............... 237
   9. Conclusion and future outlooks 239
   Abbreviations and symbols 240
   References 241

Section III Adsorption techniques for natural gas sweetening

Chapter 11 Swing technologies for natural gas sweetening: Pressure, temperature, vacuum, electric, and mixed swing processes 261

Meisam Ansarpour and Masoud Mofarahi

    1. Introduction 261
    2. Swing adsorption processes 264
    3. Conclusion and future outlooks 313
  Abbreviations and symbols 313
  References 314

Chapter 12 Zeolite sorbents and nanosorbents for natural gas sweetening 329

Maryam Koohi-Saadi and Mohammad Reza Rahimpour

    1. Introduction 329
    2. Natural gas purification techniques 332
    3. Conclusion and future outlooks 352
  Abbreviations and symbols 352
  References 353

Chapter 13 Porous metal structures, metal oxides, and silica-based sorbents for natural gas sweetening 359

Mohammad Rahmani and Fatemeh Boshagh

    1. Introduction 359
    2. Metal-based sorbents 362
    3. Silica-based sorbents 375
    4. Conclusion and future outlooks 382
  Abbreviations and symbols 383
  References 384

Chapter 14 Natural gas CO$_2$-rich sweetening via adsorption processes........395
Syed Ali Ammar Taqvi, Durreshehwar Zaeem and Haslinda Zabiri

    1. Introduction 395
    2. Adsorption processes in natural gas sweetening 396
    3. Adsorbent material selection 401
    4. Conclusion and future outlooks 409
    Abbreviations and symbols 413
    References 413

Section IV Membrane technology for natural gas sweetening
Chapter 15 Polymeric membranes for natural gas sweetening........ 419
Abdul Latif Ahmad, Muhd Izzudin Fikry Zainuddin and
Meor Muhammad Hafiz Shah Buddin

    1. Introduction 419
    2. Fundamentals of polymeric membrane gas separation 420
    3. Ideal membrane 428
    4. Current application 430
    5. Challenges 438
    6. Conclusion and future outlooks 442
    Abbreviations and symbols 443
    References 444

Chapter 16 Natural gas sweetening by ionic liquid membranes 453
Girma Gonfa and Sami Ullah

    1. Introduction 453
    2. Ionic liquid membranes 455
    3. Conclusion and future outlooks 465
    Abbreviations and symbols 465
    References 466

Chapter 17 Application of electrochemical membranes for natural gas sweetening 471

Fatemeh Haghighatjoo, Behnaz Rahmatmand and Mohammad Reza Rahimpour

    1. Introduction 471
    2. Removal of H_2S through an electrochemical membrane separator 473
    3. Removal of CO_2 through an electrochemical membrane separator 475
    4. Electrode preparation 480
    5. Membrane preparation 481
    6. Conclusion and future outlooks 482
    Abbreviations and symbols 483
    References 483

Chapter 18 Membrane technology for CO_2 removal from CO_2-rich natural gas 487

Shaik Muntasir Shovon, Faysal Ahamed Akash, Minhaj Uddin Monir, Mohammad Tofayal Ahmed and Azrina Abd Aziz

    1. Introduction 487
    2. Fundamentals of membrane gas separation for CO_2 removal ... 488
    3. Membrane processes for efficient CO_2 removal 489
    4. Current application and cases 493
    5. Conclusion and future outlooks 499
    Abbreviations and symbols 502
    References 502

Index 509

Contributors

Abdul Latif Ahmad
School of Chemical Engineering, Universiti Sains Malaysia Engineering Campus, Nibong Tebal, Pulau Pinang, Malaysia

Mohammad Tofayal Ahmed
Department of Petroleum and Mining Engineering, Jashore University of Science and Technology, Jashore, Bangladesh; Energy Conversion Laboratory, Department of Petroleum and Mining Engineering, Jashore University of Science and Technology, Jashore, Bangladesh

Faysal Ahamed Akash
Department of Petroleum and Mining Engineering, Jashore University of Science and Technology, Jashore, Bangladesh

Meisam Ansarpour
Department of Chemical Engineering, Faculty of Petroleum, Gas and Petrochemical Engineering, Persian Gulf University, Bushehr, Iran

Azrina Abd Aziz
Faculty of Civil Engineering Technology, Universiti Malaysia Pahang Al-Sultan Abdullah, Kuantan, Pahang, Malaysia

Fatemeh Boshagh
Department of Chemical Engineering, Amirkabir University of Technology (Tehran Polytechnic), Tehran, Iran

Aisha Ellaf
Department of Chemical Engineering, NED University of Engineering and Technology, Karachi, Pakistan

Babak Emdadi
Nanotechnology Laboratory, School of Science and Engineering, Khazar University, Baku, Azerbaijan

Eleonora Erdmann
Instituto de Investigaciones para la Industria Química INIQUI (CONICET-UNSA), Salta, Argentina; Facultad de Ingeniería—Consejo de Investigación, Universidad Nacional de Salta, Salta, Argentina

Nayef Ghasem
Department of Chemical and Petroleum Engineering, United Arab Emirates University, Al-Ain, United Arab Emirates

Girma Gonfa
Department of Chemical Engineering, Addis Ababa Science and Technology University, Addis Ababa, Ethiopia; Biotechnology and Bioprocess Centre of Excellence, Addis Ababa Science and Technology University

Juan Pablo Gutierrez
Instituto de Investigaciones para la Industria Química INIQUI (CONICET-UNSA), Salta, Argentina; Facultad de Ingeniería—Consejo de Investigación, Universidad Nacional de Salta, Salta, Argentina

Fatemeh Haghighatjoo
Department of Chemical Engineering, Shiraz University, Shiraz, Iran

Nadia Khan
Department of Polymer and Petrochemical Engineering, NED University of Engineering and Technology, Karachi, Pakistan

Maryam Koohi-Saadi
Department of Chemical Engineering, Shiraz University, Shiraz, Iran

Maryam Meshksar
Department of Chemical Engineering, Shiraz University, Shiraz, Iran

Masoud Mofarahi
Department of Chemical Engineering, Faculty of Petroleum, Gas and Petrochemical Engineering, Persian Gulf University, Bushehr, Iran; Department of Chemical and Biomolecular Engineering, Yonsei University, Seoul, South Korea

Minhaj Uddin Monir
Department of Petroleum and Mining Engineering, Jashore University of Science and Technology, Jashore, Bangladesh; Energy Conversion Laboratory, Department of Petroleum and Mining Engineering, Jashore University of Science and Technology, Jashore, Bangladesh

Rasoul Moradi
Nanotechnology Laboratory, School of Science and Engineering, Khazar University, Baku, Azerbaijan

Sina Mosallanezhad
Department of Chemical Engineering, Shiraz University, Shiraz, Iran

Abdul Rahim Nihmiya
Department of Civil and Environmental Technology, University of Sri Jayewardenepura, Nugegoda, Sri Lanka

Emeka Emmanuel Okoro
Department of Petroleum and Gas Engineering, University of Port Harcourt, Choba, Rivers State, Nigeria

Babalola Aisosa Oni
Department of Chemical Engineering, Covenant University, Ota, Ogun State, Nigeria; Department of Energy Engineering, University of North Dakota, Grand Forks, ND, United States

Moloud Rahimi
Department of Chemical Engineering, Shiraz University, Shiraz, Iran

Mohammad Reza Rahimpour
Department of Chemical Engineering, Shiraz University, Shiraz, Iran

Hamid Reza Rahimpour
Department of Chemical Engineering, Shiraz University, Shiraz, Iran

Mohammad Rahmani
Department of Chemical Engineering, Amirkabir University of Technology (Tehran Polytechnic), Tehran, Iran

Behnaz Rahmatmand
Department of Chemical Engineering, Shiraz University, Shiraz, Iran

Samuel Eshorame Sanni
Department of Chemical Engineering, Covenant University, Ota, Ogun State, Nigeria

Meor Muhammad Hafiz Shah Buddin
School of Chemical Engineering, Universiti Sains Malaysia Engineering Campus, Nibong Tebal, Pulau Pinang, Malaysia; School of Chemical Engineering, College of Engineering, Universiti Teknologi MARA, Shah Alam, Selangor, Malaysia

Shaik Muntasir Shovon
Department of Petroleum and Mining Engineering, Jashore University of Science and Technology, Jashore, Bangladesh

Syed Ali Ammar Taqvi
Department of Chemical Engineering, NED University of Engineering and Technology, Karachi, Pakistan

Fabiana Belén Torres
Instituto de Investigaciones para la Industria Química INIQUI (CONICET-UNSA), Salta, Argentina

Sami Ullah
Department of Chemistry, College of Science, King Khalid University, Abha, Saudi Arabia

Ali Behrad Vakylabad
Department of Materials, Institute of Science and High Technology and Environmental Sciences, Graduate University of Advanced Technology, Kerman, Iran

Haslinda Zabiri
Department of Chemical Engineering, Universiti Teknologi PETRONAS, Bandar Seri Iskandar, Perak, Malaysia; CO_2RES, Institute of Contaminant Management (ICM), Universiti Teknologi PETRONAS, Bandar Seri Iskandar, Perak, Malaysia

Durreshehwar Zaeem
Department of Chemical Engineering, NED University of Engineering and Technology, Karachi, Pakistan

Muhd Izzudin Fikry Zainuddin
School of Chemical Engineering, Universiti Sains Malaysia Engineering Campus, Nibong Tebal, Pulau Pinang, Malaysia

About the editors

Mohammad Reza Rahimpour
Department of Chemical Engineering, Shiraz University, Shiraz, Iran

Prof. Mohammad Reza Rahimpour is a Professor of Chemical Engineering at Shiraz University, Iran. He received his Ph.D. in Chemical Engineering from Shiraz University joint with the University of Sydney, Australia 1988. He started his independent career as an Assistant Professor in September 1998 at Shiraz University. Prof. M.R. Rahimpour was a Research Associate at the University of California, Davis, from 2012 to 2017. During his stay in the University of California, he developed different reaction networks and catalytic processes such as thermal and plasma reactors for upgrading of lignin biooil to biofuel with the collaboration of UCDAVIS. He has been a Chair of the Department of Chemical Engineering at Shiraz University from 2005 to 2009 and from 2015 to 2020. Prof. M.R. Rahimpour leads a research group in fuel processing technology focused on the catalytic conversion of fossil fuels such as natural gas and renewable fuels such as biooils derived from lignin to valuable energy sources. He provides young distinguished scholars with perfect educational opportunities in both experimental methods and theoretical tools in developing countries to investigate in-depth research in various fields of chemical engineering including carbon capture, chemical looping, membrane separation, storage and utilization technologies, novel technologies for natural gas conversion, and improving the energy efficiency in the production and use of natural gas industries.

Mohammad Amin Makarem
Department of Chemical Engineering, Shiraz University, Shiraz, Iran

Dr. Mohammad Amin Makarem is a Research Associate at Methanol Institute, Shiraz University. His research interests are gas separation and purification, nanofluids, microfluidics, catalyst synthesis, reactor design, and green energy. In gas separation, his focus is on experimental and theoretical investigation and optimization of pressure swing adsorption process, and in the gas purification field, he is working on novel technologies such as microchannels. Recently, he has investigated the methods of synthesizing biotemplate nanomaterials and catalysts. Besides, he has collaborated in writing and editing various books and book chapters for famous publishers such as Elsevier, Springer, and Wiley.

Maryam Meshksar
Department of Chemical Engineering, Shiraz University, Shiraz, Iran

Dr. Maryam Meshksar is a Research Associate at Shiraz University. Her research has focused on gas separation, clean energy, and catalyst synthesis. In gas separation, she is working on membrane separation process, and in the clean energy field, she has worked on different reforming-based processes for syngas production from methane experimentally. She has also synthesized novel catalysts for these processes which are tested in for the first time. Besides, she has reviewed novel technologies like microchannels for energy production. Recently, she has collaborated in writing and editing various books and book chapters for famous publishers such as Elsevier, Springer, and Wiley.

Preface

Natural gas, with its abundance, cleanliness, and economic advantages, has emerged as a crucial energy source that is extensively utilized worldwide. Composed mainly of methane, natural gas can be found in underground reservoirs or associated with crude oil. The exploration, production, transportation, and utilization of natural gas require multidisciplinary efforts and technological advancements to ensure efficient and sustainable operations. To fully exploit the potential of natural gas, a thorough understanding of its properties, extraction, processing, transportation, storage, and applications is essential.

The *Advances in Natural Gas* book series serves as a comprehensive and up-to-date reference for researchers, engineers, professionals, students, and academics involved in the natural gas industry. It provides a valuable resource for those interested in delving into the realm of natural gas science and technology.

Natural gas sweetening is a crucial step in the processing of natural gas that ensures the removal of unwanted impurities, such as hydrogen sulfide (H_2S) and carbon dioxide (CO_2), that can cause corrosion, decrease the heating value of the gas, and pose health and safety risks. Sweetening technologies have evolved over the years, ranging from absorption and adsorption techniques to membrane technology. The selection of a suitable sweetening technology depends on various factors, including the composition of the feed gas, the desired product specifications, the availability of resources, and the economic and environmental considerations.

This volume of the *Advances in Natural Gas* series, titled *Natural Gas Sweetening*, offers a comprehensive overview of the recent advances in sweetening technologies. It comprises a collection of chapters that provide insights into the concepts, standards, policies, and regulations associated with natural gas sweetening. It also explores the economic assessments and environmental challenges of natural gas sweetening technologies.

The volume is structured into four sections, each delving into different aspects of natural gas sweetening technologies. The first section serves as an introduction to natural gas sweetening, covering the basic concepts and technologies involved in sweetening processes. It highlights the importance of sweetening technologies in ensuring the quality and safety of natural gas for various applications. This section also provides an overview of the economic and environmental considerations associated with sweetening technologies.

The second section of the volume explores absorption techniques for natural gas sweetening, including the use of amines, physical and hybrid solvents, and solvents modified with nanoparticles. The section also delves into the use of encapsulated liquid sorbents and cryogenic-fractionation for natural gas sweetening. Additionally, it provides an overview of the absorption processes for CO_2 removal from CO_2-rich natural gas.

The third section of the volume focuses on adsorption techniques for natural gas sweetening, including swing technologies and the use of zeolite sorbents, nanosorbents, metal oxides, and silica-based sorbents. This section also explores the adsorption processes for CO_2-rich natural gas sweetening.

The fourth and final section of the volume addresses membrane technology for natural gas sweetening, including the use of polymeric membranes, ionic liquid membranes, dense metal membranes, and electrochemical membranes. This section also explores membrane technology for CO_2 removal from CO_2-rich natural gas.

In summary, this volume of the *Advances in Natural Gas* series provides a comprehensive overview of natural gas sweetening technologies. It not only highlights the potential of sweetening technologies in ensuring the quality and safety of natural gas but also addresses the challenges associated with their production and extraction. This volume is structured to provide valuable insights to researchers, engineers, professionals, students, and academics involved in the natural gas industry. We invite you to delve into the fascinating world of natural gas sweetening and explore the immense potential of these technologies for sustainable energy practices.

Mohammad Reza Rahimpour
Mohammad Amin Makarem
Maryam Meshksar

Reviewer acknowledgments

The editors feel obliged to appreciate the dedicated reviewers (listed below) who were involved in reviewing and commenting on the submitted chapters and whose cooperation and insightful comments were very helpful in improving the quality of the chapters and books in this series.

Dr. Fatemeh Haghighatjoo
Department of Chemical Engineering, Shiraz University, Shiraz, Iran

Ms. Parvin Kiani
Department of Chemical Engineering, Shiraz University, Shiraz, Iran

Ms. Fatemeh Zarei-Jelyani
Department of Chemical Engineering, Shiraz University, Shiraz, Iran

Ms. Fatemeh Salahi
Department of Chemical Engineering, Shiraz University, Shiraz, Iran

Natural gas sweetening concepts

Introduction to natural gas sweetening methods and technologies

Syed Ali Ammar Taqvi and Aisha Ellaf
Department of Chemical Engineering, NED University of Engineering and Technology, Karachi, Pakistan

1. Introduction

Natural gas (NG) is a clean energy alternative, and the qualities of being cheap, giving fewer emissions, and not producing any ash make it a cleaner source of energy compared to other fossil fuels. Of course, NG cannot beat renewable energy sources in this run, but it is considered a stopgap energy solution for the world until renewable energy sources are fully developed [1]. With the advancement and research on purification techniques, consumption of NG is expected to increase by another 15% by 2030, eventually replacing oil and coal [2]. NG is used in electricity generation and as fuel for process heating in combined heat and power systems, cooking, drying, refrigeration purposes, and operating compressors [3].

The composition of NG varies in different regions and reservoirs based on different underground conditions and extraction procedures. NG occurs as associated and nonassociated gas from reservoirs with and without crude oil, respectively. A major chunk of NG comprises methane, and it furthermore consists of small proportions of higher hydrocarbons, e.g., C2−C7 with impurities namely N_2, H_2O, H_2S, and CO_2. A subquality standard for NG is defined as one having $CO_2 \geq 2\%$, $N_2 \geq 4\%$, or $H_2S \geq 4$ ppm [4,5].

Removal of N_2 from NG increases its gross calorific value and provides ease in gas processing. The most efficient method for N_2 removal is cryogenic distillation. Dehydration separates water from NG to avoid corrosion, hydrate formation, and oversaturation of NG and thus meet the pipeline specifications [6,7].

H_2S and CO_2 together makeup the acid gas and must necessarily be removed from NG to:
- Meet the pipeline specifications to avoid choking, corrosion, and pipe damage, thus resulting in pollution- and corrosion-free gas processing.
- Comply with environmental standards, thus making it less harmful for human beings.
- Increase the calorific value of NG because calorific value depends on methane content; lesser the impurities and greater the methane content, higher will be the calorific value.
- Decrease the transportation and processing costs [6,8].

The process of removal of H_2S, CO_2, and mercaptans from sour NG is termed "natural gas sweetening." Acidic emissions being such a global concern have boosted the number of research studies and experimentation on the development and advancement of NG sweetening techniques to come up with such techniques that would result in efficient, ecofriendly, and maximum removal of these impurities, keeping in view the cost constraints [9].

2. Sweetening techniques

A number of NG sweetening technologies are now available and summarized in Fig. 1.1; the main technologies are categorized as:
- Membrane separation
- Absorption
- Adsorption
- Cryogenics
- Chemical looping

Research for the absorption process is primarily focused on the production, combination, and improving performance of various solvents. The focus of adsorption technology is on novel and altered materials. For improved performance, membrane-based research employs membranes made of various materials, such as composite and hybrid membrane [10,11].

2.1 Membrane separation

Membranes are semipermeable barriers employed for bulk removal of impurities from sour NG. In membranes, separation is based on the permeation rate difference of different components of sour NG through the membranes and the selectivity of

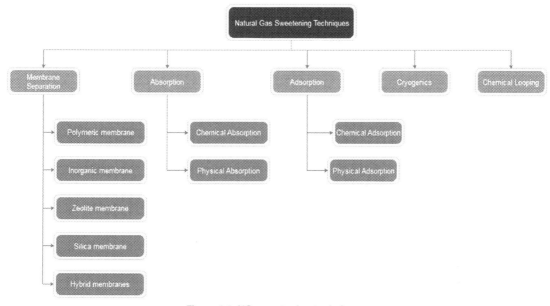

Figure 1.1 NG sweetening techniques.

the membrane to pass one component over the other. The permeation rate of each component in turn depends on [12]:
- Component characteristics
- Membrane characteristics
- Partial pressure difference or concentration difference of components across the membrane

Components that pass through the membrane are called "permeate" and those retained are called "retentate or residue." Depending on the requirements and the availability, either permeate or the residue can be the desired product; in our case of NG sweetening, the residue, that is, the sweet NG, is the desired product [13].

Membrane technology must fulfill certain requirements in order to be employed for NG sweetening [13]:
- High permeation rate and high selectivity for species to be removed in comparison with other species
- High membrane stability in the presence of contacting gas components
- Membranes must be uniform and free of defects
- Small thickness of the membrane's active region for high penetration rates
- Capable to withstand tough operating conditions

Membrane separation takes place by the following mechanisms [12]:
- Knudsen diffusion
- Molecular sieving
- Solution–diffusion separation
- Surface diffusion
- Capillary condensation

2.1.1 Knudsen diffusion

This type of diffusion occurs when the mean free path of gas molecules is larger than the size of the pores through which they are traveling, causing the molecules to collide more with the walls rather than with each other as shown in Fig. 1.2 [14]. Knudsen diffusivity is given by Eq. (1.1):

$$D_K = \left(\frac{2}{3}\right) r_p v_T \quad (1.1)$$

where

r_p = radius of the passage

$v_T = \left(\frac{8RT}{\pi M}\right)^{1/2}$ = average velocity of the molecules by their thermal energy

T = temperature (K)
M = molecular weight

The reduced form of Eq. (1.1) is shown in Eq. (1.2):

$$D_k = 9700 r_p (T/M)^{1/2}, \text{ in } \frac{cm^2}{s}; \text{ where } r_p \text{ is in cm} \quad (1.2)$$

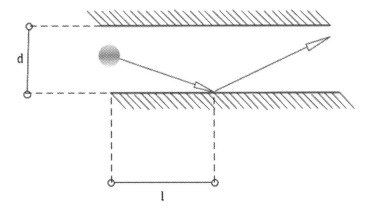

Figure 1.2 Knudsen diffusion [15].

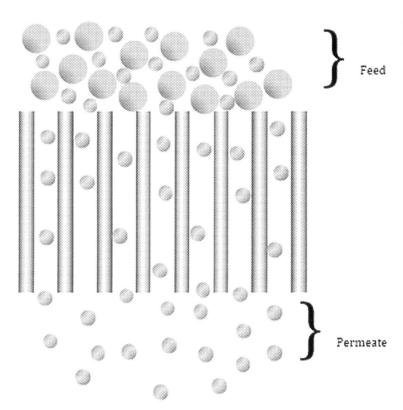

Figure 1.3 Molecular sieving [16].

2.1.2 Molecular sieving

In molecular sieving as can be observed from Fig. 1.3, the size of the molecules is equal to the size of pores. Selectivity depends on the molecular size; the molecules with larger or equal size pores are retained, while the smaller ones pass with high diffusion rates [16].

2.1.3 Solution–Diffusion separation

Unlike the other mechanisms, as shown in Fig. 1.4 [17], in the solution–diffusion mechanism, the transfer of molecules does not take place through pores but rather by dissolution and motion through dense membranes. The process takes place through three steps [17].

i. Depending on the selectivity and absorbability, a single type of component of the gas dissolves into the dense membrane.

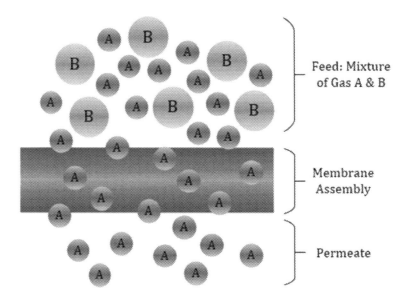

Figure 1.4 Solution–diffusion separation [17].

    ii. The dissolved component moves from the feed–membrane interface to the permeate–membrane interface based on the concentration gradient.
    iii. On the opposite end, desorption takes place due to pressure drop.

2.1.4 Surface diffusion

Surface diffusion takes place by the hopping or leaping phenomenon of the adsorbed molecules on the solid surface from one active site to another after absorbing enough energy to overcome the energy barrier as shown in Fig. 1.5 [18]. Flux due to surface diffusion is given by Eq. (1.3):

$$J_s = -D_s \frac{dC_s}{dz} \quad (1.3)$$

where
    D_s = surface diffusion coefficient, in m^2/s
    C_s = surface concentration of the adsorbed molecules, in kmol/m^2

2.1.5 Capillary condensation

Gases with vapor condensation below the saturation vapor pressure of the liquid fraction of its pure form condense on the

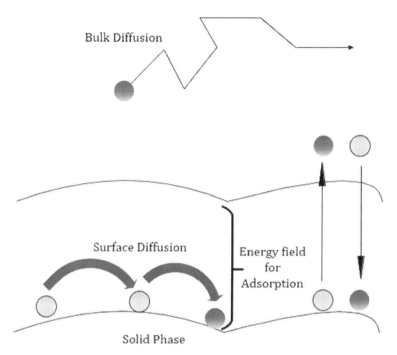

Figure 1.5 Surface diffusion [18].

surface of pores and constrict the noncondensing gas to pass through the pores, allowing the more condensed gas to permeate selectively through the pores [19]. This process is called capillary condensation. Condensation initiates at saturation pressure, P_{rel}, according to the Kelvin's equation as given by Eq. (1.4):

$$\ln P_{rel} = -\frac{2\gamma V_m \cos\theta}{RT}\frac{1}{r_k} \qquad (1.4)$$

γ = surface tension (N/m)
V_m = molar volume of the gas (m³/mol)
θ = contact angle (°)
r_k = Kelvin radius (m)

Fig. 1.6 shows the diagrammatic comparison of different membrane separation mechanisms.

2.1.6 Types of membranes

Types of membranes include [12]:
a) Polymeric membrane: These membranes perform separation by solution–diffusion mechanism. They are either rubbery membranes or glass membranes. For operations above the

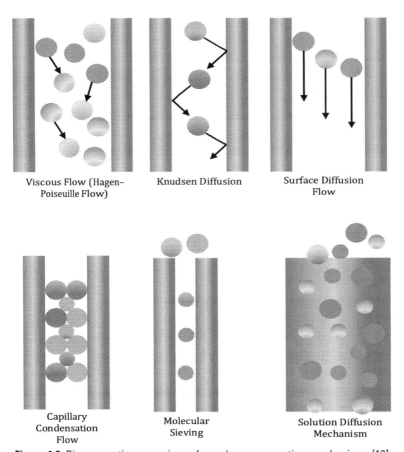

Figure 1.6 Diagrammatic comparison of membrane separation mechanisms [16].

glass transition temperature, rubbery membranes are employed, and for the ones below the glass transition temperature, glass membranes are employed.

b) Inorganic membrane: These membranes are either porous or nonporous. For high selective separation, nonporous membranes are a better option compared to porous ones.

c) Zeolite membrane: In these membranes, separation is done based on competitive adsorption on the inorganic crystalline structures.

d) Silica membrane: In these membranes, amorphous silica with pore size <1 nm is responsible for high selectivity. Further research and developments are being made to increase performance and selectivity and decrease the time required for diffusion.

e) Membranes in conjunction with chemical absorption.

2.2 Absorption

Absorption is divided into chemical absorption and physical absorption.

Chemical absorption is the process of absorption of carbon dioxide from acid gas by forming weak bonds between the molecules of carbon dioxide and the solvent used for absorption. After absorption, CO_2 is desorbed in the stripper/regenerator section. Chemical absorption is carried out at low partial pressure (or low concentration) of carbon dioxide in feed gas and low absorber temperature and absorber pressure [12].

Physical absorption of carbon dioxide in solvents is based on Henry's law. It states that dissolved gas in a liquid is proportional to its partial pressure above the liquid. Physical absorption depends on the solubility of carbon dioxide in the employed solvent. It is enhanced by the high partial pressure of carbon dioxide in the feed gas and low temperature. Physical solvent processes in detail are present in Table 1.1 [20].

Chemical absorption is favored for gases having a low partial pressure of carbon dioxide. It is more challenging to remove carbon dioxide present in less concentration as its molecules will be dispersed long distances, thus these molecules cannot be separated based on physical absorption, rather these molecules require to be attracted by forces of strong chemical bonds, which would bring the dispersed molecules together for fine purification of acid gas. For chemical absorption, we use basic solvents because carbon dioxide is an acid; thus, the bond formation requires acid—base neutralization. The overall process flow diagram of whole chemical absorption process is shown in Fig. 1.7 [12].

Chemical absorption techniques include amine absorption and aqua ammonia process.

2.2.1 Amine absorption

In the amine absorption process, amines are used to purify sour NG from CO_2. The amine is contacted with the acidic NG in the absorber, where CO_2 is absorbed into the amine by chemical absorption. Rich amine is then moved to regenerator to remove the absorbed CO_2 by heating and breaking the bonds using steam at 100—200°C. Afterward, steam and carbon dioxide are condensed to produce a stream of carbon dioxide that is >99% concentrated and may either be stored or used commercially. Lean solvent is cooled and compressed to absorber pressure before being returned to the absorber as shown in Fig. 1.8 [22].

Table 1.1 Details of physical solvent processes [21].

Processes	Solvents used	Feed condition	Advantages	Disadvantages
Selexol process	Polyethylene glycol + dimethyl ethers	High-pressure, low-pressure, high acid gas conditions	• Solvent is noncorrosive • Nonthermal regeneration of the solvent	• Inefficient at low pressures
Retinol process	Chilled methanol	Low- and moderate-concentration CO_2 gas streams	• No foaming phenomenon	• Costly refrigeration
Ifpexol-2 process	Refrigerated methanol		• Solvent is noncorrosive • Thermally and chemically stable	• Large capital cost • Low temperatures cause the formation of amalgams
Fluor process	Propylene carbonate ($C_4H_6O_3$)	High-pressure CO_2 conditions Concentration of H_2S than 20 ppmv	• Nonthermal regeneration of the solvent • High CO_2 loading	• Requires high solvent circulation rates • Solvent is costly
Purisol process	N-methyl pyrrolidone	High-pressure, high-concentration CO_2 conditions	• No foaming • Thermally and chemically stable	• Efficiency decreases at low pressures
Sulfinol process	Mixtures of DIPA/MDEA and tetrahydrothiophene dioxide (sulfolane) in different combinations	Separation in a wide range of 50%v/v H_2S and above 20%v/v CO_2	• Requires low solvent circulation rates	• Solvent is corrosive • Causes foaming
Morphysorb process	NFM and NAM mixtures	HP and HAG conditions	• High solvent loading • Less energy consumption • Solvent is noncorrosive	

DIPA, diisopropylamine; *HAG*, high acid gas; *HP*, high-pressure; *MDEA*, methyldiethanolamine; *NAM*, N-acetylmorpholine; *NFM*, N-formylmorpholine.

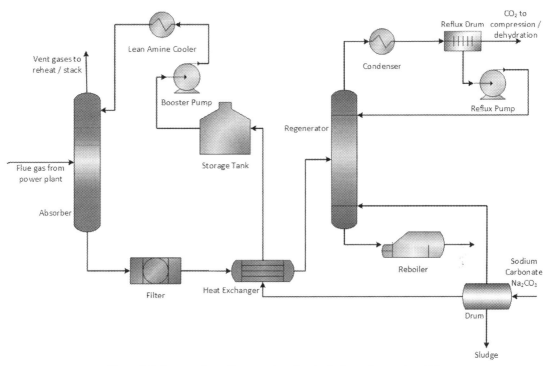

Figure 1.7 Process flow diagram of a chemical absorption system [12].

Monoethanolamine (MEA) is the most widely used solvent for acid gas separation from NG by absorption. Acid gas loadings, efficiency, energy requirement, and problems in operation, each and every part of MEA has been studied in detail as the solvent has been into operation for decades [11,24]. Hence, that is why MEA has been selected as the base case against which the solvent blends will be compared.

CO_2 separation from NG through absorption is the most studied field. These days the absolute focus in the absorption process is developing a solvent to overcome the drawbacks in the absorption process like solvent degradation and corrosion. Murtaza et al. studied the blends of methyldiethanolamine (MDEA) with MEA, diglycolamine (DGA), and diethanolamine. The simulation results indicated that the blend MDEA + DGA (40% + 5%) reduced the reboiler duty requirement from 18.99 MMBtu/hr. to 15.43 MMBtu/hr. Low corrosion and viscosity furthermore accounted for this decrease [25].

A model developed by Abotaleb et al. [26] consist of 10 cases of primary amines, secondary amines, tertiary amines, and blends. The results showed that only primary amines and blends could

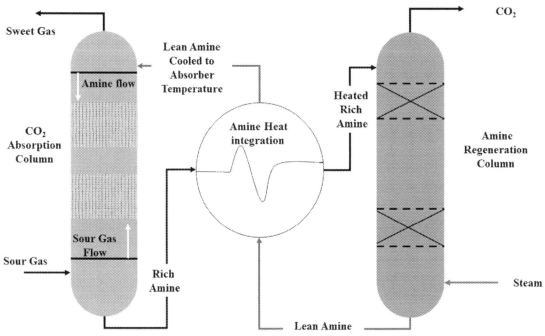

Figure 1.8 General amine absorption process [23].

meet NG pipe specifications. The MDEA/piperazine (PZ) case with wt.% (20/10) remained favorable due to reduced energy requirements than the other cases, and absorption capacity was enhanced by 67%.

MDEA with PZ as an activator with 49.8 wt.%./0.2 wt.% has a better CO_2 absorption capacity than MDEA [27]. However, an 8% increase in reboiler duty was observed, and H_2S absorption capacity was reduced. Further study is required of other operational parameters of this blend.

MDEA/MEA blends with varying compositions were analyzed by Ahmad et al. [28]. It was found that the blend with 10/20 mol% and 20/10% were most efficient because of minimized CO_2 and H_2S in the outlet.

A simulation model process optimization of Iraq's NG sweetening plant employing three solvents, MDEA, MDEA−MEA, and MDEA−sulfolane, based on CO_2 and H_2S absorption efficiency and economic perspective was developed by Khanjar and Amiriusing [29]. CO_2 and H_2S absorption capacity of MDEA (30wt.%) and MEA (15wt.%) was observed to be much higher than the other two systems. The solvent price for MDEA−MEA was reduced by 25% compared to MDEA (45 wt.%) due to low MEA price. In addition, the reboiler duty of this solvent is considerably significantly less than that of MDEA. In a similar manner, a number of other research had been and are still being performed in the field of NG sweetening using amine absorption.

Ionic liquids (ILs) based on their properties example negligible vapor pressure and high thermal stability studies by Karadas et al. [30] and Kazmi et al. [8]. Pyridinium (IL) was analyzed against MEA and DME and it was observed that IL provided more energy savings, lesser anergy involved in equipment, and higher exergy efficiency of equipment and was more economically viable compared to MEA and DME.

A case study on NG purification from CO_2 using deep eutectic solvents (DES) with greater heating value was performed by Haghbakhsh et al. [31], using two DES, namely reline and glyceline. In the results, carbon dioxide stream showed greater purity using DES compared to Selexol. Overall duties were less for plants using DES than for Selexol, but exergy destruction was higher for DES. DES were found feasible for different industries.

2.2.2 Aqua ammonia process

Aqua ammonia process is envisioned to replace the MEA process because [12]:

i. Aqua ammonia process makes the sweetening process simpler and economical by capturing all sorts of toxicities, e.g., SOx, NOx, HCl, HF, and CO_2 from sour NG
ii. Aqua ammonia process has lower heat of reaction than amine-based systems.
iii. No absorbent degradation in aqueous ammonia process.
iv. Process by-products: ammonium nitrate and ammonium sulfate can be used to form fertilizer.

On the other hand, aqua ammonia process has the following drawbacks:

i. Ammonia has high volatility, thus causing losses during the regeneration process as a result of elevated temperatures.
ii. Requirement of cooling the flue gas to 0–10°C range for high CO_2 loading of solvent and reducing ammonia slip.

The property of ammonia to cause explosion upon reaction with CO_2 in dry form has convinced the industries to use it in aqueous form. (Explosive limit for gaseous ammonia is 15%–28% (v/v).) Process flow of aqua ammonia as shown in Fig. 1.9 is similar to that of amine-based absorption, the difference comes in the requirements of cooling of flue gas by a series of direct contact coolers. From Fig. 1.9, it can be observed that cooled sour gas flows in counter-current direction with absorbent slurry

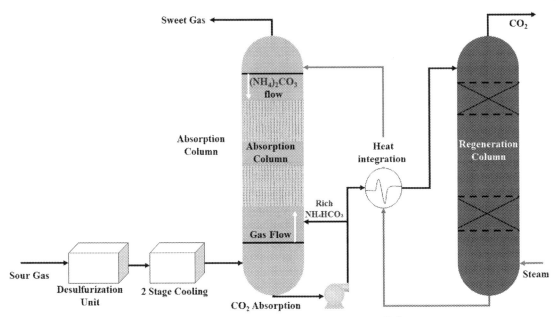

Figure 1.9 Aqua ammonia process flow diagram [32].

(ammonium carbonate and ammonium bicarbonate). Solvent is then regenerated at temperature >120°C and pressure >2 MPa; CO_2 leaves the regenerator at high pressure, causing ease in delivery for storage [32].

2.3 Adsorption

Adsorption is the process by which the gas molecules are attached to the surface of solid material. Adsorption is classified into physical adsorption (physisorption) and chemical adsorption (chemisorption).

Molecular sieves are used to separate one molecule from others based on its molecular size and weight. The efficiency of adsorption increases by increasing adsorbents' surface area and incorporating primary organic groups. Since CO_2 is acidic, anhydrous conditions on the surface produce ammonium carbamate, whereas ammonium bicarbonate is formed when water is present. Separation using molecular sieves can be observed from Fig. 1.10 [33].

The activated carbon adsorption process can be explained by heteroatoms like oxygen and nitrogen existing in the form of either acidic, basic, or neutral functional groups. By incorporating heteroatoms, the adsorption of CO_2 can be enhanced.

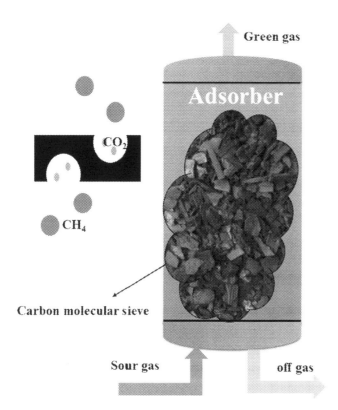

Figure 1.10 Molecular sieves [33].

The adsorption capacity of activated carbon is inversely proportional to the temperature. Fig. 1.11 shows the activated carbon adsorption mechanism [33].

Lithium zirconate is used as a high-temperature CO_2 adsorbent. Eutectic carbonate functional group increases CO_2 absorption rate. Combining alkali or alkali-earth with carbonate or halide groups enhances the CO_2 uptake rate and increases the adsorption capacity. Lithium silicate shows a much larger CO_2 adsorption capacity. Large adsorption capacity, rapid adsorption rate, operation at a vast temperature range, and stability make lithium silicate a preferable choice [12].

Research is being conducted on a suitable material for the adsorption process. The adsorption process performs best at relatively high CO_2 partial pressures, but its efficiency decreases at high temperatures of feed gas stream.

The viability of removing CO_2 from NG using the pressure swing adsorption (PSA) method was examined by Grande et al. [34]. NG with a composition of 83% CH_4, 10% CO_2, and 7% C_2H_6

Figure 1.11 Adsorption by activated carbon [33].

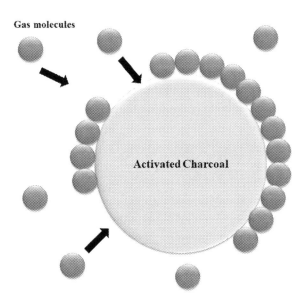

that is available at 70 bars and 313 K and with an inlet flow rate of 500,000 m³/h can be upgraded using the PSA unit.

Zhong et al. [35] used a combination method of gas hydrate production and adsorption to absorb CO_2 at a pressure range of 3.0–6.0 MPa at 274.2 K. To saturate the fixed bed and lessen the hydrate phase equilibrium conditions, 1.0 mol % tetrahydrofuran solution was utilized in place of liquid water. As the fixed bed saturation grew, the CO_2 recovery and separation factor increased.

The activated carbon adsorption to capture CO_2 postcombustion from an NG combined cycle (NGCC) power plant was conducted by Jiang et al. [36]. Due to the lower regeneration temperature of 358 K, the net efficiency employing activated carbon increases marginally from 50.8% to 51.1%. Activated carbon is more efficient and cheaper than MEA and can improve with increased heat and mass recovery.

The combination of the cyclic diamine 2-(aminomethyl)piperidine and Mg_2(dobpdc) (dobpdc^{4-} 4,4′-dioxidobiphenyl-3,3′-dicarboxylate) results in an adsorbent that is capable of capturing around 90% of the CO_2 present in a humid NG flue emission stream [37]. By using a mechanism that allows access to a significant CO_2 cycling capacity with only a minor temperature variation (2.4 mmol CO_2/g at 100°C), this material can capture CO_2.

It is not required to perform a cooling step, increasing process productivity, when using an adsorbent that is 70% zeolite and 30% binder conducting material (with a capacity greater than

1.1 mol of CO_2 per kilogramme at low CO_2 partial pressure) [38]. A concentrated stream with 80% CO_2 can be produced using the proposed electric swing adsorption (ESA) technique, according to simulations, starting from a stream with 3.5% CO_2. Energy usage per ton of CO_2 was 2.04 GJ, indicating that ESA can be included in the list of CO_2 collection systems.

Preconcentrating CO_2 from 4% to 15%–20% can be done with essentially minimal energy input by using a membrane method that uses incoming combustion air as a sweep stream in a selective exhaust gas recycle setup [39]. The minimum energy required for a CO_2 collection step can be reduced by up to 40%, thanks to the selective recycle membrane design. The amount of CO_2 that can be captured from an NGCC power plant employing an all-membrane design with a capture step and a selective recycling membrane is 90%.

Using MEA aqueous solution, CO_2 absorption was carried out in a 25 cm long, circular microchannel with an 800 m diameter was utilized by Aghel et al. [40]. Atmospheric pressure and a 10% molar CO_2 feed gas concentration were used for all experiments. Temperature (15–55°C), input solvent flow rate (0–0.04 lit/min), inlet gas flow rate (one to nine lit/min), and MEA concentration (0–40 weight percent) were the operating variables. The absorption percentage was 100% at optimal conditions.

2.4 Cryogenics

In cryogenics, separation takes place by the fractional condensation of CO_2. Cryogenics has the advantages of abolishing water consumption, application of chemicals, corrosion-related issues, and allowing the recovery of CO_2 in the form of a liquid, thus causing ease in transportation and pumping to the injection site for enhancing oil recovery over the other techniques like physical and chemical absorption, membrane separation, and adsorption methods, which require substantial compression to reach geosequestration standards. Overall cryogenics process can be observed from Fig. 1.12 [41].

2.4.1 CryoCell

One of the technologies using cryogenics is CryoCell, which utilizes the solidification property of CO_2 as the key to the separation of CO_2 from the other NG components. The vapor mixture is cooled to a temperature just above the CO_2 freezing point such that the majority of it condenses, and it is then flashed using a Joule–Thomson valve causing the mixture to separate into vapor, liquid, and gas. The cooling temperature and flash pressure drop

20 Chapter 1 Introduction to natural gas sweetening methods and technologies

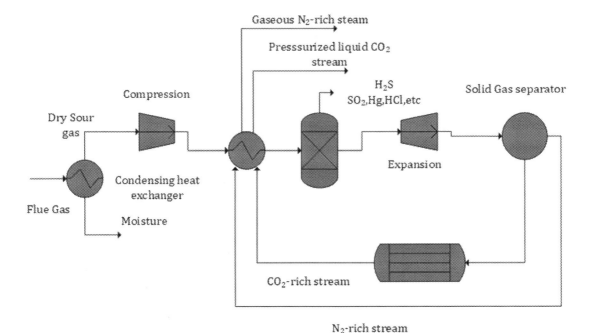

Figure 1.12 Illustration of cryogenic process [42].

are selected such that the vapor phase CO_2 composition is minimal and the liquid phase methane composition is minimal Fig. 1.13 [41].

The current field experience and test outcomes from the Western Australian CryoCell demonstration plant was operated by Cool Energy [41]. It was demonstrated that CryoCell technology can remove CO_2 from NG containing 60 mol percent down to 26 mol percent, from 40 mol percent to 14 mol percent, from 21 mol percent down to 4 mol percent, and from 13 mol percent to 3 mol percent. The CryoCell vessel heater was kept between −50 and −60°C throughout the experiments, which were conducted at inlet feed gas temperatures ranging from −50 to −65°C at feed gas pressures of 5500−6500 kPag.

A unique cryogenic switched packed bed technology for the cryogenic CO_2 capture from NG was devised by Babar et al. [43]. The pure CO_2 feed, NG sample 1, and NG sample 2 each had a saturation time of 300, 500, and 600 s, respectively. A 200-second switching time was discovered for CO_2 recovery rates of greater than 98%.

A Stirling cooler−based cryogenic CO_2 collecting system was created by Song et al. [44]. It is interesting that the procedure enables solid CO_2 capture, avoiding the need for solvents and

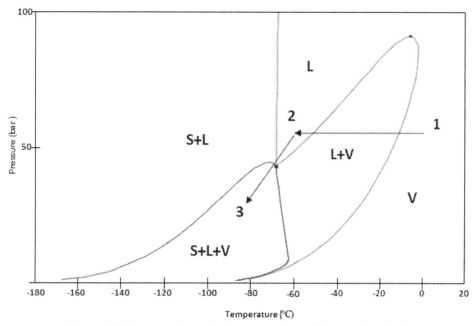

Figure 1.13 Phase envelope of a lean natural gas–CO_2 gas mixture [41].

pressure considerations in conventional approaches. According to the findings, CO_2 recovery did not increase as temperature decreased. The CO_2 capture efficiency could reach 96% when the feed gas flow rate was 1.5 L/min and the SC-1 temperature threshold was 30°C.

A packed bed high-pressure cryogenic hybrid network was constructed by Maqsood et al. [45]. The feed used had a CO_2 content of 30%, 50%, and 70%. In the cryogenic separator, measurements of separation characteristics and vapor–liquid isothermal flash were made at pressures of 20, 30, and 40 bar and temperatures ranging from −20 to 60°C. With 90.6% to 97.3% CH_4 purity and 2.65% to 12.39% methane losses in various configurations, the hybrid cryogenic network demonstrated a 37% energy saving compared to the conventional cryogenic distillation network.

A hybrid NH_2–MIL–101/CA cryogenic packed bed column to test the effectiveness of CO_2 capture was utilized by Babar et al. [46]. In comparison to spherical glass beads and monofilament fibers, it was discovered that hollow fibers significantly lower pressure drop by a factor of 61 and 33, respectively. The hollow fibers offered 230% and 122% more specific surface area than glass beads and monofilament fibers, respectively. The CO_2 capture efficiency of composite hollow fibers was noted to be 141.9%, which is 9.5% higher than the efficiency of pure CA hollow fibers. It takes less energy to cool the composite CA/NH$_2$−MIL−101(Al) hollow fibers. The NH$_2$−MIL−101(Al) hollow fiber was shown to increase CO$_2$ capture efficiency while lowering capital costs and pressure drop.

The possibility for an efficient postcombustion carbon capture using a hybrid technique combining membrane and cryogenic separation was shown by Belaissaoui et al. [47]. For a needed CO$_2$ purity of 0.9 and a required capture ratio greater than 85%, the effects of three different CO$_2$ feed contents, 5%, 15%, and 30%, and two different membrane selectivities, 50 and 100, have been examined. The hybrid process exhibits lower energy consumption than the reference technology for a CO$_2$ feed concentration between 15% and 30%, chemical absorption in MEA, and CO$_2$ purity above 89%.

Surface-area-to-volume ratio of the column is increased by adopting dynamically operated, inexpensive fiber beds, which moreover help to greatly lower the sweep gas pressure drop through the column [48]. Using the dry-jet, wet-quench phase inversion spinning technique, porous hollow fibers made of cellulose acetate were shown to be quite permeable, showing minimal diffusion resistance in the fiber wall. Compared to columns filled with conventional spherical beads, they led to up to a 61% reduction in flue gas pressure drop and a 100% increase in total CO$_2$ collected.

2.5 Chemical looping

Chemical looping combustion (CLC) was first introduced in 1983 by Knoche and Richter. It splits up combustion into intermediate redox reactions having no direct contact among air and fuel for which it is also called unmixed combustion. Instead of mixing air with fuel, oxygen is carried by solid oxygen carriers, mostly oxides of transition metals like Fe$_2$O$_3$, CuO, and NiO [12].

A CLC process as shown in Fig. 1.14 consists of two fluidized bed reactors, one for air called air reactor and the other for fuel called fuel rector, between which the solid oxygen carrier circulates. The oxygen carrier gets oxidized inside the air reactor, and it is then sent to the fuel reactor where the fuel gets oxidized to CO$_2$ and H$_2$O and the oxygen carrier gets reduced and is circulated again inside the system. The reactions are given below [10].

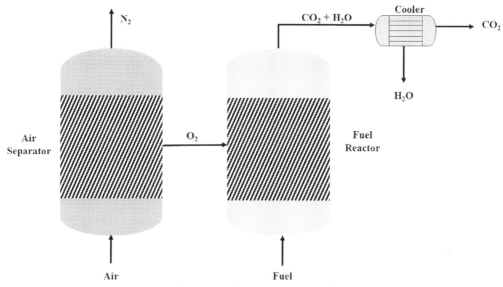

Figure 1.14 Chemical looping [49].

Air reactor:

$$M_yO_{x-1} + \frac{1}{2}O_2 \rightarrow M_yO_x \quad (1.5)$$

Fuel reactor:

$$C_nH_{2m} + (2n+m)M_yO_x \rightarrow nCO_2 + mH_2O + (2n+1)M_yO_x \quad (1.6)$$

CLC has numerous advantages over conventional combustion processes; a harmless exhaust stream from an air reactor mostly consists of N_2 gas left after oxygen is separated from the air. A well-designed system can have no formation of NO_x because there is no flame involved in the process. The fuel reactor exhaust stream from the fuel reactor comprises CO_2 and H_2O, which can be separated by a simple condensation system. The utmost advantage of CLC is that, unlike conventional amine scrubbing processes, it requires less energy for CO_2 capture, saving a huge amount in terms of operational cost [10].

In a study, Erlach et al. [50] achieved a 2.8% higher thermal efficiency for a CLC−IGCC (integrated coal gasification combined cycle) power plant compared to an IGCC using physical absorption; CLC allowed 100% carbon capture, while physical absorption only gave 85%.

For a CLC process, the energy requirement can be as low as 400 kJ/kg CO_2, which is far less than conventional absorption processes [51].

CLC has certain challenges like the design of reactors having interconnected fluidized beds and a solid oxygen carrier traveling between reactors with no gas stream leaking into one another. Erlach et al. [50] developed the prototype of this reactor, which showed 99.5% fuel conversion efficiency at ambient temperature.

A rotating wheel model where air and fuel were to be fed in different compartments of the wheel was proposed by Shimomura et al. [52]. Another vital research area for CLC is finding a suitable oxygen carrier having high stability, a large capacity to transport oxygen, and a high fuel conversion ratio.

Oxides of iron copper or nickel should be used with inerts like alumina and silica [50].

Johansson et al. [53] achieved 99% conversion efficiency using NiO with $MgAl_2O_4$ as oxygen carrier. Another vital area of research can be the temperature of air reactors.

A multistage CLC was studied by Naqvi et al. [54]; it was found that at 1200°C reaction temperature and no reheating, the CLC combined cycle achieved 535 efficiencies. Moreover, it was further concluded that using a single stage and reheating the same power plant achieved an efficiency of 51% at 1000°C and 53% at 1200°C.

Table 1.2 provides the overall summarized comparison of the discussed NG purification technologies.

3. Selection of the most suitable sweetening technique

The following factors decide the selection of the most suitable technique used to remove CO_2 from contaminant NG [4,56]:
- The partial pressure of carbon dioxide in the sour NG
- Volume of acid NG
- Impurity types and concentrations in acidic NG
- The required degree of carbon dioxide sequestration and air pollution regulations
- Energy and costs affiliated with the regeneration of the solvent
- Desired purity of sweet gas and product specifications
- Sensitivity of the solvent against the impurities present in acid gas
- Total capital and operating cost of the process

Tables 1.3–1.6 present the ranges of process parameters for the selection of the most suitable separation technique.

Table 1.2 Overall comparison of natural gas purification technologies [55].

Process	Basis of separation	Advantages	Disadvantages
Membranes	• Knudsen diffusion • Molecular sieving • Solution–diffusion separation • Surface diffusion • Capillary condensation	• Low impact on the environment • Stable at high pressures • High recovery of products • Low capital investment • Simple and versatile • Efficient in terms of space and weight • No regeneration energy required	• Moderate purity • Recompression of permeate
Absorption	• Chemical absorption: Acid–base neutralization by chemical solvent • Physical absorption: Henry's law	• Most widely used for removal of CO_2 and H_2S from natural gas	In chemical absorption: • Due to low partial pressure in the feed gas, more time is required for separation • Process is expensive due to the high duty required for amine regeneration In physical absorption: • Hydrocarbons are coabsorbed in physical solvents, resulting in reduced product revenue and often requiring recycling compression
Adsorption	• Attachment to the surface of adsorbents	• High CO_2 loading	• High regeneration energy required
Cryogenics	• Fractional condensation of CO_2	• Separated CO_2 in liquid form is ready to transport • No requirement for chemical reagents	• Refrigeration requires high energy • High capital expenditure • Prior removal of components like SO_x and NO_x due to freezing and blockage problems
Chemical looping	• Combustion of fuel in O_2 rather than air	• Air reactor exhaust gas is harmless as it mainly consists of N_2 • Less energy duty required, hence the low operational cost	Not commercially implemented

Table 1.3 Process selection table for CO_2 removal with no H_2S present [4].

Partial pressure of acid gas in product P (psia)	Partial pressure of acid gas in feed P*(psia)	Sweetening technique
0.1–0.2	1–100	Amine
0.2–1	1–10	
1 psia < P < 10 psia	P<=P*<10	
0.2–0.3	10–100	Activated hot potassium carbonate or
0.3–10	10–15	amine
0.1–0.4	100–1000	Hybrid
0.4–1.9	100–1000	Physical solvent, hybrid, or hot potassium carbonate
1.9–100	80–1000	Physical solvent
0.3–0.8	15–100	Activated hot potassium carbonate or
0.8–17	15–50	inhibited concentrated amine
17 psia < P < 50 psia	P<=P'<50 psia	
0.8–18	50–100	Physical solvent or activated hot
18–60	50–80	potassium carbonate

Table 1.4 Process selection table for H_2S removal with no CO_2 present [4].

Partial pressure of acid gas in product P (psia)	Partial pressure of acid gas in feed P*(psia)	Sweetening technique
0.01–0.1	1–3	Stetford
0.01–0.1	3–10	Amine, Stetford, or hybrid
0.1–0.25	0.01–10	
0.01–0.25	10–60	ADIP Process, hybrid, or high
0.25–1	0.01–60	loading diethanolamine (DEA)
1–10	10–60	
1 psia < P < 10 psia	P<=P*<10 psia	
0.01–0.1	60–1000	High loading DEA, hybrid, or physical solvent
0.1–1	60–100	Physical solvent or hybrid
0.1–1	100–1000	Physical solvent
1–10	60–1000	

Table 1.5 Process selection table for simultaneous H_2S and CO_2 removal [4].

Partial pressure of acid gas in product P (psia)	Partial pressure of acid gas in feed P* (psia)	Sweetening technique
0.1–0.2	1–75	Hybrid or amine
0.2–1	1–10	
1 psia < P < 10 psia	P<=P*<10	
0.1–1	75–1000	High loading diethanolamine or physical solvent
1–100	75–1000	Physical solvent
0.2–0.3	10–100	Activated hot potassium carbonate, hybrid, or amine
0.3–10	10–50	
10 psia < P < 50 psia	P<=P*<50	
0.3–50	50–75	Physical solvent

Table 1.6 Process selection table for selective H_2S removal with CO_2 present [4].

Partial pressure of acid gas in product P (psia)	Partial pressure of acid gas in feed P* (psia)	Sweetening technique
0.1–1	1–3	Stretford, tertiary amine, or mixed amine
1 psia < P < 3 psia	P<=P'<3	
0.1–3	3–60	Tertiary amine, sterically hindered amine, or mixed amine
3 psia < P < 60 psia	P<=P*<60	
0.1–60	60–1000	Physical solvent
60–100	100–1000	
60 psia < P < 100 psia	P<=P*<100	

4. Conclusion and future outlooks

The research on NG purification techniques is primarily focused on enhancing the efficiency of these techniques. For instance, although membrane separation is less energy intensive, more research is required to study the behavior of membranes on low partial pressures of acid gas, and a lot more research should be focused on developing new materials for membranes to achieve a high selectivity toward acid gas. The absorption process is a lot more mature than other separation techniques, but it does require quite an amount of attention on issues like corrosion in equipment and a large amount of energy requirement for solvent regeneration. The areas to conduct further studies can be the development of new solvents that can solve the issues with the process. An adsorption process requires further studies to develop new adsorbents that could enhance the adsorption capacity and adsorption rate and are more stable than existing adsorbents at different operating conditions and sour gas compositions. For the cryogenic process, research can be done to design a suitable cryogenic network with process parameters, gas flow rates, and sour gas composition. Several studies were focused on designing columns for cryogenic separation to ensure maximum recovery of CO_2. The cryogenic technique can also be used in combination with other techniques like membrane separation in order to have a much better separation efficiency. CLC has certain challenges like the design of reactors having interconnected fluidized beds and a solid oxygen carrier traveling between reactors with no gas stream leaking into one another. Several designs have been proposed to deal with this issue; however, the process still requires a more suitable model. Further multistage CLC can also be studied for better acid gas separation.

Abbreviations and symbols

CLC	Chemical looping combustion
Cs	Surface concentration of the adsorbed molecules, in $kmol/m^2$
DEA	Diethanolamine
DES	Deep eutectic solvents
DGA	Diglycolamine
DIPA	Diisopropylamine
D_K	Knudsen diffusivity (cm^2/s)
DME	1,2-dimethoxyethane
Ds	Surface diffusion coefficient, in m^2/s
HAG	High acid gas
IGCC	Integrated coal gasification combined cycle
IL	Ionic liquid
M	Molecular weight
MDEA	Methyldiethanolamine
MEA	Monoethanolamine
NAM	N-acetylmorpholine
NFM	N-formylmorpholine
NG	Natural gas
NGCC	Natural gas combined cycle
P	Partial pressure of acid gas in product (psia)
P*	Partial pressure of acid gas in feed (psia)
PFD	Process flow diagram
PZ	Piperazine
rk	Kelvin radius (m)

rp	Radius of the passage (cm)
T	Temperature (K)
V$_m$	Molar volume of the gas (m^3/mol)
vT	Average velocity of the molecules by their thermal energy
γ	Surface tension (N/m)
θ	Contact angle (°)

References

[1] Administration USEI. International energy outlook 2019 with projections to 2050. 2019.
[2] Qyyum MA, Qadeer K, Ahmad A, Lee M. Gas−liquid dual-expander natural gas liquefaction process with confirmation of biogeography-based energy and cost savings. Applied Thermal Engineering 2020;166:114643.
[3] U.S Energy information Administration.
[4] Kidnay AJ, Parrish WR. Fundamentals of natural gas processing. CRC Press; 2006.
[5] Ellaf A, Taqvi SAA, Zaeem D, Siddiqui FUH, Kazmi B, Idris A, et al. Energy, exergy, economic, environment, exergo-environment based assessment of amine-based hybrid solvents for natural gas sweetening. Chemosphere 2023;313:137426.
[6] Rufford TE, Smart S, Watson GC, Graham BF, Boxall J, Da Costa JD, et al. The removal of CO_2 and N_2 from natural gas: a review of conventional and emerging process technologies. Journal of Petroleum Science and Engineering 2012;94:123−54.
[7] Taqvi SAA, Zabiri H, Singh SKM, Tufa LD, Naqvi M. Investigation of control performance on an absorption/stripping system to remove CO_2 achieving clean energy systems. Fuel 2023;347:128394.
[8] Kazmi B, Raza F, Taqvi SAA, Ali SI, Suleman H. Energy, exergy and economic (3E) evaluation of CO_2 capture from natural gas using pyridinium functionalized ionic liquids: a simulation study. Journal of Natural Gas Science and Engineering 2021;90:103951.
[9] Islam MS, Yusoff R, Ali BS, Islam MN, Chakrabarti MH. Degradation studies of amines and alkanolamines during sour gas treatment process. International Journal of the Physical Sciences 2011;6(25):5883−96.
[10] Sifat NS, Haseli Y. A critical review of CO_2 capture technologies and prospects for clean power generation. Energies 2019;12(21):4143.
[11] Kazmi B, Taqvi SAA, Ali SI. Ionic liquid assessment as suitable solvent for biogas upgrading technology based on the process system engineering perspective. ChemBioEng Reviews 2022;9(2):190−211.
[12] Olajire AA. CO_2 capture and separation technologies for end-of-pipe applications−a review. Energy 2010;35(6):2610−28.
[13] Kohl AL, Nielsen R. Gas purification. Elsevier; 1997.
[14] Dutta BK. Principles of mass transfer and separation processes. Wiley Online Library; 2009.
[15] Gilron J, Soffer A. Knudsen diffusion in microporous carbon membranes with molecular sieving character. Journal of Membrane Science 2002;209(2):339−52.
[16] Rackley SA. Carbon capture and storage. 2nd ed. 2017.
[17] Ji G, Zhao M. Membrane separation technology in carbon capture. Recent advances in carbon capture and storage. 2017. p. 59−90.

[18] Hubbe MA, Azizian S, Douven S. Implications of apparent pseudo-second-order adsorption kinetics onto cellulosic materials: a review. Bioresources 2019;14(3).
[19] Bum Park H. Gas separation membranes. Encyclopedia of membrane science and technology. 2013. p. 1–32.
[20] Selvan KK, Panda RC. Mathematical modeling, parametric estimation, and operational control for natural gas sweetening processes. ChemBioEng Reviews 2018;5(1):57–74.
[21] Vega F, Cano M, Camino S, Fernández LMG, Portillo E, Navarrete B. Solvents for carbon dioxide capture. Carbon dioxide chemistry, capture and oil recovery. IntechOpen; 2018. p. 142–63.
[22] Peters L, Hussain A, Follmann M, Melin T, Hägg MB. CO_2 removal from natural gas by employing amine absorption and membrane technology—a technical and economical analysis. Chemical Engineering Journal 2011; 172(2–3):952–60.
[23] MacDowell N, Florin N, Buchard A, Hallett J, Galindo A, Jackson G, et al. An overview of CO_2 capture technologies. Energy and Environmental Science 2010;3(11):1645–69.
[24] Sultan T, Zabiri H, Taqvi SAA, Shahbaz M. Plant-wide MPC control scheme for CO_2 absorption/stripping system. Materials Today: Proceedings 2021;42: 191–200.
[25] Murtaza A, Qureshi K, Unar IN. Energy optimization for amine gas sweetening process by mixed amines using simulations. Engineering Science and Technology, an International Journal 2019;3:35–40.
[26] Abotaleb A, El-Naas MH, Amhamed A. Enhancing gas loading and reducing energy consumption in acid gas removal systems: a simulation study based on real NGL plant data. Journal of Natural Gas Science and Engineering 2018;55:565–74.
[27] Etoumi A, Alhanash M, Almabrouk M, Emtir M. Performance improvement of gas sweetening units by using a blend of MDEA/PZ. Chemical Engineering Transactions 2021;86:1057–62.
[28] Ahmad Z, Kadir NNA, Bahadori A, Zhang J. Optimization study on the CO_2 and H_2S removal in natural gas using primary, secondary, tertiary and mixed amine. In: AIP conference proceedings. vol. 2085(1). AIP Publishing LLC; 2019.
[29] Khanjar JM, Amiri EO. Simulation and parametric analysis of natural gas sweetening process: a case study of Missan Oil Field in Iraq. Oil & Gas Science and Technology–Revue d'IFP Energies nouvelles 2021;76:53.
[30] Karadas F, Atilhan M, Aparicio S. Review on the use of ionic liquids (ILs) as alternative fluids for CO_2 capture and natural gas sweetening. Energy & Fuels 2010;24(11):5817–28.
[31] Haghbakhsh R, Raeissi S. Deep eutectic solvents for CO_2 capture from natural gas by energy and exergy analyses. Journal of Environmental Chemical Engineering 2019;7(6):103411.
[32] Sabouni R, Kazemian H, Rohani S. Carbon dioxide capturing technologies: a review focusing on metal organic framework materials (MOFs). Environmental Science and Pollution Research 2014;21(8):5427–49.
[33] Siriwardane RV, Shen MS, Fisher EP, Poston JA. Adsorption of CO_2 on molecular sieves and activated carbon. Energy & Fuels 2001;15(2):279–84.
[34] Grande CA, Roussanaly S, Anantharaman R, Lindqvist K, Singh P, Kemper J. CO_2 capture in natural gas production by adsorption processes. Energy Procedia 2017;114:2259–64.

[35] Zhong D-L, Wang JL, Lu YY, Li Z, Yan J. Precombustion CO_2 capture using a hybrid process of adsorption and gas hydrate formation. Energy 2016;102: 621–9.
[36] Jiang L, Gonzalez-Diaz A, Ling-Chin J, Roskilly, AP, Smallbone AJ. Post-combustion CO_2 capture from a natural gas combined cycle power plant using activated carbon adsorption. Applied Energy 2019;245:1–15.
[37] Siegelman RL, Milner PJ, Forse AC, Lee JH, Colwell KA, Neaton JB, et al. Water enables efficient CO_2 capture from natural gas flue emissions in an oxidation-resistant diamine-appended metal–organic framework. Journal of the American Chemical Society 2019;141(33):13171–86.
[38] Grande CA, Ribeiro RP, Rodrigues AE. CO_2 capture from NGCC power stations using electric swing adsorption (ESA). Energy & fuels 2009;23(5): 2797–803.
[39] Merkel TC, Wei X, He Z, White LS, Wijmans JG, Baker RW. Selective exhaust gas recycle with membranes for CO_2 capture from natural gas combined cycle power plants. Industrial and Engineering Chemistry Research 2013; 52(3):1150–9.
[40] Aghel B, Heidaryan E, Sahraie S, Nazari M. Optimization of monoethanolamine for CO_2 absorption in a microchannel reactor. Journal of CO2 Utilization 2018;28:264–73.
[41] Hart A, Gnanendran N. Cryogenic CO_2 capture in natural gas. Energy Procedia 2009;1(1):697–706.
[42] Liu B, Yang X, Chiang PC, Wang T. Energy consumption analysis of cryogenic-membrane hybrid process for CO_2 capture from CO_2-EOR extraction gas. Aerosol and Air Quality Research 2020;20(4):820–32.
[43] Babar M, Mukhtar A, Mubashir M, Saqib S, Ullah S, Quddusi AHA, et al. Development of a novel switched packed bed process for cryogenic CO_2 capture from natural gas. Process Safety and Environmental Protection 2021;147:878–87.
[44] Song C-F, Kitamura Y, Li SH, Ogasawara K. Design of a cryogenic CO_2 capture system based on Stirling coolers. International Journal of Greenhouse Gas Control 2012;7:107–14.
[45] Maqsood K, Ali A, Nasir R, Abdulrahman A, Mahfouz AB, Ahmed A, et al. Experimental and simulation study on high-pressure VLS cryogenic hybrid network for CO_2 capture from highly sour natural gas. Process Safety and Environmental Protection 2021;150:36–50.
[46] Babar M, Bustam MA, Ali A, Maulud AS, Shafiq U, Shariff AM, et al. Efficient CO_2 capture using NH_2–MIL–101/CA composite cryogenic packed bed column. Cryogenics 2019;101:79–88.
[47] Belaissaoui B, Moullec YL, Willson D, Favre E. Hybrid membrane cryogenic process for post-combustion CO_2 capture. Journal of Membrane Science 2012;415:424–34.
[48] Lively RP, Koros WJ, Johnson J. Enhanced cryogenic CO_2 capture using dynamically operated low-cost fiber beds. Chemical engineering science 2012;71:97–103.
[49] Penthor S, Pröll T, Hofbauer H. Chemical-looping combustion using biomass as fuel. 2011.
[50] Erlach B, Schmidt M, Tsatsaronis G. Comparison of carbon capture IGCC with pre-combustion decarbonisation and with chemical-looping combustion. Energy 2011;36(6):3804–15.
[51] Brandvoll Oy, Bolland O. Inherent CO_2 capture using chemical looping combustion in a natural gas fired power cycle. In: Turbo expo: power for land, sea, and air; 2002.

[52] Shimomura Y. The CO_2 wheel: a revolutionary approach to carbon dioxide capture. Modern Power Systems 2003;23.
[53] Johansson E, Mattisson T, Lyngfelt A, Thunman H. Combustion of syngas and natural gas in a 300 W chemical-looping combustor. Chemical Engineering Research and Design 2006;84(9):819−27.
[54] Naqvi R, Bolland O. Multi-stage chemical looping combustion (CLC) for combined cycles with CO_2 capture. International Journal of Greenhouse Gas Control 2007;1(1):19−30.
[55] Shimekit B, Mukhtar H. Natural gas purification technologies-major advances for CO_2 separation and future directions. Advances in natural gas technology, vol 2012; 2012. p. 235−70.
[56] Bahadori A. Natural gas processing: technology and engineering design. Gulf Professional Publishing; 2014.

2

Natural gas sweetening standards, policies, and regulations

Nadia Khan[1] and Syed Ali Ammar Taqvi[2]

[1]Department of Polymer and Petrochemical Engineering, NED University of Engineering and Technology, Karachi, Pakistan; [2]Department of Chemical Engineering, NED University of Engineering and Technology, Karachi, Pakistan

1. Introduction

Natural gas, oil, and coal are mostly found underground and formed naturally. The reservoirs of natural gas have been discovered onshore and offshore, from which natural gas is produced as oil-associated gas or oil-free gas. According to an analysis, the gas reserves were estimated to be 6972 million cubic feet, against an annual consumption of 78.4 trillion cubic feet [1]. Like crude oil, natural gas is also a hydrocarbon, which is considered as the primary energy source of energy at domestic and commercial level for generating heat and electricity. It also served as a fuel for vehicles and petrochemical feedstock for the manufacturing of plastics and other commercial chemicals. Raw natural gas is a mixture of gases consisting of 80% hydrocarbon and 20% nonhydrocarbons [2]. The nonhydrocarbons are CO_2 and H_2S, which are considered as contaminates and account for the largest percentage among all the nonhydrocarbons. The natural gas is known as "sour" if it contains more than 5.7 mg of H_2S per one cubic meter of natural gas, and it is considered as "sweet" if it contains only CO_2 [3]. In the first step, the natural gas needs to be clean from these acid gases, due to their corrosive and toxic nature and to increase the heating value of the natural gas by removing CO_2 [4]. The removal process of acid gases from a natural gas is called gas sweetening process, which is explicitly designed to remove these acidic components and is a vital step for the following reasons:

- Health hazards: Breathing issues have been witnessed and can cause death in minutes at 500 ppm of H_2S.
- According to sales contracts up to 0.25 grain of H_2S per 100 scf of gas are allowed.
- The minimization of corrosion effects is considered as the key benefit along with the recovery of sulfur, which is considered as another commercial incentive [5,6].

Due to growing severe environmental rules and limitations on emissions and requirement for good-quality natural gas, the acid gas removal has become indispensable [6,7].There are numerous gas sweetening methods such as chemisorption through reactive chemical solvent, physical separation which includes membranes, cryogenic separation, and molecular sieve. The most commonly utilized process is absorption and the most commonly used solutions are amines and ionic liquids (ILs) [8–10]. Scrutiny of possible impact of any material on health and environment is essentially required before applying it. Amines and ILs have been proved as potential candidates for capturing CO_2 and H_2S, but they have some environmental and health issues. It is evident from several studies that amine degradation products have adverse effects on aquatic and territorial organisms [11]. They can cause irritation to skin and can be toxic to wild and aquatic life, but they have not been found to be carcinogenic except triethanolamine. Among all amines, monoethanolamine (MEA) has been widely used due to high biodegradability and due to no harmful and adverse effects on human health, animals, and aquatic life. Other amines such as aminomethyl propanol, methyldiethanolamine, and piperazine have been found to show adverse effects on human, animal, and aquatic life. These amines degrade to other products, when released to the atmosphere [12–16]. Nitrosamines are a product produced as a result of amine degradation and have been proved as the most fatal degradation product. Nitrosamine is cancer causing and pollutes drinking water. Similarly, ILs are also used as solvents in the gas sweetening process. In several studies, ILs have been reported as green solvents due to their nontoxic and biodegradable nature and proven to show better performance compared to conventional solvents. It has also been observed that frequently used solvents have a certain level of toxicity, which needs to be addressed [17,18].

Natural gas plays a vital role in the economy of various countries and certain targets are set by them to minimize climate change due to the processing of natural gas. Methane, being a main constituent of natural gas, has the potential of global warming and change of average global temperature over the next 30 years [19,20].

The standards, policies, and regulations regarding natural gas sweetening can vary depending on the country or region.

In the United States, the Environmental Protection Agency (EPA) has established the National Emission Standards for Hazardous Air Pollutants (NESHAP) for the oil and gas industry. These standards set limits on the amount of H_2S emissions that can be released during natural gas sweetening operations [21].

In Canada, the Canadian Council of Ministers of the Environment has developed the Canadian Ambient Air Quality Standards for H_2S emissions. The standards vary by province and territory and are designed to preserve human health and safeguard the environment [22].

The European Union (EU) has established the Industrial Emissions Directive (IED) to regulate emissions from industrial activities, including natural gas sweetening. The IED sets emission limits for H_2S and other pollutants and requires operators to use best available techniques (BAT) to minimize emissions [23].

Other countries and regions may have their own standards, policies, and regulations regarding natural gas sweetening. For example, in Australia, the National Greenhouse and Energy Reporting Act 2007 requires companies to report their greenhouse gas (GHG) emissions, including those from natural gas sweetening operations [24].

In addition to government regulations, many companies have their own internal standards and policies regarding natural gas sweetening. These may include guidelines for minimizing emissions, ensuring worker safety, and meeting environmental and sustainability goals.

2. Environmental policies and regulations

An environmental policy and regulation is a legal requirement that is established by a government or other regulatory body to protect the environment, including air, water, land, and ecosystems, from harmful impacts caused by human activities. Environmental standards and regulations set limits on the release of pollutants and waste into the environment and establish requirements for environmental management and monitoring. This section provides brief information regarding policy, standards, and regulation related to environmental issues caused by the natural gas sweetening process [25].

2.1 Regulating history

The Clean Air Act necessitates that the EPA establish New Source Performance Standards (NSPS) for industries that produce, or greatly contribute to, air pollution that may put public health or welfare at risk. In 1985, the United States established a regulation and emission control standard, 40 CFR part 60, subpart KKK, for the oil and natural gas source category for volatile organic compound (VOC), after which more comprehensive regulations including fugitive and GHG emissions were regulated in 2016 along with NSPS and named as 40 CFR part 60, subpart OOOO and 40 CFR part 60, subpart OOOOa. Particularly CFR part 60, subpart OOOOa deals with GHG and methane emissions [26,27]. The two subparts OOOO and OOOOa contain similar but identical limits regarding natural gas processing [28]. Moreover, in August 2020, the EPA further emended the rules of 2016 NSPS. The EPA gathered the data and estimated the environmental emissions from gas processing facility to track the emission trends over decades and to assess the effectiveness of the developed OOOO and OOOOa regulations [28].

The regulations are relevant to the range of emission sources that belong to the natural gas category, including production, processing, transmission, and storage. Despite being mainly associated with rural areas, the natural gas sector has witnessed a surge in the development of wells and facilities, resulting in emissions occurring in or around populated areas throughout the nation.

2.2 Air emissions

The main issue in natural gas sweetening is the control of air emissions from the process. Air emissions include H_2S and CO_2 emissions, fugitive emissions, VOC, SO_2, and GHG. The summary of regulations related to H_2S and CO_2 emissions, fugitive emissions, GHG, and venting and flaring is provided below.

2.2.1 H₂S and CO₂ emission

The allowable limits for CO_2 and H_2S emissions from the natural gas sweetening plants can vary depending on the country or region. However, there are some general guidelines that are commonly followed in the industry.

For CO_2 emissions, the allowable limits can range from 0.5% to 2% by volume depending on the regulations and requirements of the specific region. For example, EPA sets limits on GHG emissions, including CO_2 emissions, from natural gas processing plants. The EPA's NSPS also set a limit of 2% by volume of CO_2 emissions for new or modified plants [29].

For H_2S emissions, the allowable limits can range from 10 to 250 parts per million (ppm) by volume depending on the country or region. For example, in the United States, the Occupational Safety and Health Administration sets a permissible exposure limit of 10 ppm of H_2S for workers in the petroleum sector [30,31].

It is worth emphasizing that these are general guidelines and specific regulations can vary by country or region.

2.2.2 Regulations for fugitive emissions

The term fugitive emission in natural gas sweetening facilities represents unintended release of emissions from the leaks in tubing, valves, connections, flanges, packings, pumps, and compressor seals and storage tank leaks. The primary source of pollutants of concern includes VOC emissions.

The EPA has developed regulations that require natural gas processing facilities to monitor and control fugitive emissions of methane and other pollutants [32]. In 1979, crude and natural gas processing facilities became priority for EPA. Since 1979, EPA has been providing the standards and regulation for VOC emissions and is now broadcasting for GHG as well. The year 1985 provided the VOC emission limit of 40 CFR part 60, subpart KKK and SO_2 emission limit of 40 CFR part 60, subpart LLL. Initially, the sources of VOC emissions were limited; subsequently, EPA amended its regulation and endorsed new regulations for controlling VOC releases from hydraulic fracturing gas wells, compressors of both centrifugal and reciprocating types, controllers that operate through air pressure, and vessels designed for storage purposes. During 2016, EPA finalized standards for VOC emissions and amended in 2020. 40 CFR part 60, subpart OOOO addresses only pneumatic controllers, centrifugal compressors, and reciprocating compressors, which are commonly used equipment in industries. Moreover, EPA has issued NSPS to regulate VOC from gas processing facilities and along with that a control technique guideline document has also been published by EPA, which provides recommendation for controlling VOC emissions [26,27,32,33].

There are several standards for fugitive emissions from natural gas processing, some of which are:

EPA Method 21: This method is used to detect leaks of VOC machineries employed in the treatment of natural gas. It uses portable gas analyzer to measure emissions from equipment such as valves, pumps, and compressors.

ISO 15848-1: This standard provides guidelines for the testing and qualification of valves used in natural gas processing, including requirements for fugitive emissions. It specifies methods for measuring and reporting valve emissions and sets maximum allowable leakage rates for different types of valves.

API 622: This standard provides guidelines for the testing and qualification of packing and gasket materials used in natural gas processing. It includes requirements for fugitive emissions testing and sets maximum allowable leakage rates for different types of packing and gaskets.

ANSI/ISA-7.33-01: This standard provides guidelines for the design and installation of instrumentation systems used in natural gas processing. It includes requirements for fugitive emissions and specifies methods for detecting and repairing leaks in instrumentation systems [34].

Fugitive emissions policies and regulations vary country to country. Few are mentioned below.

- Canada: In Canada, the federal government has developed regulations to decrease methane release from natural gas treatment plant. The regulations require companies to report their emissions, implement measures to reduce fugitive emissions, and use new technologies to monitor emissions. The regulations aim to control the methane release by 2025, stating that emissions should be 40%–45% lower than 2012 levels [35].
- Australia: In Australia, the federal government has developed regulations to reduce fugitive release from natural gas treatment plant. These regulations necessitate companies to critically control and report their emissions, implement measures to reduce fugitive emissions, and use the BAT to prevent and reduce emissions. The regulations aim to control methane releases from gas treatment plant by 2030, which should be 26%–28% lower than 2005 levels [36].
- United Kingdom: The Environmental Permitting (England and Wales) Regulations 2016 and the Pollution Prevention and Control (Scotland) Regulations 2012 are the main regulations monitoring the fugitive emissions from natural gas processing in the United Kingdom. Under these regulations, operators of natural gas treatment facilities must obtain an environmental permit from the relevant environmental agency in their area (either the Environment Agency in England and Wales or the Scottish Environment Protection Agency in Scotland). The permit sets out the conditions that the operator must comply with to operate their facility in an environmentally responsible manner, including requirements for the control of fugitive

emissions. The permit requires operators to implement a leak detection and repair program to identify and repair leaks from equipment and pipelines. Operators are also required to monitor and report their emissions and to use the BAT to control their emissions.

In addition to these regulations, the UK government has also set targets for reducing emissions, including emissions from the natural gas sector. The UK Climate Change Act 2008 requires the government to set legally binding targets for reducing emissions in the United Kingdom, and the government has set a target of net zero emissions by 2050 [37–39].

2.2.3 Regulations for GHGs

In natural gas sweetening processes, a significant amount of CO_2 may be released. EPA has provided regulations related to GHG emissions for such processes. In 2016, along with VOC emission regulations, GHG emission regulation was also finalized, 40 CFR part 60, subpart OOOOa [27]. GHG emission regulations vary by country; permissible limits from natural gas sweetening plant are mentioned in Table 2.1. In addition, states have their own regulations governing the management of air quality with regard to the production and processing of oil and gas and can be influenced by factors such as political priorities, economic goals, and international agreements. Here are some examples of GHG emission regulations in different countries:

Table 2.1 Permissible limits for air emissions [43–46].

Air emission limits for natural gas sweetening facility		
Pollutants	**Units**	**Permissible limits**
Nitrogen compounds	Mg/Nm^3	150b, 50c
Sulfur compounds	Mg/Nm^3	75
Particulate matter PM10	Mg/Nm^3	10
Volatile organic compounds	Mg/Nm^3	150
Carbon monoxide	Mg/Nm^3	100

a. Dry gas at 15% oxygen
b. The limit of 150 mg/NM^3 for nitrogen compounds applies to installations that have a combined heat input capacity of up to 300MWth.
c. The limit of 50 mg/NM^3 for nitrogen compounds applies to installations with a combined heat input capacity exceeding 300MWth.

- EU: The EU has set a goal of reducing GHG emissions by at least 40% by 2030 compared to 1990 levels. The EU Emissions Trading System is a cap-and-trade system that covers various industries, while regulations also exist for the transportation and energy sectors [40].
- China: China is the world's largest GHG emitter and has set a goal of achieving carbon neutrality by 2060. The country has implemented various measures such as carbon pricing, renewable energy subsidies, and emission standards for vehicles and industries [41].
- Canada: Canada has set a goal of reducing GHG emissions by 30% below 2005 levels by 2030. The country has implemented various regulations such as a carbon pricing system, methane regulations, gas treatment plant, and emissions standards for vehicles [35].
- Japan: Japan has set a goal of reducing GHG emissions by 26% below 2013 levels by 2030. The country has implemented various measures such as a carbon pricing system, emissions standards for vehicles, and energy efficiency regulations for buildings and appliances [42].

2.2.4 Regulations for venting and flaring

Venting and flaring are critical operational and safety measures essential in natural gas sweetening process when some gases that are not needed for further processing or that exceed the capacity of the processing equipment may be generated. In a gas sweetening process, venting may occur when excess or unwanted gases are simply released from the system through a vent pipe or other openings. Flaring involves the burning of gases, typically in a controlled manner, to convert them into less harmful substances, such as carbon dioxide and water vapor [47]. In the gas sweetening process, flaring may occur when excess or unwanted gases are burned off in a flare stack, which is designed to safely combust the gases and release them into the atmosphere. It is important to acknowledge that the flaring of associated natural gas is influenced by multiple factors that are influenced by federal, state, and local legislation and regulations, which go beyond those specifically associated with flaring or air quality. Among these are measures aimed at limiting or promoting the construction of natural gas pipelines (the lack of pipeline infrastructure or delays in its construction can increase the need for flaring) and local regulations aimed at mitigating noise and light pollution [47,48].

The regulations implemented by the federal government concerning oil and natural gas production equipment and flaring encompass the Quad O (40 CFR part 60, subpart OOOO) and Quad Oa (40 CFR part 60, subpart OOOOa) standards. These guidelines concentrate on controlling hydrocarbon discharges from onshore installations, including storage vessels such as tanks, controllers that bleed continuously, compressors of both reciprocating and centrifugal types, hydraulically fractured wells, mechanisms for leak detection and repair, units used for SO_2 sweetening, and glycol dehydrators and of VOCs and GHGs, respectively. The summary of OOOO and OOOOa is mentioned in Table 2.2.

Table 2.2 Summary of subpart OOOO and OOOOa [26,27].

Source	Final standards of performance for air emissions
H_2S discharge	Recordkeeping and reporting obligations must be fulfilled by installations whose acid gas contains hydrogen sulfide (H_2S) with a design capacity of less than 2 long tons per day (LT/D), expressed in terms of sulfur.
Volatile organic compound (VOC) discharge from one storage vessel	A storage vessel facility, comprising a single storage vessel, may generate VOC emissions of 6 tons or more annually. The emissions from the facility must be kept under 4 tons per year, without factoring in any control measures.
Wet seal centrifugal compressors (excluding those installed at well sites)	95% reduction
Reciprocating compressors (excluding those installed at well sites)	Either replacing the rod packing within 36 months or 26,000 operating hours or channeling the emissions from the rod packing into a closed vent system under negative pressure, which directs them to a process
Pneumatic controllers used in natural gas treatment facilities	Natural gas bleed rate should be zero.
Pneumatic pumps used in natural gas treatment facilities	Natural gas release should be zero.
Equipment leaks used in natural gas treatment facilities	Leak detection and repair program reflecting the leak definitions and monitoring frequencies established for 40 CFR part 60, subpart VVa.
Sweetening units at onshore	Achieve, at a minimum, an SO_2 emission reduction efficiency that is determined by the information specified in Tables 2.1 and 2.2 of 40 CFR part 60, subpart OOOOa.

Venting and flaring regulations vary by country. States have their own regulations for managing air quality that pertain to the production and processing of gas and can be influenced by factors such as political priorities, economic goals, and international agreements. Multiple state agencies gather information about flaring and venting operations. Nevertheless, in numerous instances, the information is furnished on a voluntary basis, and there are no consistent guidelines for reporting. Here are some examples of venting and flaring regulations in different countries:

In Canada, the federal government has developed guidelines to control methane discharge from the oil and gas processing, which includes natural gas sweetening processes. The regulations require facilities to monitor and report their emissions, implement measures to reduce venting and flaring, and use the BAT to prevent and minimize emissions. The regulations aim to reduce methane emissions from the oil and gas sector by 40%–45% below 2012 levels by 2025 [22].

The EU has developed regulations to reduce venting and flaring in the oil and gas sector, which includes natural gas sweetening processes. The regulations require facilities to capture or reuse natural gas that would otherwise be vented, and to use BAT to prevent and minimize emissions. The EU's regulations aim to reduce methane emissions from the oil and gas sector by at least 40% below 2015 levels by 2030 [40]. Moreover, Norway has set a goal to reduce GHG emissions by 40% below 1990 levels by 2030 [49].

3. Regulations for controlling emissions from equipment

The regulation for controlling emissions from equipment at natural gas sweetening plants is provided by the NSPS and NESHAP. This section covers the details for the requirements of compressors, pneumatic controller, and pneumatic pump and vessels.

3.1 Regulation for centrifugal compressor

A single centrifugal compressor with wet seal degassing system is required to reduce the methane and VOC emissions by 95%. For the affected facility, an operator is allowed to use a cover and closed vent system to provide a path for emissions. Each closed vent system must be initially inspected according to the defined procedures for the detection of emissions, after which annual visual inspections are also required to identify the defects such as visible cracks, holes, or gaps in piping and leaks. Each connection and component must be monitored closely to ensure there are no detectable emissions. The same protocol will be utilized for covering the vent, with the exception that during visual inspections, faults in the separator wall; broken, cracked, or otherwise impaired seals or gaskets on closure devices; and missing or broken hatches, access covers, caps, or other closure devices must be monitored. An initial performance test of every equipment should be conducted within 3 months of initial startup as mentioned in Standard 60.5413a. Moreover, quarterly assessment should be conducted within 60 months after performing initial assessment. The same guidance is provided for control devices, and the standard provides design specifications and performance testing requirements. For all devices, reporting and recordkeeping are essential.

To control discharge from reciprocating compressors, it is mandatory to change compressor's rod packing within 26,000 h of operation. In case of VOC discharge from the rod packing, collect the emissions using rod packing emission collector system and direct it toward the process through a closed vent system. For compliance, it is necessary to continuously monitor the working period of rod packing and its times from the last replacement. It is also required to conduct annual inspections and engineering assessments of the system. Moreover, it is also essential to keep the record of inspection and performance evaluations [26,27].

3.2 Regulation for pneumatic controller

The standard is applicable for nonstop bleeding, natural gas–driven pneumatic controllers that each pneumatic controller must have bleed rate of zero for natural gas treatment facility. Each controller must be tagged, mentioning the month and year of installation and modification. According to 60.5410a(d)(1), manage the record to determine the efficiency of pneumatic controller. An initial inspection of the device is required, after which the annual inspections will be conducted. For the compliance of standards, annual report must be submitted according to standard 60.5410a(d)(5) [26,27].

3.3 Regulation for pneumatic pump

The pneumatic pump located at natural gas sweetening plant must have natural gas negligible or zero discharge rate according to 60.5393a(a). For sweetening unit, it is required to comply with the standards for SO_2, according to which it is required to calculate and compare the SO_2 emission reduction efficiency (Z) and SO_2 emission reduction efficiency achieved by sulfur recovery technology (R) as mentioned in 60.5405a(a). If $R > Z$, system is in obligingness, and if $R < Z$, system is not in obligingness. The analysis is required to be documented and submitted annually. Demonstrate compliance by reducing emission up to 95% by fulfilling the requirements of covered and closed vent by using existing control devices. In case if the monitoring device is not capable to regulate the emissions, it is required to maintain the record and submit the report [26,27].

4. Operational and design standards for various sweetening processes

Numerous standards have been used in the design and operation of the natural gas sweetening process. These standards have been established for many years and mostly engineers follow them and consider them important. These standards and regulation are set to provide the guidelines for the safe and secure operation [50].

Amines are widely and most commonly used solvents in gas sweetening process. During the amine sweetening process, standards for temperature approach, steam ratio, regenerator pressure, and outlet temperature of rich side are set to prevent the subsequent issues in the operation. The temperature approach is the temperature difference between incoming acid gas and lean amine feed, and it is recommended to maintain minimum temperature approach of 5°C. Similarly, steam plays a vital role in the quality of lean amine; therefore, in stripper, it is recommended to set 0.12 kg/L as ratio of steam to amine circulation [51].

Moreover, outlet temperature on the rich side of the lean/rich exchanger is another major parameter, which is set at 99°C over so many years. In fact, a study even proves that considering the temperature higher than 104°C is not beneficial to process. It is also usually suggested to operate the amine regenerator in the range of 2.1–2.2 bar. Normally, stripping is conducted at low pressure and high temperature, but, here in this case, increasing the regenerator pressure actually enhances the stripping [5,51].

The process engineer is required to analyze the heat exchanger operation during the process. For gas sweetening process, allowable pressure drop on shell side and tube is 10 psi and Tubular Exchanger Manufacturers Association (TEMA) standards

Table 2.3 Heat exchanger requirements [51].

Equipment	Duty (Btu/hr) (GPM)	Area (Sq ft) (GPM)
Reboiler	72,000	11.30
Lean amine	45,000	11.25
Air cooler	15,000	10.20
Condenser	30,000	5.20

are followed for designing [5,52]. Typical heat exchanger requirements are mentioned in Table 2.3.

The main equipment of gas sweetening is the amine contactor. This contactor can be a plate column or a packed column. Mostly, tray columns are used as amine contactor; however, packed columns are also used for small applications. The standard design of tray column includes 20 valve trays with mist pad below and above the trays, the trays distance should be 18 to 24 in, and space between mist pad and tray should be 3–4 ft. Amine velocity should be kept as 0.25 ft/s in downcomer [52,53].

In the case of packed column, different types of packing have been used in process, and pressure drop is reported according to the type of packing being used. The standard pressure drop is 0.2–0.6 in H_2O/ft of packing height with 1 in H_2O. The height of packing is generally 12–36in. The diameter of contactor depends on the type of packing being used. The packings less than 1 inch size are usually used for 1 ft diameter contactor, 1.5 in packings are preferred for 1–3 ft diameter, and 2–3 in packings are used for 3 ft or more diameter [52,53].

The standard gas loadings in contactor are mentioned below for MEA and diethanolamine (DEA) [51].

MEA: The acid gas loading is usually limited to 0.3–0.35 mol acid gas/mole MEA.

DEA: The acid gas loading with DEA is typically 0.35–0.82 mol acid gas/mole DEA.

Stripper is used to strip out the absorbed gas from the solvent, and like absorber, it is also a tray or packed column. Generally, it consists of 20 V grid trays with 24 in spacing. The foam factor of 0.75 and liquid and jet rates of 65%–75% should be considered in designing [5,51]. The reflux cooler is air-cooled forced draft heat exchanger in which air recirculation is advisable when

Table 2.4 Power requirements [51].

Main amine inlet pump	HP = GPM. psig. 0.00065
Amine booster pump	HP = GPM. 0.06
Amine reflux pump	HP = GPM. 0.06
Cooler	HP = GPM 0.0.36

ambient air temperature is below 10°F. The reflux ratio varies between 1.2 and 3 for tertiary and primary amines. Amine booster pumps are typically single-stage/multistage centrifugal pumps, which are selected depending upon the requirement [5,51,53]. Typical power requirements for the process are mentioned in Table 2.4.

The iron sponge process is a batch process that contains the moist ferric oxide wood flakes in a single vessel. Gas inlet nozzle is at the top section having size of approximately 12 in. (0.3 m). The ferric flakes are dispersed on heavy metal sieve plate and contain scrap pipe thread protectors and two to three in. sections of small diameter pipe. When considering a good combination, to eliminate H_2S, the bed height must be at least 10 feet (3 m), while a minimum height of 20 feet (6 m) is necessary for mercaptan removal. The infused amount of iron oxide on wood flakes is normally stated in pounds (lbs) of iron oxide (Fe_2O_3) per bushel with applicable grades of 6.5, 9, 15, or 20 lbs Fe_2O_3/bushel.

Membranes are primarily applied for bulk CO_2 removal. The membranes are modular, and the required elemental area is directly proportional to the flow rate of gas. In membranes, it is required to keep the temperature moderate because increase in temperature would give good permeation, but selectivity would be compromised. The greater driving force across the membrane is created by increasing the feed pressure, which decreases the permeability and selectivity. The size of the membrane is calculated by knowing the amount of CO_2 removal required [5,51,53]. Here are some general guidelines for the membrane pore size, permeability, temperature, pressure, flow rate, diameter, and other requirements for the membrane gas sweetening process. The membrane pore size should be optimized to achieve high selectivity for acid gases while maintaining high permeability for natural gas. Typical pore sizes range from 0.1 to 0.5 microns. The membrane permeability should be optimized to achieve high gas flow rates while maintaining high separation efficiency. The permeability depends on the specific membrane material, pore size, and thickness. The membrane gas sweetening process operates at temperatures typically ranging from 25 to 60°C [5,51]. The temperature should be selected to ensure adequate separation efficiency while minimizing membrane fouling and degradation. The membrane gas sweetening process operates at high pressures, typically between 10 and 60 bar. The pressure should be sufficient to maintain the gas in a compressed state and to ensure adequate separation efficiency. The gas flow rate should be optimized to achieve high separation efficiency while minimizing membrane fouling and degradation. The flow rate depends on the specific design and operating conditions. The membrane diameter should be optimized to achieve high gas flow rates while maintaining high separation efficiency. The diameter depends on the specific membrane material, pore size, and thickness.

The other processes are physical solvent processes such as Sulfinol and Rectisol process. The Sulfinol process utilizes both chemical and physical solvents, with solution concentration range between 25% and 40% sulfolane, 40%–55% diisopropanolamine (DIPA), and 20%–30% water. Standard loadings are 1.5 mol of acid gas/mol of Sulfinol solution [18,54].

The Sulfinol process typically uses a packed column for the absorption of H_2S, CO_2, and mercaptans from natural gas. The column should be designed to provide efficient mass transfer, high capacity, and good contact between the gas and the liquid phase [55].

The temperature of the Sulfinol process is typically between 25 and 60°C. The exact temperature depends on the contents of the gas stream and the specific Sulfinol solvent being used. The pressure of the Sulfinol process depends on the upstream and downstream conditions of the gas stream. The pressure should be sufficient to maintain the gas in a liquid phase and to prevent flashing of the Sulfinol solvent. The column diameter depends on the gas flow rate and the packing material used. The diameter should be selected to ensure adequate gas and liquid distribution and sufficient residence time for efficient mass transfer. The column height depends on the efficiency of the packing material, the amount of the impurities in the gas stream, and the desired removal efficiency. The height should be sufficient to achieve the target contaminant removal efficiency. The Sulfinol process typically uses a reflux stream to enhance the separation of the absorbed contaminants from the Sulfinol solvent. The reflux ratio depends on the specific design and operating conditions [55,56].

Similarly, the Rectisol process is a cryogenic gas sweetening process that uses methanol as the solvent for the removal of

CO_2, H_2S, and other sulfur-containing compounds. Some general guidelines for the column, temperature, pressure, flow rate, diameter, height, and reflux requirements for the Rectisol process are discussed.

The Rectisol process uses a packed column for the absorption of acid gases. The column should be designed to provide efficient mass transfer, high capacity, and good contact between the gas and the liquid phase. This process operates at low temperatures, typically between −50 and −80°C. The exact temperature depends on the contents of gas stream and the specific methanol solvent being used [55].

The Rectisol process operates at high pressures, typically between 20 and 30 bar. The pressure should be sufficient to maintain the gas in a liquid phase and to prevent flashing of the methanol solvent. The flow rate of the gas and liquid phase should be optimized to achieve high efficiency in contaminant removal while minimizing solvent degradation. The flow rate depends on the specific design and operating conditions. The column diameter depends on the gas flow rate and the packing material used. The diameter should be selected to ensure adequate gas and liquid distribution and sufficient residence time for efficient mass transfer. The column height depends on the efficiency of the packing material, the amount of impurities in the gas stream, and the desired removal efficiency. The height should be sufficient to achieve the target contaminant removal efficiency. It typically uses a reflux stream to enhance the separation of the absorbed contaminants from the methanol solvent. The reflux ratio depends on the specific design and operating conditions, accordingly. It is important to note that these general design considerations, process parameters, and condition may vary by requirement, country, or region [18,51,56].

5. Conclusion and future outlooks

A recurring theme in this chapter is that natural gas sweetening refers to the process of removing impurities, such as hydrogen sulfide and carbon dioxide, from natural gas to meet pipeline specifications and safety requirements. Standards and regulations are set for natural gas sweetening to ensure that the natural gas is safe for transportation, storage, and use. The impurities in natural gas, such as hydrogen sulfide, can be toxic and pose a health risk to workers and nearby communities. Carbon dioxide can also be dangerous in high concentrations and can lead to asphyxiation. In addition to the health risks, these impurities can also damage pipelines, equipment, and other infrastructure used to transport and store natural gas.

To address these risks and ensure the safety of workers and communities, regulatory agencies, such as the EPA in the United States, have established standards and regulations for natural gas sweetening. These regulations specify the maximum allowable levels of impurities in natural gas, as well as the methods and technologies that can be used to remove these impurities. In future, by complying with these standards and regulations, natural gas producers and processors can ensure safety of their operations and the communities they serve. Moreover, strict regulations would reduce the emissions and promote the production of clean energy that minimizes environmental impact and improves overall sustainability.

Abbreviations and symbols

AMP	Aminomethyl propanol
BAT	Best available techniques
CAA	Clean Air Act
CAAQS	Canadian Ambient Air Quality Standards
DEA	Diethanolamine
EPA	Environmental Protection Agency
ETS	Emissions Trading System
EU	European Union
GHG	Greenhouse gas
IED	Industrial Emissions Directive
MDEA	Methyldiethanolamine
MEA	Monoethanolamine
NESHAP	National Emission Standards for Hazardous Air Pollutants
NSPS	New Source Performance Standards
OSHA	Occupational Safety and Health Administration
PEL	Permissible exposure limit
PZ	Piperazine
TEA	Triethanolamine

References

[1] Alcheikhhamdon Y, Hoorfar M. Natural gas quality enhancement: a review of the conventional treatment processes, and the industrial challenges facing emerging technologies. Journal of Natural Gas Science and Engineering 2016/08/01;34:689−701.

[2] Taheri M, Mohebbi A, Hashemipour H, Rashidi AM. Simultaneous absorption of carbon dioxide (CO2) and hydrogen sulfide (H2S) from CO2−H2S−CH4 gas mixture using amine-based nanofluids in a wetted wall column. Journal of Natural Gas Science and Engineering 2016/01/01;28: 410−7.

[3] Kelley BT, Valencia J, Northrop S, Mart CJ. Controlled Freeze Zone (TM) for developing sour gas reserves. Energy Procedia 2011;4:824−9.

[4] Abkhiz V, Heydari I. Comparison of amine solutions performance for gas sweetening. Asia-Pacific Journal of Chemical Engineering 2014;9:656–62.
[5] Abdel-Aal HK, Aggour MA, Fahim MA. Petroleum and gas field processing. 2nd ed. 2015.
[6] Ababneh H, Al-Muhtaseb S. A review on the solid-liquid-vapor phase equilibria of acid gases in methane. Greenhouse Gases: Science and Technology 2022;12.
[7] Rezakazemi M, Ebadi Amooghin A, Montazer-Rahmati M, Ismail A, Matsuura T. State-of-the-art membrane based CO_2 separation using mixed matrix membranes (MMMs): an overview on current status and future directions. Progress in Polymer Science 2014;39:817–61.
[8] Chapter 7–natural gas sweetening. In: Mokhatab SP, Speight WA, editors. Handbook of natural gas transmission and processing. Boston, MA, USA: Gulf Professional Publishing; 2012. 2012.
[9] Gutierrez JP, Ale Ruiz EL, Erdmann E. Energy requirements, GHG emissions and investment costs in natural gas sweetening processes. Journal of Natural Gas Science and Engineering 2017/02/01;38:187–94.
[10] Ibrahim AY, Ashour FH, Gadalla MA, Farouq R. Exergy study of amine regeneration unit for diethanolamine used in refining gas sweetening: a real start-up plant. Alexandria Engineering Journal 2022/01/01;61:101–12.
[11] Kumar S, Cho JH, Moon I. Ionic liquid-amine blends and CO2BOLs: prospective solvents for natural gas sweetening and CO_2 capture technology—a review. International Journal of Greenhouse Gas Control 2014;20:87–116.
[12] Ellaf A, Ali Ammar Taqvi S, Zaeem D, Siddiqui FUH, Kazmi B, Idris A, et al. Energy, exergy, economic, environment, exergo-environment based assessment of amine-based hybrid solvents for natural gas sweetening. Chemosphere 2023/02/01;313:137426.
[13] Antonini C, Pérez-Calvo J-F, Van Der Spek M, Mazzotti M. Optimal design of an MDEA CO_2 capture plant for low-carbon hydrogen production—a rigorous process optimization approach. Separation and Purification Technology 2021;279:119715.
[14] Chakma A, Lemonier J, Chornet E, Overend R. Absorption of CO_2 by aqueous triethanolamine (TEA) solutions in a high shear jet absorber. Gas Separation & Purification 1989;3:65–70.
[15] El-Maghraby R, Salah A, Shoaib A. Carbon dioxide capturing from natural gas using di-glycol amine and piperazine—A new solvent mixture. International Journal of Recent Technology and Engineering 2019;8: 11378–83.
[16] Kazmi B, Taqvi SAA, Raza F, Haider J, Naqvi SR, Khan MS, et al. Exergy, advance exergy, and exergo-environmental based assessment of alkanol amine-and piperazine-based solvents for natural gas purification. Chemosphere 2022;307:136001.
[17] Ahmad Z, Kadir NNA, Bahadori A, Zhang J. Optimization study on the CO_2 and H_2S removal in natural gas using primary, secondary, tertiary and mixed amine. In: AIP conference proceedings; 2019. p. 020060.
[18] Bahadori A. Natural gas processing: technology and engineering design. Gulf Professional Publishing; 2014.
[19] Chapter CC. OECD environmental outlook to 2050. 2011.
[20] Kleinberg R. Methane emission controls: redesigning EPA regulations for greater efficacy. Columbia Center on Global Energy Policy (Commentary); 2021.
[21] Available: https://www.epa.gov/stationary-sources-air-pollution.

[22] Available: https://www.canada.ca/en/environment-climate-change/services/air-pollution.html.
[23] Industrial Emissions Directive (IED). Available: https://ec.europa.eu/environment/industry/stationary/index.htm.
[24] National greenhouse and energy reporting act 2007. Available: https://www.legislation.gov.au/Details/C2017C00358.
[25] EPA Standards. Available: https://www.epa.gov/stationary-sources-air-pollution/crude-oil-and-natural-gas-production-transmission-and-distribution.
[26] EPA Code of Federal Government, 40 CFR part 60 subpart OOOO. Available: https://www.ecfr.gov/current/title-40/chapter-I/subchapter-C/part-60/subpart-OOOO.
[27] Code of federal regulations, 40 CFR part 60 OOOOa.
[28] Kentucky Division of Compliance Assistance. Compliance guide: standards of performance for crude oil and natural gas production, transmission and distribution (40 CFR 60, subpart OOOO) and crude oil and natural gas facilities for which construction, modification, or reconstruction commenced after September 18, 2015 (40 CFR 60, SUBPART OOOOa). 2016. Available: https://eec.ky.gov/Environmental-Protection/Compliance-Assistance/DCA%20Resource%20Document%20Library/CrudeOilandNaturalGasTransmissionDistribution.pdf.
[29] EPA New Source Performance Standards for Natural Gas Processing Plants. Available: https://www.epa.gov/stationary-sources-air-pollution/new-source-performance-standards-nsps-oil-and-natural-gas.
[30] OSHA Hydrogen Sulfide Standard. Available: https://www.osha.gov/laws-regs/regulations/standardnumber/1910/1910.1000.
[31] American gas association best practices for H2S control in natural gas. Available: https://www.aga.org/contentassets/54c394dc7de84f14bb47d79b137239e4/bp_h2s_control_natural_gas_feb2019.pdf.
[32] United States Environmental Protection Agency. Oil and natural gas sector: emission standards for new, reconstructed, and modified sources. 2022. Available: https://www.epa.gov/controlling-air-pollution-oil-and-natural-gas-industry/oil-and-natural-gas-sector-emission.
[33] United States Environmental Protection Agency. Oil and natural gas sector: fugitive emissions. 2022. Available: https://www.epa.gov/controlling-air-pollution-oil-and-natural-gas-industry/oil-and-natural-gas-sector-fugitive-emissions.
[34] U.S. Environmental Protection Agency (EPA. NSPS OOOOa: standards of performance for crude oil and natural gas facilities for which construction, modification or reconstruction commenced after September 18, 2015 - fact sheet. 2016. Available: https://www.epa.gov/sites/production/files/2016-08/documents/fact_sheet_final_nov_2015_nsps_oil_gas.pdf.
[35] Government of Canada. Regulations for the prevention or control of emissions of volatile organic compounds (VOC) from petroleum sector sources. 2021. Available: https://www.canada.ca/en/environment-climate-change/services/managing-pollution/sources-industry/petroleum-sector/volatile-organic-compounds/volatile-organic-compounds-regulations.html.
[36] Australian Government Department of the Environment and Energy. National greenhouse and energy reporting Act 2007. 2020. Available: https://www.environment.gov.au/protection/national-greenhouse-energy-reporting/publications/national-greenhouse-energy-reporting-act-2007.

[37] UK Government. Net zero. 2021. Available: https://www.gov.uk/government/publications/net-zero-review-report.
[38] Scottish Government. Pollution prevention and control (Scotland) regulations 2012. 2012. Available: https://www.legislation.gov.uk/ssi/2012/360/contents/made.
[39] UK Government. Environmental permitting (England and Wales) regulations 2016. 2016. Available: http://www.legislation.gov.uk/uksi/2016/1154/contents/made.
[40] European Union. Available: https://ec.europa.eu/clima/policies/international/negotiations/paris_en.
[41] China. Available: https://unfccc.int/news/china-pledges-to-reach-carbon-emissions-peak-before-2030-and-achieve-carbon-neutrality-before-2060.
[42] Ministry of economy, trade and Industry. Available: https://www.meti.go.jp/english/policy/energy_environment/global_warming/pdf/Japans_Long-Term_Strategy.pdf.
[43] US EPA. 40 CFR Part 60. Standards of performance for new stationary, sources. Subpart QQQ—standards of performance for VOC emissions from petroleum refinery wastewater systems. Washington, DC: US EPA.
[44] US EPA. 40 CFR Part 63. National Emissions Standards for Hazardhous Air Pollutants. Subpart CC—National Emission Standards for Hazardous Air Pollutants from Petroleum Refineries. Washington, DC: US EPA; Available: https://www.epa.gov/stationary-sources-air-pollution/petroleum-refineries-national-emission-standards-hazardous-air.
[45] US EPA. 40 CFR Part 63. National Emissions Standards for Hazardhous Air Pollutants. Subpart HHH—National Emission Standards for Hazardous Air Pollutants From Natural Gas Transmission and Storage Facilities. Washington, DC: US EPA; Available: https://www.epa.gov/stationary-sources-air-pollution/petroleum-refineries-national-emission-standards-hazardous-air.
[46] US EPA. 40 CFR Part 63. National emissions standards for hazardhous air pollutants. Subpart VV—national emission standards for oil-water separator and organic-water separators. Washington, DC: US EPA.
[47] A Voluntary Standard for Global Gas Flaring and Venting Reduction. World Bank; Available: https://www.ccacoalition.org/en/resources/global-gas-flaring-reduction-public-%E2%80%93-private-partnership-voluntary-standard-global-gas.
[48] Kleinberg R. Greenhouse gas footprint of oilfield flares accounting for realistic flare gas composition and distribution of flare efficiencies. Authorea Preprints; 2022.
[49] Norwegian Ministry of Climate and Environment. Climate and environmental action plan for 2021-2030. 2020. Available: https://www.regjeringen.no/contentassets/a78ecf5ad2344fa5ae4a394412ef8975/en-gb/pdfs/stm202020210013000engpdfs.pdf.
[50] Shimekit B, Mukhtar H. Natural gas purification technologies-major advances for CO2 separation and future directions. Advances in natural gas technology 2012:235−70.
[51] M. F. S. T. R. E. Natural gas, Oilfield processing of petroleum, vol. 1. PennWell Books.; 1991.
[52] GPSA engineering data book. GSAP; 2004.
[53] Mokhatab S, Poe WA, Mak JY. Handbook of natural gas transmission and processing: principles and practices. Elsevier Science; 2015.

[54] Kazmi B, Haider J, Taqvi SAA, Ali SI, Qyyum MA, Nagulapati VM, et al. Tetracyanoborate anion–based ionic liquid for natural gas sweetening and DMR-LNG process: energy, exergy, environment, exergo-environment, and economic perspectives. Separation and Purification Technology 2022;303: 122242.
[55] N. R, Kohl AL. Gas purification. 5th ed. Gulf Professional Publishing; 1997.
[56] Stewart M, Arnold K. Gas sweetening and processing field manual. Gulf Professional Publishing; 2011.

Economic assessments and environmental challenges of natural gas sweetening technologies

Nadia Khan[1] and Syed Ali Ammar Taqvi[2]

[1]*Department of Polymer and Petrochemical Engineering, NED University of Engineering and Technology, Karachi, Pakistan;* [2]*Department of Chemical Engineering, NED University of Engineering and Technology, Karachi, Pakistan*

1. Introduction

The increased emission of greenhouse gases (GHGs) has become a major challenge globally [1]. The use of natural gas has been a great interest in the energy sector, and its demand will increase by 60% through 2040, but it contains some impurities such as CO_2, H_2S, mercaptans, and CO [2,3]. All these gases are acidic in nature and cause considerable damage to human health and industrial infrastructure. CO_2 and H_2S presence can cause stress, and the poisonous nature of H_2S can cause eye irritation, cough, nausea, and fatigue in long exposure. High-level exposure to H_2S can even be fatal [4,5]. This is a basic chicken-and-egg-problem because CO_2 is corrosive in nature and the transportation of natural gas is carried out via pipelines from gas wells to consuming area with different climates. In cold areas, CO_2 may get frozen and plug lines, whereas H_2S and mercaptans are extremely hazardous, are toxic, and cause corrosion in pipelines as well as in equipment [6,7]. In recent decades, the environment is considered as a significant factor to analyze the performance of the industry and the world has to move toward energy conservation methods that are safe, inexpensive, and eco-friendly [8]. The level of challenges has been increased and an integrated set of solutions is required. Hence, the controlled emission of CO_2 and H_2S from natural gas is essential for the clean energy production through the development of innovative technology, for the safer operation, transportation of gas, and environmental performance [9–11]. Environmental analysis is carried out to measure the emissions from the industrial process by incorporating environmental concerns as a constraint or objective function on the flow or concentration of hazardous components in a particular process [12]. The flow or concentration of hazardous components should be kept as low according to the environmental regulation [13–15]. Sometimes, incorporation of environmental concerns as constraints may not fulfill the requirement [16]. Hence, it is suggested to treat these concerns as objective functions. The case where the emission of a single component is an environmental concern, the concentration of pollutants released into the environment may be considered in environmental analysis [17,18]. Several studies have been conducted to make the operation environmentally safer by improving the H_2S removal and sulfur recovery via proposed split-loop configuration. Environmental analysis is normally carried out by measuring the acidification potential (AP), GHG emissions, venting, flaring, and global warming potential (GWP) [19–21].

Usually, the most common method to reduce the pollutant release is to upgrade the technology to bring the emission according to the defined standards such as total sulfur and CO_2 content for clean and safe fuel must be less than 0.5 to 2.0 grain/100 scf, which includes H_2S ranging less than 2.5 to 1.0 grain/100 scf and mercaptan less than or equal to 0.25 to 1.0 grain/100 scf, respectively [22]. One consequence of it would be the allocation of a large budget for the design, installation, and operation of the equipment.

A detailed economic analysis of various natural sweetening technologies has been published, which includes details of process design, mass and energy emissions variables, results, implications, and economic analysis [23,24]. These studies can provide the basis of comparison and can support collaborators and economists in evaluating the economic and environmental feasibility of these industrial processes. Economic analysis deals with factors that cause variations in cost estimates, such as size of equipment, quantity of energy required, and utilities. Economic analysis is used to associate key features of process feasibility among multiple alternatives containing plant performance indicators, cost, and emissions. It provides ease in making decisions for selecting design and design variables. Economic assessment model is being used to keep track of process stream changes also used to quantify changes in equipment and operation cost [25]. The economic viability and technical development can be easily assessed and attained by identifying the key variables and setting the processing milestones. The methodology and concepts of economic assessment are mentioned below along with summaries of case studies related to the natural gas sweetening process.

2. Methods of economic analysis

Economic assessment has been utilized for financial assessment and comparison of conventional and advanced processes. Economic assessment of any process requires technical analysis and cost analysis. Technical analysis is used to perform mass and energy balance over the whole process, determine equipment size, and estimate plant performance, including process efficiency and utility requirement. With respect to the process economics of the natural gas sweetening process, the circulation rate is a most important influential parameter. Solvent circulation rate puts a direct impact on size of pump, pipe sizing, heat exchangers, and regeneration tower and hence is considered to be a highly influential parameter for the costing of gas treating plant [26]. Moreover, it is also affected by the energy demand for solvent recovery as the reboiler heat duty is directly associated with liquid flow rate. Solution corrosion capability also plays an essential role in gas sweetening economics. This factor is used to select the material of construction of units due to high solution acidity and temperature. Cost analysis plays a crucial role in selecting alternative designs. Capital cost, operating cost, and other expenses give an insight regarding the feasibility of recommended changes to an existing gas treating unit. Some of the basic concepts for the estimation of cost are mentioned below.

2.1 Cash flow diagram

A cash flow diagram is beneficial to analyze and communicate economic problems related to engineering. Its structure consists of a horizontal time axis with vertical arrows to indicate inflow and outflow of cash. Cash flow diagram represents the flow of money in and out of the project but does not provide any estimation of the net balance. Another tool which is normally utilized along with cash flow diagram is account balance diagram. It represents the net balance at the end of the time period from the borrower's point of view.

2.2 Time value of money

Another major concept to be considered is the change in the value of money with respect to time depending upon the time spent or received. The present value of money is normally represented as P and future value is represented as F, if the interest rate is known, then this future value can be calculated using Eq. (3.1) [27].

$$F = (1+i)^n P \qquad (3.1)$$

where (1+i) is the compounded interest rate over "n" period. In an engineering project, the economic value can be evaluated by the time value of money compared to the do nothing scenario in which money is left in zero interest bank account.

2.3 Cost of equipment and operation

Cost estimation for different processes typically comprises capital, annual, and leveled costs. Initial cost is estimated by capacity exponent or other models to evaluate the direct cost of process as function of flow rate of streams or other important parameters such as design temperature and pressure. The capital cost is the summation of direct and indirect capital expenses. The direct cost consists of section direct cost and facility charges. The indirect cost consists of multiple expenses such as operation and utility charges, administration cost, sales tax, and health and environmental regulation charges. Moreover, contingency expenses are also sometimes considered for estimation of expected additional cost. Peters and Timmerhaus [28] provided a particular model, named as capacity-exponent model, for the estimation of direct cost of a specific process area given in Eq. (3.2) [28].

$$\text{Equipment cost} = \text{Base cost} \times \frac{\text{Size}_i}{\text{Size}_o}^{\text{scaling factor}} \qquad (3.2)$$

where the base cost narrates the baseline value of the sizing parameter ($Size_0$) and the equipment price is assessed at any random value of sizing parameter ($Size_i$). The scaling factor is used to consider any variation in cost due to change of size. Generally, the scaling factor is taken as 0.6 as a rule of thumb.

The total cost of the plant includes capital expenditures of the process, maintenance and construction cost, tax, allowances, payments, startup costs, and any contingencies. The annual debt and equity costs of capital funds are essential as funding for construction of new facilities. Total capital requirement (TCR) contains total plant infrastructure, prepaid royalties, equipment inventory, startup expenses, chemical and catalyst costs, and land costs. The annual cost is the summation of fixed operating cost (FOC) and variable operating cost (VOC). VOC contains consumables, fuels, slag, and disposals. The cost of energy production is also calculated by TCR, FOC, and VOC via Eq. (3.3) [28].

$$C_{elec} = \frac{\left[1000 f_{cr} TCR + f_{velf}(FOC + VOC)\right]\left(\frac{1000 \text{mills}}{\text{dollar}}\right)}{MW_{net} 8760 C_f} \quad (3.3)$$

where
C_{elect} = Cost of electricity in millions/kWh
TCR = Total capital requirement in $1000
FOC = Fixed operating cost in dollars
VOC = Variable operating cost in dollars
MW_{net} = Net power output in MW
f_{cr} = Fixed charge factor
f_{velf} = Variable levelization cost factor
C_f = Capacity factor

The production cost and capital cost are calculated for the same year in order to interpret the economic trends. However, the adjustment of cost of equipment and consumable in a present year to another year is carried out by using an appropriate price index provided in the *Chemical Engineering* Plant Cost Index (CEPCI) mentioned in Eq. (3.4) [28].

$$\text{Equipment cost in year i} = \text{Base cost} \times \left(\frac{CEPCI_{\text{year i}}}{CEPCI_{\text{base year}}}\right) \quad (3.4)$$

3. Techno-economic analysis of natural gas sweetening process

Process industries are mostly exposed to process uncertainties and abrupt changes in fuel prices, as a consequence of which the process safety, economic viability, energy saving capabilities, and environmental concerns become more critical, and it needs to be addressed properly while designing the chemical process plants [28,29]. Primarily, all the industrial processes are complex in nature and the decision-makers wish to discover the feasible solution in terms of environment and finance. Several studies have been reviewed, showing the best possible economic solution with less energy requirement and environmental emissions.

The basic step in economic analysis is to define the design criteria on which performance, cost, and emissions are to be evaluated [30,31]. The conventional amine process is economically analyzed with the Benfield HiPure process used in Abu Dhabi Gas

Liquefaction Company (ADGAS) liquefied natural gas unit. The plant operations were evaluated based on product concentration, energy requirement, and overall financial condition and performance. The simulation of an amine sweetening process was completed using ProMax V3.2. The CO_2 and H_2S absorption in the simulation was predicted by using electrolytic property packages in potassium carbonate and amine units of Benfield HiPure process. In terms of absorption, a simple amine-based process showed good results, which were comparable with the ADGAS process.

The cost was predicted for the Benfield HiPure process and its methyldiethanolamine (MDEA)-based alternatives. The design parameters of scrubber and solvent recovery columns, flash vessels, and pumps were acquired from the datasheets and the remaining equipment were designed in ProMax. Table 3.1 shows the assumption taken during this analysis.

The cost of equipment for ADGAS potassium carbonate process, which includes absorber, regenerator, feed/sweet gas HEX, lean carbonate filter, lean carbonate pumps, acid gas condenser, lean carbonate vessel, reflux drum, lean carbonate cooler, kettle reboiler, and rich carbonate vessel, was calculated as 117.65 M$, whereas for MDEA process, it was 38.67 M$, which contains absorber, regenerator, kettle reboiler, lean/rich heat exchanger, lean amine pump and cooler, rich amine flash vessel, lean amine solution filter, gas condensation drum, pure gas flash,

Table 3.1 Estimated cost for the Benfield HiPure process [28].

Economic assumptions	
Project life(yrs.)	22
Equipment salvage value	0
Construction period (yrs.)	3
Plant operating time (hr./yr.)	7920
Interest rate (%)	5
K_2CO_3 cost ($/kg)	3.0
DEA cost ($/kg)	3.8
MDEA cost ($/kg)	2.6
DGA cost ($/kg)	4.06
DIPA cost ($/kg)	3.0
Natural gas price ($/MMBTU)	3.5
Tax rate (%)	

DEA, diethanolamine; *DGA*, diglycolamine; *DIPA*, diisopropylamine; *MDEA*, methyldiethanolamine.

process gas cooler, and lean amine vessel. Working capital investment was estimated as 43.49 M$ and startup cost was 15.53 M$; hence, the total capital expense was 214.53 M$. The total operating cost (TOC) was calculated as 45.27M$, which includes insurance, taxes, fixed charges, plant overhead cost, and general expenses [30]. Moreover, the economic potential (EP) was calculated by using Eq. (3.5) mentioned below in Table 3.2 and was found to be 16.43 M$/year.

Table 3.2 Equations for economic assessment.

Eq. No.	Parameters	Calculation method	References
3.5	Economic potential	$EP = (R - OPEX)(1 - t)$	[30]
3.6	Purchase cost of project	$\log_{10} Cp^0 = 2.2891 + 1.3604 \log_{10}(Q) - 0.1027[\log_{10}(Q)]^2$	[32]
3.7	Cost adjustment in equipment	$C_{TM} = 1.18 \times 15.9 \times \frac{567.5}{397} \times \sum_{i=1}^{n} Cpi^0$	[32]
3.8	Annual capital—related cost	$CRC = 0.2 \times (CTM + C_M)$	[32]
3.9	Annual operating expenditure	$OPEX = 0.04 \times Q \times 8000$	[32]
3.10	Economic feasibility	$C^s = \frac{CRC + OPEX}{\text{Annual total NG production}}$	[32]
3.11	Fixed cost	$FC = MC + CC$	[33]
3.12	Base plant cost	$BPC = 1.12 \times FC$	[33]
3.13	Project contingency	$PC = 0.2 \times BPC$	[33]
3.14	Total facility investment	$TFI = BPC + PC$	[33]
3.15	Contract and material maintenance cost	$CMC = 0.05 \times TFI$	[33]
3.16	Local taxes and insurance	$LTI = 0.015 \times TFI$	[33]
3.17	Direct labor cost	$DL = 21.1$ US dollar/man hour	[33]
3.18	Labor overhead cost	$LOC = 1.15 \times DL$	[33]
3.19	Membrane replacement cost	$MRC = 0.6 \times MC$	[33]
3.20	Annual variable operating and maintenance cost	$VOM = CMC + LTI + DL + LOC + MRC$	[33]
3.21	Normalized GWP	$\text{Normalized GWP} = \frac{GWP}{GWP(\text{reference})}$	[34]
3.22	Acidification potential	$AP = \Sigma\, EFi \times m_i$	[34]
3.23	Equivalency factor	$EF_i = \frac{n}{2MW} \times 64.06$	[34]
3.24	Net profit	$NP = (1 - t) \times (S - C)$	[34]
3.25	Damage index	$DI = f_2(x_i) = \sqrt{\left(\sum_{j=1}^{n} FEDI_j\right)^2 + \left(\sum_{j=1}^{n} TDI_j\right)^2}$ Where $i = 1-7$; $j = 1-N$	[34]
3.26	Profit before tax	$PBT = f_3(xi) = S - C$ Where $i = 1-7$	[34]
3.27	H$_2$S concentration	$X_{H_2S,\text{sweet}} = -0.0022T^3 + 0.4178T^2 - 19.504T + 273.82$	[35]

The comparison of capital cost of Benfield HiPure process of ADGAS was carried out and it was evident from the results that three alternatives such as MDEA/diethanolamine (DEA), MDEA/diisopropylamine (DIPA), and MDEA/diglycolamine will reduce the capital cost by 50%. Moreover, in order to analyze the operating cost, two important criteria were considered such as cost of desorption of acid gas and the annual spending on energy consumption. It was evident from the results that MDEA/DIPA displayed lowest economic output among all other alternatives due to high costs required for stripping of acid gases compared to the Benfield HiPure process. Whereas, in case of power consumption, MDEA-based alternatives showed considerably lower annual costs, and replacement of Benfield HiPure with MDEA-based alternative would save approximately 48% of the annual energy expenditures. In terms of EP, MDEA-based alternatives were more preferred as they can improve the economic capacity up to 37%, and 16.7% on the net profit (NP). Hence, this study confirmed that MDEA-based alternatives are more attractive and efficient considering quality of product, capital cost, and operating cost [30].

The oxidative desulphurization process was selected for the removal of sulfur contents due to its efficient performance as it removes most of the sulfur contents and reduces it from 8500 to 700ppm as per regulations [36]. The sour gas condensate is considered as hazardous and raises many environmental and operational issues. Moreover, it also improves the odor of the condensate by removing volatile sulfur contents. Furthermore, this study also discussed the techno-economic aspect of the selected process and evaluation confirmed the feasibility of the process in terms of economic and environmental standpoint. The techno-economic evaluation of recommended design was carried out, which includes pumps, agitators, and vessels. All the prices were reported using appropriate cost indexes. The required total capital investment (TCI) was $2,444,000 and annual variable cost was $2,236,000 for the production of 3000 bbl/day sweet condensate for 330 active days. Hence, the proposed sweetening process required 2.26$/bbl to process a sour condensate. Furthermore, it should be accentuated that health, safety, and environment are the major concerns, not economics [36].

The energy requirement for the increased production was also assessed using the ProMax simulation model. The model indicated that solvent or MDEA rate, MDEA concentration, inlet temperature, steam flow rate, and MDEA entering position are the highly influential variables in natural gas sweetening process, but among these variables, the rate of MDEA is the most important factor that directly affects the energy consumption [37]. The process was optimized to increase the yield by using the same model. After optimization, the increased yield was attained, that is, 0.5%, which reduces the energy requirement by 19.1% under full load condition. Moreover, the analysis was extended by considering decreased feed gas load and pressure and increased H_2S concentration. The results provided evidence that under low load, energy can be conserved by reducing the steam generation of the regeneration unit and altering the position of amine inlet in secondary absorption tower, hence reducing the annual variable cost [38].

Other factors that affect the energy consumption are the amine temperature, condenser heat load, reboiler heat exchanger, and heat load of rich/lean amine gas sweetening unit. The optimizations of these parameters are essential for reducing the energy requirement. The effect of amine temperature was investigated to increase the profitability and reduce the operating cost [39]. The amine sweetening unit was simulated using PRO II commercial simulation software and Aspen HYSYS software. The PRO II provided the most promising results comparable to field results. The effect of change in rich amine temperature was observed by executing the economic comparison on the main heat transfer and amine transfer equipment with the loading of 28%. The cost was computed by using JCG procedure for a temperature ranging from 80 to 120°C, and it was confirmed that 100°C is the optimum temperature, which reduces the operating cost to over $97,704/yr. It has been reported in literature that CO_2 concentration also affects the capital and operating cost of amine gas sweetening process. The scale-up, fixed, and operating cost of the sweetening process was analyzed using Aspen HYSYS for treating high concentration of acid gases in natural gas. The process was simulated using Aspen HYSYS V 8.8, and optimization was carried out using amine circulation rate, lean amine temperature, and reboiler duty and amine composition as input parameters. The total cost was estimated using the central composite experimental design model. The results demonstrated that the absorber temperature was adjusted by reducing the temperature of lean amine and solvent flow rate [32]. The solvent flow rate was targeted to increase the CO_2 percentage; as a result, CO_2 concentration increased from 4.13% to 25%, which ultimately increased the cost of operation and the entire utility requirement. Moreover, due to increase in % CO_2, heat exchanger surface area and size of regenerator vessels were also increased. This linear relationship was well demonstrated by LieMather model, which did not indicate any significant difference compared to Kent–Eisenberg model [32]. The economic assessment of CO_2 removal from natural gas was also executed using two-stage membrane gas sweetening process and simulated on Aspen HYSYS, and the cost of compressor and membrane was evaluated by the cost model reported in literature [40]. The project time was 20 years, and the purchase cost was estimated using Eq. (3.6) from Table 3.2. Moreover, the cost adjustments were made by using the CEPCI for the equipment. Hence, the total cost of equipment was calculated by using Eq. (3.15) from Table 3.2. In addition, the membrane life time was considered as 5 years, and the annual capital–related cost was estimated using Eq. (3.16) from Table 3.2. The annual operating expenditures were calculated by using Eq. (3.17) from Table 3.2, in which only electrical cost was included. Hence, the natural gas processing cost can be estimated by using Eq. (3.18), where the Cs was employed to assess the economic feasibility of the carbon membrane system.

The effect of feed gas pressure on power was analyzed by varying the pressure. It was observed that as the pressure increases, power demand of the compressor also increases, while the required area of membrane decreases because of higher driving force. Moreover, the sensitivity analysis showed that by increasing the CO_2 permeance, energy consumption would decrease, which would ultimately decrease the processing cost [41]. The effect of different types of amine on the financial side was investigated and analyzed in the natural gas sweetening process using a mixture of MDEA and TEA. The authors calculated the gross profit gained by using a mixture of solvent and using pure MDEA solvent. It was observed that by using 40% MDEA, 5% total cost reduction is $ 3.86×10^5/yr at a loading of 0.005. The model was derived as shown in Eq. (3.27), which shows that the same H_2S concentration was achieved as pure MDEA if the mixture of MDEA/TEA operates at a temperature of 60°C, and decreasing it further may lead to further reduction in energy requirement [35].

The negative impacts of sulfur on the environment can be reduced by removing the sulfur contents from the feed gas. The performance, emissions, and cost estimates are entirely dependent on technology. The technological impact was measured using three gas sweetening processes such as amine process, absorption process using recompressed vapor, and a polymeric membrane system. The processes were simulated using Aspen HYSYS V8.8, and the cost was computed using activated energy analyzer. It was observed that the capital cost of membrane process was 12% higher than amine process and 5% higher than vapor recompression process, but operating cost was lower than vapor recompression process and higher than conventional amine process due to observed methane losses. Moreover, due to inclusion of compressor despite of using high thermal energy like amine process, the membrane process can be considered an energy-efficient process that reduced the energy requirement from 77% to 72%, and its low GHG emissions and reboiler absence make it feasible in terms of environment that reduced emission from 80% to 76% [42].

Similarly, the economical evaluation of mixed amine process, Sulfinol-M process, LO-CAT process, and Shell–Paques process was also carried out using Aspen economic analyzer V 8.1. In gas sweetening process, the circulation rate of solution is considered to be the most influential parameter, and in economic analyzer, four different flow rates were considered for all four processes. The operating cost is directly related to the energy consumed in the reboiler, pumps, and heater. In terms of operating cost, the LO-CAT process is the most economical process compared to Shell–Paques and Sulfinol-M process, and in terms of capital cost, the Sulfinol process has the highest capital cost, whereas LO-CAT process has the lowest capital cost. Hence, it is evident from the study that LO-CAT process is an economically feasible option [43].

The amine membrane hybrid process for natural gas sweetening was also assessed economically [33]. The hybrid system consisted of a single-stage and two-stage membrane system. The economic evaluation was carried out by considering all factors that affect TCI and TOC, such as nitrogen presence in feed, feed flow rate, stage cut, and membrane selection. The amine hybrid process was simulated using Aspen HYSYS, and economical evaluation was performed using Aspen Icarus. For single-stage amine hybrid process, by considering the cost of membrane area, it was observed that TCI of hybrid process is higher than conventional amine sweetening process, whereas the TOC is lower. The payback was calculated, and it was found that payback of less than 3 years is appropriate for economics. Moreover, nitrogen is not environmentally hazardous, but its presence increases the exportation cost; hence, its separation is mandatory to control the exportation expenses. The separation cost of nitrogen also increases the product cost; therefore as the nitrogen percentage in the feed increases, product cost also increases; hence, higher nitrogen content is not economically favorable. In addition to this, single-stage membrane did not show significant effect on product cost enhancement, but in some cases for low flow rate, large membrane area is required with small amine unit resulting in increase in product cost. It is also evident from the results that multisection membrane required high TOC and TCI so it is not appropriate for the feed containing acid gas up to 20 mol%. Therefore, a reasonable membrane should be selected to control TOC and TCI. Eqs. (3.12)–(3.20) from Table 3.2 were used to evaluate the parameters for evaluation of membrane.

The escalating energy demand, growing global warming concerns, and rigorous emission regulations have increased the efforts to optimize the gas recovery. Therefore, the efficient removal of acid gases from nature is essential to maintain the specifications of the final product with reduced energy requirement and maximum profit. The rapid increase of population has critically increased the global warming issues; hence, sustainable gas sweetening operation requires maximum profit and environmentally safe operation. GWP and AP are the main indicators in natural gas purification. The optimization of the sweetening process was carried out by considering different parameters such as lean amine temperature and pressure, sour gas temperature and pressure, rich amine feed temperature and pressure, and gas flow rate at inlet. The criteria for environmental escalation and economic gain were selected using four cases including NP, GWP, and AP. Restrictions were imposed on H_2S and CO_2 content as per the maximum allowable limit. The environmental effects were analyzed by analyzing the trade-offs between (i) GWP and NP, (ii) AP and NP, (iii) GWP and AP, and (iv) GWP, AP, and NP to minimize the environmental effluence along with growth in the economic value. The greenhouse effect is calculated via GWP and the authors used Eq. (3.21) for computing GWP. AP is expressed as SO_2 equivalent and equivalency factor of ith component (EF_i) was computed using Eqs. (3.22) and (3.23) and NP was computed by using Eq. (3.24). The trade-offs between two objective functions were obtained. For two objective optimizations, the result from reducing the GWP and AP along with increase in NP depicted that, due to rise in NP, the GWP and AP increase, resulting in the sacrifice of environmental pollution. The trade-off between GWP and AP depicted that both should be kept as minimum to ensure the low environment pollution. Moreover, for three objective optimization, as the AP decreases NP also decreases, if it is required to increase the profit, then acidic potential would increase, resulting in environmental pollution. There is also a possibility that NP increases with increase of GWP. Hence, researchers can further study the obtained results to optimize the sweetening process and attain more NP. In the same way, another study was carried out by employing the I-MODE algorithm through MS Excel VBA and Aspen HYSYS for the economic, safety, and environmental assessment of the gas sweetening process. The AP, damage index (DI), and profit before tax (PBT) were considered as objective functions and calculated by using Eqs. (3.22), (3.25) and (3.26) mentioned in Table 3.2 [34].

Different cases were analyzed by varying different process parameters such as temperature, pressure, flow rate of gas at inlet, flow rate of rich amine, and temperature of regeneration column. In all the cases, exchange of objective function is considered, in which AP and DI were kept low and PBT was kept high. It was observed that as DI increases, PBT also increases and AP decreases. The net flow method was used to obtain the preferred operating point, which enhances DIs, PBTs, and energy savings for 98% removal of H_2S. Moreover, the energy saving and PBT were maximized when the gas sweetening process was optimized for minimum AP and PBT. From the obtained results, it is confirmed that this methodology can be utilized for any other process where commercial software is being used [44].

4. Environmental challenges

Natural gas has gained significant importance in this century and a drastic increase in its utilization has been observed globally. The growing demand of natural gas is directly linked primarily with the environment as it contains acid gases that are poisonous and corrosive in nature. Therefore, environmental regulation authorities have set some limits for the emission of these gases. To meet these regulations and control these emissions, removal of acid gases from natural gas is essential. Chemical absorption through amines has been commonly used for the removal of acid gases, but they have some environmental and health hazards for humans, animals, and aquatic life [45]. These common amines react with acid gases and deteriorate. These degraded materials reduce the absorption efficiency; cause corrosion, foaming, and fouling; and increase the release of pollutants [46].The formation of nitrosamines and nitramines by atmospheric oxidation of amine is hazardous, toxic, mutagenic, and carcinogenic in nature [47]. Hence, the formation of these compounds and accidental spills or leaks of sweetening chemicals should be closely monitored, which can lead to soil and water contamination, posing risks to ecosystems and human health [48].Researchers are actively exploring alternative technologies for achieving CO_2 specifications, and one promising approach involves the use of membrane modules. Unlike traditional methods that rely on chemical solvents, membrane-based gas separation is gaining traction due to several advantages. These include its eco-friendliness, smaller space requirements, and cost-effectiveness resulting from lower energy consumption.

The principle behind membrane separation lies in the varying diffusion rates of gas molecules within the membrane materials. Membrane technology is particularly relevant in various gas processing industries, where it can be applied to tasks such as separating acid gases (like CO_2 and H_2S), removing heavy hydrocarbons, eliminating water, extracting nitrogen (N_2), and recovering helium [38].

Many sweetening processes require significant energy input for compression, heating, and cooling. The energy consumption associated with these processes can contribute to air pollution and GHG emissions if the energy is generated from fossil fuels. Adsorption processes require energy for the regeneration of adsorbents, which can contribute to energy-related environmental concerns, and cryogenic processes involve extremely low temperatures, necessitating substantial energy consumption for cooling and refrigeration. Some cryogenic processes can result in the loss of valuable gases, which can be wasteful and environmentally detrimental. Moreover, physical solvent processes demand energy for heating and cooling. If this energy comes from fossil fuels, it can lead to emissions of GHGs and other air pollutants. In addition to this, the regeneration of solvent solutions can produce solid waste, including spent adsorbent materials. Appropriate disposal and management of these waste materials are critical to prevent environmental harm [49]. In the intervening time, the ecological and health hazards elicited by amines are controllable, but still adequate efforts are required to move toward some alternate sources. Steps should be taken by the authorities to develop new or improved amines or consider other solvents that cause low damage to the environment, device mechanisms for proper handling, and disposal of amines and to fill the research gap on environmental effects from amines [50].

5. Conclusion and future outlooks

Economic assessment is used to measure the performance, emission, and cost of natural gas sweetening technologies. The variabilities in this assessment depend on the state of developed technology and the level of details used for the estimation. In this chapter, economic analysis and environmental assessment of various natural gas sweetening technologies is carried out to illustrate the feasibility of the technology. The results imply that several variables such as solvent circulation rate, solvent temperature, and concentration are highly influential variables that affect the energy consumption and economics of the process. The environmental and economic assessment can be used as a tool to highlight the need for research in a particular area, development, and demonstration of new methodologies and technologies.

In industries, the most commonly used amines are monoethanolamine, MDEA, and DEA. These amines react with acid gases and hence degrade [46,51,52]. In order to deal with this issue, several new compounds are under consideration due to their good absorption capability and higher stability. Piperazine (PZ) is a recently developed compound showing promising results in terms of stability at higher temperature [53,54]. The limited absorption of PZ can be further improved by blending it with amines. In order to search for more future alternatives, recent theoretical developments have revealed that ionic liquids also have several advantages over conventional amines. Despite having tunable properties and efficient performance, their toxicity, biodegradability, and environmental impact need further examination [55]. Sound knowledge is still required about their behavior before applying them on a large scale. The development of hybrid material and membrane is another attractive alternative, and the physiochemical analysis and characterization of these materials are essential. Moreover, along with experimental studies, computational modeling methods must be established as an assessment tool to judge the performance of material. Unquestionably, the synthesis of material, characterization, removal efficiency, environmental impacts, and computation require extensive efforts and in-depth knowledge. Furthermore, the economic feasibility of the new material must be analyzed before applying them on a large scale and an economic model must be developed. However, primarily, the future requirement is to develop new-cost effective and environmentally friendly material for the separation of acid gases from natural gas.

Abbreviations and symbols

AEA	Activated energy analyzer
AFDC	Annual debt and equity costs of capital funds
AP	Acidification potential
DEA	Diethanolamine
DGA	Diglycolamine
DI	Damage index
DIPA	Diisopropylamine
EP	Economic potential

FOC	Fixed operating cost
GHGs	Greenhouse gases
GWP	Global warming potential
IL	Ionic Liquid
MDEA	Methyldiethanolamine
MEA	Monoethanolamine
NP	Net profit
PBT	Profit before tax
PZ	Piperazine
TCI	Total capital investment
TCR	Total capital requirement
TOC	Total operating cost
VOC	Variable operating cost in dollars

References

[1] Dyment J, Watanasiri S. Aspen Technology Inc.: Bedford, MA, USA; 2015.
[2] Kumar S, Kwon H-T, Choi K-H, Lim W, Cho JH, Tak K, et al. Applied Energy 2011;88(12):4264−73.
[3] Kohl A, Nielsen R. 1997.
[4] Guidotti TL. International Journal of Toxicology 2010;29(6):569−81.
[5] Conway W, Bruggink S, Beyad Y, Luo W, Melián-Cabrera I, Puxty G, et al. Chemical Engineering Science 2015;126:446−54.
[6] Faiz R, Al-Marzouqi M. Separation and Purification Technology 2011;76(3): 351−61.
[7] Abotaleb A, El-Naas MH, Amhamed A. Journal of Natural Gas Science and Engineering 2018;55:565−74.
[8] Låg M, Lindeman B, Instanes C, Brunborg G, Schwarze PE. 2011.
[9] Abrahamsen EB, Milazzo MF, Selvik JT, Asche F, Abrahamsen HB. Reliability Engineering and System Safety 2020;198:106811.
[10] Borhani TNG, Afkhamipour M, Azarpour A, Akbari V, Emadi SH, Manan ZA. Journal of Industrial and Engineering Chemistry 2016;34:344−55.
[11] de Koeijer G, Talstad VR, Nepstad S, Tønnessen D, Falk-Pedersen O, Maree Y, et al. International Journal of Greenhouse Gas Control 2013;18: 200−7.
[12] Karl M, Wright RF, Berglen TF, Denby B. International Journal of Greenhouse Gas Control 2011;5(3):439−47.
[13] Kikuchi Y, Papadokonstantakis S, Banimostafa A, Sugiyama H, Hungerbühler K, Hirao M. Computer aided chemical engineering. Elsevier; 2012. p. 1392−6.
[14] Therivel R. Strategic environmental assessment in action. Routledge; 2012.
[15] Wood C. Environmental impact assessment: a comparative review. Routledge; 2014.
[16] Sathre R, Chester M, Cain J, Masanet E. Energy 2012;37(1):540−8.
[17] Ruiz-Mercado GJ, Smith RL, Gonzalez MA. Industrial and Engineering Chemistry Research 2012;51(5):2309−28.
[18] Badr S, Frutiger J, Hungerbuehler K, Papadokonstantakis S. International Journal of Greenhouse Gas Control 2017;56:202−20. https://doi.org/10.1016/j.ijggc.2016.11.013.
[19] Aliff Radzuan MR, Syarina NA, Wan Rosdi WM, Hussin AH, Adnan MF. Materials Today: Proceedings 2019;19:1628−37. https://doi.org/10.1016/j.matpr.2019.11.191.

[20] Al-Lagtah NMA, Al-Habsi S, Onaizi SA. Journal of Natural Gas Science and Engineering 2015;26:367–81. https://doi.org/10.1016/j.jngse.2015.06.030.
[21] Shah D, Saparov A, Mansurov U, Amouei Torkmahalleh M. Industrial and Engineering Chemistry Research 2020;59(7):3213–20. https://doi.org/10.1021/acs.iecr.9b06845.
[22] Poe WA, Mokhatab S. Modeling, control, and optimization of natural gas processing plants. gulf professional publishing; 2016.
[23] Suleiman B, Abdulkareem A, Abdulsalam Y, Musa U, Kovo A, Mohammed I. Journal of Natural Gas Science and Engineering 2016;36:184–201.
[24] Thafseer M, Al Ani Z, Gujarathi AM, Vakili-Nezhaad GR. Journal of Natural Gas Science and Engineering 2021;88:103800.
[25] Yavini TD, Ali M-DI, Muhammad WS. International Journal of Scientific Research in Science, Engineering and Technology 2015;1(5):194–203.
[26] Li W, Zhuang Y, Zhang L, Liu L, Du J. Journal of Cleaner Production 2019;232:487–98.
[27] Maier M, Ghazisaidi N, editors. FiWi access networks. Cambridge: Cambridge University Press; 2011. p. 146–59.
[28] Peters MS, Timmerhaus KD, West RE. Plant design and economics for chemical engineers. New York: McGraw-Hill; 2003.
[29] Sarfaraz B, Kazmi B, Taqvi SAA, Raza F, Rashid R, Siddiqui L, et al. Thermodynamic evaluation of mixed refrigerant selection in dual mixed refrigerant NG liquefaction process with respect to 3E's (Energy, Exergy, Economics). Energy 2023;283:128409.
[30] Ochieng R, Berrouk A, Peters C, Slagle J. Amine-based gas-sweetening processes prove economically more viable than the Benfield HiPure process. Bryan, Texas, US: Bryan Research and Engineering Inc; 2013.
[31] Ellaf A, Taqvi SAA, Zaeem D, Siddiqui FUH, Kazmi B, Idris A, et al. Energy, exergy, economic, environment, exergo-environment based assessment of amine-based hybrid solvents for natural gas sweetening. Chemosphere 2023;313:137426.
[32] Muhammad A, GadelHak Y. Journal of Natural Gas Science and Engineering 2014;17:119–30.
[33] Mozafari A, Azarhoosh MJ, Mousavi SE, Khaghazchi T, Ravanchi MT. Chemical Papers 2019;73(7):1585–603.
[34] Tikadar D, Gujarathi AM, Guria C. Journal of Natural Gas Science and Engineering 2021;95:104207.
[35] Fouad WA, Berrouk AS. Journal of Natural Gas Science and Engineering 2013;11:12–7.
[36] Moaseri E, Mostaghisi O, Shahsavand A, Bazubandi B, Karimi M, Ahmadi J. Journal of Natural Gas Science and Engineering 2013;12:34–42.
[37] Berchiche M, Belaadi S, Leonard G. In: Computer aided chemical engineering. Elsevier; 2020. p. 67–72.
[38] Shang J, Qiu M, Ji Z. Natural Gas Industry B 2019;6(5):472–80.
[39] Koolivand Salooki M, Keshavaz Bahdori M, Esfandyari M, Beigi M, Sadeghzade Ahari J. Journal of Mineral Resources Engineering 2019;4(1):45–57.
[40] Dai Y, Peng Y, Qiu Y, Liu H. Energies 2019;12(21):4213.
[41] Chu Y, He X. Membranes 2018;8(4):118.
[42] Gutierrez JP, Ruiz ELA, Erdmann E. Journal of Natural Gas Science and Engineering 2017;38:187–94.
[43] Kazemi A, Malayeri M, Shariati A. Journal of Natural Gas Science and Engineering 2014;20:16–22.

[44] Tikadar D, Gujarathi AM, Guria C. Process Safety and Environmental Protection 2020;140:283–98.
[45] Knaak J, Leung H-W, Stott W, Busch J, Bilsky J. Reviews of Environmental Contamination & Toxicology 1997:1–86.
[46] Liang ZH, Rongwong W, Liu H, Fu K, Gao H, Cao F, et al. International Journal of Greenhouse Gas Control 2015;40:26–54.
[47] Mazari SA, Alaba P, Saeed IM. Journal of Environmental Chemical Engineering 2019;7(3):103111.
[48] Khan SN, Hailegiorgis SM, Man Z, Shariff AM, Garg S. Journal of Molecular Liquids 2017;229:221–9.
[49] Låg M, Andreassen Å, Instanes C, Lindeman B, Norwegian Institute of Public Health (FHI). Health effects of different amines relevant for CO_2 capture. NILU OR 2009:1–28.
[50] Hosseini-Ardali SM, Hazrati-Kalbibaki M, Fattahi M, Lezsovits F. Energy 2020;211:119035.
[51] Léonard G, Crosset C, Toye D, Heyen G. Computers and Chemical Engineering 2015;83:121–30.
[52] Saeed IM, Alaba P, Mazari SA, Basirun WJ, Lee VS, Sabzoi N. International Journal of Greenhouse Gas Control 2018;79:212–33.
[53] Khan AA, Halder G, Saha A. International Journal of Greenhouse Gas Control 2017;64:163–73.
[54] Khan BA, Ullah A, Saleem MW, Khan AN, Faiq M, Haris M. Sustainability 2020;12(20):8524.
[55] Baj S, Siewniak A, Chrobok A, Krawczyk T, Sobolewski A. Journal of Chemical Technology & Biotechnology 2013;88(7):1220–7.

Absorption techniques for natural gas sweetening

4

Acid gases properties and characteristics in companion with natural gas

Sina Mosallanezhad, Hamid Reza Rahimpour and Mohammad Reza Rahimpour

Department of Chemical Engineering, Shiraz University, Shiraz, Iran

1. Introduction

Natural gas is a fossil fuel mostly made of methane, with trace quantities of other hydrocarbons, including ethane, propane, and butane. It is recovered by drilling from subsurface reservoirs. Natural gas is a vital energy source for many nations and has several uses. Natural gas is widely used for heating and energy production. It is a cleaner burning fuel than coal or oil, generating less CO_2 and other pollutants. Natural gas is also employed in industrial activities such as steel manufacturing, chemical manufacturing, and fertilizer production. Moreover, natural gas is utilized as a feedstock in manufacturing various goods, including polymers, textiles, and medicines [1]. It is also used as a fuel for compressed natural gas vehicles. It is rapidly employed as a greener alternative to conventional bunker fuel in the shipping sector [2]. It has no color, form, or odor when it is pure [3]. It is a combustible gas, and burning it releases a significant quantity of energy. Compared to other fossil fuels, it is regarded as a clean, ecologically beneficial fuel (crude oil and coal). Other than natural gas, burning fossil fuels release vast quantities of chemicals and particles that harm human health [4,5]. While CO_2 and nitrous oxide emissions are reduced while burning natural gas and sulfur dioxide emissions are minimal, this helps to lessen issues with the ozone layer, acid rain, and greenhouse gases [5a]. The switch to natural gas from fossil fuels increases energy efficiency and heralds a future with fewer carbon emissions. By creating carbon capture and storage technologies, CO_2 emissions may be reduced and used for other purposes [6]. Natural gas is safe when stored, transported, and used as energy. Its use has helped residential, commercial, and industrial heating. It is also utilized to generate energy and heat. For example, it is used as a feedstock or raw material in the petrochemical industry in ethylene production. It is used to produce ammonia, which is then used in fertilizer. Natural gas may make sulfur, hydrogen, and carbon black [3,4].

The significance of natural gas is reflected in its part in worldwide energy consumption. Natural gas accounted for 24% of worldwide energy consumption in 2020, according to the International Energy Agency, making it the second largest source of energy after oil. It is also expected to play a growing role in transitioning to a low-carbon energy system since it may replace higher-carbon fuels in power production and transportation [4]. Finally, natural gas is a valuable energy source with several uses in heating, power production, industrial operations, and manufacturing. Because of its lower emissions compared to other fossil fuels, it is an appealing alternative for governments wanting to decrease their carbon impact. Fig. 4.1 shows the distribution of natural gas reserves in the world.

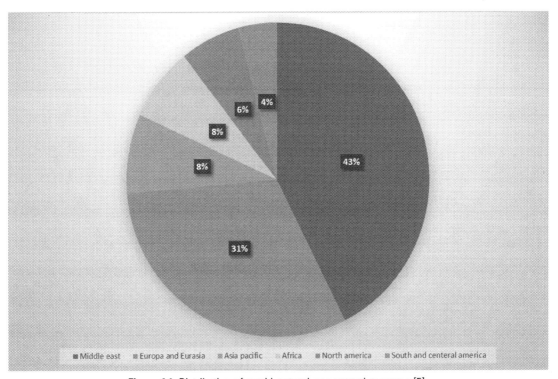

Figure 4.1 Distribution of world natural gas proved reserves [5].

2. Natural gas origins

The genesis of the various fossil fuels, mainly (in the present context) crude oil and natural gas, is the subject of several hypotheses. The most widely recognized explanation holds that organic materials or organic detritus, such as the remnants of animals or plants, are squeezed beneath the Earth over periods of geologic time that are estimated in millions of years (millennia), leading to the formation of natural gas and crude oil. The remnants of plants and animals accumulated in dense layers at the beginning of the oil and gas production eras and underwent transformation as silt, muck, and other detritus accumulated on top of them over time. Consequently, the organic stuff came under strain as the silt, mud, and other inorganic detritus transformed into rock. The biological stuff was squeezed as the pressure increased, and when coupled with other underground forces, the parts of the organic matter broke down into crude oil and natural gas [6].

3. Natural gas composition

When extracted from the ground, natural gas contains a number of different chemicals, the most important of which is methane. Natural gas is composed predominantly of methane, with smaller amounts of higher hydrocarbons like propane, ethane, butane, as well as CO_2, nitrogen, H_2S, oxygen, and trace amounts of rare gases like argon and krypton, making up the rest of the gas [7]. After being withdrawn from the reservoir, raw gas is processed to remove impurities and separate heavier hydrocarbons, mainly leaving methane. CO_2 and H_2S, which may be used to produce sulfur, are eliminated [8,9].

The initial processing step will vary depending on the kind of natural gas. Natural gas might dissolve with oil when it is recovered from wells that are connected to it. It has to be released in this situation, and then it needs to be removed from the oil [6]. The natural separation of crude oil underground due to high pressure causes gas to escape from the reservoir ultimately leading to pressure reduction of the reservoir and preservation of the remaining oil [10]. But sometimes, particularly when the gas is mixed in a light crude oil or a natural gas condensate, specialist separation equipment is needed. If separation is required, it is followed by a procedure to eliminate any water linked to the gas. However, some water could be in solution in the natural gas. Liquid water can be easily separated from the gas. Either an adsorption method or an absorption approach is used to eliminate this [11,12].

The higher hydrocarbons must be removed in the next step. Once separated, they are often referred to as natural gas liquids. Once again, two methods are often used: the cryogenic expander process and the absorption technique. The first one is more efficient in separating lighter hydrocarbons from natural gas, primarily ethane [6]. After the liquid components of natural gas have been separated, the liquids are treated to separate the various hydrocarbons they contain. Pentane, butane, propane, and ethane are often found in these. A liquid known as natural gasoline may also be produced during separation [10,13]. They are marketable for use as fuels, in chemical reactions, and in oil refineries. Due to H_2S and CO_2 acidic properties, natural gas is processed after hydrocarbon separation to remove them from the residual gas stream [14].

4. Removing acid gases from natural gas

Natural gas is harmful and corrosive due to H_2S, CO_2, and other sulfur compounds. Designing gas-handling infrastructure, particularly pipelines, to handle acid gases may not be economically feasible given the high cost of constructing corrosion-resistant equipment [14a]. In order to lower the acid gas concentration to acceptable levels, acid gas removal devices are placed at the beginning of the gas handling process [15].

The avoidance of CO_2 frost formation in cryogenic units, the improvement of process gas calorific value, and the lowering of process gas corrosivity all contribute to CO_2 removal [16]. The amount of CO_2 in the crude gas is typically decreased to the pipeline-required requirements, which is the concentration range needed for safe gas distribution at a suitable heating value and corrosivity levels [17]. To avoid the development of CO_2 frost in the static and rotational machinery of the cryogenic unit, a larger degree of CO_2 removal is needed if the gas must subsequently travel through cryogenic conditions [18]. The cryogenic unit process parameters (P, T) determine the amount of removal, and it should be noted that low temperatures and high pressures promote the development of CO_2 frost. For instance, the CO_2 level of liquefied natural gas unit feed gas must be kept to a maximum of 50 ppmv [19].

H_2S, after CO_2, is the second largest cause of process gas corrosivity. H_2S toxicity and environmental laws are what essentially drive natural gas purification from its H_2S presence. Due to its severe toxicity even at ppm levels, H_2S in natural gas necessitates particular gas handling and processing concerns. The symptom experienced after human exposure depends on the air concentrations and the amount of time spent exposed. When a person is exposed to an environment polluted with H_2S, instantaneous mortality is proven at a concentration of 2000 ppm, but extended exposure is also capable of becoming lethal at a concentration of 500 ppm (1 h). Lower exposure levels are still dangerous but may not result in instant death [20].

The process of converting sour gas (H_2S rich) into sweet gas (H_2S free) by H_2S removal is referred to as "sweetening." Since the removal of both acid gases (CO_2 and H_2S) occurs during the same procedure, industry may more easily refer to sweetening as the process of removing acid gases in general. The desiccant regeneration in the sweetening unit causes the acid gas removal units to create H_2S-rich gas. The H_2S-rich gas cannot be discharged into the environment due to its toxicity. Instead, sulfur oxides (SO_x), which are less harmful than H_2S, are often produced by flaring the poisonous gas. H_2S flaring may be avoided during regular plant operations because of the conservative environmental standards that local authorities have set for operating enterprises [6,21].

5. Definition of physical and chemical properties

5.1 Boiling point

The boiling point is a fundamental physical property of a substance that refers to the temperature at which it changes from liquid to gaseous state. At the boiling point, the vapor pressure of the liquid is equal to the external pressure applied to the surface of the liquid. When this occurs, bubbles of vapor form throughout the liquid, and the liquid vaporizes rapidly [22,23].

The boiling point is an essential property of a substance, as it can be used to identify and characterize a pure substance. Each pure substance has a specific boiling point that remains constant as long as the pressure remains constant. Therefore, the boiling point can be used to distinguish between different compounds and determine a sample's purity. The boiling point is also affected by external factors, such as pressure, which can alter the boiling point of a substance. For example, at higher altitudes where the atmospheric pressure is lower, the boiling point of water is lower than at sea level. This is why cooking times and temperatures may need to be adjusted when cooking at higher altitudes [24].

5.2 Melting point

A substance's melting point (or, more rarely, liquefaction point) is the temperature at which it transitions from solid to liquid. The liquid and solid phases are in balance at the melting point. A substance's melting point is pressure dependent and is generally expressed at a standard pressure such as 1 atm or 100 kPa. The freezing point or crystallization point is defined as the temperature of the reversal shift from liquid to solid. Because of the tendency of substances to supercool, the freezing point may seem to be lower than it really is. When determining a substance's "characteristic freezing point," the real approach is nearly invariably the concept of detecting the disappearance rather than the production of ice, that is, the melting point [25].

5.3 Density

The density of a material (specific mass or volumetric mass density) is its mass per unit volume. The density of a pure material has the exact numerical value as its mass concentration. Varied materials have different densities, and density may be important in terms of buoyancy, purity, and packing. At normal temperature and pressure settings, iridium and osmium are the densest known elements. Temperature and pressure affect a material's density. For solids and liquids, this variance is often minimal, but for gases, it is significantly bigger. Increasing the pressure on an item reduces its volume and hence increases its density. With a few exceptions, raising the temperature of a material reduces its density by increasing its volume. Heating the bottom of a fluid leads to convection of heat from the bottom to the top in most materials because the density of the heated fluid decreases, causing it to rise compared to denser unheated material [26].

5.4 Vapor density

Vapor density is an essential concept in the study of gases. It measures the mass of a gas about the volume it occupies compared to the mass and volume of a standard gas. This standard gas is typically hydrogen, oxygen, or air. Using hydrogen as the reference substance, vapor density is calculated as the ratio of the mass of a given volume of gas to the mass of an equal volume of hydrogen at the same temperature and pressure. This ratio is equal to half the relative molecular mass of the gas when the density of hydrogen is taken as 1 [27].

5.5 Vapor pressure

Vapor pressure is the pressure exerted by a vapor when the vapor is in equilibrium with the solid or liquid form of the same substance, like when conditions are such that the substance can exist in both or all three phases. Vapor pressure is a measure of the tendency of a material to change into the gaseous or vapor state, and it increases with temperature. The temperature at which the vapor pressure at the surface of a liquid becomes equal to the pressure exerted by the surroundings is called the boiling point of the liquid [28].

5.6 Viscosity

Viscosity refers to a fluid's resistance, which can be either a liquid or a gas, to change its shape or allow neighboring portions to move relative to one another. It essentially signifies the opposition to flow. Fluidity, conversely, is the reciprocal of viscosity and measures how easily a fluid can flow. For instance, molasses has a higher viscosity compared to water. Viscosity can be thought of as the internal friction between the molecules of a fluid, hindering the development of differences in velocity within the fluid as parts of it are forced to move. Viscosity plays a significant role in determining the forces that must be overcome when using fluids for lubrication or transporting them through pipelines. It also regulates the flow of liquids in processes like spraying, injection molding, and surface coating. For many fluids, the shear stress responsible for inducing flow is directly proportional to the shear strain or deformation rate. In simpler terms, this ratio of shear stress to shear strain is constant for a specific fluid at a fixed temperature, known as dynamic or absolute viscosity. Fluids that exhibit this behavior are called Newtonian fluids, named after Sir Isaac Newton, who first described viscosity mathematically [29].

As the temperature of a liquid rises, its viscosity decreases rapidly. This means that it becomes easier for the liquid to flow. On the other hand, gases become more viscous when the temperature increases, causing them to flow more sluggishly. It is interesting how heating can affect the flow properties of different substances in such contrasting ways [29].

5.7 pH

pH is a numerical indicator of how acidic or basic aqueous or other liquid solutions are. The phrase, which is often used in chemistry, biology, and agronomy, converts the hydrogen ion concentration, which typically varies between 1 and 1014 g-equivalents per liter, into numbers between 0 and 14. The hydrogen ion concentration in pure water, which has a pH of 7, is 107 g-equivalents per liter, making it neutral (neither acidic nor alkaline). A solution with a pH below 7 is referred to as acidic, while one with a pH over 7 is referred to as basic, or alkaline [30].

Typically, a pH meter is used to measure pH, which converts the difference in electromotive force between appropriate electrodes put in the solution under test into pH measurements. A pH meter's primary components are a voltmeter connected to an electrode that responds to pH and a constant electrode. The reference electrode is typically a mercury–mercurous chloride (calomel) electrode; however, a silver–silver chloride electrode is sometimes used. The pH-responsive electrode is often made of glass. The two electrodes function as a battery when submerged in a solution. The hydrogen ion activity in the solution immediately affects the electric potential (charge) that the glass electrode generates. The voltmeter detects the difference in potential between the glass and reference electrodes. The meter may have a digital reading or an analog display (scale and deflected needle). Although analog readouts provide more significant signals of the rate of change, digital readouts have the benefit of being more precise. Portable pH meters with batteries are often used to measure the pH of soils in the field. Litmus paper or combining indicator dyes in liquid suspensions and comparing the resultant colors to a color chart with a pH calibration may also be used to test pH. However, these methods could be more precise [31].

As it indicates which crops will grow well in the soil and what modifications need to be done to make the soil suitable for growing any other crops, the pH of the soil is the most significant single feature of the moisture linked with it in agriculture. While conifers and many members of the Ericaceae family, including blueberries, will not survive on alkaline soil, acidic soils are often considered infertile, and this is true for most conventional crops. It may be "sweetened" or neutralized by applying lime to acidic soil. Aluminum and manganese are more soluble in soil when soil acidity rises, and many plants, especially crops, can only tolerate tiny amounts of those metals. The microbial breakdown of organic matter, fertilizer salts that hydrolyze or nitrify, draining salt marshes for agricultural use, sulfur compound oxidation, and other factors contribute to increased soil acidity [32].

5.8 Heat of vaporization

The heat of vaporization is the quantity of heat necessary to convert 1 g of a liquid into vapor, all without raising the temperature of the liquid [33]. The heat of vaporization is a latent heat, with "latent" originating from the Latin word "latere," meaning to be hidden or concealed. Latent heat represents the additional heat needed to transition a substance from a solid to a liquid at its melting point or from a liquid to a gas at its boiling point while keeping the temperature of the substance constant. It is important to note that latent heat is associated with a change in state rather than a temperature change. Due to its substantial heat of vaporization, the water evaporation process leads to a noticeable cooling effect, while condensation results in a warming effect [34].

5.9 Ionization potential

Ionization potential, also known as ionization energy, is an important physical property of an atom or molecule that describes the energy required to remove an electron from the outermost shell of an atom or molecule [35]. This process is known as ionization, and the energy required to perform this process is referred to as the ionization potential. The ionization potential is a fundamental property of atoms and molecules that plays an essential role in various areas of physics and chemistry. The ionization potential is typically measured in units of electron volts (eV) and is often listed in physical and chemical data tables for different elements and compounds. The ionization potential of an atom or molecule is determined by the electronic structure of the atom or molecule, including the number of electrons in the outermost shell, the electron configuration of the atom or molecule, and the strength of the electrostatic interactions between the electrons and the positively charged nucleus. The ionization potential is a critical parameter in many areas of physics and chemistry. For example, in solid-state physics, the ionization potential is used to determine the electrical conductivity of materials since the electrical conductivity is related to the ease with which electrons can be removed from the material. In analytical chemistry, the ionization potential is used to identify and quantify different molecules in a sample since the ionization potential of a molecule is related to its chemical structure and can be used as a fingerprint for the molecule [36].

6. Physical and chemical properties of CO_2 and H_2S

6.1 Physical properties

A physical property refers to a trait or attribute of a substance that can be detected or quantified without altering the substance's fundamental nature. For instance, silver is glossy as a metal and possesses exceptional electrical conductivity. It also can be shaped into thin sheets, a quality known as malleability. Conversely, salt lacks luster, is brittle in its solid form, and only conducts electricity when dissolved in water, a process it readily undergoes. Examples of physical properties in matter encompass characteristics such as color, hardness, malleability, solubility (ability to dissolve), electrical conductivity, density (mass per unit volume), melting point (the temperature at which it changes from solid to liquid), and boiling point (the temperature at which it transitions from liquid to vapor). These properties provide valuable insights into the nature and behavior of substances without causing any alteration in their chemical identity [37]. For example, Co_2 and H_2S melting points are −56.558 [38] and −85.490 [39], respectively.

6.2 Chemical properties

Chemical properties of matter refer to its inherent potential to undergo a chemical transformation or reaction due to its specific composition. These properties are determined by the elements present, the arrangement of electrons, and the types of chemical bonds within the matter. Defining a chemical property often involves referencing a potential change, as chemical properties describe how a substance can react or change when exposed to specific conditions. With sufficient knowledge of chemistry, one can examine the formula of a compound and deduce specific chemical properties. For instance, hydrogen possesses the chemical property capable of ignition and explosion under suitable conditions—this exemplifies a chemical property. Similarly, metals, as a general category, exhibit the chemical property of reacting with acids. For instance, zinc has the chemical property of reacting with hydrochloric acid to produce hydrogen gas—this reaction signifies a chemical property [37].

Tables 4.1 and 4.2 show the physical and chemical properties of CO_2 and H_2O.

Table 4.1 Properties of CO_2.

Property	Value	References
Boiling point	−78.464°C	[38]
Melting point	−56.558°C	[38]
Density	1.799 g/L	[38]
Vapor density	1.53	[40]
Vapor pressure	10.5 mm Hg at −120°C	[41]
Viscosity	21.29 uPa-sec at 300 K	[38]
pH	The pH of saturated CO_2 solutions varies from 3.7 at 101 kPa (1 atm) to 3.2 at 2370 kPa (23.4 atm)	[42]
Heat of vaporization	83.12 cal/g	[41]
Ionization potential	13.77 eV	[43]

Table 4.2 Properties of H_2S.

Property	Value	References
Boiling point	−60.33°C	[39]
Melting point	−85.49°C	[39]
Density	1.5392 g/L at 0°C at 760 mm Hg	[39]
Vapor density	1.189	[40]
Vapor pressure	17.6 atm	[44]
Viscosity	Gas at 101.325 KPa at 25°C; 0.012 8 m Pa S; 0.012 8 cP	[45]
pH	4.5	[39]
Heat of vaporization	Molar enthalpy of vaporization: 14.08 kJ/mol at 25°C; 18.67 kJ/mol at −59.55°C	[46]
Ionization potential	10.46eV	[44]

7. Conclusion and future outlooks

Natural gas is a gas that is abundant in hydrocarbons, with methane being its primary component. The significant amount of energy released during the combustion process of natural gas makes it advantageous for use as a fuel source. However, acid gases (CO_2 and H_2S) occur in raw natural gas in varied amounts from ppm to a few percent, producing corrosiveness and poisoning of the gas; in this chapter, we discussed their physical and chemical features. For this reason, CO_2 and H_2S must be removed from natural gas in the acid gas removal due to their toxicity, environmental damage, and economic aspects [8]. The primary objective of the acid gas removal unit is to effectively eliminate the acidic constituents to comply with the required sulfur and CO_2 criteria for the sales gas.

Therefore, systems that, in addition to sweetening natural gas, can store acid gases and use them in various processes should be designed; for example, CO_2 can be used in methanol production. The design of the systems should be in such a way that it is the least environmentally harmful and economical.

Abbreviations and symbols

AGRU Acid gas removal unit
CCUS Carbon capture, utilization, and storage
CNG Compressed natural gas
CO₂ Carbon dioxide
H₂S Hydrogen sulfide
IEA International Energy Agency

References

[1] Carroll JJ. Acid gas injection and carbon dioxide sequestration. John Wiley & Sons; 2010.
[2] Aslam MU, Masjuki HH, Kalam MA, Abdesselam H, Mahlia TMI, Amalina MA. An experimental investigation of CNG as an alternative fuel for a retrofitted gasoline vehicle. Fuel 2006;85(5−6):717−24.
[3] Speight J. Liquid fuels from natural gas. Handbook of alternative Fuel technologies, vol. 153; 2007.
[4] U. S. EIA. Natural gas 1998: issues and trends. April 1999. DOE/EIA-0560 (98), http://www.eia.doe.gov/pub/oil_gas.
[5] Faramawy S, Zaki T, Sakr A-E. Natural gas origin, composition, and processing: a review. Journal of Natural Gas Science and Engineering 2016; 34:34−54.
[5a] Haghighatjoo F, Rahimpour MR, Farsi M. Techno-economic and environmental assessment of CO_2 conversion to methanol: direct versus indirect conversion routes. Chemical Engineering and Processing - Process Intensification 2023. https://doi.org/10.1016/j.cep.2023.109264.
[6] Breeze P. Chapter 2 - the natural gas resource. In: Breeze P, editor. Gas-turbine power generation. Academic Press; 2016. p. 9−19.
[7] Snowdon LR. Natural gas composition in a geological environment and the implications for the processes of generation and preservation. Organic Geochemistry 2001;32(7):913−31.
[8] Rufford TE, et al. The removal of CO2 and N2 from natural gas: a review of conventional and emerging process technologies. Journal of Petroleum Science and Engineering 2012;94:123−54.

[9] Datta AK, Sen PK. Optimization of membrane unit for removing carbon dioxide from natural gas. Journal of Membrane Science 2006;283(1−2):291−300.
[10] Baker RW, Lokhandwala K. Natural gas processing with membranes: an overview. Industrial & Engineering Chemistry Research 2008;47(7):2109−21.
[11] Baker RW. Future directions of membrane gas separation technology. Industrial & Engineering Chemistry Research 2002;41(6):1393−411.
[12] Bernardo P, Drioli E, Golemme G. Membrane gas separation: a review/state of the art. Industrial & Engineering Chemistry Research 2009;48(10):4638−63.
[13] Watler KG. Process for separating higher hydrocarbons from natural or produced gas streams. Google Patents; 1989.
[14] Scholes CA, Stevens GW, Kentish SE. Membrane gas separation applications in natural gas processing. Fuel 2012;96:15−28.
[14a] Mosallanezhad S, Kiani P, Makarem MA, Rahimpour MR. Application of disaster engineering in oil, gas, and petrochemical industries. In: Rahimpour MR, Abrishami Shirazi N, Omidvar B, Makarem MA, editors. Crises in oil, gas and petrochemical industries. vol. 1. Elsevier; 2023. p. 47−70.
[15] Muller C, et al. A review of the practical application of micro-aeration and oxygenation for hydrogen sulfide management in anaerobic digesters. Process Safety and Environmental Protection 2022;165:126−37.
[16] Alcheikhhamdon Y, Hoorfar M. Natural gas purification from acid gases using membranes: a review of the history, features, techno-commercial challenges, and process intensification of commercial membranes. Chemical Engineering and Processing - Process Intensification 2017/10/01;120:105−13. https://doi.org/10.1016/j.cep.2017.07.009.
[17] Fouladi N, Makarem MA, Sedghamiz MA, Rahimpour HR. CO_2 adsorption by swing technologies and challenges on industrialization. In: Advances in carbon capture. Elsevier; 2020. p. 241−67.
[18] Sridhar S, Smitha B, Aminabhavi TM. Separation of carbon dioxide from natural gas mixtures through polymeric membranes—a review. Separation and Purification Reviews 2007;36(2):113−74.
[19] Zarezadeh F, Vatani A, Palizdar A, Nargessi Z. Simulation and optimization of sweetening and dehydration processes in the pretreatment unit of a mini-scale natural gas liquefaction plant. International Journal of Greenhouse Gas Control 2022;118:103669.
[20] Ma Y, et al. Hydrogen sulfide removal from natural gas using membrane technology: a review. Journal of Materials Chemistry A 2021;9(36):20211−40.
[21] Stewart M, Arnold K. Gas sweetening and processing field manual. Gulf Professional Publishing; 2011.
[22] Theodore L, Dupont RR, Ganesan K. Pollution prevention: the waste management approach to the 21st century. CRC Press; 1999.
[23] Goldberg D. 3,000 solved problems in chemistry. McGraw-Hill Professional; 1988.
[24] Jackson CM. Chemical identity crisis: glass and glassblowing in the identification of organic compounds: essay in honour of Alan J. Rocke. Annals of Science 2015;72(2):187−205.
[25] Karthikeyan M, Glen RC, Bender A. General melting point prediction based on a diverse compound data set and artificial neural networks. Journal of Chemical Information and Modeling 2005;45(3):581−90.
[26] T. E. O. E. D. Britannica. Encyclopedia Britannica. https://www.britannica.com/science/density. [Accessed].

[27] Daintith J. A dictionary of chemistry. Oxford University Press; 2008. in English.
[28] Pitzer KS, Lippmann DZ, Curl Jr RF, Huggins CM, Petersen DE. The volumetric and thermodynamic properties of fluids. II. Compressibility factor, vapor pressure and entropy of vaporization1. Journal of the American Chemical Society 1955;77(13):3433−40.
[29] T. E. O. E. Britannica. Viscosity. Encyclopedia Britannica, https://www.britannica.com/science/viscosity. [Accessed].
[30] Pye HOT, et al. The acidity of atmospheric particles and clouds. Atmospheric Chemistry and Physics 2020;20(8):4809−88.
[31] Bates RG. Definitions of pH scales. Chemical Reviews 1948;42(1):1−61.
[32] Buck RP, et al. Measurement of pH. Definition, standards, and procedures (IUPAC Recommendations 2002). Pure and Applied Chemistry 2002;74(11):2169−200.
[33] Guralnik DB. Webster's new world dictionary of the American language. No Title. 1970.
[34] Kramer PJ, Boyer JS. Water relations of plants and soils. Academic press; 1995.
[35] T. E. O. E. Britannica. Ionization energy. Encyclopedia Britannica, https://www.britannica.com/science/ionization-energy. [Accessed].
[36] Zhan C-G, Nichols JA, Dixon DA. Ionization potential, electron affinity, electronegativity, hardness, and electron excitation energy: molecular properties from density functional theory orbital energies. The Journal of Physical Chemistry A 2003;107(20):4184−95.
[37] LibreTextsChemistery. Differences in matter- physical and chemical properties. https://chem.libretexts.org/Bookshelves/Introductory_Chemistry/Introductory_Chemistry/03%3A_Matter_and_Energy/3.05%3A_Differences_in_Matter-_Physical_and_Chemical_Properties#:~:text=Physical%20properties%20include%20color%2C%20density,look%20for%20a%20chemical%20change. [Accessed].
[38] Haynes WM. CRC handbook of chemistry and physics. CRC press; 2016.
[39] O'Neil MJ, Smith A, Heckelman PE. The merck index: an encyclopedia of chemicals, drugs, and biologicals. Whitehouse Station, NJ, USA: Merck and Co," *Inc.*; 2006. p. 1204.
[40] Lewis RJ, Sax NI. Sax's dangerous properties of industrial materials. 11th ed. Hoboken, N.J.: J. Wiley & Sons; 2004 Available: https://archive.org/details/saxsdangerouspro0001lewi_f8v7.
[41] O'Neil MJ. The Merck index: an encyclopedia of chemicals, drugs, and biologicals. RSC Publishing; 2013.
[42] Kroschwitz JI, Howe-Grant M, Kirk RE, Othmer DF. Encyclopedia of chemical technology. John Wiley & Sons; 1996.
[43] O. S. A. H. A. (OSHA). Carbon dioxide. https://www.osha.gov/chemicaldata/183. [Accessed].
[44] O. S. A. H. A. (OSHA). Hydrogen sulfide. https://www.osha.gov/chemicaldata/652. [Accessed].
[45] Book MGD. Inc. Matheson gas products, vol. 7094; 1980. Secaucus, NJ.
[46] Haynes WM. CRC handbook of chemistry and physics. Internet Version 2011. Boca Raton, FL: Taylor Francis Group; 2011.

5

Application of amines for natural gas sweetening

Abdul Rahim Nihmiya[1] and Nayef Ghasem[2]
[1]Department of Civil and Environmental Technology, University of Sri Jayewardenepura, Nugegoda, Sri Lanka; [2]Department of Chemical and Petroleum Engineering, United Arab Emirates University, Al-Ain, United Arab Emirates

1. Introduction

Fossil fuels are a nonrenewable energy source that has been fulfilling human energy requirements and powering economies for long years. Regardless of the boom in renewable energy sources, fossil fuels still supply more than half of the global energy requirement. Fossil fuels are formed underground because of the decomposition of carbon-rich remains of animals and plants for hundreds of years under high pressure and temperature. The primary fossil fuels are coal, crude oil, and natural gas. Natural gas, the gaseous form of fossil fuel, is considered the Earth's cleanest burning hydrocarbon because it causes less environmental pollution than coal and oil [1–3]. Natural gas is the lowest-carbon hydrocarbon, odorless, colorless, and nontoxic but highly flammable. Also, natural gas is cheaper than other types of fossil fuels and extremely reliable.

Furthermore, natural gas has more efficient storage and transportation features compared to other renewable energy sources. Hence, it is a favorably used energy source to provide warmth for cooking and heating and to fuel power stations that provide electricity to homes and businesses [4]. Moreover, natural gas is utilized as fuel for vehicles. Natural gas is associated with other hydrocarbon reservoirs in coal bed sand as methane clathrates, or it is found in deep underground rock formations. There are three primary categories of natural gas wells: oil wells, gas wells, and condensate wells [5]. The principal category is an oil well with related gas. Natural gas wells are explicitly drilled for natural gas and have no or little oil. Condensate wells comprise liquid condensate along with natural gas. Either at the wellhead or during the natural gas processing, the condensate liquid, a hydrocarbon mixture, is separated from the natural gas.

Raw natural gas is a combination of light hydrocarbons comprising methane, ethane, propane, butanes, and pentanes. Furthermore, raw natural gas consists of other compounds such as carbon dioxide (CO_2), hydrogen sulfide (H_2S), nitrogen, helium, and other compounds, including earthy impurities. Natural gas composition is certainly not fixed and varies depending on several factors like the origin, location of the deposit, and geological structure [4]. However, methane is the key element present in natural gas, as shown in Table 5.1 [1,6]. Based on the composition of natural gas, it can be categorized as given in Table 5.2 [1].

Hence, natural gas often contains significant amounts of CO_2 and H_2S on top of other gaseous contaminants. Before it is utilized or liquefied, natural gas must be purified by removing the major nonhydrocarbon acidic gases, CO_2 and H_2S. The process of acid gas separation from natural gas has become a necessary process, and it is known as natural gas sweetening. As per pipeline standards, CO_2 level in the natural gas must be less than 2% (vol), whereas, for the liquefaction process, CO_2 and H_2S level in the natural gas must be no more than 50 and 2 ppm, respectively. This is exclusively true with the more and more rigorous environmental considerations, natural gas economic value, and operational issues. The combustion of sulfur compounds emits harmful air pollutants and in due course causes acid rain when in contact with water. Many operational difficulties, pipeline corrosion, or even pipeline break can be caused by natural gas

Table 5.1 Natural gas typical composition [1,6].

Constituent	Chemical formula	Composition (%)
Methane	CH_4	70—90
Ethane, propane, and butane	C_2H_6, C_3H_8, C_4H_{10}	0—20
Carbon dioxide	CO_2	0—8
Oxygen	O_2	0—0.2
Nitrogen	N_2	0—5
Hydrogen sulfide	H_2S	0—5
Rare gases	He, Ne, Ar, Xe	Trace

Table 5.2 Natural gas categorization based on composition [1].

Natural gas category	Composition
Lean gas	Principally methane
Wet gas	A substantial amount of heavy hydrocarbons
Sour gas	Substantial amount of H_2S
Sweet gas	Slight H_2S
Residue gas	Natural gas (heavy hydrocarbons extracted)
Casing head gas	Derivative from petroleum

that does not comply with standard pressures, heating value, specific gravities, or moisture levels [6,7]. For example, the existence of acid gases like CO_2 and H_2S can create problems such as pipelines and equipment oxidization. Moreover, in liquefied natural gas processing plants, when the natural gas temperature drops to a certain temperature in cooling, the frozen CO_2 may cause transportation drawbacks and pipeline systems get block [8]. Additional difficulties triggered by water and other impurities are atmospheric contamination, foaming in gas dryers and CO_2 removal units and burners, and catalyst damage. H_2S removal eliminates toxicity and improves the odor of the gas. Also, the presence of contaminants upsets the selling price of natural gas as it decreases the energy content of the gas.

Therefore, natural gas extracted at the wellhead must be processed. That is, natural gas is processed as it generally comprises natural gas and liquids contaminants beforehand, which is securely distributed to the elevated pressure, lengthy pipes that carry it to the consumers. The processed natural gas passed into the central gas transportation network should satisfy certain quality measures. The requirement for the natural gas standards in the pipeline for the pipeline grid to function properly is given in Table 5.3 [6,7].

Hence, removing acid gases is one of the essential practices carried out in natural gas treatment processes. Many strategies have been utilized to optimally function the sweetening plant of natural gas through gas recovery enhancement, satisfying the specific discharge emission standards. To keep up the final product specifications with improved exploitation of energy resources, it is necessary to make the acid gas removal process more effective. Profit is the main criterion of an operating plant,

Table 5.3 Natural gas delivery pipeline standards (US) [6,7].

Components	US pipeline specification
CO_2	<2 mol%
H_2S	<4 ppm
H_2O	<120 ppm
C_{3+}	950–1050 Btu/scf dew point −20°C
Total inerts	<4 mol%

and it is desired to increase the plant efficiency over the period. An extensive range of techniques are existing for acid gas removal, and they comprise absorption, adsorption, membrane contactors, and cryogenics. The overall details and assessment of the broadly used gas removal techniques are extensively available in the literature [6,9–12].

The process technologies that use amines for natural gas sweetening have been used for decades and emerging with innovative modern techniques. The following sections have emphatically presented the application of amines for natural gas sweetening.

2. Amine-based techniques for acid gas removal

Natural gas sweetening using amine solutions is based on a weak base (alkaline) chemical reaction with weak acid. The solution of amines is based on nature, and H_2S and CO_2 existing in natural gas are acidic. Acid gases react with the elements in the liquid and get absorbed into the liquid. Typically, the desorption happens by elevating the liquid temperature to an upper value. Amines are the extensively used chemical absorbents for CO_2 and H_2S separation from process gas streams [13–16]. Amines that are extensively utilized in the industry today are primary amines such as monoethanolamine (MEA) and diglycolamine (DGA); secondary amines such as diethanolamine (DEA) and diisopropanolamine (DIPA); and tertiary amines such as methyldiethanolamine (MDEA) and triethanolamine (TEA).

Amines are chemical compounds containing hydroxyl (-OH) and amino (-NH₂, -NHR, and -NR₂) functional groups on an alkane backbone. The molecular arrangement of the particular amine governs how it reacts with acid gases and its acid gas absorption efficiency. Chemical structures of generally used amines are given in Fig. 5.1. The chemical and physical characteristics of generally used amines are listed in Table 5.4 [13,17]. Depending on chemical structure, dissimilar kinds of amines have dissimilar kinetics and reaction mechanisms. Primary and secondary amines follow different mechanisms than tertiary amines, and sterically hindered amines vary from nonsterically hindered amines.

When a liquid amine solution contacts a gas stream comprising H_2S and CO_2, the gaseous acid compounds react with the amine to produce a salt, which is a soluble acid−base complex in the treating solution. The reaction between acid gases and amines is exothermic, and a significant quantity of heat is released in the absorber. H_2S reacts rapidly with the primary, secondary, or tertiary amine, irrespective of the structure of the amine to form the amine hydrosulfide. Eq. (5.1) shows the direct proton transfer reaction:

$$R_1R_2R_3N + H_2S \rightarrow R_1R_2R_3NH + HS \qquad (5.1)$$

The reaction among CO_2 and amines is a little more multifaceted. The zwitterion mechanism is used to typically define the reaction of CO_2 with primary, secondary, and sterically hindered amines, while the base-catalyzed hydration mechanism is used

Primary amines

Monoethanolamine (MEA) $\qquad NH_2 - CH_2 - CH_2 - OH$

Diglycolamine (DGA) $\qquad NH_2 - (CH_2)_2 - O - (CH_2)_2 - OH$

Secondary Amines

Diethanolamine (DEA) $\qquad NH_1 - (CH_2 - CH_2 - OH)_2$

Tertiary Amines

Triethanolamine (TEA) $\qquad N - (CH_2 - CH_2 - OH)_3$

Methyldiethanolamine (MDEA) $\qquad CH_3 - N - (CH_2 - CH_2 - OH)_2$

Hindered Amines

2-amino-2-methyl-1-proponol (AMP) $\qquad NH_2 - C - (CH_3)_2 - CH_2 - OH$

Figure 5.1 Molecular structures of generally used amines.

Table 5.4 Physical and chemical properties of commonly used amines [13,17].

Amine	Other names	Chemical formula	Physical properties	Chemical properties
MEA	• Monoethanolamine • Glycinol	C_2H_7NO	• Viscous colorless liquid • Unpleasant ammonia-like odor • Density 1.0117 g/cm^3 • Boiling point 170°C	• Molar mass 61.084 g/mol • Pickup of acid gas 0.33—0.4 mol/mol • Heat duty of reboiler 280—335 kJ/L • Reboiler temperature 107—127°C • Heat of reaction 1445—1630 kJ/kg CO_2
DGA	• Diethylene glycol amine • Diglycolamine	$C_4H_{11}NO_2$	• Colorless, slightly viscous liquid • Mild amine odor • Density 1.054 g/cm^3 • Boiling point 221.3°C	• Molar mass 105.14 g/mol • Pickup of acid gas 0.25—3 mol/mol • Heat duty of reboiler 300—360 kJ/L • Reboiler temperature 121—127°C • Heat of reaction 2000 kJ/kg CO_2
DEA	Diethanolamine	$C_4H_{11}NO_2$	• Colorless crystals • Ammonia odor • Density 1.097 g/cm^3 • Boiling point 269.2°C	• Molar mass 105.14 g/mol • Pickup of acid gas 0.35—0.65 mol/mol • Heat duty of reboiler 245—280 kJ/L • Reboiler temperature 110—121°C • Heat of reaction 1350—1515 kJ/kg CO_2
MDEA	Methyldiethanolamine	$C_5H_{13}NO_2$	• Colorless liquid • Ammoniacal • Density 1.038 g/cm^3 • Boiling point 247.4°C	• Molar mass 119.17 g/mol • Pickup of acid gas 0.20—0.55 mol/mol • Heat duty of reboiler 220—335 kJ/L • Reboiler temperature 110—127°C • Heat of reaction 1325—1390 kJ/kg CO_2

to define the reaction of CO_2 with tertiary amines [18]. Primary, secondary, and hindered alkanolamine chemical reactions related to CO_2 absorption are [19]:

$$R_1R_2NH + CO_2 \leftrightarrow R_1R_2NH^+COO^- \text{ (zwitterion)} \quad (5.2)$$

$$R_1R_2NH^+COO^- + B \leftrightarrow BH^+ + R_1R_2NCOO^- \text{(carbamate)} \quad (5.3)$$

$$R_1R_2NCOO^- \leftrightarrow R_1R_2NH + HCO_3^- \quad (5.4)$$

$$HCO_3^- \leftrightarrow CO_2 + OH^- \quad (5.5)$$

where B is any base, that can be amine, or OH^{-1} in an aqueous solution. Generally, it is assumed that primary, secondary, or hindered amine reaction with CO_2 is first order in both CO_2 and amine. As per the intermediate zwitterion steady-state principle, reaction rate of CO_2 in the aqueous solutions can be written as [19,20]:

$$r = \frac{k_1(CO_2)(R_1NH_2)}{1 + \dfrac{k_{-1}}{k_B(B)}} \quad (5.6)$$

where k_1 and k_{-1} are correspondingly the forward and backward reaction rate constants. k_B is the backward reaction rate constant. Considering a few adjustments and assumptions like the amine concentration staying at a fixed value for a typical case, Eq. (5.6) has been additionally streamlined to have only one constant, k_{app}, and the below pseudo-first-order rate expression results [20].

$$r_{CO_2} = k_{app}([CO_2] - [CO_2]_e) \quad (5.7)$$

where $[CO_2]_e$ is the CO_2 concentration that is in equilibrium with the other ionic and nonionic species present in the solution. The importance of Eq. (5.7) is that it can be useful for both absorption and desorption.

The chemical reaction involved in tertiary alkanolamine is:

$$R_1R_2R_3N + CO_2 + H_2O \leftrightarrow R_1R_2R_3NH^+ + HCO_3^- \quad (5.8)$$

The rate expression for Eq. (5.8) can be written as:

$$r_{CO_2} = -k_3[R_1R_2R_3N][CO_2] + \frac{k_3}{K_3}[R_1R_2R_3NH^+][HCO_3^-] \quad (5.9)$$

where k_3 is the equilibrium reaction rate constant.

Considering a few adjustments and assumptions like the concentration of amine does not vary significantly throughout one specific process and the forward reaction is the foremost, Eq. (5.9) can be streamlined as Eq. (5.7) [20].

It is essential to consider and take into account that the reaction of CO_2 with primary, secondary, and hindered amine produces carbamate. In contrast, the reaction of CO_2 with a ternary amine produces bicarbonate. The bicarbonates dissociate more quickly than the carbamates [21]. Consequently, in terms of solvent regeneration, the CO_2 absorption with ternary amines is high energy demanding than the CO_2 absorption with primary amines. Moreover, from Eq. (5.4), it can be perceived that carbamate can be changed to free amine molecules for advanced reaction with CO_2. In Eq. (5.3), free amine release enthalpy of dissociation is determined by the stability of the carbamate [14,21]. The stability of the produced carbamate reduces because of the steric character of amines. Thus, sterically hindered amines such as 2-amino-2-methyl-1-propanol (AMP) can be regenerated easily for further CO_2 absorption.

The amine most considered for acid gas absorption is MEA. The benefits of MEA are its instantaneous reaction with acid gases and the preferred thorough separation of acid gases from natural gas streams. The CO_2 reactivity with amines follows the order of primary, secondary, and ternary amines; for instance, the reactivity of CO_2 with MEA, DEA, and MDEA at 25°C is 7,000, 1,200, and 3.5 $m^3s^{-1} kmol^{-1}$, respectively [21]. The existence of the hydroxyl group allows MEA to dissolve into polar solvents such as water freely, and many industrial applications use MEA solutions with concentrations of 20–30% w/w. Commonly, the MEA concentrations in the solution is restricted to 30 wt.% due to operational difficulties such as degradation products causing equipment corrosion [22]. The reaction of MEA with CO_2 produces stable carbamates; therefore, it has to be heated up to 120°C for regeneration and wants 165 kJ per mole of CO_2 [21,23]. The regeneration process can be attributed to almost 80% of the overall cost of absorption and desorption processes, regardless of the efficient waste heat integration [23].

Therefore, several researchers considered tertiary amines like MDEA, which remove CO_2 by formation of bicarbonate [21,22]. Even though the absorption rate is slower than that of the primary and secondary amines, they have lower absorption heat. Accordingly, to resolve this challenging trade-off between energy consumption and regeneration capture rate, hindered amines and adaptation of mixture of various groups of amines are being considered. Sterically hindered amines, for example, AMP is having an amine functional group bounded by a crowded steric environment. Because of this sterically hindering chemical structure, the carbamate produced by CO_2 absorption is not very stable. Similar kind of this steric hindrance is usually introduced to the amine group, which has been found to enhance CO_2 solubility while the absorption of heat is reduced [21]. Hence, amines that have sterically hindered structure have been promoted as encouraging preference for CO_2 removal by a number of scientists.

The concept of blended amines is to mix a fast-reacting solvent with an amine, to have a low heat of absorption [21]. Most originations of the blended amines are based with tertiary amines. Due to its small energy necessities for regeneration and equivalent CO_2 solubility, MDEA is attaining interest as the significant constituent of blended amines [14]. Adding a comparatively lesser quantity of a primary amine like MEA or a secondary amine like DEA into the MDEA solution improves the capturing rate while retaining the benefits of MDEA [14,24]. In this process, the fast-reacting amine is identified as a promoter. One of the widely used promoters for CO_2 capture is piperazine (PZ), a cyclic diamine due to its fast foundation of carbamate with CO_2 [21]. As a great CO_2 capture promoter, PZ has been considered by a number of investigators [25–28].

Although removal of acid gases by amine absorption is a very mature technique and for many decades has been used, it has various weaknesses like poor CO_2 loading capacity, increased energy intake during high-temperature absorbent regeneration, equipment corrosion rate is high, degradation of amines, and large equipment size [15,21,29]. The advantages and disadvantages of commonly used amines are compared in Table 5.5 [10,13,17,30]. Thus, several researchers are going on to discover effective and efficient absorbent liquids for CO_2 capture.

3. Amine-based absorption process

Absorption is the finely developed method for extracting acid gases from natural gas. In absorption, an element existing in gas phase is taken by a liquid phase in which it is favorably soluble. The widespread process arrangement for sweetening operations has an absorption unit and a regeneration unit operation together, as shown in Fig. 5.2. In the absorption tower, the natural gas with contaminants is contacted by the lean solvent, where it absorbs the CO_2 and H_2S confined in the gas. Then the solvent concentrated with acid gases known as a rich solvent is taken from the absorber bottom part and is flashed in the flash drum. Then it is in the stripping tower regenerated by heating. Generally,

Table 5.5 Advantages and disadvantages of commonly used amines [10,13,17,30].

Amine	Advantage	Disadvantage
MEA	• Low solvent cost • Thermal stability is high • Because of its primary amine character, good reactivity with CO_2	• High solvent vapor pressure • Higher corrosion • Because of high heat of reaction with acid gases, regeneration energy requirements are high • In a mixed acid gas system, nonselective separation • Development of irreversible degradation compounds
DEA	• Possibly lower solvent losses due to low solvent vapor pressure • With relation to MEA, less corrosive • Comparatively cheap	• Related to MEA and DGA agent, less reactivity • Principally nonselective separation in mixed acid gas systems because of the reduced ability to slip a considerable amount of CO_2 • Greater circulation desires • Nonreclaimable through conventional reclaiming methods
DGA	• Capital and functional cost savings because of its lower circulation requirements • High reactivity • Low freeze point • Excellent thermal stability	• In a mixed acid gas system, nonselective separation • From inlet, gas absorbs aromatic elements, which possibly confounds the sulfur recovery unit design • Compared to MEA and DEA, solvent cost is high
MDEA	• In mixed acid gas applications, selectivity of H_2S over CO_2 • Possibly lower solvent losses due to low solvent vapor pressure • Less corrosive • High resistance to degradation • Effective energy utilization (low capital and functional cost)	• Compared to MEA, DEA, and DGA agent, solvent cost is high • Comparative reactivity is less • Nonreclaimable through conventional reclaiming methods

DEA, diethanolamine; *DGA*, diglycolamine; *MDEA*, methyldiethanolamine; *MEA*, monoethanolamine.

the lean solvent processed at $P < 2$ bar gauge and around 125°C is withdrawn from the regeneration column and is cooled in the cooler and sent back to the absorber. The sweet gas removed from the absorber top part is directed to the dehydration process. The stream of acid gases taken off from the regenerator is carried to a sulfur recovery unit.

Absorption generally takes place in a countercurrent tower known as an absorption column. Through the column, liquid passes from bottom to top and gas passes from top to bottom. The absorption column can be built-in with necessary trays,

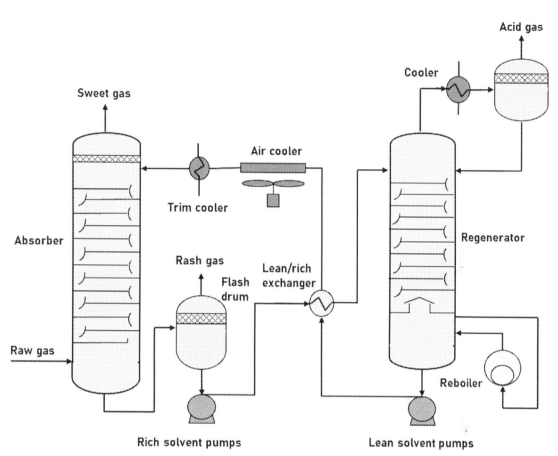

Figure 5.2 Regenerative acid gas removal process.

compacted with packing, or fixed with sprays or other necessary internals as per the required surface area for gas—liquid contact. Either physical or chemical solvents are utilized as absorption liquids. Whether physical or chemical solvents have to be utilized as absorption liquids is based on the feed gas, acid gas's partial pressures, temperature, and necessary percentile of acid gas removal. The most broadly used physical solvents for acid gas removal are Purisol, Selexol, and Rectisol. The most extensively used chemical solvents are amines, the mixture of amines, formulated amines, and inorganic carbonates [13,15,31]. Amine units are in general more economical than other treating methods since they regenerate amine for repeated use, reducing cost of chemicals.

The correct choice of amine is vitally crucial as it improves equipment dimensions and cost and reduces plant operating

costs, comprising regeneration costs. During the choice of the amine for design or prevailing plant assessment, to make the correct choice, some of the following factors have to be considered [32]:
- Categories of contaminants in the unprocessed natural gas
- The concentration of acid gases in the raw natural gas and the required amount of removal
- Selectivity of acid gas to react with the solution (acid gas preference)
- The bulk volume of raw natural gas to be processed
- The economic value of sweetening agents and plant costs
- Natural gas treatment process parameters
- The amine circulation rate volume
- The reboiler/condenser duty
- Amine acid gas loading capacity

The critical problems in conventional absorption columns are due to disperse gas phase and liquid phase, so that the liquid and gas phase cannot be separated independently. The liquid and gas phase connection permits across the phase boundary momentum transfer. Therefore, the interfacial area and the mass transfer coefficients cannot be controlled freely as both rely on the operating and process conditions and are therefore coupled. Under certain operating conditions, this interdependence of the interfacial area and the mass transfer coefficient can cause unsteady and unproductive operation of the dispersive type of liquid–gas contactors. For instance, operational difficulties and limitations like flooding, channeling, foaming, and entrainment are caused by excessive gas and liquid flow ratios [33,34]. Moreover, in some instances, corrosion of the units is caused by the solvent, such as amines used in the absorption process. Moreover, the chemical absorption processes are considered as the comparatively high heat of acid gas absorption and need considerable heat for regeneration. So, a high amount of energy consumption is required in solvent regeneration processes. Further, owing to high-temperature solvents, evaporation is a huge problem. Furthermore, as recycling of all the solvents back to the absorber columns is not possible, the discarding of the absorption liquids creates environmental issues [7].

Since the 1980s, separation of gases by using membranes has developed into a commercially feasible technique; currently membrane technology has been adopted by many plants to remove acid from raw natural gas [11,35]. Principally nonporous polymer membranes that have selectivity have technologically advanced several gas separation processes. Here the membrane provides the selectivity based on molecule size to be separated, diffusivity, and/or differences in insolubility. Chemicals are not necessary for the separation with a selective membrane. As the development of membrane technology progressed, technically advanced membranes with pores were established to be utilized in gas–liquid membrane contactors (GLMCs), which associate the benefits of membrane technology with conventional absorption technology.

In the GLMC, liquid and gas pass through the two different sides of the microporous membrane. Therefore, without dispersing one phase to another, the GLMC membrane only works as a wall between gas and liquid phases. In common, in membrane contactors if microporous membranes are hydrophobic in nature, at the opening of the pores of the microporous membrane, the liquid–gas interface is immobilized by cautious regulation of the pressure variance among the two phases. The concentration gradient is the driving force for gas–liquid absorption/desorption applications in membrane contactors. Through the membrane's pores, from the rich phase, the gas elements that have to be separated move to the gas–liquid boundary. Then on the other side, they interact with the diluted phase. For example, in the case of CO_2 separation from natural gas, as presented in Fig. 5.3, from the feed gas side CO_2 molecules diffuse via the membrane and are then absorbed by the other side flowing selective absorption liquid. The concept of CO_2 absorption by sodium hydroxide in a hollow fiber membrane contactor (HFMC) was first proposed by Qi and Cussler [36]. After that, GLMC, as an acid gas separation unit, has been broadly considered by many scientists, and stimulated practical and theoretical outcomes have been conveyed [37–41].

In GLMC, the absorbent liquid provides the selectivity for acid gas more reasonably than the porous membrane. Hence, proper selection of absorption liquids is critical in defining the effectiveness of GLMC for acid gas separation. Absorption liquid selection should be centered on the same principles that are applied to conventional absorption units. Therefore, usually used amines have been extensively studied as absorbent liquids in GLMC applications [30,42,43]. However, in GLMC, typically, the absorbent liquid should not wet the membrane. Therefore, absorbent liquid with high surface tension has to be utilized. Even though hydrophobic membranes are utilized for acid gas removal in membrane contactors, absorbent liquids with inadequate surface tension have more affinity to penetrate into the membrane pores and wet the membrane. The mass transfer resistance from the thick liquid layer in the pores is higher in a wetted membrane and regulates the overall mass transfer resistance [44]. Due to their low

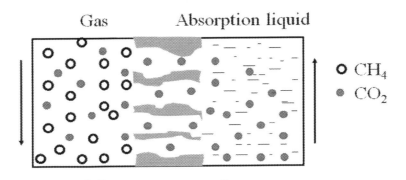

Figure 5.3 CO$_2$ molecule diffusion through a microporous membrane.

surface tension, amines are inherently problematic, as the amine solutions can simply seek into the membrane pores and slowly with time wet the membrane [45,46]. Consequently, absorption liquids with additional promising features for acid gas removal are accordingly being demanded.

4. Current applications and cases

Acid gas absorption using amines is a much developed technology that has been utilized for long periods of time. Darani et al. [47] studied the CO$_2$ removal efficiency in NISOC (National Iranian South Oilfields Company), which is one of the gas sweetening plants consuming aqueous DEA solution. As the significant operational parameters, amine concentrations, temperatures, and lean amine circulation rates were considered. It was observed that the optimum CO$_2$ removal efficiency was attained for 30 wt.% amine concentration at a temperature of 40°C and 260 m^3/h circulation rate. Furthermore, the studies concluded that in an aqueous solution, amine concentration is the most influential parameter on CO$_2$ removal efficiency, and the lean amine temperature has minimal effect. The outcomes of this study could help better understand the impact of the natural gas sweetening process operative parameters and attain their suitable values to accomplish maximum efficiency.

One of the major fossil fuel markets worldwide is situated in the Gulf of Mexico area. The natural gas extracted in the Gulf of Mexico comprises elevated amounts of CO$_2$, about 1.47%, and little amounts of H$_2$S. Aspen HYSYS V.8.4 program was used by Sulaiman et al. [48] to simulate and optimize the potential gas sweetening process. The simulation study was performed by engaging the PZ-activated MDEA solution, achieving high CO_2 removal efficiency. Moreover, optimization of the gas sweetening process is attained by using various amine types and mixes of amine such as MEA and DEA. Furthermore, operation variables such as amine concentration, circulation rate, and additional operating conditions are studied for all amine types. They concluded that the mixture of MDEA 30 wt.% and PZ 2.5 wt.% is the utmost efficient and suggested amine technique.

Jamekhorshid et al. [49], in a prevailing industrial process plant in Phase I of the South Pars Gas Complex positioned in Persian Gulf region, examined the efficiency of two amines, DEA and MDEA, and blended them with various mass concentrations and circulation rates to absorb CO_2 and H_2S from natural gas. Efficiency studies were conducted for circulation flow rates between 140 and 350 m^3/h for MDEA and between 300 and 520 m^3/h for DEA. MDEA concentrations varied from 30 to 65 wt.%, and DEA concentrations varied from 30 to 40 wt.%. Various MDEA: DEA ratio performances were also examined. The influence of these factors has been distinctly analyzed on the factors such as acid gas loadings, acid gas content in the sweetened gas, pH, amine carryover, and reboiler duty. The outcomes proved that the circulation rate and elevated solvent concentration favor the natural gas sweetening process. To regenerate CO_2-rich loaded solutions in the reboiler, they need more equivalent energy, which uplifts operational costs. Corresponding to the outcomes, optimal flow rate and concentration of solvents of DEA and MDEA were attained at 32 and 40 wt.% and 350 and 150 m^3/h, respectively.

Khurmala field is deliberated as a central fuel source for the Iraqi Kurdistan region. Projects were initiated to recover and utilize it as power station feedstock or sell it in the international market. However, it has been confirmed from the quality checks that the natural gas from Khurmala has high concentrations of CO_2 (4.4%) and H_2S (5.3%). Hence, Abdul Rahman et al. [50] aimed to use the newest version of the Aspen HYSYS V.7.3 program to simulate the probable Khurmala gas sweetening process. Furthermore, the DEA solution was used by researchers to simulate the amine gas sweetening process, and it attained high acid gas removal. For example, H_2S concentration in the sweet gas stream was about 4.0 ppm at a 400 m^3/h amine volumetric flow rate.

Furthermore, the mockup work also accomplished process standardization by utilizing various types of amines and their mixtures, for example, MDEA and MEA. It also studied a few important amine process variables such as amine volumetric flow rate and amine concentration for each amine type. Additionally, the optimization work confirmed that using 35% w/w DEA possibly will be the most suggested variable. Abdul Rahman et al. [51] also considered the North Gas Company's (NGC) raw natural gas. The NGC natural gas consists of H_2S 2.95% and CO_2 2.54%. A DEA amine system was consumed to reduce concentration of acid gases below 5 ppm for H_2S and 2% for CO_2. Bryan Research and Engineering's ProMax process simulation software was used by researchers to optimize the amine sweetening system by varying types of amine, that is, DEA and MDEA. Concentrations of amines and solvent volumetric flow rates were also optimized. It was determined as the optimum operating condition at which MDEA solution with 50 wt.% concentration circulated at flow rate of 414 m^3/h. This design together satisfied sweet gas standards and reduced steam consumption to 30.9% and the reboiler duty to 38 MW.

Fouad et al. [16] studied the utilization of mixed tertiary amines, namely MDEA and TEA, for gas sweetening energy requirement reduction at the Habshan gas sweetening unit in Abu Dhabi, UAE. Outcomes proved that up to a 3.0% drop in the unit operational cost could be attained using the 40% wt. MDEA +5% wt. TEA while achieving the sweet gas standards regarding H_2S and CO_2 concentrations. Results for the blend were related to the performance of the regular MDEA 45% w/w concentration solvent used in Habshan and other probable primary/tertiary and secondary/tertiary amine blends. The cost saving was accomplished by reducing the cost of plant raw materials. Further energy requirements for both regenerator reboiler and trimmed cooler were also reduced.

In Iraq, Missan Oil Company/Buzurgan Oil Field of Natural Gas Processing Plant simulation and parametric studies were considered by Khanjar et al. [52]. As per the simulation results, solvent concentration, feed flow rate, and temperature were optimized. Outcomes revealed that the concentration of H_2S and CO_2 in the sweet gas stream increases with flow rate and feed temperature. At that point, the effect of mixture solvents was considered as the next step. As a physical–chemical mixture solvent, sulfolane–MDEA was selected, whereas as chemical mixture solvent, MDEA–MEA was selected. The simulation outcomes confirmed that via using a mixture solvent, reboiler duty, the solvent price, and cooling duty can be decreased.

Abkhiz et al. [53], in Fajr Jam Gas Refining Company, have compared the various types of amine characteristics with mixed or activated amine characteristics. The study examined the furthermost significant variables in gas sweetening refineries, including amine volumetric flow rate, consumption of steam, and CO_2 removal efficiency, for sweetening trains applying DEA, MDEA, DEA/MDEA, and activated MDEA (aMDEA). In the case of consumption of steam and volumetric flow rates, MDEA and aMDEA reveal the smaller amount of required rates to satisfy lean amine loadings. For CO_2 removal, aMDEA has the greatest effectiveness.

Sahl et al. [54], to lessen the CO_2 level of sweet natural gas products from amine sweetening plants, presented a two-step approach. Step 1 is amine blending, and step 2 is a slight process modification. In step 1, they simulated an industrial natural gas sweetening plant by Aspen HYSYS, and the model results were confirmed compared to the plant data. Subsequently, various mixtures of MDEA−MEA and MDEA−DEA were explored. Then the ideal amine mixture of MDEA 28 wt.% and MEA 10 wt.% was reported. Compared to the base case (plant data), the ideal amine mixture attained a noteworthy removal (99.9%) in CO_2 concentration of sweet natural gas. In step 2, two types of amine stream split (lean amine stream split and semilean amine stream split) were analyzed. The work considered split stream flow rate, absorber recycles stage, and withdrawal regenerator stages. Compared to step 1, both types of stream split achieved a considerable decrease in CO_2 concentration of sweet natural gas products and amine circulation rate. However, with 63.6% and 69.1% reduction in lean amine circulation rate and CO_2 concentration of sweet natural gas, respectively, the semilean amine stream split was superior to the lean amine split.

Karimi et al. [55] compare the pros and cons of each amine solution using the analytic hierarchy process (AHP). Corresponding to the AHP procedure, four operational conditions and seven other possibilities were considered. Then, the simulation of the natural gas sweetening process was developed, and eventually, process parameters were optimized. The acid gas contents' reduction and reboiler duty drop were taken as target functions. The simulation outcomes specified that the ideal temperature of feed gas and concentration of MDEA are 30°C and 39 wt.%, respectively. As one of the optimization targets of the gas sweetening process, the concentration of CO_2 and H_2S attained 88 ppm and 2.4 in the optimal condition. Consequently, 5% reduction in the MDEA solution feeding and approximately 0.04% reboiler duty drop were observed related to the conventional sweetening process.

Asma et al. [56] investigated three gas sweetening units in Libya: the Sahel gas plant, Mellitah Complex, and Al-Estiklal. For natural gas sweetening, these units primarily use MDEA as a solvent. In the plants, without being advanced, handling higher gas production was very challenging. So the work aimed to decrease operating costs and upsurge throughput while continuing the quality of the product. The process was simulated using HYSYS simulation software. The possible benefits of consuming a mixture of MDEA and PZ were assessed. Outcomes revealed that the mixture of MDEA and PZ offers better gas treatment. In the Mellitah Complex, the raw feed gas (1.98 vol % CO_2 and 0.9915 ppm H_2S) flow rate can be improved to 18,820 kmol/h from 16,208 kmol/h. In the Al-Estiklal gas plant, the raw feed (0.975 mol% CO_2) flow rate can be increased to 8750 kmol/h from 8012.79 kmol/h. However, the Sahel gas plant revealed an unlikely performance, where the mixing of PZ caused H_2S deviation from the requisite standards. The concentration of H_2S attained 25.944 ppm, and the concentration of CO_2 was 1.347 mol%. However, the drop in the amine temperature decreases CO_2 to 1.276 mol% and H_2S to 13.76 ppm. The outcomes validated the possibility for important developments in increasing throughput through the activation of MDEA by usage of PZ.

In the natural gas sweetening processes, deep eutectic solvents (DESs) have extended substantial courtesies as new green solvents. ParisaJahanbakhsh-Bonab et al. [57], by molecular dynamics models, studied the effectiveness of 6: 1 M ratio of MDEA and choline chloride (ChCl) (an amine-based DES) for acid gas removal. The CO_2 and H_2S solubility outcomes in the DES phase revealed that CO_2 and H_2S solubilities are increasing with pressure. The maximum values of 0.04 and 0.225 are obtained for CO_2 and H_2S solubility, respectively, at 310 K and 2 MPa. Moreover, solubility selectivity parameter calculations at various pressures revealed monotonic characteristics for the solubility selectivity of H_2S over CH_4. The H_2S over CH_4 solubility selectivity has a larger value (10–20 times) than the CO_2 over CH_4 solubility selectivity at 0.1–2 MPa pressure range. It can be decided compared to aqueous MDEA solvent that the MDEA-based DES consumed here has greater effectiveness for H_2S removal and better selectivity toward CH_4 and CO_2. Outcomes point out that MDEA-based DES is a better replacement for the usually adopted amine solvents in the natural gas sweetening plants.

High-performance polyamines, namely, Tetraethylenepentamine (TEPA), PZ, and branched polyethylenimine (PEI-B), at concentrations of 0.3, 0.5, and 0.1 M, were studied by Jaffary et al. [58] for MDEA activation for natural gas sweetening, mainly for CO_2 separation. A gas stream containing 50% CO_2 and 50% N_2 mimicked the natural gas. The effectiveness assessment was evaluated considering the absorption and desorption rate, rich and lean loading, cyclic capacity, and heat duty. Outcomes revealed that PEI-B demonstrated good performance in all assessment variables. Moreover, ^{13}C nuclear magnetic resonance study carried out on 0.1 M polyamines at equilibrium-rich loadings revealed that PEI-B formed and desorbed the great volumes of CO_2 products as related to those from PZ. Moreover, for the absorption process, two pseudo-first-order models with respect to amine and CO_2 were established. Equally in two models, even though PZ showed the high value for rate constant per amino group than PEI-B, the overall rate constant was greatest. Moreover, to adequately describe the absorption, first-order exponential rise was established. Whereas to describe the desorption processes, decay expressions were established.

Cao et al. [59] employed GLMC consisting of polyvinylidene fluoride porous hollow fibers for CO_2 absorption. MDEA-based nanofluid was used as an absorbent liquid. A mechanistic model was developed by performing both theoretical and experimental works. The model considered the mass transfer of components in whole subdomains of the contactor module. The expected results of the established model and simulations revealed that the MDEA-based solvent spread with carbon nanotube (CNT) increases CO_2 absorption percentage more than the fresh MDEA solvent. Moreover, the performance of CO_2 absorption for nanofluid-based MDEA was improved with increasing liquid flow rate, MDEA concentration, and membrane porosity. However, the improvement of the membrane tortuosity and gas velocity caused a reduction in CO_2 removal effectiveness of the module. Furthermore, it was exposed that the CNT's 'influence on CO_2 removal is greater in the solvent existence of lesser MDEA content (5%). The model was confirmed by relating it with the experimental data, and significant correlation was observed.

Magnone et al. [60] employed ceramic HFMC, which is hydrophobically modified to recognize the associations among the CO_2 absorption characteristics of five single absorbents (amine-based). They are MEA, DEA, MDEA, AMP, and PZ. Their 16 binary mixtures in the blended amine-based absorbents were also considered. Maintaining all other variables constant, CO_2 absorption properties via ceramic HFMCs that are hydrophobically modified greatly depend on concentration and chemical nature on single and blended amine-based absorbents. MEA and DEA revealed the top CO_2 absorption flux among the single amine solutions. In the blended amine-based absorbents for the CO_2 absorption, replacing MDEA with 20 wt.%, DEA improved the CO_2 chemical absorption to about $7 \times 10^4 \, mol/m^2 \, s$ from $1 \times 10^4 \, mol/m^2 \, s$. Aqueous MDEA mixed with DEA showed a percentage of CO_2 absorption flux over 500% compared with the equivalent aqueous solution of a single MDEA.

5. Conclusion and future outlooks

Raw natural gas from wells comprises substantial concentrations of H_2S and CO_2 referred to as acid gases. Before it is consumed or liquefied, natural gas must be refined by removing the acid gases, which is known as natural gas sweetening. Sulfur compounds in natural gas can be hazardous, even toxic, to inhale by a human. Acid gases can also be highly corrosive to pipes. When in contact with water, CO_2 produces carbonic acid. Pipeline standards need CO_2 and H_2S to be below particular levels to safeguard the pipe's integrity and avoid corrosion that can cause drips or breaks. Therefore, the separation of acid gases is essential.

Since the inception of the gas processing industry, various types of sweetening technologies have been technologically advanced. Among them, chemical absorption is the most extensively employed gas treating process. In the chemical absorption process, in the contactor tower, a chemical solvent extracts acid gases from natural gas via chemical reactions, either a conventional absorption tower or GLMC. Mostly used chemical solvents in the contactor towers are amines. Amines remove CO_2 and H_2S from natural gas by employing chemical reactions in the contactor tower by combining a water solution of a weak base, known as an alkanolamine, with the weak acids formed by acid gases. This reaction produces amine carbamate or amine hydrosulfide, which are soluble salts in a water solution. With the addition of heat, these reactions are completely reversible. Amine units are commonly more economical compared to alternate treating systems since they regenerate amine for repeated use, decreasing chemical costs. Currently, the amines that are regularly used in industry are:
- Primary amines (eg: MEA, DGA)
- Secondary amines (eg: DEA, DIPA)
- Tertiary amines (eg: MDEA, TEA
- Mixed amines
- Formulated amines (eg: aMDEA, UCARSOL, FLEXSORB)

Formulated amine solvents have been explicitly formulated to carry out a straightforward task. There are various types of formulated amines that contain various amine blends and various reaction promoters.

Various factors have to be negotiated when deciding on a more suitable amine for a sweetening process. The elements that should be considered are the acid gas elements of the raw natural gas, the composition, the pressure of available gas, flow rate, temperature of the gas to be processed, and the concentration requirement of the sales gas. Further, it is essential to think through the loading ability of amine besides its rate of circulation and cost of regeneration. As diverse operation parameters have been considered and demonstrated with each amine, they have become extensively known in the natural gas sweetening process industry. Considering all the research work and process parameters, MEA is typically not the best amine choice because of its elevated heat of reaction and poorer acid gas carrying capacity per gallon of solution. But, many chemical plants still use MEA as an absorption liquid when the feed gas pressure is small, and pipeline standards of gas or complete removal of the acid gases are wanted. As functional expertise widened in the early 1970s for DEA, it came to be the "workhorse" of the chemical industry because of its greater acid gas loading ability, lesser heat of reaction, and subsequent lower energy necessities. DEA could also have the probable selective H_2S removal from gas feeds comprising CO_2 under typical conditions. MDEA has some exceptional abilities based on the application. For bulk acid gas removal in pressure swing plants, MDEA can be utilized due its low heat of reaction. At present, MDEA is fairly recommended because while absorbing the maximum amount of CO_2, it is suitable to slip as much H_2S as possible.

With a lot of research and development on the natural gas sweetening process, there has been a swing in solvent selection from conventionally used amines to proprietary amines advanced with improved selectivity as a crucial part of this advancement. For very much selective H_2S removal, solvents by Union Carbide (UCARSOL), BASF (aMDEA), ExxonMobil (Flexsorb), the DOW Chemical Company (Gas Spec), and others have been industrialized, which show better selectivity and H_2S removal to lower treated gas standards. But these solvents are MDEA-based solvents. These solvents have other uses, such as H_2S removal from CO_2-enhanced oil recovery enrichment processes. Developments of all these novel amines with selective kinetics have contributed the way to successful gas sweetening with more significant prominence on environmental emission than before. The objective of natural gas sweetening has always been the removal of undesirable contaminations safely and economically. A steady advancement must be undertaken to decrease cost while still satisfying preferred product standards, reducing corrosion, and minimizing operation issues.

Abbreviations and symbols

AHP	Analytic hierarchy process
AMP	2-Amino-2-methyl-1-propanol
CO_2	Carbon dioxide
DEA	Diethanolamine
DESs	Deep eutectic solvents
DGA	Diglycolamine
DIPA	Diisopropanolamine
GLMC	Gas–liquid membrane contactors
H_2S	Hydrogen sulfide
MDEA	Methyldiethanolamine
MEA	Monoethanolamine
PEI-B	Branched polyethylenimine
PZ	Piperazine
TEA	Triethanolamine

References

[1] Viswanathan B. In: Viswanathan BBT-ES, editor. Chapter 3—natural gas. Amsterdam: Elsevier; 2017. p. 59–79.
[2] Speight JG. In: Speight JGBT-NGSE, editor. 4—composition and properties. Boston: Gulf Professional Publishing; 2019. p. 99–148.
[3] Speight JG. The chemistry and technology of petroleum. CRC Press; 2006.
[4] Liang F-Y, Ryvak M, Sayeed S, Zhao N. The role of natural gas as a primary fuel in the near future, including comparisons of acquisition, transmission and waste handling costs of as with competitive alternatives. Chemistry Central Journal 2012;6(1):S4.
[5] Stewart M, Arnold K. Gas sweetening and processing field manual. Gulf Professional Publishing; 2011.
[6] Baker RW, Lokhandwala K. Natural gas processing with membranes: an overview. Industrial and Engineering Chemistry Research 2008;47(7): 2109–21.
[7] Shimekit B, Mukhtar H. Natural gas purification technologies-major advances for CO2 separation and future directions. Europe: INTECH Open Access Publisher; 2012.
[8] Mokhatab S, Poe WA. Handbook of natural gas transmission and processing. Gulf Professional Publishing; 2012.
[9] Baker RW. Membrane technology and applications. 2nd ed. John Wiley and Sons; 2004.
[10] Rahim NA. Modeling and experimental study of carbon dioxide absorption/stripping in gas liquid membrane contactor. Al Ain: United Arab Emirates University; 2015. PhD thesis.

[11] Ghasem NM, Rahim NA, Al-Marzouqi M. Carbon capture from natural gas via polymeric membranes. In: Encyclopedia of information science and technology. 4th ed. IGI Global; 2018. p. 3043—55.
[12] Ghasem N, Al-Marzouqi M, Rahim NA. Absorption of CO_2 form natural gas via gas-liquid PVDF hollow fiber membrane contactor and potassium glycinate as solvent. Jurnal Teknologi 2014;69(9).
[13] Polasek J, Bullin J. Selecting amines for sweetening units. Energy Progress 1984;4(3):146—9.
[14] Mandal BP, Guha M, Biswas AK, Bandyopadhyay SS. Removal of carbon dioxide by absorption in mixed amines: modelling of absorption in aqueous MDEA/MEA and AMP/MEA solutions. Chemical Engineering and Science 2001;56(21):6217—24.
[15] Mohamadirad R, Hamlehdar O, Boor H, Monnavar AF, Rostami S. Mixed amines application in gas sweetening plants. Chemical Engineering Transactions 2011;24(20):265—70.
[16] Fouad WA, Berrouk AS. Using mixed tertiary amines for gas sweetening energy requirement reduction. Journal of Natural Gas Science and Engineering 2013;11:12—7.
[17] Ghasem N. In: Rahimpour MR, Farsi M, Makarem CC, editors. Chapter 21—CO2 removal from natural gas. Woodhead Publishing; 2020. p. 479—501. M. A. B. T.-A.
[18] Awais M. Determination of the mechanism of the reaction between CO_2 and alkanolamines. 2013.
[19] Liu X. Rate based modelling of CO_2 removal using alkanolamines. 2014.
[20] Jamal A. Absorption and desorption of CO_2 and CO in alkanolamine systems. 2002.
[21] Yu C-H, Huang C-H, Tan C-S. A review of CO_2 capture by absorption and adsorption. Aerosol and Air Quality Research 2012;12(5):745—69.
[22] Benedict ML. CO2 capture from flue gas using amino acid salt solutions. 2012.
[23] Yeh JT, Pennline HW, Resnik KP. Study of CO_2 absorption and desorption in a packed column. Energy and Fuels 2001;15(2):274—8.
[24] Idem R, Wilson M, Tontiwachwuthikul P, Chakma A, Veawab A, Aroonwilas A, et al. Pilot plant studies of the CO_2 capture performance of aqueous MEA and mixed MEA/MDEA solvents at the University of Regina CO_2 capture technology development plant and the Boundary Dam CO2 capture demonstration plant. Industrial and Engineering Chemistry Research 2006;45(8):2414—20.
[25] Zhang Z, Yan Y, Zhang L, Chen Y, Ju S. CFD investigation of CO_2 capture by methyldiethanolamine and 2-(1-piperazinyl)-ethylamine in membranes: part B. Effect of membrane properties. Journal of Natural Gas Science and Engineering 2014;19:311—6.
[26] Lin S-H, Chiang P-C, Hsieh C-F, Li M-H, Tung K-L. Absorption of carbon dioxide by the absorbent composed of piperazine and 2-amino-2-methyl-1-propanol in PVDF membrane contactor. Journal of the Chinese Institute of Chemical Engineers 2008;39(1):13—21.
[27] Freeman SA, Dugas R, Van Wagener DH, Nguyen T, Rochelle GT. Carbon dioxide capture with concentrated, aqueous piperazine. International Journal of Greenhouse Gas Control 2010;4(2):119—24.
[28] Barker C. Chapter 2 origin, composition and properties of petroleum. In: Donaldson EC, V Chilingarian G, editors. Yen PS, editor. Enhanced oil recovery, I, vol 17. Elsevier; 1985. p. 11—45. T. F. B. T.-D.

[29] Mozafari A, Azarhoosh MJ, Mousavi SE, Khaghazchi T, Ravanchi MT. Amine–membrane hybrid process economics for natural gas sweetening. Chemical Papers 2019;73(7):1585–603.
[30] N. A. Rahim, "Overview of absorbents used in gas liquid membrane contactor for CO_2 absorption.".
[31] Mitra S. A technical report on gas sweetening by amines, vol. 1. Petrofac Engineering India Ltd; 2015.
[32] Khan SY, Yusuf M, Malani A. Selection of amine in natural gas sweetening process for acid gases removal: a review of recent studies. Petroleum Petrochemical Engineering Journal 2017;1(3):1–7.
[33] Leung DYC, Caramanna G, Maroto-Valer MM. An overview of current status of carbon dioxide capture and storage technologies. Renewable and Sustainable Energy Reviews 2014;39:426–43.
[34] Metz B, Davidson O, De Coninck H, Loos M, Meyer L. Carbon dioxide capture and storage. 2005.
[35] N. Ghasem, M. Al-Marsouqi, and N. A. Rahim, "Simulation of gas/liquid membrane contactor via COMSOL Multiphysics®."
[36] Drioli E, Criscuoli A, Curcio E. Membrane contactors: fundamentals, applications and potentialities: fundamentals, applications and potentialities, vol 11. Elsevier; 2011.
[37] Ghasem N, Al-Marzouqi M, Abdul Rahim N. Modeling of CO_2 absorption in a membrane contactor considering solvent evaporation. Separation and Purification Technology 2013;110:1–10.
[38] Rahim NA, Ghasem N, Al-Marzouqi M. Stripping of CO_2 from different aqueous solvents using PVDF hollow fiber membrane contacting process. Journal of Natural Gas Science and Engineering 2014;21:886–93.
[39] Ghasem N, Al-Marzouqi M, Rahim NA. Effect of polymer extrusion temperature on poly (vinylidene fluoride) hollow fiber membranes: properties and performance used as gas–liquid membrane contactor for CO_2 absorption. Separation and Purification Technology 2012;99:91–103.
[40] Bougie F, Iliuta I, Iliuta MC. Absorption of CO_2 by AHPD–Pz aqueous blend in PTFE hollow fiber membrane contactors. Separation and Purification Technology 2014;138(0):84–91.
[41] Boributh S, Rongwong W, Assabumrungrat S, Laosiripojana N, Jiraratananon R. Mathematical modeling and cascade design of hollow fiber membrane contactor for CO_2 absorption by monoethanolamine. Journal of Membrane Science 2012;401:175–89.
[42] Li J-L, Chen B-H. Review of CO_2 absorption using chemical solvents in hollow fiber membrane contactors. Separation and Purification Technology 2005;41(2):109–22.
[43] Mansourizadeh A, Ismail AF. Hollow fiber gas–liquid membrane contactors for acid gas capture: a review. Journal of Hazardous Materials 2009;171(1–3):38–53.
[44] Mosadegh-Sedghi S, Rodrigue D, Brisson J, Iliuta MC. Wetting phenomenon in membrane contactors – causes and prevention. Journal of Membrane Science 2014;452(0):332–53.
[45] Kumar PS, Hogendoorn JA, Feron PHM, Versteeg GF. New absorption liquids for the removal of CO_2 from dilute gas streams using membrane contactors. Chemical Engineering and Science 2002;57(9):1639–51.
[46] Lu J-G, Zheng Y-F, Cheng M-D. Wetting mechanism in mass transfer process of hydrophobic membrane gas absorption. Journal of Membrane Science 2008;308(1):180–90.

[47] Darani NS, Behbahani RM, Shahebrahimi Y, Asadi A, Mohammadi AH. Simulation and optimization of the acid gas absorption process by an aqueous diethanolamine solution in a natural gas sweetening unit. ACS Omega May 2021;6(18):12072−80.

[48] Abdul-Wahab S, Fadlallah S, Al-Rashdi M. Evaluation of the impact of ground-level concentrations of SO2, NOx, CO, and PM_{10} emitted from a steel melting plant on Muscat, Oman. Sustainable Cities and Society 2018; 38:675−83.

[49] Jamekhorshid A, Davani ZK, Salehi A, Khosravi A. Gas sweetening simulation and its optimization by two typical amine solutions: an industrial case study in Persian Gulf region. Natural Gas Industry B 2021; 8(3):309−16.

[50] Abdulrahman RK, Sebastine IM. Natural gas sweetening process simulation and optimization: a case study of Khurmala field in Iraqi Kurdistan region. Journal of Natural Gas Science and Engineering 2013;14:116−20.

[51] Abdulrahman RK, Zangana MHS, Ali KS, Slagle JC. Utilizing mixed amines in gas sweetening process: a Kirkuk field case study and simulation. 2017 international conference on environmental impacts of the oil and gas industries: Kurdistan region of Iraq as a case study (EIOGI). 2017. p. 5−8.

[52] Khanjar JM, Amiri EO. Simulation and parametric analysis of natural gas sweetening process: a case study of Missan Oil Field in Iraq. Oil and Gas Science and Technology—Revue d'IFP Energies Nouvelles 2021;76:53.

[53] Abkhiz V, Heydari I. Comparison of amine solutions performance for gas sweetening. Asia-Pacific Journal of Chemical Engineering 2014;9(5):656−62.

[54] Bin Sahl A, Siyambalapitiya T, Mahmoud A, Sunarso J. Towards zero carbon dioxide concentration in sweet natural gas product from amine sweetening plant. IOP Conference Series: Materials Science and Engineering 2021; 1195(1):12038.

[55] Karimi A, Sadeghi A. AHP-based amine selection in sour gas treating process: simulation and optimization. Iranian Journal of Chemistry and Chemical Engineering 2022;41(11):3772−85.

[56] Limami H, Manssouri I, Cherkaoui K, Khaldoun A. Recycled wastewater treatment plant sludge as a construction material additive to ecological lightweight earth bricks. Cleaner Engineering and Technology 2021;2: 100050.

[57] Jahanbakhsh-Bonab P, Sardroodi JJ, Avestan MS. The pressure effects on the amine-based DES performance in NG sweetening: insights from molecular dynamics simulation. Fuel 2022;323:124249.

[58] Jaffary B, Jaafari L, Idem R. CO_2 capture performance comparisons of polyamines at practical concentrations for use as activators for methyldiethanolamine for natural gas sweetening. Energy and Fuels 2021; 35(9):8081−94.

[59] Cao Y, Rehman ZU, Ghasem N, Al-Marzouqi M, Abdullatif N, Nakhjiri AT, et al. Intensification of CO_2 absorption using MDEA-based nanofluid in a hollow fibre membrane contactor. Scientific Reports 2021;11(1):1−12.

[60] Magnone E, Lee HJ, Shin MC, Park JH. A performance comparison study of five single and sixteen blended amine absorbents for CO_2 capture using ceramic hollow fiber membrane contactors. Journal of Industrial and Engineering Chemistry 2021;100:174−85.

6

Physical and hybrid solvents for natural gas sweetening: Ethers, pyrrolidone, methanol and other sorbents

Samuel Eshorame Sanni[1], Babalola Aisosa Oni[1,3] and Emeka Emmanuel Okoro[2]

[1]*Department of Chemical Engineering, Covenant University, Ota, Ogun State, Nigeria;* [2]*Department of Petroleum and Gas Engineering, University of Port Harcourt, Choba, Rivers State, Nigeria;* [3]*Department of Energy Engineering, University of North Dakota, Grand Forks, ND, United States*

1. Introduction

Natural gas (NG) is an energy source that is useful both domestically and industrially owing to its earthly abundance and low-cost involvement during production [1]. NG is a nonpolluting fuel as opposed to other fuels (solid/liquid fuels) [2]. In 2017, NG consumption rose to 1.3% [3] and is expected to increase by 15% (by approximately 70–83 quadrillion Btu) by 2030 [4]. Despite the merits associated with NG, it is usually entrained by contaminants (H_2S, CO_2, and other trace gases), which impose severe consequences including low heating value/low energy efficiency, safety risks, damage to pipelines, as well as process equipment; hence, there is need to rid the gas of these impurities [5]. Numerous methods have been proposed for NG treatment, and these include absorption, adsorption, and membrane separation [6,7]. CO_2 and H_2S removal from NG via absorption reduces the annual total cost of producing the gas. Consequently, novel technologies have ensued for the effective separation of these contaminants from NG [8,9]. Some traditional methods of gas sweetening include the Rectisol process, which uses cold methanol to trap acid gases from NG [10–12], and the Fluor process, in which propylene carbonate (PC) serves as a solvent, which takes advantage of CO_2 partial pressure during gas sweetening. In the Selexol process, dimethyl ether/polyethylene glycol (PEG) or their mixture is employed as solvent for acid gas removal within pressure limits of 2.07–13.8 MPa.

NG sweetening is an essential means of meeting gas–liquid product specs/heating value requirements; this helps in abating catalyst poisoning tendencies in downstream facilities, equipment corrosion, and toxicity, thus promoting environmental safety. Amine and its hybrid solvents (monoethanolamine [MEA], diglycolamine [DGA], diethanolamine [DEA], methyldiethanolamine [MDEA] [13,14], and diisopropanolamine), alongside other proprietary and mixed amines, have known significant use in the sweetening of NG. Physical solvents for NG sweetening usually thrive under acid gas partial pressure >50 psi, high acid gas solubility in the solvent, low heavy hydrocarbon concentration, and bulk presence of gaseous impurities. Based on the literature, 90%–97% NG purification is feasible [15–22], and according to Fang and Zhu [23], amines/carbonates/aqueous ammonia/enzymes, polymers, and ionic liquids (ILs) are new waves of materials to be considered for NG sweetening. In NG purification, the end use determines the permissible contaminant gas concentration; for instance, a CH_4 concentration of more than 90% is acceptable for use in internal combustion engines [24]; high amounts of CO_2 cause a decline in engine performance [25]/lowers the heating capacity of the fuel [26]; and H_2S limits of ~3500 ppm can result in the corrosion of the internal parts of engines [27]. During pipeline transportation of NG, significant concentrations of H_2S can lead to pipeline leakage/corrosion, fire explosions, as well as losses related to aquatic and human lives [28]; these in turn necessitate the treatment of the gas before it is transported to end users [29]. Based on literature, the sulfur threshold of NG is 0.1 ppm [30]. Furthermore, NG liquefaction ensues at 161°C and 1 atm, hence the need for CO_2 removal [31,32]. In order to ensure an apt selection of a suitable solvent for NG sweetening, sorbent property considerations including viscosity, corrosivity, thermal resistance, density, solute solubility, cost of solvent, and spent sorbent recovery are very crucial [33,34].

2. Physical and hybrid sorbents for NG sweetening

2.1 Acid gas absorption in amines and alcohols

One of the most prominent acid gas impurities in NG is carbon dioxide, which is released in large quantities during gas combustion for the generation of heat and electricity in power plants. Another prominent acid gas found in NG is hydrogen sulfide, a highly toxic gas that naturally coexists with CH_4/light hydrocarbons as well as CO_2 in oil and gas reservoirs. CO_2 in NG causes a decline in its heating value, whereas the presence of H_2S may induce corrosion of transmission lines and process equipment, hence the need to strip the aforementioned gases from NG. Physical and chemical absorption are viable approaches for stripping CO_2 and H_2S from NG [12]. During absorption, both gases are physically taken up by a choice-solvent, that is, by solvents such as methanol/alkanol amines [35–38], 1-methyl-2-pyrrolidone, etc., or by chemical dissolution of the contaminated NG in an aqueous solution of 2-aminoethanol (monoethanolamine)/2,2'-iminodiethanol (DEA) or 2,2'-(methylimino)diethanol (N-methyldiethanolamine) [12]. Thereafter, desorption of the trapped gases then occurs via a reversible reaction within a spent solvent regenerator to give a mixture of CO_2 and H_2S/sour gas for further treatment as done in the Claus process; lean solvent recycling also takes place at the absorption stage. In lieu of the acid gas stripping potentials of chemical solvents, they are somewhat associated with high energy costs/high heat capacities of the diluent mixed with the absorbents [39], solvent loss during desorption caused by vaporization, and solvent degradation, which yield toxic/corrosive by-products. Trapped H_2S from NG can be converted to elemental sulfur via the Claus process prior to storage and transportation [40]. Although CO_2 is usually associated with H_2S in NG formations, absorbents for sweetening NG must possess the ability to selectively separate H_2S from CH_4 and CO_2, which in turn imparts on the Claus process efficiency, hence the need to consider the use of hindered amines (N-methyldiethanolamine and tert-butylaminoethoxyethanol), which are very useful for the selective sweetening of NG [41,42].

Among the existing chemical absorption technologies for NG sweetening, amine scrubbers such as monoethanolamide (MEAm)/DEA/triethanolamine/DGA/MDEA and piperazine are the most popularly adopted solvents for CO_2. In lieu of the aforementioned, amine degradation is a major limitation during gas scrubbing operations [43,44]. In addition, due to the formation of corrosive products during the degradation reaction of CO_2-entrained amines [45], inorganic hydroxides (NaOH, KOH, and Ca(OH)$_2$) have proven to be good absorbents for CO_2 stripping from NG [46]. However, solvent regeneration is a major limitation when this technique is used.

Some physical solvents for CO_2 capture from NG include the Selexol (Honeywell UOP, Des Plaines, IL, USA) and Rectisol (Air Liquide, Houston, TX, USA) with good CO_2 absorptive capacity and selectivity; however, they are relatively less energy intensive because their applications span below room temperature. The Selexol process makes use of a solvent/mixture of dimethyl ethers of polyethylene glycol (DEPG) or polyethylene glycol dimethyl ether, owing to their high stability and low vapor pressure [47]. Furthermore, DEPG is capable of removing H_2S and CO_2 in two-stage absorbers connected to multiple regeneration flash tanks. In lieu of the fact that DEPG offers the advantage of dehydrating NG, at low temperatures, DEPG has a very high viscosity, which in turn impedes mass transfer. Although the Rectisol process makes use of methanol for purifying syngas/NG for high CO_2 recovery, the solvent used has been observed to have a higher affinity/selectivity for H_2S over CO_2. According to Jansen et al. [48], high H_2S and CO_2 removal from NG can be achieved by the use of methanol as solvent within the temperature limit of -15 to $-60°C$ where its viscosity is somewhat moderate. The recovery process for the spent methanol is quite expensive, owing to the inherent high vapor pressure of CH_3OH, which leads to the introduction of expensive solvents as a viable means of combating solvent loss under cryogenic conditions [49]. Some establishments including the Great Plains Synfuels Plant (Saskatchewan, Canada) and Alberta Carbon Trunk Line (Alberta, Canada) [50–52] as well as the Sasol's synfuel plants in South Africa [53], have recorded successful applications of the Rectisol process. Other common commercially viable physical solvents for CO_2 capture from NG include PC that is usually employed at low temperatures during CO_2 capture (Fluor process) and N-methyl-2-pyrrolidone (NMP), which is used commercially in the Purisol process owing to its high selectivity for H_2S over CO_2 within the temperature range of $15-40°C$. However, the high vapor pressure of NMP makes it susceptible to high water requirement during washing of the gaseous stream for high solvent recovery [49].

2.2 Modified membrane composites for gas sweetening

A new class of adsorbent that have gained alarming interest in stripping CO_2 from NG is polymeric membranes, this is, as a result of the simplicity in their design/fabrication, scale-up flexibility, low thermal resistance, and low energy costs [54]. Unfortunately, most membranes are not durable/span through a short service life during NG purification as a result of saturation/pore plugging, which in turn results in low membrane performance with time; hence, they require frequent replacements during the sorption step [55]. Therefore, to improve the service life of membranes, some adjustments in the chemical properties of the materials are made by imposing surface/intricate characteristics in them so as to improve their thermal as well as chemical/hydrological stability.

The CO_2, CH_4, and N_2 adsorptive potentials of a poly(vinyl trimethylsilane) (PVTMS) membrane composite–modified hydrogel chitosan (CS) and PEG (CS + PEG) polymer blend were studied by Kunalan et al. [56]. The physicochemical/surface morphology of the pristine and modified polymer blend was examined. At optimum conditions (transmembrane pressure = 76 cmHg/reaction temperature = 30°C), the effect of membrane wetting, number of sequential coats, and properties of the pristine/modified membranes on gas adsorption were determined. CO_2 permeability was estimated to be 153 barrer, while gas pair selectivity of 33 and 23 were recorded for CO_2–N_2 and CO_2–CH_4 gas pairs for the unmodified and CS + PEG–modified PVTMS membranes, respectively. The CO_2 retention of the modified membrane composite was seen to approach the Robeson's upper limit CO_2 concentration, thus justifying its efficiency for CO_2 capture.

2.3 The use of carbonate and aqueous amino acid salts

Carbonate solutions, such as potassium and calcium carbonate, have been tested and proven useful for CO_2 absorption. However, they have low volatility, reactivity/corrosion rates, as well as low absorption rates relative to amines and caustic solutions [57]. Aqueous amino acid salts have great potentials for CO_2 capture, but the loss experienced in the solvent regeneration step imposed by low volatility and high molecular weight causes increased capital costs of the absorber relative to amines and caustic solutions [58]. The work of Sanni et al. [35] involves the combination of mixed concentrations of $Ca(OH)_2$ and DEA for CO_2 and H_2S capture where the retention rates for both gases in the solvent were in the range of 95%–100%.

2.4 Ionic liquids

Since the consequences associated with traditional processes for NG sweetening are unfavorable from the viewpoint of green/sustainable chemistry, concerted/ongoing efforts are in place to seek possible replacements for conventional alkanolamines using task-specific ILs [59] as acid gas strippers from NG. ILs/molten salts are stable at temperatures within ambient temperature and 100°C. ILs have negligible/very low vapor pressures and are nonvolatile/nonflammable with high thermal and electrochemical stability. Phenolic ILs, which possess strong basicities, are known to have high potentials for CO_2 capture [60]; this is due to the inherent chemical reactivity of the phenolate anion ([PhO]) with CO_2, which in turn induces equimolar absorption of CO_2. There are speculations that phenolic ILs have the ability to trap H_2S, owing to the acidic nature of the gas. ILs are green solvents with unique properties (distinct ionic environments, wide liquid range, negligible volatility, high thermal stability and structural designations), which make them suitable for several applications [61,62]. Fig. 6.1 shows the chemical structures of some phenolic ILs, while Fig. 6.2 is illustrative of the adsorption isotherms of some phenolic ILs established for CO_2 and H_2S stripping from NG at 313.2 K.

The volatile solvent loss experienced by conventional solvents/amines can be abated by ILs during NG purification, owing to the energy involvement during the solvent-regeneration step, induced by their relatively low heat capacities [64]. High H_2S solubility in ILs has been reported for liquids bearing a myriad of anions, including chloride ([Cl]), hexafluorophosphate ([PF_6]),

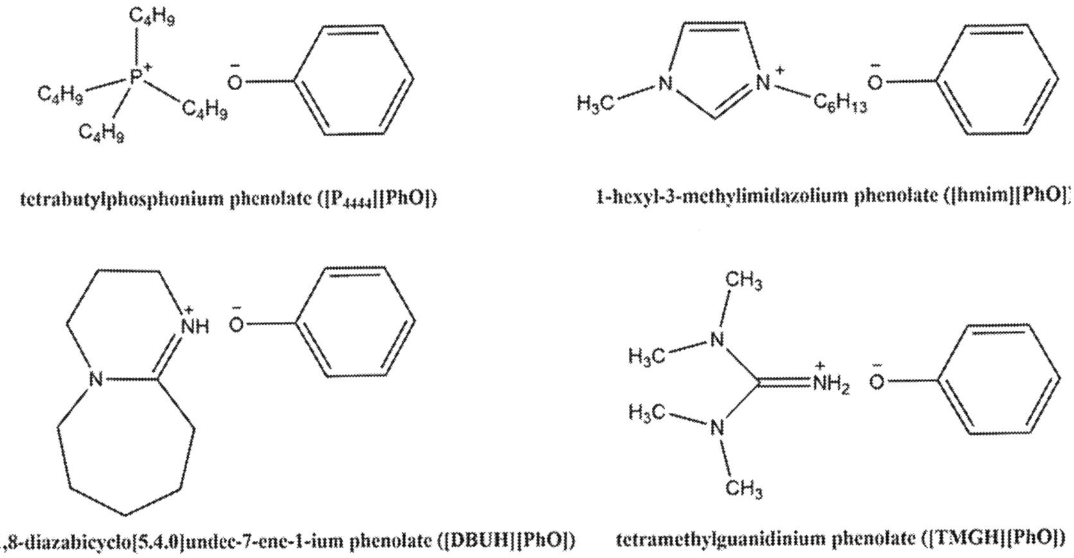

tetrabutylphosphonium phenolate ([P₄₄₄₄][PhO])

1-hexyl-3-methylimidazolium phenolate ([hmim][PhO])

1,8-diazabicyclo[5.4.0]undec-7-ene-1-ium phenolate ([DBUH][PhO])

tetramethylguanidinium phenolate ([TMGH][PhO])

Figure 6.1 Chemical structures of some synthetic phenolic ionic liquids. Adopted from Huang et al. [63] with copyright and reprint permission from Elsevier.

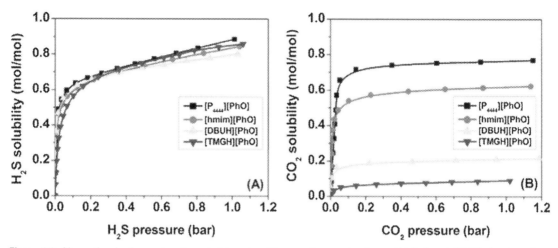

Figure 6.2 Absorption isotherms/profiles of stripped acid gases from natural gas: (A) H$_2$S and (B) CO$_2$ in phenolic ionic liquids at 313.2 K. Adopted from Huang et al. [63] with copyright and reprint permission from Elsevier.

tetrafluoroborate ([BF$_4$]), trifluoroacetate ([TfA]), acetate ([Ace]), trifluoromethanesulfonate (TfO), bis(trifluoromethanesulfonyl) imide ([Tf$_2$N]), methyl sulfate ([MeSO$_4$]), ethyl sulfate ([EtSO$_4$]), propionate ([Pro]), lactate ([Lac]), tris(pentafluoroethyl)trifluorophosphate ([eFAP]), glycinate ([Gly]), and alaninate ([Ala]) [65–74].

According to an investigation, the high solubilities of H$_2$S in some synthetic phenolic ILs can be likened to the strong interactions between basic phenolate anions with H$_2$S, whereas the solubility of CO$_2$ decreases when there is an increase in cationic hydrogen/proton donation [75]. For tetramethylguanidinium phenolate ([TMGH][PhO]), with strong anion basicity/cation of strong hydrogen bond donation characteristic, a high absorptive capacity for H$_2$S (for instance, 0.56 mol/mol at 313.2 K/0.1 bar and 0.85 mol/mol at 313.2 K/1 bar) with corresponding high H$_2$S/CO$_2$ selectivity of 6.2 at 313.2 K and 0.1 bar relative to CO$_2$ solubility at 313.2 K/1 bar and 9.4 at 313.2 K/1 bar compared to CO$_2$ solubility at 313.2 K/1 bar was recorded. Due to its small molecular size/facile synthetic nature and cost-effectiveness, the absolute solubility of H$_2$S in [TMGH][PhO] is 2.68 mol/kg at 313.2 K/0.1 bar and 4.08 mol/kg at 313.2 K/1 bar, which is way higher than that of other absorbents reported in the literature. [TMGH][PhO] can be facilely produced via one-step neutralization of 1,1,3,3-tetramethylguanidine and phenol; it is a promising candidate for the selective sweetening of NG. Three aqueous multiple Lewis base functionalized protic ionic liquids (MLB-PILs) were synthesized and tested for their H_2S adsorption capacities [75]. The ILs were tethered with tertiary amine as cations and with [Ace] as anions. Impressively, the viscosity of the MLB-PILs was less than 25 cP compared to those reported in literature at ambient condition. Moreso, the H_2S solubility (0.65–1.92 mol/mol at 313.2 K/1 bar) of the MLB-PILs was higher relative to those of other ILs.

Fig. 6.3 is an illustration of a decision support system for making an apt IL selection for acid gas removal from NG, while Fig. 6.4 showcases different categories of ILs and their striking properties, which make them highly susceptible to acid gas trapping.

Fig. 6.5 is an illustration of some acid gas stripping technologies from NG as well as their associated challenges and advantages with key prospects indicated for IL systems.

2.5 Methanol

Methanol being a physical solvent has found use in two popular processes (Rectisol and Ifpexol processes). In the Rectisol process, impurities such as aromatics, organic compounds of sulfur, and hydrogen cyanide from coal gasification are being stripped off gaseous streams. The methanol used in the Rectisol process helps to abate ice and hydrate formation at low temperatures, that is, at temperatures of −59.5–73.3°C, which helps to boost its acid gas (H_2S/CO_2) stripping tendency while taking into consideration specific requirements and feed conditions alongside nonselective and selective standard procedures [12]. The Ifpexol process is one that employs methanol in carrying out NG dehydration, NG recovery, and acid gas stripping in an entire process/loop. It has a known record of high acid gas/H_2S removal from contaminated NG streams [12].

Sun and Smith [77] conducted a single- and two-stage Rectisol simulation study using an equation of state, also known as the perturbed chain-statistical associating fluid theory (PC-SAFT) with considerations for heat recovery, acid gas removal capacity, equipment requirement, power demand, environmental emission, and process economics.

In a related study, amid recalibrating the PC-SAFT equation, the Rectisol process was reviewed in terms of its configuration and applications [53]; a detailed process simulation, optimized heat integration, utility design, as well as alternative configurations for CO_2 capture were proposed.

The prediction of the thermodynamic behavior of the Rectisol process via the SAFT equation of state was also carried out by Gao et al. [78], whereas the work of Sharma et al. [79] bothers on the

Chapter 6 Physical and hybrid solvents for natural gas sweetening **123**

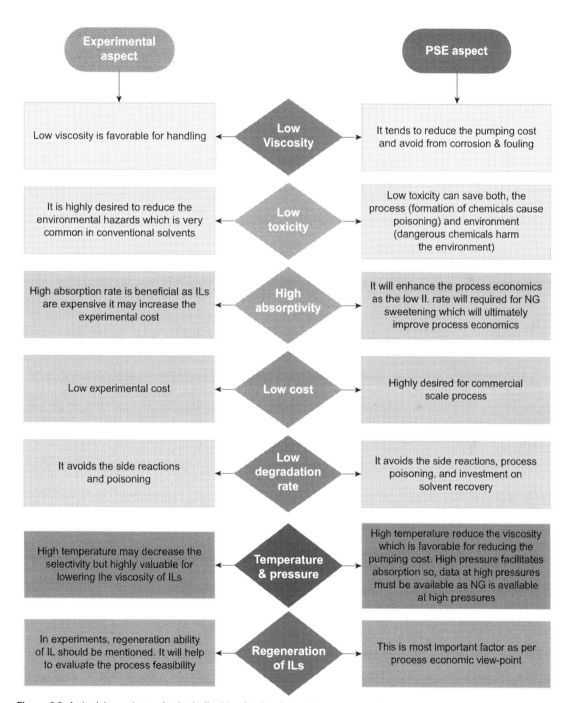

Figure 6.3 A decision scheme for ionic liquid selection for acid gas removal from natural gas. Adopted from Haider et al. [76] with preprint and copyright permission obtained from Elsevier.

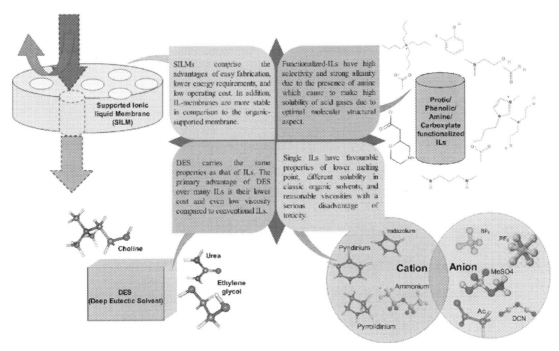

Figure 6.4 Different categories of ionic liquids and their properties. Adopted from Haider et al. [76] with preprint and copyright permission obtained from Elsevier.

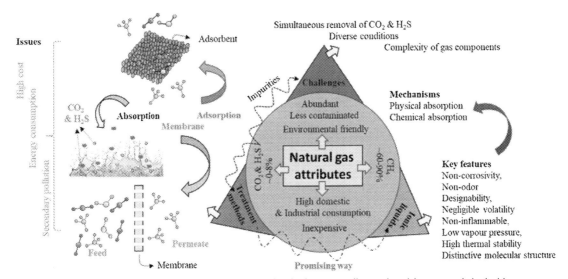

Figure 6.5 Key features, impurities, some treatment technologies, as well as related issues and desirable properties of several sorbents for natural gas purification. Adopted from Haider et al. [76] with preprint and copyright permission obtained from Elsevier.

optimization of the energy penalty and CO_2 retention in a standalone Rectisol process in a bid to ascertain the most suitable conditions for improved CO_2 retention rates.

2.6 The use of polyethylene glycol methyl isopropyl ethers

A commercially viable method that employs polyethylene glycol dialkyl ethers (methyl isopropyl ethers, MPE) is the Sepasolv-MPE; it is similar to the Selexol process in terms of the solvent used and mode of operation. The solvent has proven worthy of the selective removal of H_2S and CO_2 from natural and synthesis gases [80]. However, studies involving the use of this approach are not readily available in literature.

2.7 Propylene carbonate

The process that employs PC in purifying gas streams is the Flucor process, which takes advantage of CO_2 partial pressures above 60 psig with temperatures within the range of −17 to 65°C. The solvent has high CO_2 solubility and can be employed in carrying out simultaneous dehydration and sweetening of NG. It has the capacity of stripping very low amounts (20 ppmv) of H_2S at low temperatures with good mass transfer without significantly altering the fluid's viscosity [81].

2.8 The use of NMP

The Purisol process employs NMP at ambient conditions/low temperatures, that is, about −15°C. NMP possesses higher vapor pressure relative to DEPG and PC, which is the reason it is often necessary to wash it off with water after use. Despite its high selectivity for H_2S, this method can be employed in purifying high-pressure CO_2 containing synthesis gas released from Integrated coal gasification combined cycle gas turbine systems.

2.9 Glycerol/glycerine

Glycerol, as a polyol, is a colorless, odorless, nontoxic, sweet, and viscous liquid. Nunes et al. [82] documented CO_2 dissolution in glycerol and other sorbents of similar chain length at different temperatures (80, 120, and 150°C) and at pressures within the confines of 32 MPa. CO_2 solubility was seen to increase with increase in pressure at reduced temperatures, whereas the study by Medina-Gonzalez et al. [83] confirmed the viability of the in situ Fourier-transform infrared spectroscopy approach in measuring CO_2 solubility in glycerol within a temperature range of 40–200°C and pressures of about 35 MPa.

2.10 Tetramethylene sulfone/2,3,4,5-tetrahydrothiophene-1,1-dioxide (($CH_2)_4SO_2$)

Tetramethylene sulfone/2,3,4,5-tetrahydrothiophene-1,1-dioxide is a colorless organic sulfur-based liquid. CO_2 solubility in sulfolane solution at 25–130°C and pressures up to 7.6 MPa have been documented [84]. In another study, N_2O and CO_2 solubilities and diffusivities in aqueous and pure sulfolane solutions were recorded and correlated within a temperature range of 20–85°C [85]. As a way of understanding/determining the microscopic characteristics/behaviors that ensue when utilizing sulfolane as solvent for CO_2 and H_2S capture, the density functional theory and molecular dynamics computational chemistry were employed [86].

3. Mechanism of solute take-up by physical and hybrid sorbents

As previously mentioned, for the physical sorbents/solvents used in the Rectisol, Flucor, and Selexol processes or those used in monotonic membranes (one material type membrane), absorption/adsorption is mainly by physical means and involves mainly one type of solvent, where the molecules coexist via forces of cohesion and the attractive force between the solvent molecule and that of the absorbate is by physical attraction/weak van der Waals forces, whereas, for the case of hybrid solvents or sorbents, fluid–solid molecules/liquid–solid molecules of different coexisting molecules interact via forces of adhesion and the molecules become covalently bonded to offer a synergistic effect of the individual molecules that make up the entire system/adsorbent/absorbent. Hence, the interactive forces between hybrid sorbents and target molecules are stronger than those offered by single/monoatomic sorbents and their target molecules; hence, the adsorbate/absorbate molecular attraction is higher for the latter relative to the former.

4. Conclusion and future outlooks

Commercialized conventional NG sweetening systems have been employed over the years; however, owing to some of the major limitations associated with the use of existing physical solvents, attention has drifted in the direction of other more suitable solvents such as ILs and membrane systems, which have the potentials to overcome the limitations associated with physical solvents, thus giving high CO_2 and H_2S absorption/adsorption rates. In recent times, researchers have found that ILs are systems with high CO_2 capture potentials. Some of them are integrated in MOFs and membrane systems or coupled with zeolites in order to improve their CO_2 retention rates. By using hybrid or mixed systems, the individual components contribute their respective characteristics, which in turn gives a synergistic or an overall effect for high acid gas selectivity/retention. Such systems including [PF$_6$], [Pro], [Lac], [Tf$_2$N], [BF$_4$], [EtSO$_4$], [eFAP], [TfA], [Ace], TfO, [Gly], [MeSO$_4$], and [Ala] have been projected as suitable acid gas absorbers, hence the need to consider integrating them in MOFs, molecular sieves, and membranes as a new wave in NG sweetening/acid gas capture and storage technologies. Furthermore, membrane–IL systems alongside membranes impregnated with nanoparticles for induced thermal stability and gas trapping ability are being tested/embraced as novel systems for gas sweetening. However, it is thus necessary to take cognizance of the capabilities of hybrid systems (deep eutectic solvents/ILs) with a good measure of compatibility prior to blending, because this not only helps to maximize their synergistic effect toward high acid gas stripping relative to conventional gas sweetening solvents but also helps to abate the disadvantages associated with the latter.

Abbreviations and symbols

([EtSO$_4$])	Ethyl sulfate
(MLB-PILs)	Multiple Lewis base functionalized protic ionic liquids
[Ace]	Acetate
[Ala]	Alaninate
[BF$_4$]	Tetrafluoroborate
[eFAP]	Tris(pentafluoroethyl)trifluorophosphate
[Gly]	Glycinate
[Lac]	Lactate
[MeSO$_4$]	Methyl sulfate
[PF$_6$]	Hexafluorophosphate
[Pro]	Propionate
[Tf$_2$N]	Bis(trifluoromethanesulfonyl)imide

[TfA]	Trifluoroacetate
[TMGH][PhO]	Tetramethylguanidinium phenolate
CS + PEG	Chitosan and polyethylene glycol
DEA	Diethanolamine
DEPG	Dimethyl ethers of polyethylene glycol
DGA	Diglycolamine
DIPA	Diisopropanolamine
IL	Ionic liquid
MDEA	Methyldiethanolamine
MEAm	Monoethanolamide
MOF	Metal–organic framework
NG	Natural gas
NMP	N-methyl-2-pyrrolidone
PC	Propylene carbonate
PEGDME	Polyethylene glycol dimethyl ether
PVTMS	Poly(vinyl trimethylsilane)
TEA	Triethanolamine
TfO	Trifluoromethanesulfonate
TSILs	Task-specific ionic liquids

References

[1] Pospísil J, Charvat P, Arsenyeva O, Klimes L, Spilacek M, Klemes JJ. Energy demand of liquefaction and regasification of natural gas and the potential of LNG for operative thermal energy storage. Renewable and Sustainable Energy Reviews 2019;99:1–15. https://doi.org/10.1016/j.rser.2018.09.027.

[2] Haghtalab A, Afsharpour A. Solubility of CO_2 + H_2S gas mixture into different aqueous N-methyldiethanolamine solutions blended with 1-butyl-3-methylimidazolium acetate ionic liquid. Fluid Phase Equilibria 2015;406:10–20. https://doi.org/10.1016/j.fluid.2015.08.001.

[3] British Petroleum. BP statistical review of world energy. 2018.

[4] International Energy Agency (IEA). International energy outlook. 2019. https://doi.org/10.1080/01636609609550217.

[5] Wang Y, Liu X, Kraslawski A, Gao J, Cui P. A novel process design for CO_2 capture and H_2S removal from the syngas using ionic liquid. Journal of Cleaner Production 2019;213:480–90. https://doi.org/10.1016/j.jclepro.2018.12.180.

[6] Xu H-J, Zhang C-F, Zheng Z-S. Selective H_2S removal by nonaqueous methyldiethanolamine solutions in an experimental apparatus. Industrial & Engineering Chemistry Research 2002;41(12):2953–6. https://doi.org/10.1021/ie0109253.

[7] Song C, Liu Q, Ji N, Deng S, Zhao J, Li Y, et al. Alternative pathways for efficient CO_2 capture by hybrid processes—a review. Renewable and Sustainable Energy Reviews 2018;82:215–31. https://doi.org/10.1016/j.rser.2017.09.040.

[8] Rochelle GT. Amine scrubbing for CO_2 capture. Science 2009;325:1652–4. https://doi.org/10.1126/science.1176731.

[9] Abdeen FRH, Mel M, Jami MS, Ihsan SI, Ismail AF. A review of chemical absorption of carbon dioxide for biogas upgrading. Chinese Journal of Chemical Engineering 2016;24(6):693–702. https://doi.org/10.1016/j.cjche.2016.05.006.

[10] Bolland O. CO$_2$ capture in power plants. Norwegian University of Science and Technology; 2013. http://www.ivt.ntnu.no/ept/fag/fordypn/tep03/innhold/EP03_Part4_Absorption.pdf.
[11] Salako AE. Removal of carbon dioxide from natural gas for lng production. Norway: Trondheim, Norwegian University of Science and Technology Institute of Petroleum Technology; 2005. https://pdfs.semanticscholar.org/37ea/1e656cc310e4797ded37b8781dd06d97694c.pdf.
[12] Kohl A, Nielsen R. Gas purification. 5th. Houston Texas: Gulf Publishing Company; 1997.
[13] Nobel A. Functional chemicals. USA: Alliance Chemicals; 2017. https://alliancechemicals.com/wpcontent/uploads/2011/09/DEAspecification.pdf.
[14] Perry C. Activated carbon filtration of amine and glycol solutions. In: Gas conditioning conference. Oklahoma, USA: University of Oklahoma Extension Division; 1994.
[15] Tatin R, Moura L, Dietrich N, Baig S, Hebrard G. Physical absorption of volatile organic compounds by spraying emulsion in a spray tower: experiments and modelling. Chemical Engineering Research and Design 2015;104:409−15. https://doi.org/10.1016/j.cherd.2015.08.030.
[16] Horikawa MS, Rossi ML, Gimenes ML, Costa CMM, da Silva MGC. Chemical absorption of H$_2$S for biogas purification. Brazilian Journal of Chemical Engineering 2004;21(3):415−22. https://doi.org/10.1590/S0104-66322004000300006.
[17] Zucca T, Pellegrini LA, Nava C, Bellasio R, Bianconi R. Reduction of the flare system in the installations of natural gas. Chemical Engineering Transactions 2005;6:629−34.
[18] Pellegrini LA, Gamba S, Moioli S. Using an adaptive parameter method for process simulation of nonideal systems. Industrial & Engineering Chemistry Research 2010a;49:4923−32. https://doi.org/10.1021/ie901773q.
[19] Pellegrini LA, Moioli S, Gamba S. Energy saving in a CO$_2$ capture plant by MEA scrubbing. Chemical Engineering Research and Design 2010b;89:1676−83.
[20] Scholes CA, Stevens GW, Kentish SE. Membrane gas separation applications in natural gas processing. Fuel 2012;96:15−28. https://doi.org/10.1016/j.fuel.2011.12.074.
[21] Palomeque-santiago JF, Guzman J, Zuniga-mendiola AJ. Simulation of the natural gas purification process with membrane technology. technical and economic aspects. Revista Mexicana de Ingenieria Quimica 2016;15(2):611−24.
[22] Belaissaoui B, Favre E. Novel dense skin hollow fiber membrane contactor based process for CO$_2$ removal from raw biogas using water as absorbent. Separation and Purification Technology 2018;193:112−26. https://doi.org/10.1016/J.SEPPUR.2017.10.060.
[23] Fang M, Zhu D. Chemical absorption. In: Handbook of climate change citigation. New York Dordrecht Heidelberg London: Springer; 2012. ISBN 978- 1-4419-7990-2.
[24] Harasimowicz M, Orluk P, Zakrzewska-Trznadel G, Chmielewski A. Application of polyimide membranes for biogas purification and enrichment. Journal of Hazardous Matererials 2007;144(3):698−702. https://doi.org/10.1016/j.jhazmat.2007.01.098.
[25] Vijay VK, Chandra R, Subbarao PMV, Kapid S. Biogas purification and bottling into CNG cylinders: producing Bio-CNG from biomass for rural automotive applications. In: 2nd joint international conference on sustainable energy and environment (SEE), 2006; 2006 [Bangkok, Thailand].

[26] Huo D. The global sour gas problem, energy resource engineering. Stanford University; 2012.
[27] China National Corporation. Sour natural gas corrosion protection and purification. Beijing, China: Dongzhimen North Street; 2012.
[28] Popoola LT, Grema SA, Latinwo GK, Gutti B, Balogun AS. Corrosion problems during oil and gas production and its mitigation. International Journal of Industrial Chemistry 2013;4:1–15. https://doi.org/10.1186/2228-5547-4-35.
[29] Lee J. Soft systems. 2015. Retrieved March 24, 2017 from Oil and Gas Blog.
[30] National Energy Laboratory. Syngas contaminant removal and conditioning. Retrieved March 27, 2017. https://www.netldoe.gov/research/coal/energysystems/gasifification/gasifififpedia/agr.
[31] Rivera-Tinoco R, Bouallou C. Comparison of absorption rates and absorption capacity of ammonia solvents with MEA and MDEA aqueous blends for CO_2 capture. Journal of Cleaner Production 2010;18:875–80. https://doi.org/10.1016/j.jclepro.2009.12.006.
[32] Rodriguez N, Mustati S, Scenna N. Optimization of post-combustion CO_2 process using DEA–MDEA mixtures. Chemical Engineering Research and Design 2011;89:1763–73. https://doi.org/10.1016/j.cherd.2010.11.009.
[33] Treybal RE. Operaciones de Transferencia de Masa. 2nd ed. Mexico: Mc Graw Hill; 1996.
[34] Huertas JI, Giraldo N, Izquierdo S. Removal of H_2S and CO_2 from biogas by amine absorption. China: Intech. Open; 2011.
[35] Sanni SE, Agboola O, Fagbiele O, Yusuf EO, Emetere ME. Optimization of natural gas treatment for the removal of CO2 and H2S in a novel alkaline-DEA hybrid scrubber. Egyptian Journal of Petroleum 2020;29(1):83–94. https://doi.org/10.1016/j.ejpe.2019.11.003.
[36] Lawson JD, Garst AW. Gas sweetening data: equilibrium solubility of hydrogen sulfifide and carbon dioxide in aqueous monoethanolamine and aqueous diethanolamine solutions. Journal of Chemical & Engineering Data 1976;21(1):20–30.
[37] Lee JI, Otto FD, Mather AE. Measurement and prediction of solubility of mixtures of carbon dioxide and hydrogen sulfifide in a 2.5 N monoethanolamine solution. Canadian Journal of Chemical Engineering 1976;54(3):214–9.
[38] Sada E, Kumazawa H, Butt MA, Hayashi D. Simultaneous absorption of carbon dioxide and hydrogen sulfifide into aqueous monoethanolamine solutions. Chemical Engineering Science 1976;31(9):839–41.
[39] Galán Sánchez LM, Meindersma GW, de Haan AB. Translation IChemE, Part A: Chemical Engineering Research and Design 2007;85:31–9.
[40] Pieplu A, Saur O, Lavalley JC, Legendre O, Nedez C. Claus catalysis and H_2S selective oxidation. Catalysis Reviews: Science and Engineering 1998;40(4):409–50.
[41] Mandal BP, Biswas AK, Bandyopadhyay SS. Selective absorption of H_2S from gas streams containing H_2S and CO_2 into aqueous solutions of N-methyldiethanolamine and 2-amino-2-methyl-1-propanol. Separation and Purification Technology 2004;35(3):191–202.
[42] Lu G, Zheng YF, He DL. Selective absorption of H_2S from gas mixtures into aqueous solutions of blended amines of methyldiethanolamine and 2-tertiarybutylamino-2-ethoxyethanol in a packed column. Separation and Purification Technology 2006;52(2):209–17.
[43] Lepaumier H, Picq D, Carrette PL. New amines for CO_2 Capture. I. mechanisms of amine degradation in the presence of CO_2. Industrial &

Engineering Chemistry Research 2009;48(20):9061−7. https://doi.org/10.1021/ie900472x.

[44] Gouedard C, Picq D, Launay F, Carrette PL. Amine degradation in CO_2 capture. I. A review. International Journal of Greenhouse Gas Control 2012;10:244−70. https://doi.org/10.1016/j.ijggc.2012.06.015.

[45] Veawab A, Tontiwachwuthikul P, Chakma A. Corrosion behavior of carbon steel in the CO_2 absorption process using aqueous amine solutions. Industrial & Engineering Chemistry Research 1999;38(10):3917−24. https://doi.org/10.1021/ie9901630.

[46] Kenarsari SD, Yang D, Jiang G, Zhang S, Wang J, Russell AG, et al. Review of recent advances in carbon dioxide separation and capture. Royal Society of Chemistry Advances 2013;3(45):22739−73. https://doi.org/10.1039/c3ra43965h.

[47] Porter RTJ, Fairweather M, Kolster C, Mac Dowell N, Shah N, Woolley RM. Cost and performance of some carbon capture technology options for producing different quality CO_2 product streams. International Journal of Greenhouse Gas Control 2017;57:185−95.

[48] Jansen D, Gazzani M, Manzolini G, Dijk EV, Carbo M. Pre-combustion CO_2 capture. International Journal of Greenhouse Gas Control 2015;40:167−87.

[49] Burr B, Lyddon L. A comparison of physical solvents for acid gas removal, digital refining. Bryan Research & Engineering; 2008.

[50] Folger P. Carbon capture: a technology assessment. CRS report for Congress; 2013.

[51] Heal K, Kemp T. North West sturgeon refinery project overview - carbon capture through innovative commercial structuring in the canadian oil sands. Energy Procedia 2013;37:7046−55.

[52] Loria P, Bright MBH. Lessons captured from 50 years of CCS projects. The Electricity Journal 2021;34:106998.

[53] Gatti M, Martelli E, Marechal F, Consonni S. Review, modeling, heat Integration, and improved schemes of Rectisol®-based processes for CO_2 capture. Applied Thermal Engineering 2014;70:1123−40.

[54] Brunetti A, Scura F, Barbieri G, Drioli E. Membrane technologies for CO_2 separation. Journal of Membrane Science 2010. https://doi.org/10.1016/j.memsci.2009.11.040.

[55] Bernardo P, Clarizia G. 30 years of membrane technology for gas separation. Chemical Engineering Transactions 2013. https://doi.org/10.3303/CET1332334.

[56] Kunalan S, Palanivelu K, Syrtsova DA, Teplyakov VV. Thin-film hydrogel polymer layered polyvinyltrimethylsilane dual-layer flat-bed composite membrane for CO2 gas separation. Journal of Applied Polymer Science 2022;139(17):52024. https://doi.org/10.1002/app.52024 (in press).

[57] Cents AGH, Brilman DWF, Versteeg GF. CO_2 absorption in carbonate/bicarbonate solutions: the Danckwerts-criterion revisited. Chemical Engineering Science 2005. https://doi.org/10.1016/j.ces.2005.05.020.

[58] Song HJ, Park S, Kim H, Gaur A, Park JW, Lee SJ. Carbon dioxide absorption characteristics of aqueous amino acid salt solutions. International Journal of Greenhouse Gas Control 2012. https://doi.org/10.1016/j.ijggc.2012.07.019.

[59] Bates ED, Mayton RD, Ntai I, Davis JH. Journal of the American Chemical Society 2002;124:926−7.

[60] Wang C, Luo H, Li H, Zhu X, Yu B, Dai S, et al. Tuning the physicochemical properties of diverse phenolic ionic liquids for equimolar CO_2 capture by the substituent on the anion. Chemistry - A European Journal 2012;18(7):2153−60.

[61] Brennecke JF, Magin EJ. Ionic liquids: innovative fluids for chemical processing. AIChE Journal 2001;47(11):2384−9.
[62] Lei Z, Dai C, Chen B. Gas solubility in ionic liquids. Chemical Review 2014; 114(2):1289−326.
[63] Huang K, Zhang X-M, Zhou L-S, Tao D-J, Fan J-P. Highly efficient and selective absorption of H_2S in phenolic ionic liquids: a cooperative result of anionic strong basicity and cationic hydrogenbond donation. Chemical Engineering Science 2017;173:253−63. https://doi.org/10.1016/j.ces.2017.07.048.
[64] Fredlake CP, Crosthwaite JM, Hert DG, Aki S, Brennecke JF. Thermophysical properties of imidazolium-based ionic liquids. Journal of Chemical & Engineering Data 2004;49(4):954−64.
[65] Ahmadi MA, Pouladi B, Javvi Y, Alfkhani S, Soleimani R. Connectionist technique estimates H_2S solubility in ionic liquids through a low parameter approach. The Journal of Supercritical Fluids 2015;97:81−7.
[66] Haghtalab A, Kheiri A. High pressure measurement and CPA equation of state for solubility of carbon dioxide and hydrogen sulfiide in 1-butyl-3-methylimidazolium acetate. Journal of Chemical Thermodynamics 2015;89: 41−50.
[67] Hamzehie ME, Fattahi M, Najibi H, Van der Bruggen B, Mazinani S, Mazinani S. Application of artifificial neural networks for estimation of solubility of acid gases (H_2S and CO_2) in 32 commonly ionic liquid and amine solutions. Journal of Natural Gas Science and Engineering 2015;24: 106−14.
[68] Sanchez-Badillo J, Gallo M, Alvarado S, Glossman-Mitnik D. Solvation Thermodynamic properties of hydrogen sulfifide in [C4mim][PF6], [C4mim][BF4], and [C4mim][Cl] ionic liquids, determined by molecular simulations. The Journal of Physical Chemistry B 2015;119(33):10727−37.
[69] Wang A, Zhang K, Ren S, Hou Y, Wu W. Effificient capture of low partial pressure H_2S by tetraethyl ammonium amino acid ionic liquids with absorption-promoted solvents. RSC Advances 2016;6(103):101462−9.
[70] Zhao Y, Gao H, Zhang X, Huang Y, Bao D, Zhang S, et al. Hydrogen sulfifide solubility in ionic liquids (ILs): an extensive database and a new ELM model mainly established by imidazolium-based ILs. Journal of Chemical & Engineering Data 2016;61(12):3970−8.
[71] Afsharpour A, Haghtalab A. Simultaneous measurement absorption of CO_2 and H_2S mixture into aqueous solutions containing diisopropanolamine blended with 1-butyl-3-methylimidazolium acetate ionic liquid. International Journal of Greenhouse Gas Control 2017;58:71−80.
[72] Soltani Panah H. Modeling H2S and CO2 solubility in ionic liquids using the CPA equation of state through a new approach. Fluid Phase Equilibria 2017; 437:155−65. https://doi.org/10.1016/j.flfluid.2017.01.023 (in press).
[73] Huang J, Wu YT, Hu XB. Effect of alkalinity on absorption capacity and selectivity of SO_2 and H_2S over CO_2: substituted benzoate-based ionic liquids as the study platform. Chemical Engineering Journal 2016;297: 265−76.
[74] Huang K, Cai DN, Chen YL, Wu YT, Hu XB, Zhang ZB. Dual lewis-base functionalization of ionic liquids for highly effificient and selective capture of H_2S. ChemPlusChem 2014;79(2):241−9.
[75] Zheng W, Wu D, Feng X, Hu J, Zhang F, Wu Y-T, et al. Low viscous protic ionic liquids functionalized with multiple lewis base for highly efficient capture of H_2S. Journal of Molecular Liquids 2018;263:209−17. https://doi.org/10.1016/j.molliq.2018.04.129.

[76] Haider H, Saeed S, Qyyum MA, Kazmi B, Ahmad R, Muhammad A, et al. Simultaneous capture of acid gases from natural gas adopting ionic liquids: challenges, recent developments, and prospects. Renewable and Sustainable Energy Reviews 2020;123:109771. https://doi.org/10.1016/j.rser.2020.109771.
[77] Sun L, Smith R. Rectisol wash process simulation and analysis. Journal of Cleaner Production 2013;39:321–8.
[78] Gao N, Zhai C, Sun W, Zhang X. Equation oriented method for rectisol wash modeling and analysis. Chinese Journal of Chemical Engineering 2015;23:1530–5.
[79] Sharma I, Hoadley AFA, Mahajani SM. A. GaneshMulti-objective optimisation of a Rectisol™ process for carbon capture. Journal of Cleaner Production 2016;119:196–206.
[80] Wolfer W, Schwartz E, Vodrazka W, Volkamer K. Solvent shows greater efficiency in sweetening of gas. Oil & Gas Journal 1980;78.
[81] Mokhatab S, Poe WA. Handbook of natural gas transmission and processing. Elsevier Science; 2012.
[82] Nunes AVM, Carrera GVSM, Najdanovic-Visak V, Nunes da Ponte M. Solubility of CO_2 in glycerol at high pressures. Fluid Phase Equilibria 2013;358:105–7.
[83] Medina-Gonzalez Y, Tassaing T, Camy S, Condoret JS. Phase equilibrium of the CO_2/glycerol system: experimental data by in situ FT-IR spectroscopy and thermodynamic modeling. The Journal of Supercritical Fluids 2013;73:97–107.
[84] Jou FY, Deshmukh RD, Otto FD, Mather AE. Solubility of H_2S, CO_2, CH_4 and C_2H_6 in sulfolane at elevated pressures. Fluid Phase Equilibria 1990;56:313–24.
[85] Xu S, Wang Y-W, Otto FD, Mather AE. Solubilities and diffusivities of N2O and CO_2 in aqueous sulfolane solutions. Journal of Chemical Technology and Biotechnology 1991;51:197–208.
[86] Gutiérrez A, Atilhan M, Aparicio S. Microscopic characterization of CO_2 and H_2S removal by sulfolane. Energy Fuels 2017;31:9800–13.

Natural gas sweetening by solvents modified with nanoparticles

Moloud Rahimi, Maryam Meshksar and Mohammad Reza Rahimpour
Department of Chemical Engineering, Shiraz University, Shiraz, Iran

1. Introduction

Natural gas (NG) is clearly destined to play a key role in the future global energy development as it approximately has half carbon dioxide (CO_2) polluting effects vis-à-vis other fossil fuels [1,2]. Actually, NG as a combustible gas that is colorless, amorphous, and odorless in its pure state. It is considered the most environmentally friendly and hydrogen-rich source of energy among all hydrocarbon energy sources. Moreover, it boasts excellent efficiency for converting energy into power [3]. Besides, NG is an essential feedstock for not only energy generation but also production of a wide range of chemical products including methanol, ammonia, formaldehyde, insulation materials, varnishes, paints, glues, fuel additives, and acetic acid [4,5].

NG is a mixture of different components, each of which has advantages and disadvantages. As tabulated in Table 7.1, while the main component of NG is a mixture of hydrocarbons, mainly methane (CH_4), it also includes various impurities such as water vapor (H_2O), sulfurated components, carbon dioxide, nitrogen (N_2), helium (He), and mercury (Hg) [6]. Except for methane, which is a hydrocarbon composed of one carbon and four hydrogen atoms, other alkanes such as ethane, butane, and propane are also found in NG. Despite having different molecular structures, these compounds have many similarities, all of them are colorless, odorless, flammable, and combustible gases.

In addition to helium, the presence of most contaminants increases the expenses associated with gas field exploration. At

Table 7.1 Natural gas impurities with their amounts in various references.

Hydrocarbon components		Composition [7]	Composition [8]
Methane	CH_4	0.75–0.99	>85%
Ethane	C_2H_5	0.01–0.15	3%–8%
Propane	C_3H_8	0.01–0.10	1%–2%
Butane	C_4H_{10}	0.00–0.02	<1%
Pentane	C_5H_{12}	0.00–0.01	<1%
Nonhydrocarbon components			
Water vapor	H_2O	Inert	Inert
Carbon dioxide	CO_2	0.00–0.30	1%–2%

Table 7.1 Natural gas impurities with their amounts in various references.—*continued*

Hydrocarbon components		Composition [7]	Composition [8]
Nitrogen	N_2	0.00–0.15	1%–5%
Helium	He	0.00–0.05	0
Hydrogen sulfide	H_2S	0.00–0.30	<1%

times, these costs become so high that even for massive fields like Natuna in the South China Sea, the undertaking becomes financially unviable. Moreover, the process of liquefying NG for transportation necessitates a minimal amount of CO_2, specifically less than 50 parts per million (ppm), to prevent frosted CO_2 from clogging flow lines and causing other operational issues [9]. Fig. 7.1 shows the amount of NG reserves extracted by countries in 2022 based on the data provided on the World Natural Gas Statistics, Worldometer website.

NG or any other gases containing significant measurable amount of acidic gases such as CO_2, H_2S, or other acid gas components are referred to as sour gas and always put up a lot of cost to development field [9]. Based on the International Energy Agency report, the world's NG reserves in various countries, including North America, contain 2580 Tcf of sour gas, which is about 43% of these reserves [10]. As the largest sour gas reserves in the world, the Middle East has 60% of these reserves, while Russia's share is 34%. Amount of CO_2 in more than 30 huge gas fields that have been identified worldwide is estimated to be more than

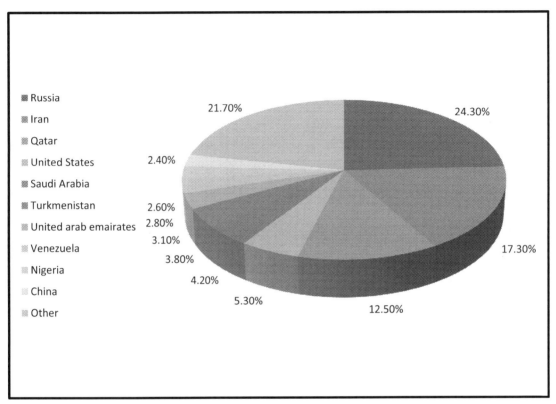

Figure 7.1 Amount of natural gas reserves extracted in 2022.

5 MMt. Based on calculations and evaluations that occur, the amount of CO_2 gas in the world's largest gas field, South Pars/North Dome Field in the Gulf, is estimated to be around 400 MMt [9]. A part of this gas is used for reinjection to help support the field pressure and the rest is released. The increasing tendency to developing sour gas indicates more reservoir CO_2 dispersing to the atmosphere unless extensive geological reserves become adopted [9].

Due to the fact that the exact composition of NG can be vary widely based on its location, each NG well has a different gas composition and so different amount of each component. The interesting thing to note is that even the wells in the same formation could have variations to the compositions, and these differences can be from partial to somewhat different value. The difference in composition and amount of components in NG extracted from different places can be observed in Table 7.2.

H_2S is a corrosive impurity found in NG that can cause damage to pipelines by forming iron sulfide when it reacts with H_2O and

Table 7.2 Composition of natural gas in different locations [11].

	Canada	Kansas	Texas
C_1	77.1	73.0	65.8
C_2	6.6	6.3	3.8
C_3	3.1	3.7	1.7
C_4s	2.0	1.4	0.8
C_5s^+	3.0	0.6	0.5
H_2S	3.3	0.0	0.0
CO_2	1.7	0.0	0.0
N_2	3.2	14.7	25.6
He	0.0	0.5	1.8

O_2 at normal temperatures. Additionally, H_2S can be changed into elemental sulfur with a small amount of O_2 and can also undergo polymerization. It is important to note that the buildup of elemental sulfur can lead to blockages in pipelines, resulting in reduced pressure and increased operational expenses. Therefore, the regulation and control of H_2S levels in NG present potential challenges for transportation in pipelines and as a feed gas for liquefied NG plants. Basically, the presence of a large amount of CO_2 in NG causes a decrease in the calorific value of the mixture compared to the pure state, which leads to a decrease in the burning speed. In addition, the presence of this component causes corrosion of equipment and pipelines. Considering that the amount of CO_2 and H_2S in gas fields should reach an acceptable level for pipeline, and the negative effects mentioned above, it is very important to remove these two impurities.

2. NG sweetening techniques

As discussed above, the existed impurities in the extracted NG not only end the reduction in the NG efficiency but also cause damage and corrosion of equipment and pipelines. Therefore, they must be removed from the gas mixture to prevent their harmful effluents. The process of separation and removal of H_2S, CO_2, and other acid gases like carbonyl sulfide (COS), mercaptan, and carbon disulfide (CS_2) from NG in order to improve its quality for transfer, consumption, and sale before reaching the end user is called gas sweetening [12]. To separate and remove H_2S and CO_2 from NG, various physical or chemical

140 Chapter 7 Natural gas sweetening by solvents modified with nanoparticles

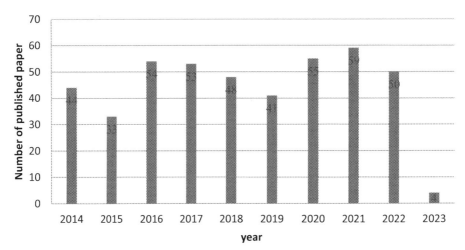

Figure 7.2 Number of published papers from January 2014 to January 2023 on natural gas sweetening techniques according to Scopus statistics.

separation methods are used, which include physical absorption, membrane permeation, cryogenic distillation, plasma technology, and adsorption process, among which NG sweetening using different absorbents is described in detail in the following, and as membrane and cryogenic methods have not yet been developed to a practical stage, more adsorption and absorption methods are used to remove CO_2, H_2S, and other impurities from NG. However, the adsorption method is not suitable for large-scale systems because it requires a large amount of energy for the removal process. The total number of publications from January 2014 to January 2023 on NG sweetening derived from Scopus with the key words of "NG sweetening, CO_2 capture from NG, and H_2S removal from NG" is illustrated in Fig. 7.2.

2.1 Absorption process for NG sweetening

Absorption process is one of the established methods to remove impurities from NG in which the gas mixture comes into a contact with a liquid solvent for transferring the sour gases into the liquid phase. In this process, a gas–liquid contactor and a series of flash tanks are needed to regenerate the solvent. The impurity is absorbed in the physical solvent in the high-pressure gas liquid contactor and then it is removed in the medium- and low-pressure flash tank without adding heat. One of the advantages of using a physical solvent is that the impurity is absorbed without any chemical reaction and can be removed by methods such as pressure reduction, inert gas passage from the solvent, and mild thermal regeneration.

In the case of absorbent selection for NG sweetening, at the first of the 21st century, the first solvent used in absorption process was carbonate solution, which was consumed in dry ice plants to separate CO_2 from flue gas. After alkanolamines were introduced, sodium carbonate solutions were quickly removed, because in addition to quick removal of CO_2, they have extremely high absorption CO_2 efficiency [13]. However, carbonate systems have been used for CO_2 and H_2S removal from gas treating plants for many decades. Recently, the usage of carbonates for postcombustion CO_2 capture gained renewed desire because of the potentially low energy of the process. Since 1930, the development of primary alkanolamine solvents such as monoethanolamine (MEA) and diglycolamine, which provide high chemical reaction, proper kinetics, moderate absorption capacity, and acceptable sustainability, was carried out in order to remove CO_2 [13–15]. The use of resistant amines, blends, and inhibitors has been suggested to reduce solvent loss and operating costs. Afterward, secondary alkanolamine solvents such as diethanolamine (DEA) and diisopropanolamine (DIPA) were produced and used as alternatives to MEA [16]. The difference between molecular structure of primary and secondary is that in primary alkanolamines, hydrogen atoms are directly bonded to nitrogen. Frazier and Kohl suggested methyldiethanolamine (MDEA) as a tertiary alkanolamine solvent, which is characterized by having a high equivalent weight to increase selectivity of H_2S [16]. Each of these alkanolamines has at least one hydroxyl group and one amino group. In 1983, ExxonMobil Research & Engineering Company developed 2-amino-2-methyl-1-propanol (AMP), which hindered amines [17]. Since 1995, sterically hindered amines widely have been investigated as potential alkanolamine absorbents. Many studies have shown that the promoters, such as piperazine, potassium carbonate, MEA, and DEA, increase the rate of absorption process by combination with amine solvent, such as MDEA [16]. From 2005, the desire to use of amino acid for acid gas capture has started to investigate using sodium or potassium salt glycine, the simplest primary amino acid for CO_2 removal [18]. In 2003, ionic liquids (ILs) were noticed as "solvents of the future" due to their potential to remove acid gases as an alternative solvent [19,20]. Generally, the solvents used in absorption process are divided in two categories, physical and chemical, as follows:

A. Chemical absorption solvents
  i. Conventional amine-based solvents
  ii. Sterically hindered amine solvents

iii. Non–amine-based solvents
iv. Solvent blends
v. ILs
vi. New generation solvents for CO_2 capture
B. Physical absorption solvents
   i. Selexol
   ii. Rectisol
   iii. Ifpexol
   iv. Fluor
   v. Purisol
   vi. Sulfinol
   vii. Morphysorb

2.2 Adsorption process for NG sweetening

Adsorption is a physical or chemical phenomenon or process in which atoms, molecules, or ions of a gas or a liquid adsorb on a solid surface, creating a film of absorbent material on the solid surface. The phenomenon of adsorption process, like surface tension, is based on surface energy, in which the atoms on the adsorbent surface form a bond with the adsorbent atoms; nature of this bond depends on the adsorbent and adsorbate composition. As a result, adsorption process is classified into two general categories: physical adsorption with weak van der Waals forces and chemical adsorption having covalent bonds [21,22].

Considering that the majority of natural resources consist of sour gases, they must be removed and sweetened. In 1773, the first systematic study on adsorption usage for adsorption of air using charcoal by Schelee began [23]. For the first time, adsorption process commercially was used for purification of white sugar. In 1783, charcoal was used as adsorbent to remove impurities from sugar [24]. In the 19th century, at the same time as the industrial revolution, efforts have been made to improve the adsorbent properties focusing on the surface area and porosity of the adsorbents [25,26]. After Chappuis measured the isotherm of the adsorbent layers, the improvement of the adsorbent was investigated based on the porosity and surface area [27]. During 30 years, because of its unique characteristics, adsorption has been widely used in various fields of industry, including purification of gas mixtures, mainly in the petrochemical, environmental, electronic, and medical industries [28]. Various types of adsorbents with different pore sizes and different selectivities that created flexible designs for the separation and purification of gas mixtures for the intended purposes were available. In 1932, pressure swing adsorption technology was first registered by Charles Skarkstrom for oxygen enrichment, in which the cyclic adsorption included four stages of feed, blowdown, purge, and pressurization [29]. In this process, the flow of acid gas with a high amount of CO_2 comes into contact with packed spherical adsorbers, which are generally placed in parallel for higher energy efficiencies [30]. After entering the column, the feed gas goes under higher atmospheric pressure and causes the adsorption of CO_2 molecules on the adsorbent surface. CO_2 was selectively adsorbed by adsorbent surface at low temperature and high pressure until equilibrium is reached. When the surface is saturated with adsorbed CO_2, regeneration of the adsorbent is done by reducing the column pressure and restricting the gas flow to separate CO_2 from adsorbent surface [31]. Various range of adsorbents are available, which are used to remove specific pollutants based on their selective adsorption level compared to impurities. In order to improve the adsorbent, synthesis and impregnation of the surface with other compounds can be carried out. It should be noted that the adsorption rate can be increased by functionalizing the adsorbent surface with various compounds. In many cases, impregnated silica xerogels with amine molecules are used industrially for simultaneous CO_2 and H_2S adsorption from NG. The main differences between the two methods of adsorption and absorption are summarized in Table 7.3. Besides, Fig. 7.3 schematically shows the difference between these two methods.

Table 7.3 Differences between adsorption and absorption processes.

Adsorption	Absorption
• Substances such as gas, liquids, or dissolved solids adhere to the surface of another material that can be a solid or liquid.	• It is a situation in which any substance (atoms, ions, or molecules) is taken or absorbed by another substance, especially a solid or liquid substance.
• It is a surface phenomenon.	• It is a bulk phenomenon.
• It is an exothermic reaction.	• It is an endothermic reaction.
• It is a temperature-dependent phenomenon.	• It is not affected by temperature.
• It steadily increases and reaches equilibrium.	• It occurs at uniform rate.
• Concentration on the surface of adsorbent is different from that in bulk.	• Concentration of absorption is the same throughout the material.

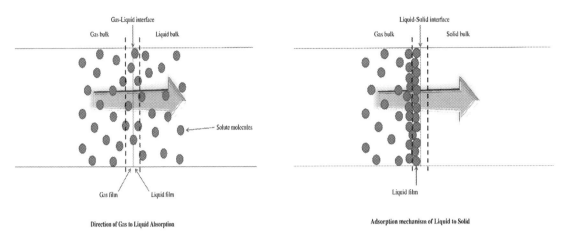

Figure 7.3 Difference between adsorption and absorption.

2.3 Absorbents modified with nanoparticles for NG sweetening

As previously discussed, removal of impurities is of particular importance in NG processing, biological methane processing, crude oil refining, and synthesis gas purification. Since 1950s, the amine-based chemical absorption process has been used to remove and absorption of acid gases such as CO_2 and H_2S from gas mixture in gas treatment plants, and has been greatly developed by attracting the attention of researchers [32]. As in the interface of a gas phase and a liquid phase, mass transfer resistance typically is higher, and reducing this resistance using practical solutions including mechanical, chemical, acoustic, and electromagnetic treatments in addition to the use of nanotechnology is practical. In these methods, more surface disturbance and thus more mass transfer are induced in both phases [33].

One of the most attractive topics in the NG sweetening is the use of nanofluids due to the improvement in heat transfer and mass transfer coefficients [34]. Nanofluids made from metals, oxides, carbides, or carbon nanotubes (CNTs) are uniform colloidal suspensions of nanometer particles in a base fluid like water, ethylene glycol, and oil. For the first time, Choi introduced nanofluids in 1995 [35]. After that, with the advancement of nanotechnology and the benefit of surface engineering, nanofluid production technology becomes able to spread nanoparticles stably in a base fluid. Nanoparticles with this feature can reduce the amount of used amines in the gas sweetening process and can be consumed subsequently. As a result, many researchers have focused on the synthesis and introduction of nanoabsorbents to effectively remove CO_2 and H_2S pollutants from NG.

The implementation of nanoabsorbents not only ends the increment in the process efficiency but also reduces environmental risks and damages caused by these polluting and harmful gases, all of which decrease the removal process costs. Although the main mechanism of increasing mass transfer by nanofluid is not known, the main proposed mechanisms include Brownian motion, hydrodynamic effects, grazing effect, shuttle effect, bubble-breaking effect, reduction of surface tension, and reduction of film thickness [36]. The effect of nanoparticles on the absorption of ammonia, carbon dioxide, and oxygen in nanofluids has also been studied, results of which showed a decrease in the liquid mass transfer coefficient and absorption speed with increase in the nanoparticle's concentration. Besides, this increase caused the reduction in the absorption speed. In nanoabsorbents, both nanofluid and nanoemulsion, methanol is widely used as the base fluid because it is the most efficient physical absorber compared to other physical absorbers in terms of energy penalty. A brief review on the performance of different nanoparticles for CO_2 and H_2S absorption in various absorbents are listed in Table 7.4.

The two main groups of solvents for CO_2 absorption that have been modified using nanoparticles are solvent-based nanoparticle suspensions and organic nanoparticle hybrid materials. Kim et al. [41] first investigated the absorption rate of NH_3 in nanofluids with a size of 50 nm of CuO, Cu, and Al_2O_3 in water and observed an increase in the absorption rate with an increase in the loading of nanoparticles. Krishnamurthy et al. [42] reported that a dye in a nanofluid diffuses faster than in water, because Brownian motion induced convection and faster diffusion. CO_2 removal was evaluated in a suspension of SiO_2 nanoparticles in water with a dosage of 0.01wt% to 0.04wt% SiO_2 by Kim et al. [43]. It was concluded from the obtained results that the solution containing 0.021wt% SiO_2 caused 24% improvement in CO_2 capture. Komati and Suresh [44] studied the effect of nano-ferrofluid for absorption process in CO_2/MDEA solution in a wetted wall column, results of which demonstrated that the increase in mass transfer coefficient for magnetite dosage of about 0.39% vol.% was 92.8%. In other researches by Park et al. in 2006 [45], 2007 [46], and 2008 [47], it was observed that the increment in the SiO_2 nanoparticle concentration in aqueous solutions of DEA, MEA, and DIPA causes decrement in the absorption rate due to the solution elasticity [44]. Lee et al. [48] investigated different loadings of methanol-based fluids

Table 7.4 Summary of the application of various nanoabsorbents for acid gas removal.

Absorbent	Nanoparticles	Size and concentration	Maximum enhancement ratio	Notes	References
MeOH (22°C)	SiO_2	10–20 nm[a] (340 nm[b]), 0.005–0.5 vol%	1.097	Tray column	[37]
	Al_2O_3	40–50 nm[a] (200 nm[b]), 0.005–0.5 vol%	1.094	Tray column	
MeOH (17°C)	Al_2O_3	40–50 nm[a], 0.05 vol%	1.012 / 1.10 (with trays)	Annular contactor	[38]
	SiO_2	10–20 nm[a], 0.05 vol%	1.011 / 1.09 (with trays)	Annular contactor	
	TiO_2	<25 nm[a], 0.05 vol%	1.046 / 1.05 (with trays)	Annular contactor	
H_2O and diethanolamine (10 wt%, 25°C)	Al_2O_3	10–20 nm[a], 0.005–0.5wt%	1.33	Wetted wall column	[39]
	SiO_2	10–15 nm[a], 0.005–0.5wt%	1.40	Wetted wall column	
MeOH (21°C)	Al_2O_3	45 nm[a], 0.005–0.1 vol%	1.27	Taylor–Couette absorber	[40]

[a]Size of nanoparticles in powder state before they were combined with base fluid to form nanoabsorbents.
[b]Size of nanoparticles in suspension.

containing Al_2O_3 and SiO_2 nanoparticles to capture CO_2 in a bubble-type absorber, results of which demonstrated the maximum increase in CO_2 absorption in a solution containing 4.5% Al_2O_3 at 20°C and 5.6% SiO_2 at −20°C. Maa et al. [49] investigated the increase in heat and mass transfer for the NH_3/H_2O bubble absorption process using CNT–ammonia binary nanofluid. According to the obtained results, absorption increased with increasing initial concentration of ammonia. Kim et al. [50] studied the dispersion of nanoparticles in binary nanofluids and concluded that the concentration of SiO_2 nanoparticles regardless of the distribution stabilizer should be less than 0.01. In addition, when the concentration of SiO_2 nanoparticles was 0.005 vol.%, the maximum improvement in heat transfer rate and mass transfer rate was 46.8% and 18%, respectively. By adding SiO_2 nanoparticles to the solution, they concluded that the total absorption increased compared to the pure liquid without solid material. In the set of experiments conducted by B. Rahmatmand et al. [51], the process of carbon dioxide absorption in a high-pressure media has been evaluated. These experiments involved in a closed container in a static state where the gas and nanofluid were directly contacted. The initial test conditions had 20, 30 and 40 bar pressure and 308 K temperature. The nanofluids contained SiO_2, Al_2O_3, and Fe_3O_4 and CNTs in pure water with concentrations of 0.02%, 0.05%, and 0.1% wt.%. CNTs also dispersed in MDEA and DEA with a 0.02% wt.% concentration. They found that SiO_2 and Al_2O_3 had a greater effect at a higher concentration of nanoparticles (0.1 wt.%) and enhanced the absorption capacity up to 21% and 18%, respectively. Fe_3O_4 and CNT were more effective at lower concentrations (0.02 wt%) and increased the amount of absorption up to 24% and 34%, respectively. CNT nanoparticles in MDEA solution were more effective than in DEA solution by increasing absorption capacity up to 23%. In a study by Park et al. [45], it was observed that with increasing silica nanoparticles concentration, the volumetric mass transfer coefficient in the liquid side and chemical absorption rate had been decreased because of elasticity in the CO_2/AMP aqueous solution.

V. Irani et al. [52] investigated CO_2 and H_2S absorption capacity of MDEA modified via reduced graphene oxide solutions, results of which depicted 16.2% and 17.7% removal efficiency for CO_2 and H_2S, respectively. Al_2O_3 and SiO_2 nanoparticles with the dosages of 0.005 and 0.1 vol.% were also used in an experimental study by Pineda et al. [37], which depicted that the maximum amount of absorption compared to pure methanol has increased by 9.4% and 9.7% using Al_2O_3 and SiO_2 nanoparticles, respectively. Jung et al. [53] tested 0.005 to 0.1 vol.% Al_2O_3/methanol nanofluids for CO_2 absorption and found that the maximum increase in CO_2 absorption compared to pure methanol was for 8.3% alumina nanoparticles at 0.01 vol.% solution. In another experiment by Jong Sung Lee et al. [54], CO_2 absorption and regeneration by using SiO_2/H_2O and Al_2O_3/H_2O nanofluids were investigated, which showed an increase in absorption performance by 23.5% and 11.8%, respectively. In a study conducted by Lee and Kang [55], aqueous NaCl solution was used to absorb CO_2 and the solubility of CO_2 was measured. Their aim was to investigate the dispersion stability of nanofluids in Al_2O_3/NaCl aqueous solution and obtain the optimal concentration to increase CO_2 absorption. In these experiments, the concentration of particles has changed from 0.005% to 0.1%, and the best performance was achieved for 0.01% by volume alumina nanofluid.

Esmaili Faraj et al. [56] investigated the effect of silica nanoparticles and exfoliated graphene oxide (EGO) in water to remove H_2S in a bubble column. They concluded that silica water nanofluid has a positive effect on H_2S absorption. According to their experiences, when only 0.02 wt% of nanoparticles were used, the amount of mass transfer by EGO water nanofluid had increased up to 40% compared to the base state. Ma et al. [57] studied the effect of Cu and CuO nanoparticles on breakthrough time and mass transfer coefficient of H_2S removal from biogas using MDEA. They observed that while the viscosity of both solutions has been increased with increasing concentration of both nanoparticles, it was decreased with increasing temperature. They reported that the absorption performance of MDEA-based CuO nanofluid was better than MDEA-based Cu nanofluid. CuO nanofluids had the maximum delay time of 380 min. The breakthrough time of two nanofluids was slightly longer with increasing temperature, and after 40°C, a sharp reduction in process efficiency was observed.

3. Conclusion and future outlooks

The importance and necessity of removing and separating impurities from the NG mixture in order to use it in the industrial processes strengthened the idea of upgrading and improving the sweetening process. The unique features and great capabilities of nanoabsorbents in absorbing and separating impurities from gas mixtures have increased the tendency to use them in the NG sweetening process. The use of modern nanotechnology and the addition of nanoparticles to the base fluids, which were introduced as nanofluids, were used for this purpose. Nanofluids generally improve the adsorption of sour gases such as CO_2 and H_2S. Although previous studies showed better performance of nanoabsorbents in CO_2 and H_2S absorption and regeneration, more studies are needed to apply the industrial gas removal system via nanofluids. To clarify the mechanisms of increasing CO_2 physical absorption performance, surface mass transfer experiments between CO_2 or H_2S and nanoabsorbers should be conducted.

Abbreviations and symbols

AMP	2-Amino-2-methyl-1-propanol
C$_2$H$_5$	Ethane
C$_3$H$_8$	Propane
C$_4$H$_{10}$	Butane
C$_5$H$_{12}$	Pentane
CH$_4$	Methane
ChCl	Choline chloride
CNT	Carbon nanotube
CO$_2$	Carbon dioxide
COS	Carbonyl sulfide
DEA	Diethanolamine
DGA	Diglycolamine
DIPA	Diisopropanolamine
EGO	Exfoliated graphene oxide
H$_2$O	Water vapor
H$_2$S	Hydrogen sulfide
LNG	Liquefied natural gas
LTTM	Low transition temperature mixture
MDEA	Methyldiethanolamine
MEA	Monoethanolamine
NG	Natural gas
rGO	Reduced graphene oxide
RSH	Mercaptan

References

[1] Rahimpour MR, Makarem MA, Meshksar M. In: Rahimpour MR, Makarem MA, Meshksar M, editors. Advances in Synthesis Gas : Methods, Technologies and Applications - Volume 1: Syngas Production and Preparation. 1. Elsevier; 2023. https://doi.org/10.1016/C2021-0-00292-3.

[2] Rahimpour MR, Makarem MA, Meshksar M. In: Rahimpour MR, Makarem MA, Meshksar M, editors. Advances in Synthesis Gas : Methods, Technologies and Applications - Volume 2: Syngas Purification and Separation. 2. Elsevier; 2023. https://doi.org/10.1016/C2021-0-00379-5.

[3] Khosravani H, Meshksar M, Rahimpour HR, Rahimpour MR. Chapter 1 - Introduction to syngas products and applications. Advances in Synthesis Gas : Methods, Technologies and Applications Volume 3: Syngas Products and Usages. Elsevier; 2023. p. 3–25.

[4] Economides MJ, Wood DA. The state of natural gas. Journal of Natural Gas Science and Engineering 2009;1:1–13.

[5] Kiani MR, Meshksar M, Makarem MA, Rahimpour E. Catalytic membrane micro-reactors for fuel and biofuel processing: a mini review. Topics in Catalysis 2021:1–20.

[6] Gutierrez JP, Benitez LA, Ruiz ELA, Erdmann E. A sensitivity analysis and a comparison of two simulators performance for the process of natural gas sweetening. Journal of Natural Gas Science and Engineering 2016;31:800–7.

[7] Korpyś M, Wójcik J, Synowiec P. Methods for sweetening natural and shale gas. Chemical Science 2014;68:213–5.

[8] Mezni M. LPG recovery unit optimization. New Mexico: Institute of Mining and Technology; 2015.

[9] Alcheikhhamdon Y, Hoorfar M. Natural gas quality enhancement: a review of the conventional treatment processes, and the industrial challenges facing emerging technologies. Journal of Natural Gas Science and Engineering 2016;34:689–701.
[10] Cho JD. Global trends of sciences information on the sour gas. Economic and Environmental Geology 2015;48:89–101.
[11] Kidnay AJ, Parrish W. Fundamentals of natural gas processing. New York: CRC; 2006.
[12] Meshksar M, Sedghamiz MA, Zafarnak S, Rahimpour MR. CO2 separation with ionic liquid membranes. In: Advances in carbon capture. Elsevier; 2020. p. 291–309.
[13] Knuutila H, Svendsen HF, Juliussen O. Kinetics of carbonate based CO2 capture systems. Energy Procedia 2009;1:1011–8.
[14] Bottoms R. Process for separating acidic gases. US Patent; 1930. p. 1783901.
[15] Vega F, Sanna A, Navarrete B, Maroto-Valer MM, Cortés VJ. Degradation of amine-based solvents in CO2 capture process by chemical absorption. Greenhouse Gases: Science and Technology 2014;4:707–33.
[16] Karamé I, Shaya J, Srour H. Carbon dioxide chemistry, capture and oil recovery. BoD–Books on Demand; 2018.
[17] Shah MS, Tsapatsis M, Siepmann JI. Hydrogen sulfide capture: from absorption in polar liquids to oxide, zeolite, and metal–organic framework adsorbents and membranes. Chemical Reviews 2017;117:9755–803.
[18] Weiland RH, Hatcher NA, Nava JL. Post-combustion CO2 capture with amino-acid salts. GPA Eur 2010;22:24.
[19] Greer AJ, Jacquemin J, Hardacre C. Industrial applications of ionic liquids. Molecules 2020;25:5207.
[20] Rogers RD, Seddon KR. Ionic liquids–solvents of the future? Science 2003; 302:792–3.
[21] Ferrari L, Kaufmann J, Winnefeld F, Plank J. Interaction of cement model systems with superplasticizers investigated by atomic force microscopy, zeta potential, and adsorption measurements. Journal of Colloid and Interface Science 2010;347:15–24.
[22] Khosrowshahi MS, Abdol MA, Mashhadimoslem H, Khakpour E, Emrooz HBM, Sadeghzadeh S, et al. The role of surface chemistry on CO2 adsorption in biomass-derived porous carbons by experimental results and molecular dynamics simulations. Scientific Reports 2022;12:8917.
[23] Rouquerol J, Avnir D, Everett D, Fairbridge C, Haynes M, Pernicone N, et al. Guidelines for the characterization of porous solids. In: Studies in surface science and catalysis. Elsevier; 1994. p. 1–9.
[24] Kyzas GZ, Fu J, Matis KA. The change from past to future for adsorbent materials in treatment of dyeing wastewaters. Materials 2013;6:5131–58.
[25] Adam N. The adsorption of gases and vapours. Nature Publishing Group UK London; 1945.
[26] Gregg SJ, Sing KSW, Salzberg H. Adsorption surface area and porosity. Journal of the Electrochemical Society 1967;114:279Ca.
[27] Chappuis P. Ueber die Wärmeerzeugung bei der Absorption der Gase durch feste Körper und Flüssigkeiten. Annalen der Physik 1883;255:21–38.
[28] Sircar S. Publications on adsorption science and technology. Adsorption 2000;6:359–65.
[29] Grande CA. Advances in pressure swing adsorption for gas separation. International Scholarly Research Notices 2012;2012.

[30] Ho MT, Allinson GW, Wiley DE. Reducing the cost of CO2 capture from flue gases using pressure swing adsorption. Industrial and Engineering Chemistry Research 2008;47:4883—90.
[31] Fouladi N, Makarem MA, Sedghamiz MA, Rahimpour HR. CO2 adsorption by swing technologies and challenges on industrialization. In: Advances in carbon capture. Elsevier; 2020. p. 241—67.
[32] Kohl AL. Alkanolamines for hydrogen sulfide and carbon dioxide removal. In: Gas purification; 1985. p. 29—35.
[33] Mandal B, Bandyopadhyay S. Simultaneous absorption of carbon dioxide and hydrogen sulfide into aqueous blends of 2-amino-2-methyl-1-propanol and diethanolamine. Chemical Engineering Science 2005;60:6438—51.
[34] Meshksar M, Makarem MA, Hosseini Z-S, Rahimpour MR. Chapter 5—effect of nanofluids in solubility enhancement. In: Rahimpour MR, Makarem MA, Kiani MR, Sedghamiz MA, editors. Nanofluids and mass transfer. Elsevier; 2022. p. 115—32.
[35] Choi SU, Eastman JA. Enhancing thermal conductivity of fluids with nanoparticles. Argonne, IL (United States): Argonne National Lab.(ANL); 1995.
[36] Kiani MR, Meshksar M, Makarem MA, Rahimpour MR. Chapter 2—preparation, stability, and characterization of nanofluids. In: Rahimpour MR, Makarem MA, Kiani MR, Sedghamiz MA, editors. Nanofluids and mass transfer. Elsevier; 2022. p. 21—38.
[37] Pineda IT, Lee JW, Jung I, Kang YT. CO2 absorption enhancement by methanol-based Al2O3 and SiO2 nanofluids in a tray column absorber. International Journal of Refrigeration 2012;35:1402—9.
[38] Pineda IT, Choi CK, Kang YT. CO2 gas absorption by CH3OH based nanofluids in an annular contactor at low rotational speeds. International Journal of Greenhouse Gas Control 2014;23:105—12.
[39] Taheri M, Mohebbi A, Hashemipour H, Rashidi AM. Simultaneous absorption of carbon dioxide (CO2) and hydrogen sulfide (H2S) from CO2—H2S—CH4 gas mixture using amine-based nanofluids in a wetted wall column. Journal of Natural Gas Science and Engineering 2016;28:410—7.
[40] Pineda IT, Kang YT. CO2 absorption enhancement by nanoabsorbents in Taylor—Couette absorber. International Journal of Heat and Mass Transfer 2016;100:39—47.
[41] Kim J-K, Jung JY, Kang YT. The effect of nano-particles on the bubble absorption performance in a binary nanofluid. International Journal of Refrigeration 2006;29:22—9.
[42] Krishnamurthy S, Bhattacharya P, Phelan P, Prasher R. Enhanced mass transport in nanofluids. Nano Letters 2006;6:419—23.
[43] Kim W-g, Kang HU, Jung K-m, Kim SH. Synthesis of silica nanofluid and application to CO2 absorption. Separation Science and Technology 2008;43:3036—55.
[44] Komati S, Suresh AK. CO2 absorption into amine solutions: a novel strategy for intensification based on the addition of ferrofluids. Journal of Chemical Technology and Biotechnology: International Research in Process, Environmental & Clean Technology 2008;83:1094—100.
[45] Park S-W, Choi B-S, Lee J-W. Effect of elasticity of aqueous colloidal silica solution on chemical absorption of carbon dioxide with 2-amino-2-methyl-1-propanol. Korea-Australia Rheology Journal 2006;18:133—41.
[46] Park S-W, Choi B-S, Kim S-S, Lee J-W. Chemical absorption of carbon dioxide into aqueous colloidal silica solution containing

monoethanolamine. Journal of Industrial and Engineering Chemistry 2007; 13:133−42.
[47] Park S-W, Choi B-S, Kim S-S, Lee B-D, Lee J-W. Absorption of carbon dioxide into aqueous colloidal silica solution with diisopropanolamine. Journal of Industrial and Engineering Chemistry 2008;14:166−74.
[48] Lee JW, Jung J-Y, Lee S-G, Kang YT. CO_2 bubble absorption enhancement in methanol-based nanofluids. International Journal of Refrigeration 2011; 34:1727−33.
[49] Ma X, Su F, Chen J, Bai T, Han Z. Enhancement of bubble absorption process using a CNTs-ammonia binary nanofluid. International Communications in Heat and Mass Transfer 2009;36:657−60.
[50] Kim H, Jeong J, Kang YT. Heat and mass transfer enhancement for falling film absorption process by SiO_2 binary nanofluids. International Journal of Refrigeration 2012;35:645−51.
[51] Rahmatmand B, Keshavarz P, Ayatollahi S. Study of absorption enhancement of CO_2 by SiO_2, Al_2O_3, CNT, and Fe_3O_4 nanoparticles in water and amine solutions. Journal of Chemical and Engineering Data 2016; 61:1378−87.
[52] Irani V, Tavasoli A, Vahidi M. Preparation of amine functionalized reduced graphene oxide/methyl diethanolamine nanofluid and its application for improving the CO_2 and H_2S absorption. Journal of Colloid and Interface Science 2018;527:57−67.
[53] Jung J-Y, Lee JW, Kang YT. CO_2 absorption characteristics of nanoparticle suspensions in methanol. Journal of Mechanical Science and Technology 2012;26:2285−90.
[54] Lee JS, Lee JW, Kang YT. CO_2 absorption/regeneration enhancement in DI water with suspended nanoparticles for energy conversion application. Applied Energy 2015;143:119−29.
[55] Lee JW, Kang YT. CO_2 absorption enhancement by Al_2O_3 nanoparticles in NaCl aqueous solution. Energy 2013;53:206−11.
[56] Esmaeili Faraj SH, Nasr Esfahany M, Jafari-Asl M, Etesami N. Hydrogen sulfide bubble absorption enhancement in water-based nanofluids. Industrial and Engineering Chemistry Research 2014;53:16851−8.
[57] Ma M, Zou C. Effect of nanoparticles on the mass transfer process of removal of hydrogen sulfide in biogas by MDEA. International Journal of Heat and Mass Transfer 2018;127:385−92.

Encapsulated liquid sorbents for sweetening of natural gas

Babak Emdadi and Rasoul Moradi
Nanotechnology Laboratory, School of Science and Engineering, Khazar University, Baku, Azerbaijan

1. Introduction

The primary energy source that is widely used to meet industrial and domestic demands is natural gas (NG). Carbon dioxide (CO_2) and hydrogen sulfide (H_2S) are pollutants found in NG produced from a variety of sources. More than 50% of current acid gas elimination techniques use aqueous dilutions of alkanolamines, making the amine gas-producing fresh technique the most widely used method for the removal of sour gases. However, this process of producing fresh gas, specifically for amine recurrence, is energy centralized [1]. Therefore, enhancing amine gas production's sweet performance can lead to exceptional energy savings and unrecognized financial rewards for the present generation of gas plants. For chemical industries, it is crucial to properly dispose of acid gas pollutants like CO_2 and H_2S from gas mixtures. H_2S and CO_2 are typically the main gas contaminants in NG implementation. This crucial piece of advice must be taken into account while using NG to produce energy; H_2S and CO_2 must undoubtedly be removed from NG. The unfavorable ingredient CO_2 might reduce the energy efficiency of NG. H_2S is a toxic gas that is both deadly and caustic [2]. Therefore, it is essential to separate CO_2 and H_2S from NG because the natural world and the effects of human activity on it are important.

Numerous issues with traditional methods include storming, bottoming, bubbling, piping, increasing expenditures, and managing expenses. As a result, many researchers have looked into the possibility of increasing the effectiveness of these methods [3].

As a pure energy source that is superior to petroleum and a combustible black or dark brown rock made primarily of carbonized plant matter, found primarily in underground deposits and widely used as fuel, NG is anticipated to play a significant role as a global energy reserve in the ensuing periodic years [4–6]. However, sulfur and other acidic compounds co-exist beside CH_4 in NG are primary elements and chemicals that are able to dissolve some metals, turn litmus red, and neutralize alkalis; typically, this type of corrosive or sour-tasting substance is called acid. These dangerous vapors, such as CO_2 and H_2S, are typically also added as pollutants in weak-state biological gas, which includes biogas and trash yard gas. These undesired contaminants will cause channel erosion and a decrease in energy range. H_2S is also a positively polluting gas that, when it bursts, releases SO_2 into the environment [7–9]. Increased-effective desulfurization has therefore been a source of growing worry in relation to or defined by industry-promoting realistic gas in light of the requirement for practical energy usage and global conservation [10–13]. Nowadays, fluid-phase chemical contact by amines is one method frequently used for the separation of H_2S from NG [14,15]. However, this method suffers from a number of flaws, including amine loss, severe degradation, and an inherently high renewal price [16–18]. There have been significant efforts made to provide more sorbents for H_2S absorption and split. Ionic liquids (ILs), which are made up of a variety of positively charged ions that are related to or derived from bio-sources and/or chemicals and minerals [19,20], have received widespread attention in many fields in previous decades [21–24]. The process flow diagram for sweetening NG using amine dilution is shown in Fig. 8.1 [25].

In order to address global climate issues and achieve targets, carbon capture and storage (CCS) is fundamentally required [26,27]. For considerable financial and related to the government or public affairs of a country acceptance of CCS, however, price drops for capturing are necessary [28]. In comparison to the most cutting-edge aqueous amine-based capturing techniques, the U.S. Department of Energy wishes to see the development of carbon capture technologies that lower the cost of lightning by 30% [28].

Numerous goal-planned, nonwater sorbents have been developed to boost the energy output of carbon capture. These sorbents, which comprise ILs, phase-changing ILs (PCILs), and CO_2-binding organic liquids, have shown promise for much reduced energy consumption in the after-eruption capture as compared to a kind of liquid known as aqueous amines. Significant process innovations are needed to combine superior energy-impressive sorbents and compounds that have the ability

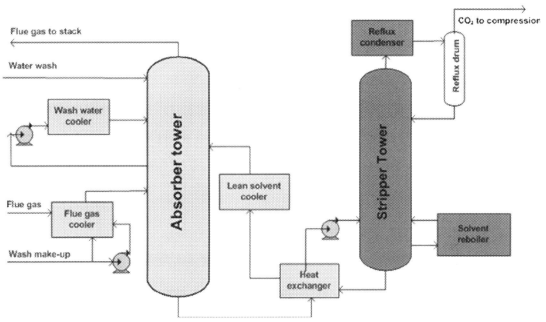

Figure 8.1 Diagram for natural gas (NG) conventional sweetening by amine dilution process [25]. Copyright 2014. Reprinted with permission from Springer.

of collecting molecules of another substance by sorption while increasing fund consumption in order to achieve transformative carbon capture (subtractive fund expense). One approach to doing this is by procedure exacerbation, in which established measurement procedures are redesigned to provide better outcomes or combine several tasks into a single machine.

CO_2 is deemed a considerable and substantial nonuseful gas for the world over the centuries, with its universal atmospheric engagement boosting specifically because of fossil fuel eruption, especially fuel plant publications [29]. One of the favorable short-term methods for decreasing the worldwide heating impacts generated by greenhouse gases includes straightforward CO_2 capturing at energy factories [30]. The foundation of conventional post-ignition CO_2 absorption technologies is amine absorption operations, which have a number of drawbacks, including corrosiveness and fast, unpredictable change, which raises operating costs and has an adverse effect on the environment [31–33]. Thus, it is vital to expand creative and expense-useful techs able to in a well-organized and competent way capturing CO_2, overpowering the difficulties connected to in a way that is concerned with buying, selling, and making a profit existing technique. ILs are being investigated for CO_2 capturing because of their single attributes, for example, an increased solvating size for various combinations, adjustability, weak fog compression, broad fluid temperature span, increased thermic durability, noninflammability [34–36].

Recently, several studies have been devoted to the topic of obtaining CO_2 using coercion [37]. For the disposal of acid gases, such as H_2S or CO_2, from flue gas in boats, chemical sorption is frequently used in conjunction with amines, such as monoethanolamine (MEA) [38,39]. This amine scrubbing method has certain drawbacks, including greater amine erosion, sorbent failure while waiting for renewal, increased energy requirements for revival, and higher operational costs [40,41]. Particularly for extensive-shape usages for example energy works, counting on the character of explosion (pre-explosion or postexplosion tech) and the engagement of CO_2 in the chimney gas creek, it is vital to progress novel and useful sorbents for CO_2 capturing [42–44]. However, we must take into account this significant issue if we hope to reduce CO_2 emissions from combustible black or dark brown rock, which is mostly composed of carbonized plant matter and is primarily found in subterranean deposits and is commonly utilized as fuel for energy plants. However, current technologies for extracting carbon from chimney gas are burdened by rising costs for resources including money, energy, and chemicals [45–48], restricting their extension to history. The customary specified strategy for capturing CO_2 obtains chimney gas in connection with an aqueous amine solution, commonly monoethanolamine (MEA), which responds with CO_2 to create carbamates [49]. Whenever MEA has a quick sorption measure and increased CO_2 transporting size, it also contains diverse deficiencies that refuse extensive usage [50].

MEA is negatively degrading, produces polluted decadence yields, and requires significant power to eliminate CO_2, a chemical that has the ability to resorb molecules of other substances [51–55]. Novel compounds with the ability to absorb molecules of other chemicals are needed to overcome these restrictions. Newly, stable sorbents with strong surface measurements and weak CO_2 connections powers, for example, zeolites [56], metal-organic frames [57–59], frustrated Lewis pairs [60], and nanoporous polymers [61], have been presented; however, they sorrow from weak CO_2 size in the existence of water (which is plentiful in the chimney gas), weak durability, on numerous sorption–desorption processes and/or increased expense of yields [62,63].

2. Encapsulated liquid sorbents

Physical sorbents are negligibly eroding in comparison to chemical sorbents permitting the utilization of carbon alloy for factory and tool building boosting expense decline in capture procedures [32]. The renewal of these mixtures is performed by compression decline in insomuch chemical sorbents that need an elevated temperature [1]. Various artful procedures, for example, CO_2 elimination from NG, hydrogen, and combination gas generation including great CO_2 contents use physical sorbents [37].

The removal of H_2S and CO_2 (sour gases) from NG before it is transitioned via channels is sensible since the presence of these components in the channel may result in technological issues (for example, the corroding of pipes) [64–66]. Rules limit the greatest levels of H_2S and CO_2 condensations in NG to 4 ppm and 2 mol%, respectively [67]. For the disposal of these sour gases from natural pipeline gas in various working environments and to separate CO_2 and H_2S condensations in the tart gas, significant processes have developed [66,68,69]. Abolgasem et al. [70] studied inherent gas constructed from one of the gas areas, with a somewhat more than half H_2S (5 mol %) deficit and fairly less than half CO_2 (1.33 mol %) thought and four of the good-progressed methods for the natural and also inherent gas making sweet are manufactured in imitation of some other material for meeting with the pipe laying characteristics.

These include the shell procedure, the flexible hydrogen sulfide removal method known as the LO CAT strategy, the blended alkanolamine method (blended methyldiethanolamine, also known as N-methyldiethanolamine and more commonly as MDEA and diethanolamine, also known as DEA), and the sulfinol-M operation. Alkanolamine resolutions are the considerable prevalent resolutions for acid gas disposal from NG and are in a commercial way utilized for NG-making sweets in so much more than the previous century [71]. Because MDEA resolution, besides needing somewhat weak power for recovering, is famous for its strong opting into H_2S, and the surplus of much fewer numbers of preliminary or subsidiary amines to MDEA, eventuates in CO_2 sorption measure of MDEA resolution to boost [72–74].

As a suitable alkanolamine dilution for those gases with elevated H_2S and low CO_2 concentration, the MDEA, DEA combination of amines has been chosen. The fact that the NG produced from this domain has an H_2S quantity of 5% and a CO_2 value of 1.33% is a significant aspect that needs our attention. In addition, this gas contains light components with weak molecular weights, such as CO, CO_2, and N_2. However, the primary components of this gas are ethane with a mole atom of 0.1021 and methane with a mole atom of 0.7203. The utilization gas's mean MW is 23.4, and its temperature and compression are comparable to 54°C and 24.5 bar [72–74].

The sorbent encapsulation is also an interesting approach for CO_2 capturing. In micro and nanoencapsulation procedures, the sorbent is enclosed inside a shell (often polymer) (Fig. 8.2) [76,77].

In nano-encapsulation processes, capsules with diameters ranging from 1 to 1000 nm are formed, whereas in microencapsulation, capsule figures with typical measurements ranging from 1 to 1000 m are used [76,78]. The operational combination of micro- and nanocapsules, which also includes the capsule cover, is covered with a different material. One or more independent components of the functional composite can be found in the core of either mononuclear or polynuclear micro- and nanocapsules (Fig. 8.3A and B).

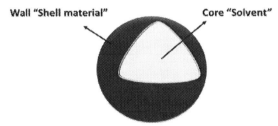

Figure 8.2 Encapsulated sorbent microcapsule system in simple illustration [75]. Copyright 2020. Reprinted with permission from Elsevier.

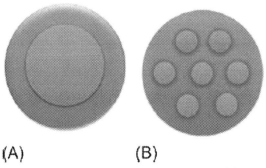

Figure 8.3 Several sorts of capsules: (A) mononuclear and (B) polynuclear [75]. Copyright 2020. Reprinted with permission from Elsevier.

Size, type, effectiveness of absorption, and micro/nanocapsule shape are all strongly influenced by starting materials and procurement techniques [79]. For sorbent encapsulation, many approaches are available. Spray drying, emulsification, in situ polymerization, and interfacial polycondensation are the main processes for the presentation of micro and nanocapsules.

2.1 Spray-drying strategy

The spray-drying process is the most widely used method for making micro and nanocapsules since it has lower manufacturing costs than other methods. The creation of an emulsion carrying the functional mixture and cachet fence material shown by emulsion atomization and atomized bit dehydration is a part of this process [80,81]. The optimization of functional characteristics can lead to encapsulation proficiency (feed temperature, air channel temperature, and current levels). Emulsion consistency is impacted by a decrease in input temperature, for example, by a simultaneous decrease in the size and fluidity of comparable spraying emulsions [82,83]. The difference in the temperatures at the air entrance and output is directly related to the level of capsule drying.

2.2 Emulsification strategy

Emulsions were employed in several applications. Two or more immiscible fluids (typically water and oil) can be mixed to create an emulsion by distributing one of the fluids in the form of microcapsules in the other. This method can be classified as W/O (water is dispersed as microcapsules in the oil ongoing phase), O/W (oil microcapsules are spread throughout the aqueous phase), W/O/W (microcapsules of water scattered in the oil phase of an O/W emulsion), or O/W/O (emulsified microcapsules synthesized using a microfluidic device to polymeric capsules) [84,85]. Typically, customary procedures to construct emulsions include glob decomposition utilizing a blending system. The shocking power is regular generating microcapsules highly polydispersed in size. Microfluidic techniques can be applied for extremely training favorably monodisperse bits as an emulsion in a microfluidic appliance to construct individual microcapsules. UV irradiance, warmth, or chemical yields can be utilized to polymerize made microcapsules utilizing a microfluidic machine [84,85].

2.3 In situ and interfacial polymerization strategies

Because in situ polymerization results in increased encapsulation efficiency and size, it is a viable method for producing microcapsules. This process involves the direct polymerization of reflective monomers on the bit surface or in microcapsules [86]. A specific type of concentration polymerization called interfacial polymerization involves making two reflexive monomers soluble in unmixed fluids. One monomer can be dissolved in the oil phase, while the second monomer can be dissolved in the water phase. Reflexive monomers react quickly by polycondensation, building a polymer, and encasing the anticipated sorbent as they reach the dilution interface.

The cover area and CO_2 absorption level may expand as a result of encapsulated CO_2 sorbents, which may result in a decrease in the absorber potential measurement. Additionally, positively sticky sorbents, weak sorption kinetics, deprecation yields, or deposition reminders can all make sorbent encapsulation a desirable option [87–90]. Although sorbent encapsulation is a useful technology, its success depends on optimal operating characteristics. To describe capsule therapy in typical functioning scenarios, analytical, numerical, and practical assessments must be written.

At the present time, various classes of substances have been utilized as covers and encapsulated mechanisms. Table 8.1 illustrates the primary sorts of substances utilized in microcapsule preparation and their effect on CO_2 absorption.

Because of amine's propensity to alter quickly and unexpectedly as well as their depreciation creations, amine-founded CO_2 capture systems have the potential to have an adverse impact on the environment. Capturing carbon dioxide with encapsulated sorbents such piperazine, MEA, potassium carbonate, and sodium carbonate at rates up to 30% by weight. The polymeric shell keeps sorbents, deposits, or depreciation products contained, where they may be heated to renew them. Encapsulated sorbents conveyed both the CO_2 sorption size of fluid sorbents (water toleration and strong absorption size and the capacity to select the best one) and the physical manners of solid substances that have the property of collecting molecules of another substance through sorption [91].

Successful silicone encapsulation of sodium carbonate dilutions (Na_2CO_3) and aqueous potassium carbonate (K_2CO_3) for CO_2 capture was achieved [91]. Synthesized microcapsules had a constant diameter of $600 \pm 6\,\mu m$ and shell thickness of $31 \pm 1\,\mu m$. In CO_2 collection operations, carbonate dilutions

Table 8.1 Comparison of the substances utilized in encapsulated forms and CO$_2$ absorption rate of them.

Microcapsule features			CO$_2$ uptake			
Outer shell	Encapsulated material	Operating pressure/ temperature (bar-Kelvin)	mg/g	mmol/g	mol/mol	References
Semicosil 949 UV	30wt% Na$_2$CO$_3$	0.1—313.15	100	—	—	[91]
Liquid silicon rubber (0—2wt% Dow Corning 749 fluid)	30wt% K$_2$CO$_3$	~1.01—293.15	—	1.6—2	—	[92]
SITRIS	NDIL0231 IL	—	—	0.488	—	[87]
C$_{CAP}$ support	[P66614][2-CNPyr] IL	1—298	—	1.40	—	[93]
PVDF-HFP	[Hmim][TF$_2$N] IL	~25—296.15	—	3.052	—	[94]
C$_{CAP}$ support	[Emim][TF$_2$N] IL	~25—296.15	—	2.464	—	[94]
C$_{CAP}$ support	[Bmim][Gly] IL	1—303	38	—	—	[95]
C$_{CAP}$ support	[Bmim][PRO] IL	1—303	31	—	—	[95]
PDMS	[Bmim][MET] IL	1—303	20	—	—	[95]
PDMS	[P2222][Bnlm] PCIL	1—333.15	—	—	~0.9	[96]
Thiolene Q	[P2228][2CNPyr] IL	1—295.15	—	—	~0.9	[96]
Thiolene Q	NDIL0230 IL	~0.1—298.15	—	2.5	—	[97]
Tegorad 2650	NDIL 0309 IL	~0.1—298.15	—	1.5	—	[97]

have been investigated as an alternative to aqueous monoethanolamine dilutions. In addition to being abundant, kind to the environment, and resistant to deterioration, carbonates also have a low propensity to alter quickly and unexpectedly. Temperature between 80 and 150°C is required for carbonate regeneration. Encapsulated fluid carbonates have been suggested as a viable alternative to carbonate solids' progressive CO$_2$ sorption level.

Heating (up to 70°C) was used to resuscitate the capsules by releasing CO$_2$ and solubilizing the deposits. Encapsulated sodium carbonate (30 wt% Na$_2$CO$_3$) had a CO$_2$ absorption size of 0.10 g CO$_2$/g sorbent at 0.1 bar. This result is comparable to MEA absorption size. The degree of CO$_2$ sorption is higher in microcapsules made of a fluid carbonate nucleus and silicone shells than in nonencapsulated sorbents. Microcapsules with a pH of m-cresol purple and a 5–30 weight percent aqueous potassium carbonate (K_2CO_3) dilution were arranged. The CO_2 sorption desorption technique is enabled by the pH indicator counted in the aqueous nucleus, as illustrated in Fig. 8.4.

To determine the quantity of CO_2 absorption, microcapsules were exposed to CO_2 (flow rate of 56 mL/h). Because of the dilution's pH (11.9), nucleus fluid has a purple color before the CO_2 sorption test. When CO_2 sorption occurs, a significant change from purple to yellow is observed in that the pH part falls below 8 [92]. Bicarbonate is formed in K_2CO_3 dilutions by the chemical sorption of CO_2:

$$K_2CO_3 + H_2O + CO_2 \rightarrow 2KHCO_3 \qquad (8.1)$$

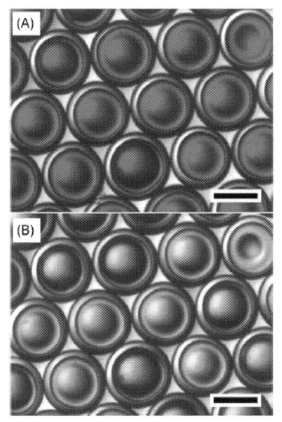

Figure 8.4 Microcapsules holding m-cresol purple (pH indicator) and 5wt% K_2CO_3 in the aqueous nucleus: (A) Before CO_2 absorption and (B) after CO_2 sorption. The ranking bar is 200 μm [92]. Copyright 2016. Reprinted with permission from ACS.

Thus, after satiation is advanced in the microcapsules, CO_2 can be desorbed by heating. Heating prefers the inverse response, releasing CO_2:

$$KHCO_3 \rightarrow K_2CO_3 + H_2O + CO_2 \qquad (8.2)$$

Microcapsules containing 30 weight percent K_2CO_3 had a CO_2 sorption quantity ranging from 1.6 to 2.0 mmol/g depending on the level and breadth of the cover.

By using double-emulsion forms developed in a microfluidic device, aqueous amine dilutions (triethylenetetramine (TETA) and diethylenetriamine (DETA)) were encapsulated in salt or ester of acrylic acid (Fig. 8.5) [98].

Techno-financial analysis is also required for the operation of capturing systems employing encapsulated CO_2 compounds that have the ability to absorb molecules of another material. The cost of collecting after-ignition CO_2 with encapsulated 30% MEA wt. MEA absorbent was approximated to 30% MEA wt. absorbent in a conventional sorption structure. Two operating formats were used to estimate the encapsulated MEA technique. Considerable fixed-bed columns (FB) were analyzed using a design that included an eruptive fluidized-bed amplifier and a circulating fluidized-bed (CFB) absorber. In a typical sorption scenario, the encapsulated MEA approach outperformed the aqueous MEA absorbent in terms of cost. In compared to the CFB approach, the disadvantages of the FB design include a higher amount of water vaporization and increased cost [99].

The implementation of the encapsulated sorbent process can be advanced by narrowing the shell to increase capsule penetrance (>3000 boundaries). However, the cover width drop has the potential to improve warmth while supporting the structure of new sorption and renewal queue. An encapsulated sodium carbonate dilution was used to complete a techno-economic analysis. Two different types of response components (concrete and carbon steel) were given an operational expenditure that was equivalent in value each year and was compared to a typical MEA absorber structure [100].

Trihexyl (tetradecyl) phosphonium 2-cyanopyrrolide ([P66614] [2-CNPyr]) encapsulation of an IL incorporating an aprotic heterocyclic anion (AHA-IL) was calculated for CO_2 sorption. [P66614][2-CNPyr] To hasten the kinetics of absorption, IL was trapped in carbonaceous submicrocapsules. Pictures of carbonaceous submicrocapsules and encapsulated AHA-IL are shown in Fig. 8.6A (AHA-ENIL). A comparable globular shape with border

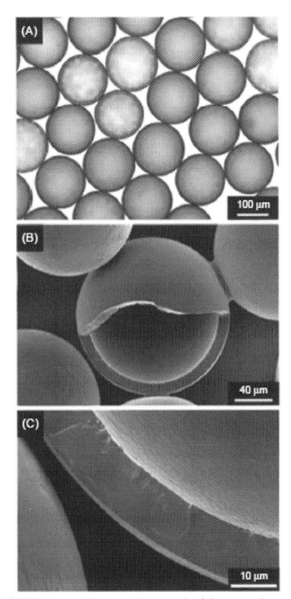

Figure 8.5 (A) Water monodisperse microcapsules (after accepting and being polymer). The moderate bit diameter is 164 ± 3.8 μm. (B) Scanning electron microscope (SEM) picture of a preserved by removal or evaporation of moisture and disconnected microcapsule. (C) Making bigger of the capsule covering [98]. Copyright 2014. Reprinted with permission from ACS.

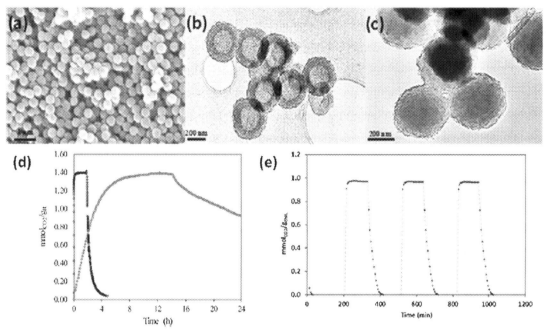

Figure 8.6 (A) Scanning electron microscopy (SEM) images and (B) transmission electron microscopy (TEM) microdiagrams of the carbonaceous submicrocapsules model, (C) microdiagrams of AHA-ENIL organized with [P66614][2-CNPyr]. (D) CO_2 sorption dynamic bends for nonencapsulated AHA-IL and AHAENIL at 298K and 1bar, utilizing [P66614] [2-CNPyr], and (E) [P66614] [2 CNPyr] AHA-ENIL CO_2 sorption–desorption processes at 1bar (sorption: 298K; desorption: 323K, 25 mLN2/min) [93]. Copyright 2018. Reprinted with permission from ACS.

width of roughly 70 nm and modest capacity diffusion of about 350 nm was provided by carbonaceous submicrocapsules assistance. Additionally, we must take into account the important tip that we illustrate in Fig. 8.6B. The [P66614][2-CNPyr] IL completely fills the carbonaceous submicrocapsules backing nucleus. For both encapsulated and unencapsulated AHA-IL, CO_2 dynamic sorption bends (see Fig. 8.6D) disclosed a decreased least duration to thermodynamic balance shape 12 h to 20 min (at 298K and 1 bar) utilizing encapsulated AHA-IL (AHA-ENIL). This treatment is because of sorbent encapsulation boosting growth in the gas-fluid surface connection space. Encapsulated AHA-ENIL is able to be readily renewed upholding the dynamic execution and CO_2 sorption for multiple periods (Fig. 8.6 C) [93].

The amine input of the polymeric cover, which prevents sorbent depreciation as well as the departure of each depreciation yield, deviates design erosion, and increases incorporated fluid sorbents by 25% or more, is the main advantage of encapsulated methods over a conventional amine-based method [101].

3. NG sweetening using encapsulated liquids

Sorbent encapsulation opens up the possibility of novel methodological options. In contrast to water-founded amine techniques employing a conventional packed overlook, an amine-founded encapsulated plan may appear to be a fluidized bed. Both chimney gas and depriving gas operated as a batch approach can be used to mix capsules. The improved surface area per volume of the sorbent as a result of enclosed designs is another advantage. Using very sticky sorbents, which can be challenging to utilize in conventional structures, is made possible by this feature [75]. First method of sweetening of NG is the use of a blended amine function. Alkanolamines are extensively utilized for the disposal of sour gases from NG. Alkanolamines are categorized (founded on the count of alkyl classes that make a connection to the N atom of the amino class) in early amines (for example, MEA), peripheral amines (for example, DEA), and third in order amines (for example, MDEA), per of which include benefits and detriments [72,102]. Preliminary amines include better response levels, while third-in-order amines require weak power for renewal [103]. The choosing of an appropriate amine dilution for making fresh is able to influence supply measuring and handling expenses. MDEA is understood for its increase electively than H_2S in the existence of CO_2 and its response with H_2S is immediate; however, sorption of CO_2 in MDEA happens at weak measure [66,104]. In addition, the revitalization of third-in-order amines needs weaker power in comparison to preliminary and peripheral amines [105], so MDEA appears to be a great option for making fresh NG with 5% H_2S and 1.33% CO_2. The earlier investigations recommend that for increasing the CO_2 sorption level of third-in-order amines (for example, MDEA), fewer quantities of preliminary or peripheral amines (for instance, DEA) can be counted to the dilution [106].

An alternative argument for blending preliminary or peripheral amines with third-in-order amines is to synthesize the increased responsibility of preliminary and peripheral amines with low power needs for the revitalization of third-in-order amines [72]. DEA likewise is able to be built by responding ethylene oxide and ammonia [107]. MDEA and DEA both are built and exist for purchase. These amines absorb sour gases via these chemical responses [108,109]:

$$\text{Peripheral amines}: R_2NH + H_2S \rightarrow (R_2NH)^+ + (HS)^- \quad (8.3)$$

Third in order amines : $R_3N + H_2S \to (R_3NH)^+ + (HS)^1$ (8.4)

Peripheral amines: $CO_2 + R_2NH \leftrightarrow R_2NH + COO^-$ (8.5)

Third in order amines: $CO_2 + R_3N + H_2O \leftrightarrow R_3NH^+ + HCO^{-3}$

(8.6)

Elevated compression and a weak temperature prefer the size of the responses whenever the inverse of these responses is selected by an increased temperature and weak compression [72]. Furthermore, [110] proposes that the absorber of the amine-making fresh factory has to work at elevated degrees of warmth and weak compression and the renewal column to perform at heightened degrees of warmth and weak compressions. Fig. 8.7 shows the graph of the process flow diagram for the mixed amine operation.

Second method of sweetening of NG is the sulfinol-M procedure. As suggested by Ref. [111], the sulfinol method is fit for the making fresh of gases containing an overview of negligible compressions of H_2S and CO_2 more increased in comparison to 1 bar. In NG, the sum of insignificant compressions of H_2S and CO_2 is calculated to be 1.4904 bar, making the sulfinol process a

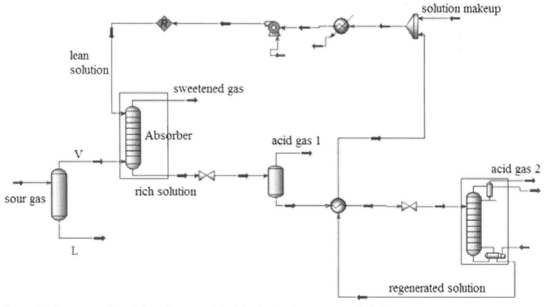

Figure 8.7 Representation of flow diagram of the blended amine operation [70]. Copyright 2014. Reprinted with permission from Elsevier.

viable alternative for producing this gas from scratch. When removing sour gas from NG, the sulfinol-M method dilutes the chemical with water, sulfolane, and MDEA. Sulfolane is an opposite organic combination that has the potential to absorb sour gases (H_2S and CO_2) [112]. It is combined by reacting to sulfur dioxide gas and butadiene to create the average yield 3-sulfolene, and then 3-sulfolene is combined with hydrogen gas to form Sulfolane, and this chemical is in a way that is focused on buying, selling, and making money that is intended and built for buying [113].

The specific sorption of H_2S from NG in the presence of CO_2 is the main application for the sulfinol-M dilution [114]. The sulfinol-M method takes advantage of the advantages of both sorbents, which are the large sorbent size of material sorbents and the good cleanliness of became sweet gas of chemical sorbents [115]. Sulfolane is responsible for the material sorption and MDEA for chemical sorption via responses (3–6) [116]. Fig. 8.8 depicts the process flow diagram used in the sulfinol-M method.

The chelated iron (LO-CAT) treatment is the third way to sweeten NG. H_2S is removed from gases using a ferric iron dilution in the LO-CAT process, which dates back to 1830 and the introduction of Fe_2O_3 [68]. Fe (III) is used as a build-up oxidant to remove H_2S from gases. The method of producing fresh gas using chelated iron dilution is inexpensive, in accordance with the state of the environment, a pure operation, and the dilution in this manner is not harmful [117]. The technology for using iron was

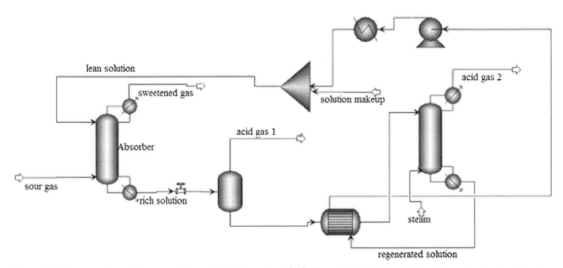

Figure 8.8 Process flow diagram of the sulfinol-M method [70]. Copyright 2014. Reprinted with permission from Elsevier.

pioneered in redox catalysts that used Fe and ethylene diamine tetraacetic (EDTA) complexes [70].

$$2Fe^{3+} + HS^- \leftrightarrow 2Fe^{2+} + S + H^+ \quad (8.7)$$

$$2Fe^{2+} + 1/2\,O_2 + H_2O \leftrightarrow 2Fe^{3+} + 2OH^- \quad (8.8)$$

In Fig. 8.9, the process flow diagram (PFD) of the LO-CAT operation is indicated.

A particular molar ratio of 4:1 [68] exists between the iron in the dilution and the H_2S in the input gas. The involvement of iron in the dilution is assessed using this ratio. Utilizing weather at the compression of the renewal column causes the dilution to be renewed. In a boxed or spray bed, the renewal reaction often takes place [118]. In this process, disposal of hydrogen sulfide operates at about 100% efficiency. The diluted stream does not need to be heated or cooled.

The Shell Paques process is the fourth way to sweeten NG. The Shell and Paques groups first developed the shell-paques technology for the provisioning procedure's order of NG, combination gas, and series gases. Similar technology is used in this method as it is in thiopaq tech, which is used in petroleum, gas, and petrochemical industries to order increased-compression tart gases (up to 1300 psig) [119]. The entire disposal of H_2S may be achieved in this function by using this technique, which is suited for H_2S spans of 100 ppm to considerably enhance H_2S condensations and compression spans of 1–75 bar. With the use of an acidic dilution made up of acidic soda, water, and sulfur bacteria, the sorption of sour gases is released throughout the

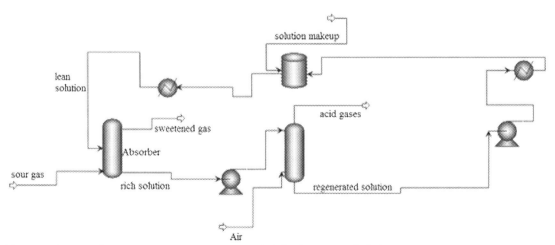

Figure 8.9 Process flow diagram of the chelated iron (LO-CAT) operation [70]. Copyright 2014. Reprinted with permission from Elsevier.

shell process. The affluent dilution enters into a bioreactor where it is introduced to air and H$_2$S is converted to basic sulfur. The dilution links the tart gas in the absorber column and the acidic soda (NaOH) absorbs H$_2$S. The reaction of H$_2$S sorption is [120].

$$NaOH \rightarrow Na^+ + OH^- \tag{8.9}$$

$$H_2S + OH^- \rightarrow HS^- + H_2O \tag{8.10}$$

It can be finalized from reactions (9) and (10) that the sorption of H$_2$S depletes the OH$^-$ ion (and therefore acidic dilution), and the response happening in the bioreactor making the depleted OH$^-$ is:

$$HS^- + 1/2\ O_2 \rightarrow \frac{1}{8} S_8 + OH^- \tag{8.11}$$

The presence of microscopic organisms in the bioreactor boosts the level of response (11), in which hydroxyl ion is produced again. The produced OH- ion can absorb further H$_2$S molecules in the absorber. The sorption level of CO$_2$ is weak, and this dilution particularly absorbs H$_2$S from the NG. According to the attaining channel characteristics for the fresh NG, as shown in an image of the procedure flow graph of the shell method is shown in Fig. 8.10.

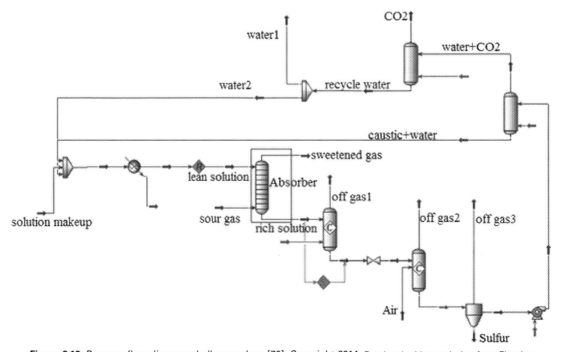

Figure 8.10 Process flow diagram shell procedure [70]. Copyright 2014. Reprinted with permission from Elsevier.

We have to pay consideration to this prominent point that one of the considerable and necessary characteristics of the sour gas disposal rates is the discharge level (gpm) of the dilution in flow and scheme expenses are robust roles of the dilution in flow. As written in Ref. [121], 70% of the enterprise of a factory is straightly connected to this character. The cause for this impact is that the dilution flow level impacts the extent of installations for example absorbers, channels, and amplifiers. In Fig. 8.11, we have shown the need for dilution circulation level because of the following transferring channels' features for several tart gas circulation levels.

Therefore, according to the information drawn out from the results of the diagrams, the shell procedure requires the most inferior dilution circulation level for sour gas disposal (due to the following channel features). The sulfinol-M and blended amine procedures also require weaker dilution circulation levels in comparison to the LO CAT procedure. The distinction among the required circulation level of the shell procedure and the different three procedures is meaningful, whenever the required circulation level of dilution for blended amine, sulfinol-M, and LO CAT procedures is near. An increased dilution circulation level eventuates the factory to need more extensive tools and hence forces the prices of the manufacturer to amplify on the other meaning, the prosperous dilution sour gas condensation will decline because of the increased circulation level of the dilution,

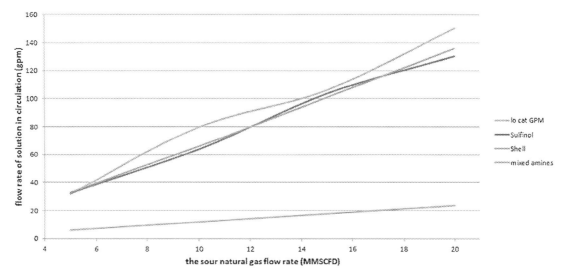

Figure 8.11 Needed dilution circulation level (gpm) because of the following transferring channels' features for several tart gas circulation levels [70]. Copyright 2014. Reprinted with permission from Elsevier.

and hence, the erosion level will decline at the channels and tool. According to these data, a procedure cannot be selected just established on the dilution circulation level.

Dilution loading, which is defined as the ratio of the moles of material used for sorption to the moles of sour gas terminated from the NG, is another crucial functional characteristic in the sour gas disposal division. The loading indicates how well 1 mole of sorbent may entrap sour gases.

When the sour gas loading is heightened, the erosion level in structures would be increased too; thus, in harmony to Fig. 8.10, the shell procedure has the most destructive efficiency since the process of corroding metal is brought in statement. Hence, the finest method in this point is the blended amine procedure, whenever proficiency of the LO CAT approach is good too. However, this character cannot be accepted in statement disconnected from the dilution circulation level, due to the prosperous amine sour gas loading growths, the dilution circulation level required (for a special feed gas) reduces, so the measure would have more elevated prices connected to erosion in structures whenever weaker improved and managing prices connected to dilution circulation level.

In Fig. 8.13, we have shown yearly performing expenses (US$/year) for diverse procedures at diverse acid gas circulation levels.

The primary functioning prices for making fresh parts are the expenses bonded to the power depleted in the reboilers of the renewal queues (or manufacture of unrestricted vapor), pumps, and stoves. As displayed in Fig. 8.10, in this situation, the most suitable procedure for sour gas disposal established on the entire yearly functioning prices is the LO CAT operation, the shell, and the sulfinol-M procedures also have more inferior functional expenses in comparison to the blended amine procedure. In Fig. 8.14, we have displayed entire financial expenses (US$) for diverse procedures at further starting gas circulation levels.

As indicated in Fig. 8.10 established on whole finances expenses, the LO-CAT procedure is the most suitable procedure for the disposal of sour gases from the NG. The sulfinol-M includes the most elevated financial expenses between the four procedures, and blended amine and shell procedures have inferior financial expenses in comparison to the sulfinol-M procedure and better in comparison to the LO CAT procedure. To select one of the four procedures, not only finances, and functioning expenses, in addition, expenses connected to elevated erosion levels for the procedures have to be brought in the statement. As portrayed in Figs. 8.11 and 8.12, the LO CAT procedure includes the

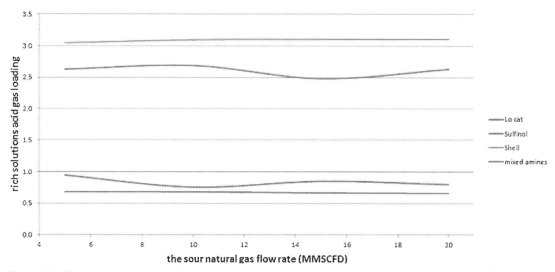

Figure 8.12 Prosperous dilution loading for various procedures at several tart gas circulation levels [70]. Copyright 2014. Reprinted with permission from Elsevier.

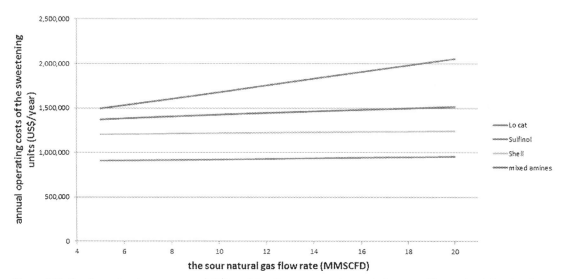

Figure 8.13 Yearly performing expenses (US$/year) for diverse procedures at diverse acid gas circulation levels [70]. Copyright 2014. Reprinted with permission from Elsevier.

most suitable financial efficiency for this feed at progressed acid gas circulation levels (5−20 MMSCFD).

Furthermore, as we can see in Fig. 8.12, it is evident that the LO CAT and the blended amine procedures are the two procedures that include good efficiency established on better dilution of

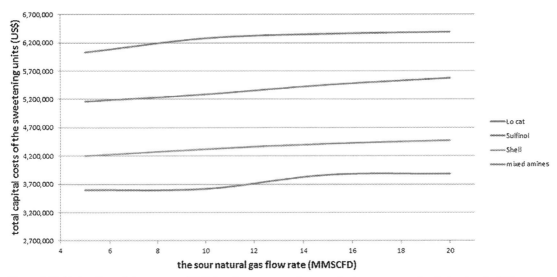

Figure 8.14 Entire financial expenses (US$) for diverse procedures at further tart gas circulation levels [70]. Copyright 2014. Reprinted with permission from Elsevier.

sour gas loading and hence, include weaker expenses correlated to erosion in structures. We have to pay attention to this considerable point that the two factors, it appears that the LO CAT procedure is the most useful procedure when finances and performing expenses along with sour gas loading are taken into account. As displayed in Figs. 8.11–8.14, the simulation and financial assessment of the four procedures have been brought out for weak (progressed) tart gas circulation levels. Nevertheless, the immense circulation levels of tart NG for artful usages are able to be a matter of coming investigations.

4. Conclusion and future outlooks

For the goal of making sweet NG with a fairly good H_2S range of 5 mol% and weak CO_2 range of 1.33 mol%, an appropriate alkanolamine dilution has been opted and afterward the procedure of making sweet with the utilization of alkanolamines is manufactured and in an economic way considered; three other procedures which are also qualified for making sweet of NG of this kind are manufactured and in an economic subject considered and afterward analogized to the alkanolamine procedure. It has been disclosed that in this possibility, the LO CAT operation is better in the economic topic in comparison to the alkanolamine function, which is the considerably famous procedure for the goal of NG-making sweets.

Sorbent encapsulation offers subjects and possibilities in the carbon capturing area. Usually, multiple encapsulated sorbents show more elevated CO_2 capturing size in comparison to nonencapsulated sorbents and can be renewed utilizing favorable functional situations.

The primary benefit of encapsulated sorbents over customary aqueous amine-based sorbents is the sorbent aloneness of the polymeric shell impeding sorbent degradation and tool erosion. However, encapsulated techniques facilitate a better surface area permitting the utilization of favorably thick sorbents, which can be hard to use on customary procedures. These outcomes emphasize the possibility of involving encapsulated sorbents in CO_2 capturing methodologies. The primary possible problem is handling to combine encapsulated liquid, a substance which has the property of collecting molecules of another substance by sorption suggesting elevated penetrance (>3000 barrer), weak warmth of response, and evaporation level. Confounding this possible problem can be a chance for progressing new substances, which have the property of collecting molecules of another substance by sorption with synergic treatment guiding to progressed implementation of CO_2 capturing with novel beneficial attributes. Further investigations and analyses can be accomplished to additional optimize the sweetening of the NG in the easiest ways and also make the best use of the plant. These kinds of examinations can include studying the influence of utilizing diverse amines or exploring the pretreatment of the acid gas and the sensitivity research of its compression, temperature, flow rate, and other specifications taking into attention the physical and chemical characteristics of the ingredients that can affect the efficiency of the gas sweetening procedure.

Abbreviations and symbols

[emmim] [B(CN)₄]	1-ethyl-2,3-dimethyl-imidazolium tetracyanoborate
[hmim][Tf₂N]	1-hexyl-3-methylimidazolium bis (trifluoromethylsulfonyl) imide
[HMIM][TFSI]	1-hexyl-3-methylimidazolium bis(trifluoromethylsulfonyl) imide
[N-bupy][BF₄]	N-butylpyridinium tetrafluoroborate
[omim][Tf₂N]	1-octyl-3-methylimidazolium bis (trifluoromethylsulfonyl) imide
[P2222][BnIm]	tetraethylphosphonium benzimidazolide
[P2228][2CNPyr]	triethyl(octyl)phosphonium 2-cyanopyrrolide
[P66614][2-CNPyr]	trihexyl(tetradecyl)phosphonium 2-cyanopyrrolide
[Tf₂N]	bis(trifluoromethylsulfonyl) imide
CCap	carbonaceous submicrocapsules

CCS	carbon capture and storage
CFB	circulating fluidized bed
CH$_4$	methane
CO$_2$	carbon dioxide
CO$_2$BOLs	CO$_2$-binding organic liquids
DBU	1,8-diazabicyclo[5.4.0]-undec-7-ene
DEA	ethanolamine
DEPG	dimethyl ether of polyethylene glycol
DETA	diethylenetriamine
EAOC	equivalent annual operating cost
ENIL	encapsulated ionic liquid
H$_2$CO$_3$	carbonic acid
H$_2$S	hydrogen sulfide
IGCC	integrated gasification combined cycle
Ils	ionic liquids
K	Kelvin
K$_2$CO$_3$	potassium carbonate
KHCO$_3$	potassium bicarbonate
Koechanol	1-((1,3-dimethylimidazolidin-2-ylidene)amin)propan-2-ol
MDEA	methyldiethanolamine
MEA	monoethanolamine
MECS	microencapsulated CO$_2$ sorbents
Na$_2$CO$_3$	sodium carbonate
NG	natural gas
nm	nanometer
NMP	N-methyl-2-pyrrolidone
O$_2$	oxygen
PCIL	phase-change ionic liquid
PDMS	polydimethylsiloxane
pH	hydrogen potential
psig	pounds per square in gauge
PVDF-HFP	poly(vinylidene fluoride-co-hexafluoropropylene)
RTILs	room-temperature ionic liquids
SEM	scanning electron microscopy
TEA	triethanolamine
TEM	transmission electron microscopy
TETA	triethylenetetramine
TGA	thermogravimetric analysis
UV	ultraviolet
wt.%	weight percent
°C	degree Celsius
μm	micrometer

References

[1] Borhani TN, Wang M. Role of solvents in CO$_2$ capture processes: the review of selection and design methods. Renewable and Sustainable Energy Reviews 2019;114:109299. https://doi.org/10.1016/j.rser.2019.109299.

[2] Faiz R, Al-Marzouqi M. Mathematical modeling for the simultaneous absorption of CO$_2$ and H$_2$S using MEA in hollow fiber membrane contactors. Journal of Membrane Science 2009;342(1−2):269−78. https://doi.org/10.1016/j.memsci.2009.06.050.

[3] Gabelman A, Hwang ST. Hollow fiber membrane contactors. Journal of Membrane Science 1999;159(1−2):61−106. https://doi.org/10.1016/S0376-7388(99)00040-X.

[4] Bae YS, Snurr RQ. Development and evaluation of porous materials for carbon dioxide separation and capture. Angewandte Chemie International Edition 2011;50(49):11586−96. https://doi.org/10.1002/anie.201101891.

[5] Vaesen S, Guillerm V, Yang Q, Wiersum AD, Marszalek B, Gil B, De Weireld G. A robust amino-functionalized titanium (IV) based MOF for improved separation of acid gases. Chemical Communications 2013;49(86):10082−4. https://doi.org/10.1039/C3CC45828H.

[6] Grande CA, Rodrigues AE. Layered vacuum pressure-swing adsorption for biogas upgrading. Industrial & Engineering Chemistry Research 2007;46(23):7844−8. https://doi.org/10.1021/ie070942d.

[7] Saiyed HN. Hydrogen sulfide: human health aspects, concise international chemical assessment document No. 53. Indian Journal of Medical Research 2006;123(1):96.

[8] Lambert TW, Goodwin VM, Stefani D, Strosher L. Hydrogen sulfide (H2S) and sour gas effects on the eye. A historical perspective. The Science of the Total Environment 2006;367(1):1−22. https://doi.org/10.1016/j.scitotenv.2006.01.034.

[9] Koech PK, Rainbolt JE, Bearden MD, Zheng F, Heldebrant DJ. Chemically selective gas sweetening without thermal-swing regeneration. Energy & Environmental Science 2011;4(4):1385−90. https://doi.org/10.1039/C0EE00839G.

[10] Barea E, Montoro C, Navarro JA. Toxic gas removal−metal−organic frameworks for the capture and degradation of toxic gases and vapours. Chemical Society Reviews 2014;43(16):5419−30. https://doi.org/10.1039/C3CS60475F.

[11] Maghsoudi H, Soltanieh M. Simultaneous separation of H2S and CO2 from CH4 by a high silica CHA-type zeolite membrane. Journal of Membrane Science 2014;470:159−65. https://doi.org/10.1016/j.memsci.2014.07.025.

[12] Sidi-Boumedine R, Horstmann S, Fischer K, Provost E, Fürst W, Gmehling J. Experimental determination of hydrogen sulfide solubility data in aqueous alkanolamine solutions. Fluid Phase Equilibria 2004;218(1):149−55. https://doi.org/10.1016/j.fluid.2003.11.020.

[13] Dhage P, Samokhvalov A, Repala D, Duin EC, Tatarchuk BJ. Regenerable Fe−Mn−ZnO/SiO 2 sorbents for room temperature removal of H 2 S from fuel reformates: performance, active sites, Operando studies. Physical Chemistry Chemical Physics 2011;13(6):2179−87. https://doi.org/10.1039/C0CP01355B.

[14] Li YG, Mather AE. Correlation and prediction of the solubility of CO2 and H2S in aqueous solutions of methyldiethanolamine. Industrial & Engineering Chemistry Research 1997;36(7):2760−5. https://doi.org/10.1021/ie970061e.

[15] Pani F, Gaunand A, Richon D, Cadours R, Bouallou C. Absorption of H2S by an aqueous methyldiethanolamine solution at 296 and 343 K. Journal of Chemical & Engineering Data 1997;42(5):865−70. https://doi.org/10.1021/je970062d.

[16] Karadas F, Atilhan M, Aparicio S. Review on the use of ionic liquids (ILs) as alternative fluids for CO2 capture and NG sweetening. Energy & Fuels 2010;24(11):5817−28. https://doi.org/10.1021/ef1011337.

[17] Shiflett MB, Yokozeki A. Separation of CO2 and H2S using room-temperature ionic liquid [bmim][PF6]. Fluid Phase Equilibria 2010; 294(1−2):105−13. https://doi.org/10.1016/j.fluid.2010.01.013.
[18] Kumar S, Cho JH, Moon IL. Ionic liquid-amine blends and CO2BOLs: prospective solvents for NG sweetening and CO2 capture technology—a review. International Journal of Greenhouse Gas Control 2014;20:87−116. https://doi.org/10.1016/j.ijggc.2013.10.019.
[19] Jalili AH, Rahmati-Rostami M, Ghotbi C, Hosseini-Jenab M, Ahmadi AN. Solubility of H2S in ionic liquids [bmim][PF6],[bmim][BF4], and [bmim] [Tf2N]. Journal of Chemical & Engineering Data 2009;54(6):1844−9. https://doi.org/10.1021/je8009495.
[20] Plechkova NV, Seddon KR. Applications of ionic liquids in the chemical industry. Chemical Society Reviews 2008;37(1):123−50. https://doi.org/10.1039/B006677J.
[21] Aparicio S, Atilhan M. Computational study of hexamethylguanidinium lactate ionic liquid: a candidate for NG sweetening. Energy & fuels 2010; 24(9):4989−5001. https://doi.org/10.1021/ef1005258.
[22] Zhang X, Zhang X, Dong H, Zhao Z, Zhang S, Huang Y. Carbon capture with ionic liquids: overview and progress. Energy & Environmental Science 2012;5(5):6668−81. https://doi.org/10.1039/C2EE21152A.
[23] Ju YJ, Lien CH, Chang KH, Hu CC, Wong DSH. Deep eutectic solvent-based ionic liquid electrolytes for electrical double-layer capacitors. Journal of the Chinese Chemical Society 2012;59(10):1280−7. https://doi.org/10.1002/jccs.201100698.
[24] Zhao Y, Zhang J, Han B, Song J, Li J, Wang Q. Metal−organic framework nanospheres with well-ordered mesopores synthesized in an ionic liquid/CO2/surfactant system. Angewandte Chemie International Edition 2011; 50(3):636−9. https://doi.org/10.1002/anie.201005314.
[25] Sabouni R, Kazemian H, Rohani S. Carbon dioxide capturing technologies: a review focusing on metal organic framework materials (MOFs). Environmental Science and Pollution Research 2014;21(8): 5427−49. https://doi.org/10.1007/s11356-013-2406-2.
[26] Gasser T, Guivarch C, Tachiiri K, Jones CD, Ciais P. Negative emissions physically needed to keep global heating below 2 C. Nature Communications 2015;6(1):1−7. https://doi.org/10.1038/ncomms8958.
[27] Unit, I. E. Carbon capture and storage: the solution for deep emissions reductions. IEA; 2015.
[28] Miller DC, Litynski JT, Brickett LA, Morreale BD. Toward transformational carbon capture systems. AIChE Journal 2016;62(1):2−10. https://doi.org/10.1002/aic.15066.
[29] Figueroa JD, Fout T, Plasynski S, McIlvried H, Srivastava RD. Advances in CO2 capture technology—the US department of energy's carbon sequestration program. International Journal of Greenhouse Gas Control 2008;2(1):9−20. https://doi.org/10.1016/S1750-5836(07)00094-1.
[30] Herzog H. What future for carbon capture and sequestration? Environmental Science and Technology-Columbus 2001;35(7):148A.
[31] Baltus RE, Counce RM, Culbertson BH, Luo H, DePaoli DW, Dai S, et al. Examination of the potential of ionic liquids for gas separations. Separation Science and Technology 2005;40(1−3):525−41. https://doi.org/10.1081/SS-200042513.
[32] Olajire AA. CO2 capture and separation technologies for end-of-pipe applications—a review. Energy 2010;35(6):2610−28. https://doi.org/10.1016/j.energy.2010.02.030.

[33] Ziobrowski Z, Krupiczka R, Rotkegel A. Carbon dioxide absorption in a packed column using imidazolium based ionic liquids and MEA solution. International Journal of Greenhouse Gas Control 2016;47:8–16. https://doi.org/10.1016/j.ijggc.2016.01.018.

[34] Kasahara S, Kamio E, Otani A, Matsuyama H. Fundamental investigation of the factors controlling the CO2 permeability of facilitated transport membranes containing amine-functionalized task-specific ionic liquids. Industrial & Engineering Chemistry Research 2014;53(6):2422–31. https://doi.org/10.1021/ie403116t.

[35] Otani A, Zhang Y, Matsuki T, Kamio E, Matsuyama H, Maginn EJ. Molecular design of high CO2 reactivity and low viscosity ionic liquids for CO2 separative facilitated transport membranes. Industrial & Engineering Chemistry Research 2016;55(10):2821–30. https://doi.org/10.1021/acs.iecr.6b00188.

[36] Gurkan BE, de la Fuente JC, Mindrup EM, Ficke LE, Goodrich BF, Price EA, Brennecke JF. Equimolar CO2 absorption by anion-functionalized ionic liquids. Journal of the American Chemical Society 2010;132(7):2116–7. https://doi.org/10.1021/ja909305t.

[37] Yu CH, Huang CH, Tan CS. A review of CO2 capture by absorption and adsorption. Aerosol and Air Quality Research 2012;12(5):745–69. https://doi.org/10.4209/aaqr.2012.05.0132.

[38] Feron P, editor. Absorption-based post-combustion capture of carbon dioxide. Woodhead Publishing; 2016.

[39] Ferrara G, Lanzini A, Leone P, Ho MT, Wiley DE. Exergetic and exergoeconomic analysis of post-combustion CO2 capture using MEA-solvent chemical absorption. Energy 2017;130:113–28. https://doi.org/10.1016/j.energy.2017.04.096.

[40] Jones CW. CO2 capture from dilute gases as a component of modern global carbon management. Annual Review of Chemical and Biomolecular Engineering 2011;2:31–52. https://doi.org/10.1146/annurev-chembioeng-061010-114252.

[41] Papatryfon XL, Heliopoulos NS, Molchan IS, Zubeir LF, Bezemer ND, Arfanis MK, Schubert TJ. CO2 capture efficiency, corrosion properties, and ecotoxicity evaluation of amine solutions involving newly synthesized ionic liquids. Industrial & Engineering Chemistry Research 2014;53(30):12083–102. https://doi.org/10.1021/ie501897d.

[42] Wang M, Lawal A, Stephenson P, Sidders J, Ramshaw C. Post-combustion CO2 capture with chemical absorption: a state-of-the-art review. Chemical Engineering Research and Design 2011;89(9):1609–24. https://doi.org/10.1016/j.cherd.2010.11.005.

[43] American Chemical Society. Recent advances in post-combustion CO2 capture chemistry. American Chemical Society; 2012. https://doi.org/10.1021/bk-2012-1097.ot001.

[44] Liu H, Huang J, Pendleton P. Tailoring ionic liquids for post-combustion CO2 capture. Recent Advances in Post-Combustion CO2 Capture Chemistry 2012;1097:153–75. https://doi.org/10.1021/bk-2012-1097.ch008.

[45] Chu S. Carbon capture and sequestration. Science 2009;325(5948). https://doi.org/10.1126/science.1181637.

[46] Haszeldine RS. Carbon capture and storage: how green can black be? Science 2009;325(5948):1647–52. https://doi.org/10.1126/science.1172246.

[47] Rochelle GT. Amine scrubbing for CO2 capture. Science 2009;325(5948):1652–4. https://doi.org/10.1126/science.1176731.

[48] Friedlingstein P, Solomon S, Plattner GK, Knutti R, Ciais P, Raupach MR. Long-term climate implications of twenty-first century options for carbon dioxide emission mitigation. Nature Climate Change 2011;1(9):457−61. https://doi.org/10.1038/nclimate1302.

[49] Choi S, Drese JH, Jones CW. Adsorbent materials for carbon dioxide capture from large anthropogenic point sources. ChemSusChem: Chemistry & Sustainability Energy & Materials 2009;2(9):796−854. https://doi.org/10.1002/cssc.200900036.

[50] Bishnoi S, Rochelle GT. Absorption of carbon dioxide into aqueous piperazine: reaction kinetics, mass transfer and solubility. Chemical Engineering Science 2000;55(22):5531−43. https://doi.org/10.1016/S0009-2509(00)00182-2.

[51] Nielsen CJ, Herrmann H, Weller C. Atmospheric chemistry and environmental impact of the use of amines in carbon capture and storage (CCS). Chemical Society Reviews 2012;41(19):6684−704. https://doi.org/10.1039/C2CS35059A.

[52] da Silva EF, Booth AM. Emissions from postcombustion CO_2 capture plants. 2013. https://doi.org/10.1021/es305111u.

[53] Voice AK, Rochelle GT. Products and process variables in oxidation of monoethanolamine for CO_2 capture. International Journal of Greenhouse Gas Control 2013;12:472−7. https://doi.org/10.1016/j.ijggc.2012.11.017.

[54] Reynolds AJ, Verheyen TV, Adeloju SB, Meuleman E, Feron P. Towards commercial scale postcombustion capture of CO_2 with monoethanolamine solvent: key considerations for solvent management and environmental impacts. Environmental Science & Technology 2012;46(7):3643−54. https://doi.org/10.1021/es204051s.

[55] Lively RP, Chance RR, Koros WJ. Enabling low-cost CO_2 capture via heat integration. Industrial & Engineering Chemistry Research 2010;49(16):7550−62. https://doi.org/10.1021/ie100806g.

[56] Banerjee R, Phan A, Wang B, Knobler C, Furukawa H, O'Keeffe M, et al. High-throughput synthesis of zeolitic imidazolate frameworks and application to CO_2 capture. Science 2008;319(5865):939−43. https://doi.org/10.1126/science.1152516.

[57] Britt D, Furukawa H, Wang B, Glover TG, Yaghi OM. Highly efficient separation of carbon dioxide by a metal-organic framework replete with open metal sites. Proceedings of the National Academy of Sciences 2009;106(49):20637−40. https://doi.org/10.1073/pnas.0909718106.

[58] Farha OK, Özgür Yazaydın A, Eryazici I, Malliakas CD, Hauser BG, Kanatzidis MG, Hupp JT. De novo synthesis of a metal−organic framework material featuring ultrahigh surface area and gas storage capacities. Nature Chemistry 2010;2(11):944−8. https://doi.org/10.1038/nchem.834.

[59] Xiang S, He Y, Zhang Z, Wu H, Zhou W, Krishna R, et al. Microporous metal-organic framework with potential for carbon dioxide capture at ambient conditions. Nature Communications 2012;3(1):1−9. https://doi.org/10.1038/ncomms1956.

[60] Voicu D, Abolhasani M, Choueiri R, Lestari G, Seiler C, Menard G, Kumacheva E. Microfluidic studies of CO_2 sequestration by frustrated Lewis pairs. Journal of the American Chemical Society 2014;136(10):3875−80. https://doi.org/10.1021/ja411601a.

[61] Patel HA, Hyun Je S, Park J, Chen DP, Jung Y, Yavuz CT, et al. Unprecedented high-temperature CO_2 selectivity in N_2-phobic

nanoporous covalent organic polymers. Nature Communications 2013; 4(1):1−8. https://doi.org/10.1038/ncomms2359.
[62] Liu Y, Wilcox J. Molecular simulation studies of CO2 adsorption by carbon model compounds for carbon capture and sequestration applications. Environmental Science & Technology 2013;47(1):95−101. https://doi.org/10.1021/es3012029.
[63] Li H, Eddaoudi M, O'Keeffe M, Yaghi OM. Design and synthesis of an exceptionally stable and highly porous metal-organic framework. Nature 1999;402(6759):276−9. https://doi.org/10.1038/46248.
[64] Campbell JM. Campbell petroleum series. 1974.
[65] Datta AK, Sen PK. Optimization of membrane unit for removing carbon dioxide from NG. Journal of Membrane Science 2006;283(1−2):291−300. https://doi.org/10.1016/j.memsci.2006.06.043.
[66] Rufford TE, Smart S, Watson GC, Graham BF, Boxall J, Da Costa JD, et al. The removal of CO2 and N2 from NG: a review of conventional and emerging process technologies. Journal of Petroleum Science and Engineering 2012;94:123−54. https://doi.org/10.1016/j.petrol.2012.06.016.
[67] Abdulrahman RK, Sebastine IM. NG sweetening process simulation and optimization: a case study of Khurmala field in Iraqi Kurdistan region. Journal of NG Science and Engineering 2013;14:116−20. https://doi.org/10.1016/j.jngse.2013.06.005.
[68] Kohl AL, Nielsen R. Gas purification. Gulf Professional Publishing; 1997.
[69] Gpsa G. Engineering data book. Gas Processors Suppliers Association 2004;2:16−24.
[70] Kazemi A, Malayeri M, Shariati A. Feasibility study, simulation and economical evaluation of NG sweetening processes−part 1: a case study on a low capacity plant in Iran. Journal of NG Science and Engineering 2014;20:16−22. https://doi.org/10.1016/j.jngse.2014.06.001.
[71] Rinker EB, Ashour SS, Sandall OC. Absorption of carbon dioxide into aqueous blends of diethanolamine and methyldiethanolamine. Industrial & Engineering Chemistry Research 2000;39(11):4346−56. https://doi.org/10.1021/ie990850r.
[72] Fouad WA, Berrouk AS. Using mixed tertiary amines for gas sweetening energy requirement reduction. Journal of NG Science and Engineering 2013;11:12−7. https://doi.org/10.1016/j.jngse.2012.07.003.
[73] Nuchitprasittichai A, Cremaschi S. Optimization of CO2 capture process with aqueous amines using response surface methodology. Computers & Chemical Engineering 2011;35(8):1521−31. https://doi.org/10.1016/j.compchemeng.2011.03.016.
[74] Rangwala HA, Morrell BR, Mather AE, Otto FD. Absorption of CO2 into aqueous tertiary amine/MEA solutions. Canadian Journal of Chemical Engineering 1992;70(3):482−90. https://doi.org/10.1002/cjce.5450700310.
[75] Einloft S, Bernard FL. Encapsulated liquid sorbents for CO2 capture. In: Advances in carbon capture. Woodhead Publishing; 2020. p. 125−50. https://doi.org/10.1016/B978-0-12-819657-1.00006-2.
[76] Campos E, Branquinho J, Carreira AS, Carvalho A, Coimbra P, Ferreira P, et al. Designing polymeric microparticles for biomedical and industrial applications. European Polymer Journal 2013;49(8):2005−21. https://doi.org/10.1016/j.eurpolymj.2013.04.033.
[77] Singh MN, Hemant KSY, Ram M, Shivakumar HG. Microencapsulation: a promising technique for controlled drug delivery. Research in Pharmaceutical Sciences 2010;5(2):65.

[78] Fang Z, Bhandari B. Encapsulation of polyphenols—a review. Trends in Food Science & Technology 2010;21(10):510—23. https://doi.org/10.1016/j.tifs.2010.08.003.

[79] Bakry AM, Abbas S, Ali B, Majeed H, Abouelwafa MY, Mousa A, et al. Microencapsulation of oils: a comprehensive review of benefits, techniques, and applications. Comprehensive Reviews in Food Science and Food Safety 2016;15(1):143—82. https://doi.org/10.1111/1541-4337.12179.

[80] Silva PTD, Fries LLM, Menezes CRD, Holkem AT, Schwan CL, Wigmann ÉF, Silva CDBD. Microencapsulation: concepts, mechanisms, methods and some applications in food technology. Ciência Rural 2014; 44:1304—11. https://doi.org/10.1590/0103-8478cr20130971.

[81] Gharsallaoui A, Roudaut G, Chambin O, Voilley A, Saurel R. Applications of spray-drying in microencapsulation of food ingredients: an overview. Food Research International 2007;40(9):1107—21. https://doi.org/10.1016/j.foodres.2007.07.004.

[82] Ferguson L, Scovazzo P. Solubility, diffusivity, and permeability of gases in phosphonium-based room temperature ionic liquids: data and correlations. Industrial & Engineering Chemistry Research 2007;46(4): 1369—74. https://doi.org/10.1021/ie0610905.

[83] Condemarin R, Scovazzo P. Gas permeabilities, solubilities, diffusivities, and diffusivity correlations for ammonium-based room temperature ionic liquids with comparison to imidazolium and phosphonium RTIL data. Chemical Engineering Journal 2009;147(1):51—7. https://doi.org/10.1016/j.cej.2008.11.015.

[84] Chen CH, Shah RK, Abate AR, Weitz DA. Janus particles templated from double emulsion droplets generated using microfluidics. Langmuir 2009; 25(8):4320—3. https://doi.org/10.1021/la900240y.

[85] Shah RK, Shum HC, Rowat AC, Lee D, Agresti JJ, Utada AS, Weitz DA. Designer emulsions using microfluidics. Materials Today 2008;11(4): 18—27. https://doi.org/10.1016/S1369-7021(08)70053-1.

[86] Jyothi NVN, Prasanna PM, Sakarkar SN, Prabha KS, Ramaiah PS, Srawan GY. Microencapsulation techniques, factors influencing encapsulation efficiency. Journal of Microencapsulation 2010;27(3): 187—97. https://doi.org/10.3109/02652040903131301.

[87] Stolaroff JK, Ye C, Oakdale JS, Baker SE, Smith WL, Nguyen DT, Aines RD. Microencapsulation of advanced solvents for carbon capture. Faraday Discussions 2016;192:271—81. https://doi.org/10.1039/C6FD00049E.

[88] Raksajati A, Ho MT, Wiley DE. Techno-economic evaluation of CO2 capture from flue gases using encapsulated solvent. Industrial & Engineering Chemistry Research 2017;56(6):1604—20. https://doi.org/10.1021/acs.iecr.6b04095.

[89] Finn JR, Galvin JE. Modeling and simulation of CO2 capture using semipermeable elastic microcapsules. International Journal of Greenhouse Gas Control 2018;74:191—205. https://doi.org/10.1016/j.ijggc.2018.04.022.

[90] Stolaroff JK, Ye C, Nguyen DT, Oakdale J, Knipe JM, Baker SE. CO2 absorption kinetics of micro-encapsulated ionic liquids. Energy Procedia 2017;114:860—5. https://doi.org/10.1016/j.egypro.2017.03.1228.

[91] Vericella JJ, Baker SE, Stolaroff JK, Duoss EB, Hardin JO, Lewicki J, Aines RD. Encapsulated liquid sorbents for carbon dioxide capture. Nature Communications 2015;6(1):1—7. https://doi.org/10.1038/ncomms7124.

[92] Nabavi SA, Vladisavljević GT, Gu S, Manovic V. Semipermeable elastic microcapsules for gas capture and sensing. Langmuir 2016;32(38): 9826−35. https://doi.org/10.1021/acs.langmuir.6b02420.

[93] Moya C, Alonso-Morales N, de Riva J, Morales-Collazo O, Brennecke JF, Palomar J. Encapsulation of ionic liquids with an aprotic heterocyclic anion (AHA-IL) for CO2 capture: preserving the favorable thermodynamics and enhancing the kinetics of absorption. The Journal of Physical Chemistry B 2018;122(9):2616−26. https://doi.org/10.1021/acs.jpcb.7b12137.

[94] Kaviani S, Kolahchyan S, Hickenbottom KL, Lopez AM, Nejati S. Enhanced solubility of carbon dioxide for encapsulated ionic liquids in polymeric materials. Chemical Engineering Journal 2018;354:753−7. https://doi.org/10.1016/j.cej.2018.08.086.

[95] Santiago R, Lemus J, Moya C, Moreno D, Alonso-Morales N, Palomar J. Encapsulated ionic liquids to enable the practical application of amino acid-based ionic liquids in CO2 capture. ACS Sustainable Chemistry & Engineering 2018;6(11):14178−87. https://doi.org/10.1021/acssuschemeng.8b02797.

[96] Song T, Avelar Bonilla GM, Morales-Collazo O, Lubben MJ, Brennecke JF. Recyclability of encapsulated ionic liquids for post-combustion CO2 capture. Industrial & Engineering Chemistry Research 2019;58(12): 4997−5007. https://doi.org/10.1021/acs.iecr.9b00251.

[97] Knipe JM, Chavez KP, Hornbostel KM, Worthington MA, Nguyen DT, Ye C, et al. Evaluating the performance of micro-encapsulated CO2 sorbents during CO2 absorption and regeneration cycling. Environmental Science & Technology 2019;53(5):2926−36. https://doi.org/10.1021/acs.est.8b06442.

[98] Chen PW, Cadisch G, Studart AR. Encapsulation of aliphatic amines using microfluidics. Langmuir 2014;30(9):2346−50. https://doi.org/10.1021/la500037d.

[99] Raksajati A, Ho MT, Wiley DE. Comparison of design options for encapsulated solvent processes for CO2 capture. Energy Procedia 2017; 114:764−70. https://doi.org/10.1016/j.egypro.2017.03.1219.

[100] Kotamreddy G, Hughes R, Bhattacharyya D, Stolaroff J, Hornbostel K, Matuszewski M, et al. Process modeling and techno-economic analysis of a CO2 capture process using fixed bed reactors with a microencapsulated solvent. Energy & Fuels 2019;33(8):7534−49. https://doi.org/10.1021/acs.energyfuels.9b01255.

[101] Aines RD, Bourcier WL, Spadaccini CM, Stolaroff JK. U.S. Patent No. 8,945,279. Washington, DC: U.S. Patent and Trademark Office; 2015.

[102] Padurean A, Cormos CC, Cormos AM, Agachi PS. Multicriterial analysis of post-combustion carbon dioxide capture using alkanolamines. International Journal of Greenhouse Gas Control 2011;5(4):676−85. https://doi.org/10.1016/j.ijggc.2011.02.001.

[103] Mudhasakul S, Ku HM, Douglas PL. A simulation model of a CO2 absorption process with methyldiethanolamine solvent and piperazine as an activator. International Journal of Greenhouse Gas Control 2013;15: 134−41. https://doi.org/10.1016/j.ijggc.2013.01.023.

[104] Haimour N, Sandall OC. Absorption of H2S into aqueous methyldiethanolamine. Chemical Engineering Communications 1987; 59(1−6):85−93. https://doi.org/10.1080/00986448708911987.

[105] Yildirim Ö, Kiss AA, Hüser N, Leßmann K, Kenig EY. Reactive absorption in chemical process industry: a review on current activities. Chemical

[106] Anufrikov YA, Kuranov GL, Smirnova NA. Solubility of CO2 and H2S in alkanolamine-containing aqueous solutions. Russian Journal of Applied Chemistry 2007;80(4):515–27. https://doi.org/10.1134/S1070427207040015.

[107] Jones C, Edens MR, Lochary JF. Alkanolamines from olefin oxides and ammonia. Kirk-Othmer Encyclopedia of Chemical Technology; 2000. https://doi.org/10.1002/0471238961.0112110105040514.a01.pub2.

[108] Cummings AL, Smith GD, Nelsen DK. Advances in amine reclaiming—why there's no excuse to operate a dirty amine system. In: Laurance reid gas conditioning conference, vol 2007; 2007.

[109] Lu JG, Zheng YF, He DL. Selective absorption of H2S from gas mixtures into aqueous solutions of blended amines of methyldiethanolamine and 2-tertiarybutylamino-2-ethoxyethanol in a packed column. Separation and Purification Technology 2006;52(2):209–17. https://doi.org/10.1016/j.seppur.2006.04.003.

[110] Ebenezer SA, Gudmunsson JS. Removal of carbon dioxide from NG for LNG production. Semester project work; 2005.

[111] Hiller H, Reimert R, Marschner F, Renner HJ, Boll W, Supp E, Driesen HE. Gas production Ullmann's encyclopedia of industrial chemistry. 2006.

[112] Jou FY, Deshmukh RD, Otto FD, Mather AE. Solubility of H2S, CO2, CH4 and C2H6 in sulfolane at elevated pressures. Fluid Phase Equilibria 1990;56:313–24. https://doi.org/10.1016/0378-3812(90)85111-M.

[113] Stewart O, Minnear L. Sulfolane technical assistance and evaluation report. Final Report. Alaska Department of Environmental Conservation; 2010.

[114] Nasir P. A mixed solvent for a low total sulfur specification. In: AIChE national meeting; August 1990 [San Diego, CA].

[115] Kunkel LV. U.S. Patent No. 3,630,666. Washington, DC: U.S. Patent and Trademark Office; 1971.

[116] Nikolic DL, Wijntje R, Rao PPH, Van Der Zwet G. Sulfinol-X*: second-generation solvent for contaminated gas treating. In: IPTC 2009: international petroleum technology conference. European Association of Geoscientists & Engineers; December 2009. https://doi.org/10.3997/2214-4609-pdb.151.iptc14017.

[117] Yong YU, Youzhi LIU, Guisheng QI. Rapid regeneration of chelated iron desulfurization solution using electrochemical reactor with rotating cylindrical electrodes. Chinese Journal of Chemical Engineering 2014;22(2):136–40. https://doi.org/10.1016/S1004-9541(14)60005-7.

[118] Yang JP, Li HT, Xiao JG. Study on desulfurization from acidic gaseous stream by chelate iron. Journal of Chemical Industry and Engineering 2002;23(2):23–4.

[119] Benschop A, Janssen A, Hoksberg A, Seriwala M, Abry R, Ngai C. The shell-Paques/THIOPAQ gas desulphurization process: successful start up first commercial unit. 2002.

[120] Sherbaniuk R, Beasley T, Abry RGF. Reducing solution gas flaring has many positive results and managing hydrogen sulphide the natural way. Energy Processing Canada 2003;96(1):8–16.

[121] Chan CW, Tontiwachwuthikul P. A decision support system for solvent selection of CO2 separation processes. Energy Conversion and Management 1996;37(6–8):941–6. https://doi.org/10.1016/0196-8904(95)00281-2.

9

Cryogenic fractionation for natural gas sweetening

Juan Pablo Gutierrez[1,2], Fabiana Belén Torres[1] and Eleonora Erdmann[1,2]

[1]Instituto de Investigaciones para la Industria Química INIQUI (CONICET-UNSA), Salta, Argentina; [2]Facultad de Ingeniería—Consejo de Investigación, Universidad Nacional de Salta, Salta, Argentina

1. Introduction

Natural gas represents a nonrenewable source due to its combustible characteristics and can be used in an extensive range of applications, including its use in residences, transport, and industries. Natural gas is used in power plants, in cement industries, in metals obtainment, and for petrochemical compounds synthesis. Numerous alkanes conform to the mixture that is called "natural gas," ranging from methane to superiors. Nevertheless, this mixture is contaminated with other types of components, reducing the calorific value of the gas. In this sense, common contaminants are carbon dioxide, hydrogen sulfide, water, nitrogen oxides, sulfur, and mercury. According to various authors, the location where the natural gas is extracted defines the specific percentages that each compound has in the total mixture [1,2].

Particularly, due to the impurities in sour natural gas, this must be treated by removing the acid gases before its transportation to the different consumers. In general, before being sold, natural gas must be treated to remove the carbon dioxide (CO_2) and hydrogen sulfide (H_2S) present in the gas and achieve the concentration adequate for its distribution. This process is known in industries as natural gas sweetening or acid gas removal. Conventional removal technologies are employed to remove the acid gases from natural gas where the operative pressure and temperature play an important role in the separation of these components. Generally speaking, natural gas needs to be treated before any application and, for this reason, it must achieve acceptable conditions. Every country has its legislation to control the specifications for the distribution and consumption of natural gas. Numerous processes exist to apply in the natural gas treatment and satisfy these bound values particularly in regard to CO_2 and H_2S compositions [1,3].

Different technologies can be applied in the process to separate carbon dioxide and hydrogen sulfide from natural gas. As mentioned before, the treatment to remove acid gases (CO_2, H_2S) is named sweetening technology. The presence of the acid gases causes some issues in the transportation like the corrosion of the pipelines. Another associated problem is the decrease in the calorific value, which affects the economic potential and energetic efficiency of the entire process [4]. Furthermore, the world's attention toward climate change has led to care for the CO_2 emissions into the atmosphere and the interests of the conditioning technologies [5,6]. Most importantly, an analysis realized by the Intergovernmental Panel on Climate Change (IPCC) and by the International Energy Agency exposed the importance of research and investment in CO_2 capture technologies. In this regard, it is important to study its transportation and use or storage to be able to afford the problems of climate change. Besides, the IPCC mentions that our planet needs to maintain the global temperature increases under 1.5°C, and the recommended technologies are associated with carbon capture and storage [5,7].

Several works and many studies related to the sweetening process have been published. In addition to this, numerous new methods appeared, and others were being improved to achieve the restrictions. From the works published by Dai et al. [8] and Rezakazemi et al. [9], different technologies are implemented for CO_2 separation such as different material-based membranes modules, chemical absorption (employing variety of solvents including alkanolamines, ionic liquids, and others), physical adsorption, low temperature process, and the combination of the aforementioned, the so-called hybrid or combined systems. Other works use cryogenic technology for CO_2 capture in biogas upgrading and liquefaction. Spitoni et al. [10] present a work that consists of a new technique for removing CO_2 at a cold load. Authors study the performance of the plant and carry out the optimization, with a sensitivity analysis, and the evaluation of the process.

Cryogenic fractionation represents a process which can be applied in the separation of CO_2 from natural gas or fuel gas. This technology uses the properties of the components to be separated from the source; the condensation and desublimation temperatures are the principal properties involved in the separation nature. This process has some advantages over others. For example, it is capable of operating a large amount of gas, it is able to work at moderate pressures, and it allows the obtainment of high purity and recovery of CO_2, 99.99% approximately in both cases. Another characteristic of this process is the absence of any chemical solvent for the separation.

Different processes were developed in industries to be applied in cryogenic fractionation. Holmes and Ryan [11] patented a new configuration for the cryogenic distillation, while Tunier et al. [12,13] presented the dynamically operated packed beds. Although Turnier et al. presented a novel method because both water and carbon dioxide can be separated at the same time, elevated energy requirement resulted to be a main disadvantage [14].

2. Thermodynamic principles of CO_2 and CH_4 separation

Pure CO_2 is present at atmospheric pressure as a gas, but different conditions affect its state. It is important to note some differences between CO_2 and CH_4. CO_2 has a sublimation point of $-78.5°C$, meanwhile for the CH_4, the melting point is $-182°C$. As mentioned previously, with adequate pressure and temperature conditions of thermodynamic equilibrium, the components of natural gas will be separated into vapor, liquid, and solid phases. As a result of this, one stream of the process contains high purity solid CO_2 and another stream contains both liquid and vapor phases, which consists of less CO_2 and hydrocarbons [3].

Fig. 9.1 shows a general description thermodynamics P–T phase envelope for the binary mixture methane/carbon dioxide. As it can be appreciated, this envelope is developed by the authors only for the coexisting compounds and, consequently, another component or contaminant will distort the curves. Triple points in the graph are represented by A and B, respectively. At the same time, liquid CO_2 vapor pressure is represented by the portion between B and C, while critical locus is represented by the portion between D and C. The curve between B and C separates the vapor/liquid phase from the vapor phase (located below). S–V phase is represented over the curve between A and B [15].

Furthermore, Fig. 9.2 illustrates the influence of the carbon dioxide composition in the P–T phase envelope for the binary mixture methane/carbon dioxide. Compositions analyzed in this

188 Chapter 9 Cryogenic fractionation for natural gas sweetening

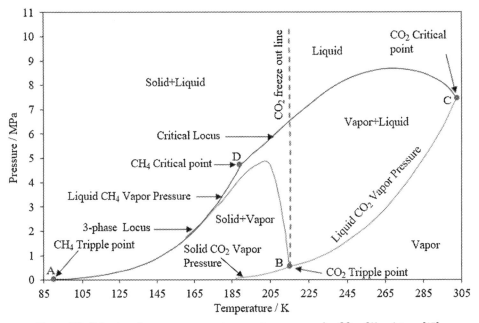

Figure 9.1 Behavior of pressure versus temperature curves, for CO_2–CH_4 mixture [15].

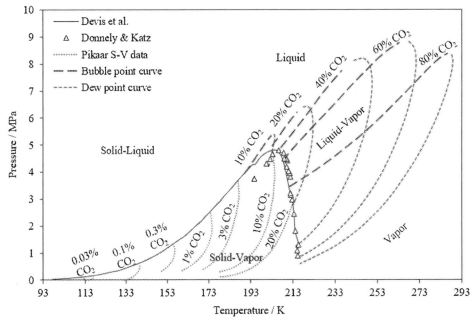

Figure 9.2 Behavior of pressure versus temperature curves, for various CO_2–CH_4 proportions [15].

graph are 10 mol %, 20 mol %, 40 mol %, 60 mol %, and 80 mol % of CO_2 in the binary mixture. Again, this analysis is only for the binary mixture excluding any contaminant. As it can be expected, the dew and bubble points vary according to the CO_2/CH_4 compositions. Consequently, every combination with a specific composition has a specific critical point with different pressure and temperature conditions. At constant pressures, decreasing CO_2 content in the mixture will imply the temperature of the dew and bubble to decrease, and conversely. Moreover, frost lines are also affected by the specific composition of the mixture.

Equilibrium data for the binary mixture methane/carbon dioxide can be found in literature according to experimental data obtained from specific CO_2–HC compositions. Nevertheless, most recent procedures predict thermodynamic data by using equation of state–based simulators [15].

Fig. 9.2 represents some important parameters by varying the CO_2 concentration, curves of dew and bubble point, and solid and vapor phases data for the mixture CO_2–CH_4. It is possible to carry out two cryogenic processes following the regions of the figure: above the curve, a V–L separation, conventional methods can be performed, and under the curve, an S–V separation, nonconventional process can be performed. The evaluation of the equilibrium curves of CO_2–CH_4 mixtures results in an important step for the extrapolation that can be carried out from them, facing the predominance of CH_4 in natural gas streams.

Following this, Fig. 9.3 presents some developing cryogenic technologies for carbon dioxide capture [17]. As it can be seen, cryogenic processes for CO_2 capture can be divided into nonconventional V–L, nonconventional V–S, and hybrid systems. In this chapter, authors mostly detail nonconventional and hybrid technologies. Due to the separation carried out for a hybrid system that offers a good recovery of CO_2 and a stream with a CO_2 concentration over 99%, this process has attracted more attention. For this reason, much research has paid attention to hybrid technologies [16].

3. Cryogenic processes for acid gas removal

The CO_2 gas separation from flue gases, natural gas, or biogas could be possible with cryogenic technology. This process uses the different condensation and desublimation mix properties [10]. As stated by Brunetti et al. [18], a high purity CO_2 stream is obtained and, consequently, the recovery is 99.99% approximately. In contrast, contaminants can produce serious

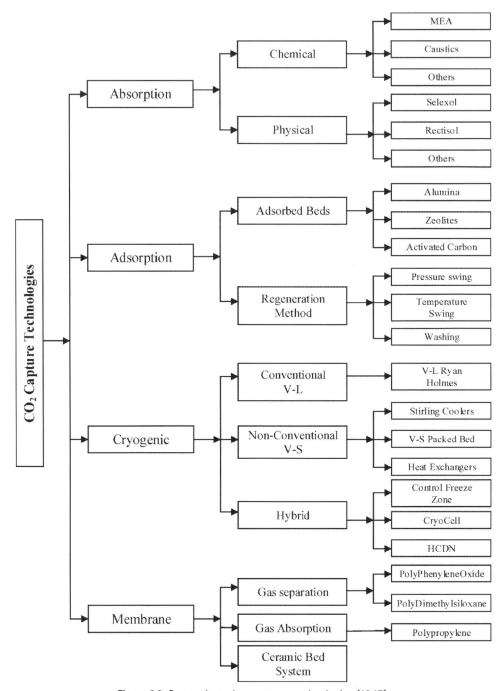

Figure 9.3 Cryogenic carbon capture technologies [16,17].

obstruction and increase productive expenditures. Recently, cryogenic-based approaches have received more attention [19], and new cryogenic CO_2 capture technologies can be found in literature.

Some solutions are detailed to address the issues of the cryogenic technology, for example, to improve the efficiency, decrease the loss of energy and exergy of this process, and moreover prevent water condensation, which could cause blockage in some equipment.

3.1 Cryogenic packed bed

Tunier et al. [12,13] designed a process for CO_2 capture, including a dynamically manipulated packed bed incorporated into the cryogenic system, as shown in Fig. 9.4. A steel monolith structure is used as the pack material, and the energy requirement is provided by using liquefied natural gas (LNG) [20]. This method is advantageous because it is possible to separate H_2O and CO_2 from flue gases streams, based on their different dew and sublimation points. Some common inconveniences of the cryogenic process could be prevented when a packed bed is used in the cryogenic system, like clogging and pressure drops. Furthermore, chemical absorbent and elevated pressure are not required [12,13]. In addition to this, Babar et al. [15] studied a continuous process to remove CO_2 from natural gas for this purpose by using

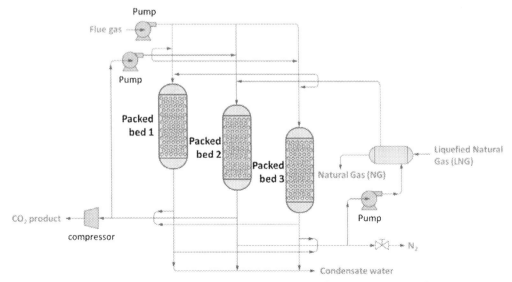

Figure 9.4 Three-stage packed bed process (Song et al. [17] and Tunier et al. [12,13]).

an alternate cryogenic packed bed. In this research, the refrigerant used was liquid nitrogen for prechilling the glass packing contained by the packed bed. Inside the cryogenic column, the packed bed was constituted by spherical glass beads with a diameter of 8 mm [21].

In addition, packed cryogenic beds can be utilized to capture CO_2 in upgraded biogas. Moreover, it is possible to remove the H_2S that can be present at different points of the bed pack. Concurrently, a stream of methane is obtained with a purity of around 99.1% without experiencing any phase change. Tuinier and Annaland [22] compared a typical vacuum pressure swing adsorption (VPSA) with a cryogenic packed bed process. Differences can be listed in Table 9.1.

Another advantage of the bed pack with respect to the VPSA process is that the methane product stream is obtained at cryogenic temperature and thus implies to be immediately liquefied and injected directly into a pipe network. In that respect, liquefaction expenditures can be omitted [22].

Cryogenic packed bed processes have demonstrated more efficiency than VPSA and alkanolamines-based absorption processes. However, it is necessary to resolve several challenges before being sold to industries. Sensible and latent heat losses are expected to be reduced due to thermal insulation of the beds. If high H_2S removal efficiency is required, packed bed temperature should be kept around $-150°C$ and, consequently, operating expenditures are elevated. As stated by Abatzoglou and Boivin [23], cold energy required in this cryogenic process is competitive when LNG is used as a cold source. Nevertheless, if this source is unavailable, the energy requirement is highly elevated because of the use of refrigerating sources and the advantages in terms of energy consumption of this process with respect to common cryogenic processes are insignificant.

Table 9.1 Comparison between packed bed and vacuum pressure swing adsorption (VPSA) processes [22].

	Packed bed	VPSA process
Methane recovery	79.7%	94.3%
Methane productivity	43.1 kg h−1 mpacking−3	350.2 kg h−1 mpacking−3
Energy requirement	2.9 MJ/kg CH_4	3.7 MJ/kg CH_4
Installation costs	Lower in packed bed due to smaller bed sizes	

Table 9.2 Cryogenic packed bed: application, cost investment, cost operation, and produced CO_2 purity [12,13].

Application		Cost investment	Cost operation	CO_2 purity
Cryogenic packed bed	Separation of CO_2 and H_2S based on the deposition of the acid gases in packed bed columns	345 M$ CO_2 captured: 450 ton/h	18,000$/h	99.9% CO_2 recovery

Some uses, cost investment, cost operation, and the CO_2 purity for the cryogenic packed bed process are shown in Table 9.2.

3.2 Cryogenic separation based on multicompression stages with intercoolers

CO_2 removal consists of energetically efficient installations. Among the different processes for acid gas removal to achieve the target performances, polymeric-based membrane modules have generated increasing interest. However, based on performances of already used materials, one single stage is not sufficient to achieve a typical CO_2 content reduction from 30%–40% to 2% mole. In this regard, multistage membrane units are needed for 15% CO_2 content or higher, common carbon dioxide compositions in flue gases or sour natural gas. Multistage membrane modules have been introduced by several researchers in recent decades [24], and, more recently, other technologies are under evaluation to improve the performance of CO_2 separation.

Therefore, cryogenic technologies based on multicompression stages using intercoolers result in an appropriate method to improve the separation of the main acid gases such as CO_2 and H_2S. As mentioned before, this method exhibits some good advantages that could be positive for air separation and flue gas conditioning. Cryogenic processes belong to the so-called low-temperature technologies. Additionally, most of these systems are implemented in different gas separation processes to separate CO_2 from different sources such as air, sour natural gas, and flue gas. However, the high energy requirements to obtain high-purity CO_2 have led the researchers to deeply analyze this application for particular cases. Hart and Gnanendran [3] present the cryogenic separation applied for a high-pressure sour natural gas, Meratla

[25] for the mixture of CO_2/O_2 in combustion cycle, and Zanganeh and Shafeen [26] for some mixture with elevated CO_2 concentration.

The cryogenic distillation process results differently from conventional distillation because the CO_2 separation is independent from boiling points, and only dew and sublimation points are considered. This method was introduced by Holmes and Ryan [11], where a distillation column is used for the removal of acid gases from sour natural gas. In cryogenic distillation, the separation of the components occurs at very low temperature ($<-73.3°C$) and elevated pressure. The final products could be liquefied CO_2 or CO_2 in the vapor phase at high pressure, and this implies an advantage to transport or storage of the gas [27].

When it is necessary to perform the separation of CO_2 from fuel gas or from another source like flue gas, which contains high carbon dioxide concentration, this separation could be made by using two cryogenic processes combined. In the first technology, the CO_2 desublimates into the fins heat exchanger surface to be finally separated at high pressure into the liquid phase. In the second technology, the CO_2 desublimates in contact with the packing material and then it is removed as a gaseous product.

One favorable characteristic is this technology is avoiding chemical agents to remove CO_2. Furthermore, it is possible to scale up the process to be used in the industries with large volumes of treating gas streams. From this process, the CO_2 is recovered as a liquid, resulting in an economical way to transport or pump the gas to its final disposition, for example, for the injection to enhanced coal bed methane (ECBM) or enhanced oil recovery (EOR).

The main advantage of the cryogenic process is the high CO_2 recovery, with around 99.95% [28]. Another advantage of this process is that the product is obtained at low temperatures in order to be reused as a cold energy source, for example, to be used for mechanical cleaning [17].

One of the biggest problems in the process is the high energy consumption to maintain the low temperature, which makes the process to turn cost noneffective. An additional operational problem is the CO_2 solidification at very low temperatures. Another cause of obstruction in the pipe is the presence of H_2O [28]. Another deficiency in the process is the presence of impurities like oxides of nitrogen and sulfur, water, and oxygen. These components need to be treated before the removal of CO_2.

As mentioned before, the operative cost can result in a serious problem for this kind of technology. Principally, the presence of water can cause a rise in pressure loss or plugging because of the ice formation. For this reason, it is necessary to implement some steps for the removal of water traces from the fuel/flue gas stream, resulting in a high investment and operative costs. Moreover, the increasing formation of a layer of CO_2 on the surface of the heat exchanger will affect the mass transfer, reducing the performance and the efficiency of the entire process [29].

The separation of CO_2 by using a cryogenic fractionation technology is simulated by Aspen HYSYS, for this purpose in the work of Liu et al. [14]. For this purpose, authors considered 100,000 Nm^3/d of natural gas with 81.4% CO_2 concentration. In this work, the main factors analyzed include the column pressure, condensation, and reboiling temperatures, and the concentration of CO_2. All these factors could affect the system cooling consumption. As a result of the simulation, the CO_2 purity is over 95%, while the recovery is around 90%, for high CO_2 concentrations. Some effects of the parameters on the complete system are noted, like a significant cooling consumption, and reasonable values are obtained. These results will improve the application of this kind of technology for CO_2 separation on a large scale. The flow sheet of the proposed configuration is presented in Fig. 9.5.

The importance of this work relies on the fact that the energy analysis provides some recommendations for the use of cryogenic fractionation at different industrial levels. Despite these advantages, several challenges exist, such as the effect of the liquefaction temperature and the treatment of noncondensable gas products.

Table 9.3 presents the application, energy consumption, cost operation, and produced CO_2 purity for the cryogenic separation based on multicompression stages with intercoolers.

3.3 External cooling loop cryogenic for CO_2 capture

The external cooling loop cryogenic is another technology used for CO_2 capture (CCCECL); it operates at cryogenic temperatures, forming solid particles of the acid gas. CCCECL is a continuous process to remove carbon dioxide in a flue gas stream and it is proven to recover over 99%, with low energy consumption of around 0.74 MJe/kg CO_2 [31]. In this technology, water must be removed before reaching the freezing temperature. After that, the flue gas needs to be cooled near 150 K; this operation occurs in a heat exchanger where the CO_2 is desublimated. Then, solid CO_2 forms in a slurry, making possible its separation from the

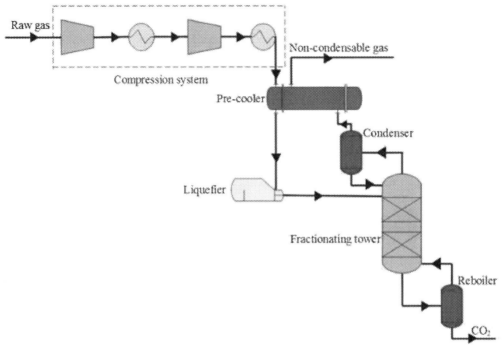

Figure 9.5 Cryogenic distillation process [14].

Table 9.3 Cryogenic separation based on multicompression stages with intercoolers: application, energy consumption, cost operation, and produced CO_2 purity [27,30].

	Application	Energy consumption	Cost operation	CO_2 purity
Cryogenic separation based on multicompression stages with intercoolers	CO_2 separation from treating gas concentration over 60%	1.9 MJ/kg CO_2	8.6 k€/ (m³/h)	High purity (>90%) 99.92% purity of CO_2 in the liquid phase

liquid phase and melting it under pressure, reaching atmospheric temperature.

There are various advantages present in this kind of system. Some of these consist of the absence of noxious chemicals, low energy requirement, energy storage potential, a simple technology, representing a low cost, multipollutant capture, and others. the main disadvantage is that CO_2 solids formation causes technical problems in this technology [31].

As mentioned by Ref. [32], the reuse of cold waste energy represents a common alternative to reduce energy consumption in industries. Some possibilities to reuse cold energy are an inertial carbon extraction system, a thermal swing process, and/or an external cooling loop. Fig. 9.6 illustrates a combined cryogenic carbon dioxide separation flowsheet with an external cooling loop (CCCECL).

In the CCCECL process, flue gas from existing systems is dried, compressed, and cooled to a temperature slightly above the CO_2 solidification point, then the gas is expanded to further cool it, while an amount of CO_2 as a solid is precipitated. Then, the gas is pressurized and reheated via the heat exchange with the incoming gases. At the end of the process, the CO_2 is separated by liquid CO_2, and a N_2-rich stream is discharged [32].

The CCCECL process has some configurations that can store energy in the form of LNG. The capability to manage the energy loss of CO_2 removal by using a stored refrigerant and driving to the process during peak demand and transferring the parasitic load to the grid is the main advantage to help meet demands

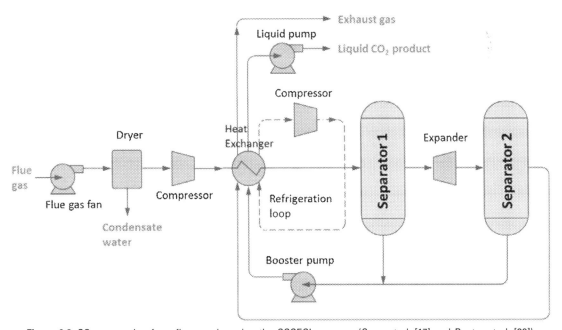

Figure 9.6 CO_2 separation from flue gas by using the CCCECL process (Song et al. [17] and Baxter et al. [32]).

Table 9.4 CCCECL technology: application, energy consumption, and produced CO₂ purity [32,33].

	Application	Energy consumption	CO$_2$ purity
External cooling loop in the cryogenic CO$_2$ separation.	Reusing some residual cold waste energy is a practical approach to reduce energy requirement. Water demand is reduced by 25%–30%.	0.98 MJelectrical/ kg CO$_2$	99% of the CO$_2$ at −135°C and 90% at −120°C

CCCECL, external cooling loop cryogenic carbon capture.

and regenerate the refrigerant during low-demand periods [32]. In addition, the rapid load change capability of CCCECL is beneficial in order to integrate conventional power generation systems with renewable intermittent power sources. CCCECL energy requirement is less energy demanding than other conventional processes, with an estimated rate of 0.98 MJelectrical/kg CO$_2$ [33].

Table 9.4 presents the application, energy consumption, and produced CO$_2$ purity by using the CCCECL technology.

3.4 Cryogenic liquid

This technology is presented by Fazlollahi et al. [34]. It puts in contact the flue gas, which contains the CO$_2$, with the cryogenic liquid. The process causes the formation of solid particles of CO$_2$ in different stages of a desublimation column. After that, the produced CO$_2$ is sent to a filtration stage to obtain purified carbon dioxide. A contact liquid is used to improve CO$_2$ separation and avoid some issues with this cryogenic process. One of the principal characteristics of these systems is having low vapor pressures to decrease losses through evaporation. The process consists of the contact of both phases and the accumulation of the CO$_2$ layer solid. Then the solid–liquid phase is sent to a separator, which is constituted by a continuous filter press. Thereafter, the contact liquid is recooled in a closed loop refrigerator step to be reused in the process again.

Table 9.5 presents the application, cost investment, cost operation, and produced CO$_2$ purity for the cryogenic liquid technology.

Table 9.5 Cryogenic liquid technology: application, cost investment, cost operation, and produced CO_2 purity [34].

Application		Cost investment	Cost operation	CO_2 purity
Cryogenic liquid	The use of a contact liquid is to prevent solid CO_2 formation on the surface of the equipment. This liquid once utilized is reconditioned for entering again in the process.	657,190 USD (1000 kmol/h—feed stream)	38,146 $/h	99% CO_2 capture

3.5 Heat exchangers

This technology is presented by Popov et al. [35]. Heat exchangers constitute an elementary unit in the cryogenic process. In the cryogenic process, there are three remarkable steps: (1) the cooling of the feed gas; (2) the process of distillation and CO_2 separation; and (3) liquefaction for the final disposal. The principal advantages of the use of heat exchangers are in their design. This equipment can operate with high effectiveness and achieve good performance. Due to the high energy requirement in the process, this equipment requires large heat transfer areas.

Cryogenic heat transfer could be applied in an extensive range of applications. To consider an economical operation, this heat exchanger needs to get an effectiveness of around 95% or even higher. This technology is up-and-coming because the involved cost in the process seems to be less than 30% compared to other similar technologies for capturing CO_2 [35]. Heat exchangers could represent at least 20%—30% of the total cost investment in some plants, for example, in air separation or natural gas treatment [36].

4. Current applications and improvements

4.1 CryoCell process

Cool Energy Ltd. developed another technology used for CO_2 capture, the CryoCell process. This CryoCell was tested in collaboration with other industrial partners such as Shell Global Solutions in Western Australia [3]. Fig. 9.7 presents a basic CryoCell design. Initially, it is necessary to dehydrate the feed gas stream to water contents of around 5 ppm. The dried gas then is sent to

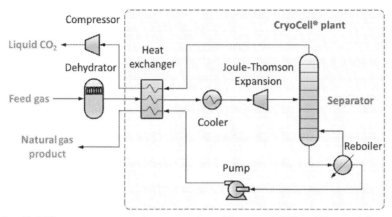

Figure 9.7 CryoCell CO_2 capture process from natural gas (Song et al. [17] and Hart and Gnanendran [3]).

a heat exchanger and a cooler where it is precooled and cooled to its freezing temperature. After that, a Joule–Thomson valve is used to expand the liquid that is sent to the CryoCell separation unit as a three-phase mixture. In the separator, the solid CO_2 is collected at the bottom, melted by a heater, and separated as a liquid phase. The gas from the top is compressed to achieve sale gas specifications, and the liquid is pumped to the required disposal pressure.

4.2 Cryogenic–membrane hybrid system case

Some conventional technologies to separate CO_2 present bottlenecks, and to overcome this, the combination of two or more single CO_2 separations methods is used to create the hybrid processes. Some of the already studied hybrid systems are amine-based absorption with polymeric membrane modules, adsorption columns with polymeric membrane modules, cryogenic separation columns with adsorption columns, and cryogenic separation columns with polymeric membrane modules [17].

Hybrid processes based on cryogenic technology are promising alternatives in CO_2 capture because of the product obtained, CO_2 in a liquid phase, and the low energy requirements of 1.163 MJ/kg CO_2 [17].

The combination of two or more single CO_2 capture methods attracts increasing attention as they can improve the performance of the individual processes [37]. In this case, combining the cryogenic and membrane technologies provides a better separation performance for feed gases with high CO_2 content and the purity required in the outlet stream is significantly elevated.

Gutierrez et al. [1] presents a steady-state simulation of the process of CO_2 removal from natural gas where a similar membrane module is used. In the work, authors present a detailed discussion of the process of chemical absorption of CO_2 separation by using methyldiethanolamine. This type of membrane module can be integrated into a cryogenic fractionation to optimize the entire system and improve the CO_2 purity and consumed energy in the reboiler of the distillation tower.

Song et al. [38] presents the steady-state simulation of a membrane based in combination with a cryogenic process at low temperature. In this technology, a reduction of 1.7 MJ/kg CO_2 in a standard coal-fired power plant represents a performance enhancement.

Fig. 9.8 shows a hybrid system: cryogenic distillation and membrane separation. In this process, the stream to be separated is sent to a cooler— compression system, where the manipulation of some parameters such as the temperature and pressure permits to obtain liquefied CO_2 [28]. Then the separation occurs in the cryogenic system, where the cryogenically cooled feed is introduced in the distillation column. The contact between both vapor and liquid phases occurs in some trays, and this number depends on the required purity grade [11].

This operation produces CH_4, which is obtained at the top column and liquid CO_2 recollected from the bottom. Then the membrane technology is used to improve the separation and the purity of the desired streams. In this case, it is possible the use of polymeric membranes, where the selectivity defines the ability to interact between the participant molecules [28]. Rich CO_2 from

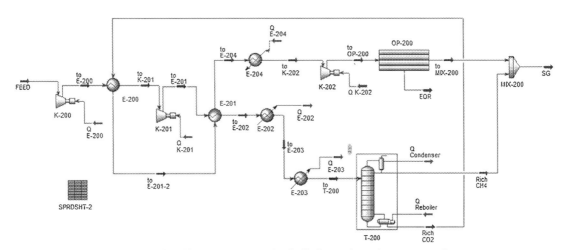

Figure 9.8 Hybrid system: cryogenic distillation and membrane separation.

the column bottom is recycled and returned to the membrane module, where two streams are obtained—one stream rich in CH_4 (to MIX-200), and the other rich CO_2 stream (EOR).

5. Conclusion and future outlooks

Cryogenic CO_2 separation technology seems to improve the energy efficiency of comparable carbon dioxide capture systems. In cryogenic CO_2 processes, some units consume intensive energy, like the compressors and pumps, as well as multistream heat exchangers for CO_2 condensation and liquefaction. However, the investment costs of cryogenic systems are lower in comparison to single absorption or adsorption processes due to their smaller equipment sizes. In addition, this implies the magnitude of the cryogenic separation processes to be smaller in terms of installation sizes. Meanwhile, the operating costs of cryogenic processes seem to be lower than that required for traditional technologies because no chemicals and solvents are employed.

Some optimization researchers proposed different ideas for the improvement of CO_2 separation technologies. New ideas for the optimization of such processes include the combination of cryogenic distillation and membrane separation, proving a low energy penalty. Characteristics of membrane systems must be defined considering the composition of the gas to feed, the purity of the products, the availability of energy supplies, and the volumes to process.

Abbreviations and symbols

ASU   Air separation units
CAPEX   Capital expenditure
CCCECL   External cooling loop cryogenic carbon capture
ECBM   Enhanced coal bed methane
EOR   Enhanced oil recovery
EoS   Equation of state
HC   Hydrocarbon
IEA   International Energy Agency
IPCC   Intergovernmental Panel on Climate Change
LNG   Liquefied natural gas
MEA   Monoethanolamine
NG   Natural gas
OPEX   Operational expenditure
P–T   Pressure–temperature
S–V   Solid–vapor
V–L   Vapor–liquid
V–S   Vapor–solid
VPSA   Vacuum pressure swing adsorption

References

[1] Gutierrez JP, Ruiz ELA, Erdmann E. Energy requirements, GHG emissions and investment costs in natural gas sweetening processes. Journal of Natural Gas Science and Engineering 2017;38:187−94.

[2] Martínez MJ. Ingeniería de Gas, Principios y Aplicaciones. Endulzamiento del Gas Natural. Venezuela: Ingenieros Consultores SRL. Maracaibo; 2000.

[3] Hart A, Gnanendran N. Cryogenic CO2 capture in natural gas. Energy Procedia 2009;1(1):697−706.

[4] Gutierrez JP, Benitez LA, Ruiz ELA, Erdmann E. A sensitivity analysis and a comparison of two simulators performance for the process of natural gas sweetening. Journal of Natural Gas Science and Engineering 2016;31:800−7.

[5] De Guido G, Pellegrini LA. Calculation of solid-vapor equilibria for cryogenic carbon capture. Computers and Chemical Engineering 2022;156: 107569. https://doi.org/10.1016/j.compchemeng.2021.107569. 2022.

[6] de Miranda Pinto JT, Mistage O, Bilotta P, Helmers E. Road-rail intermodal freight transport as a strategy for climate change mitigation. Environmental Development 2018;25:100−10. https://doi.org/10.1016/j.envdev.2017.07.005.

[7] IPCC. Summary for policymakers. Global warming of 1.5°C. An IPCC special report on the impacts of global warming of 1.5°C above pre-industrial levels and related global greenhouse gas emission pathways. In: The context of strengthening the global response to the threat of climate change, sustainable development, and efforts to eradicate poverty; 2018.

[8] Dai Z, Ansaloni L, Deng L. Recent advances in multi-layer composite polymeric membranes for CO2 separation: a review. Green Energy and Environment 2016;1:102−28. https://doi.org/10.1016/j.gee.2016.08.001.

[9] Rezakazemi M, Li J. Post-combustion CO2 capture with sweep gas in thin film composite (TFC) hollow fiber membrane (HFM) contactor. Journal of CO2 Utilization 2020;40. https://doi.org/10.1016/j.jcou.2020.101266.

[10] Spitoni M, Pierantozzi M, Comodi G, Polonara F, Alessia A. Theoretical evaluation and optimization of a cryogenic technology for carbon dioxide separation and methane liquefaction from biogas. Journal of Natural Gas Science and Engineering 2019;62:132−43. https://doi.org/10.1016/j.jngse.2018.12.007.

[11] Holmes AS, Ryan JM. U.S. Patent No. 4,318,723. Washington, DC: U.S. Patent and Trademark Office; 1982.

[12] Tuinier MJ, Annaland MVS, Kuipers JAM. A novel process for cryogenic CO2 capture using dynamically operated packed beds-an experimental and numerical study. International Journal of Greenhouse Gas Control 2011;5: 694−701.

[13] Tuinier MJ, Hamers HP, Van Sint Annaland M. Techno-economic evaluation of cryogenic CO2 capture—a comparison with absorption and membrane technology. International Journal of Greenhouse Gas Control 2011;5:1559−65.

[14] Liu B, Zhang M, Yang X, Wang T. Simulation and energy analysis of CO2 capture from CO2-EOR extraction gas using cryogenic fractionation. Journal of the Taiwan Institute of Chemical Engineers 2019;103:67−74.

[15] Babar M, Bustam MA, Ali A, Maulud AS, Shafiq U, Mukhtar A, et al. Thermodynamic data for cryogenic carbon dioxide capture from natural gas: a review. Cryogenics 2019;102:85−104.

[16] Font-Palma C, Cann D, Udemu C. Review of cryogenic carbon capture innovations and their potential applications. C—Journal of Carbon Research 2021;7(3):58.

[17] Song C, Liu Q, Deng S, Li H, Kitamura Y. Cryogenic-based CO2 capture technologies: state-of-the-art developments and current challenges. Renewable and Sustainable Energy Reviews 2019;110. https://doi.org/10.1016/j.rser.2018.11.018.
[18] Brunetti A, Scura F, Barbieri G, Drioli E. Membrane technologies for CO2 separation. Journal of Membrane Science 2010;359:115–25.
[19] Berstad D, Anantharaman R, Neks P. Low-temperature CO2 capture technologies – applications and potential. International Journal of Refrigeration 2013;36:1403–16.
[20] Tuinier MJ, Annaland MVS, Kramer GJ, Kuipers JAM. Cryogenic CO2 capture using dynamically operated packed beds. Chemical Engineering and Science 2010;65:114–9.
[21] Babar M, Mukhtar A, Mubashir M, Saqib S, Ullah S, Hassan A, et al. Development of a novel switched packed bed process for cryogenic CO2 capture from natural gas. Process Safety and Environmental Protection 2021;147:878–87. https://doi.org/10.1016/j.psep.2021.01.010.
[22] Tuinier MJ, Annaland MVS. Biogas purification using cryogenic packed-bed technology. Industrial and Engineering Chemistry Research 2012;51:5552–8.
[23] Abatzoglou N, Boivin S. A review of biogas purification processes. Biofuels, Bioproducts and Biorefining 2009;3:42–71.
[24] Belaissaoui B, Le Moullec Y, Willson D, Favre E. Hybrid membrane cryogenic process for post-combustion CO2 capture. Journal of Membrane Science 2012;415:424–34.
[25] Meratla Z. Combining cryogenic flue gas emission remediation with a CO2O2 combustion cycle. Energy Conversion and Management 1997;38: S147–52.
[26] Zanganeh KE, Shafeen A, Salvador C. CO2 capture and development of an advanced pilot-scale cryogenic separation and compression unit. Energy Procedia 2009;1(1):247–52.
[27] Maqsood K, Ali A, Shariff A, Ganguly S. Synthesis of conventional and hybrid cryogenic distillation sequence for purification of natural gas. Journal of Applied Sciences 2014;14(21):2722–9.
[28] Aaron D, Tsouris C. Separation of CO2 from flue gas: a review. Separation Science and Technology 2005;40(1–3):321–48. https://doi.org/10.1081/SS-200042244.
[29] Mukhtar A, Saqib S, Mellon NB, Babar M, Rafiq S, Ullah S, Chawla M. CO2 capturing, thermo-kinetic principles, synthesis and amine functionalization of covalent organic polymers for CO2 separation from natural gas: a review. Journal of Natural Gas Science and Engineering 2020;77:103203.
[30] Shen M, Tong L, Yin S, Liu C, Wang L, Feng W, et al. Cryogenic technology progress for CO2 capture under carbon neutrality goals: a review. Separation and Purification Technology 2022:121734. https://doi.org/10.1016/j.seppur.2022.121734.
[31] Jensen MJ, Russell CS, Bergerson D, Hoeger CD, Frankman DJ, Bence CS, et al. Prediction and validation of external cooling loop cryogenic carbon capture (CCC-ECL) for full-scale coal-fired power plant retrofit. Faculty Publications; 2015. 1731, https://scholarsarchive.byu.edu/facpub/1731.
[32] Baxter L, Baxter A, Burt S. Cryogenic CO2 capture as a cost-effective CO2 capture process. In: Proceedings of the international Pittsburgh coal conference, Pittsburgh, PA, USA; September 20–23, 2009; 2009.
[33] Safdarnejad SM, Hedengren JD, Baxter LL. Plant-level dynamic optimization of cryogenic carbon capture with conventional and renewable power

sources. Applied Energy 2015;149:354−66. https://doi.org/10.1016/j.apenergy.2015.03.100.

[34] Fazlollahi F, Saeidi S, Safdari MS, Sarkari M, Klemeš JJ, Baxter LL. Effect of operating conditions on cryogenic carbon dioxide removal. Energy Technology 2017;5(9):1588−98. https://doi.org/10.1002/ente.201600802.

[35] Popov D, Fikiin K, Stankov B, Alvarez G, Youbi-Idrissi M, Damas A, Brown T. Cryogenic heat exchangers for process cooling and renewable energy storage: a review. Applied Thermal Engineering 2019;153:275−90. https://doi.org/10.1016/j.applthermaleng.2019.02.106.

[36] Coyle DA, Durr CA, Hill DK. Cost optimization, the contractor's approach. In: International conference on liquefied natural gas, Perth, Australia; May 1998. p. 4−7.

[37] Younas M, Rezakazemi M, Daud M, B Wazir M, Ahmad S, Ullah N, et al. Recent progress and remaining challenges in post-combustion CO_2 capture using metal-organic frameworks (MOFs). Progress in Energy and Combustion Science 2020;80. https://doi.org/10.1016/j.pecs.2020.100849.

[38] Song C, Liu Q, Ji N, Deng S, Zhao J, Li Y, et al. Reducing the energy consumption of membrane-cryogenic hybrid CO_2 capture by process optimization. Energy 2017;124:29−39.

Further reading

[1] Xu G, Li L, Yang Y, Tian L, Liu T, Zhang K. A novel CO_2 cryogenic liquefaction and separation system. Energy 2012;42:522−9. https://doi.org/10.1016/j.energy.2012.02.048.

10

Absorption processes for CO_2 removal from CO_2-rich natural gas

Ali Behrad Vakylabad

Department of Materials, Institute of Science and High Technology and Environmental Sciences, Graduate University of Advanced Technology, Kerman, Iran

1. Introduction

Absorption processes for CO_2 removal from CO_2-rich natural gas have become increasingly important due to the growing concerns about climate change and the need to reduce greenhouse gas emissions [1–5]. Carbon capture and storage (CCS) technologies, including absorption processes, are seen as a potential solution to mitigate the impact of CO_2 emissions from various industrial processes [6,7]. The CO_2-rich natural gas is typically produced from sources such as shale gas, coal-bed methane, and biogas [8]. The gas typically contains high levels of CO_2, which must be removed before it can be used for various applications [9].

Absorption processes are one of the most effective methods for removing CO_2 from natural gas. Absorption processes work by using a solvent to capture the CO_2 from the gas stream. The solvent is then regenerated by heating it, releasing the CO_2, and allowing the solvent to be reused. The most common solvents used for CO_2 absorption are aqueous solutions of amines, such as monoethanolamine (MEA), diethanolamine (DEA), and methyl diethanolamine (MDEA). The absorption process involves passing the CO_2-rich gas through an absorption column where the solvent is introduced [10,11]. The solvent absorbs the CO_2 from the gas, producing a CO_2-rich solvent stream. The solvent is then sent to a regeneration column where it is heated to release the CO_2. The CO_2 is then compressed and stored or transported for further use or disposal. Here is a detailed tabulated table that describes the various processes and equipment used in a typical plant for CO_2 removal from natural gas (Table 10.1) [9,25].

One of the advantages of absorption processes is their ability to capture CO_2 from gas streams with high concentrations of CO_2 [26]. This makes them particularly suitable for use in CO_2-rich natural gas streams. Additionally, absorption processes are well-established and have been used in various industrial applications for many years [20]. However, there are some disadvantages to absorption processes. One of the main drawbacks is the high energy consumption required for the regeneration of the solvent. The process also requires a large amount of solvent, which can be expensive and difficult to handle. The solvent can also degrade

Table 10.1 Typical unit operations and equipment of CO_2 absorption from natural gas.

Process step	Equipment used	Description	References
Gas treatment	Inlet separator	Separates any liquids or solids from the incoming natural gas stream	[12]
	Gas filter	Removes any solids or fine particles from the gas stream	[13]
	Gas heater	Raises the temperature of the gas stream to the desired level	[14]
Dehydration	Glycol dehydrator	Removes any water vapor from the gas stream	[15]
CO_2 removal	Absorber column	Uses an amine solvent to selectively absorb CO_2 from the gas stream	[13]
	Rich solvent tank	Collects the solvent rich in CO_2 from the absorber column	[16]
	Heat exchanger	Preheats the lean solvent before it enters the absorber column	[17]
Regeneration	Stripper column	Uses heat and/or reduced pressure to remove CO_2 from the rich solvent	[18]
	Reboiler	Provides heat to the stripper column to facilitate CO_2 removal from the solvent	[19,20]
	Lean solvent tank	Collects the solvent stripped of CO_2 from the stripper column	[21]
Acid gas treatment	Scrubber column	Removes impurities such as hydrogen sulfide (H_2S) and mercaptans from the acid gas stream	[12]
	Amine regenerator	Regenerates the amine solvent used in the scrubber column	[18]
	Acid gas flare	Disposes of the treated acid gas by burning it	[22]
CO_2 product	CO_2 compressor	Compresses the CO_2 product to the desired pressure	[23]
	CO_2 storage	Stores the compressed CO_2 for further use or transport	[24]

Note: This table is not exhaustive and does not include every possible piece of equipment or process used in CO_2 removal from natural gas. The specific equipment used and the sequence of processes may vary depending on the plant design, operating conditions, and other factors.

over time, leading to decreased efficiency and increased maintenance costs [27,28]. Despite these challenges, absorption processes remain a promising technology for CO_2 removal from CO_2-rich natural gas [29,30]. Ongoing research and development efforts aim to improve the efficiency and reduce the cost of these processes, making them more competitive with other carbon capture technologies. As the world continues to focus on reducing greenhouse gas emissions, absorption processes are likely to play an increasingly important role in achieving these goals [31,32].

1.1 Advances in the absorption of CO_2

New advanced ideas and technologies are being developed for the absorption of CO_2 from CO_2-rich natural gas [33,34]. Some of these technologies include:

- Membrane-based separation: This technology uses thin membranes to separate CO_2 from natural gas. The membranes have tiny pores that allow only CO_2 molecules to pass through while blocking other gases. This process is energy-efficient and has low operating costs [35].
- Chemical absorption: This process involves using chemicals to absorb CO_2 from natural gas. The absorbed CO_2 is later released and captured for storage. This technology is widely used in industrial applications and has been proven to be effective [33,36,37].
- Cryogenic separation: This process involves cooling the natural gas to very low temperatures to separate CO_2 from the gas. The CO_2 is then captured and stored. This technology is energy-intensive but has high efficiency [33].
- Adsorption: This process involves using solid materials to adsorb CO_2 from natural gas. The adsorbed CO_2 is later released and captured for storage. This technology is still in the research and development phase but has the potential to be a low-cost and energy-efficient solution [36].

In brief, these advanced technologies offer promising solutions for the absorption of CO_2 from CO_2-rich natural gas. The most widely used industrial process for CO_2 absorption from natural gas is amine-based carbon capture technology [38]. This process involves using a solution of amines (organic compounds that contain nitrogen) to capture the CO_2 from the natural gas stream. The amine solution reacts with the CO_2 to form a stable compound that can be separated from the natural gas. This process has been implemented on a large scale in various industries such as oil refining, chemical production, and natural gas processing [39].

1.2 Natural gas purification and processing

Natural gas is considered a cleaner energy source for heating, cooking, and electricity generation. Natural gas can also be used as fuel for transportation and as a chemical feedstock for manufacturing plastics and other commercially important chemicals. It is generally transported via pipelines or in the form of liquefied natural gas (LNG) [40–42]. Various technologies have been utilized to improve the fuel properties of natural gas by eliminating CO_2, such as cryogenic separation, membrane separation, and adsorption. Cryogenic separation involves cooling natural gas to such a low temperature that CO_2 becomes liquefied and separated. According to Li et al. [43], this method can decrease CO_2 concentration by up to 85% by maintaining the temperature at $-60°C$. However, this process has high investment and operational costs due to the necessary heat duty and special equipment. Membrane separation, on the other hand, provides a secure and effective separation performance based on the membrane's selective permeability. However, the membrane has a limited lifespan, and replacing it is expensive. Additionally, hydrocarbons can also seep through the membrane, resulting in hydrocarbon losses. To eliminate CO_2 through adsorption, porous solid materials like activated carbon, zeolites, silica gels, and carbon molecular sieves are utilized as solid adsorbents to isolate CO_2 from gas mixtures. As solid adsorbents have limited capacity, they require frequent regeneration, making pressure swing adsorption a common process for continuous operation. These methods have been developed to enhance the economic viability of the process [44–46].

In the process of capturing CO_2, chemical absorption is commonly used. This method involves a chemical reaction between CO_2 and an absorbent in liquid form. Various chemical solutions can be utilized as absorbents, such as MEA, DEA, MDEA, ammonia (NH_3), and sodium hydroxide (NaOH). Yincheng et al. achieved a CO_2 removal efficiency of over 90% using aqueous ammonia (NH_3) and sodium hydroxide (NaOH) as absorbents. However, NaOH can cause equipment corrosion, while NH_3 requires low temperatures to reduce vapor losses. Although MEA is a popular solvent due to its high reactivity and CO_2 absorption rate, it is expensive and potentially harmful to the environment. Therefore, it is necessary to consider cost-effective and eco-friendly processes that use alternative absorbents [47–49].

Water is a widely used and environmentally friendly solvent for eliminating CO_2 through a process known as water scrubbing. This method is recommended as an affordable and eco-friendly way to absorb CO_2, although its performance in CO_2 absorption is not particularly high. However, by introducing microchannels as an absorber, the process can be enhanced by generating larger interfacial areas, which promote the transfer of CO_2 from gas to liquid. The use of water or chemical solutions to remove CO_2 from natural gas can result in additional costs for offshore natural gas plants, depending on their location. Seawater, which is easily accessible in coastal regions and offshore, is a promising absorbent due to its richness in alkali ions such as Na^+, Mg^{2+}, and Ca^{2+}, which react directly with CO_2. Therefore, it has the potential to be an economical and environmentally friendly solvent [50,51].

To achieve high removal efficiency of CO_2 using seawater, a rather high flow rate of seawater is necessary, which may require a flow rate 250 times greater than that of the amine solution. Li et al. used steel slag containing metal oxides like CaO and MgO to enhance CO_2 removal using seawater. The dissolution of these components increased the alkalinity of seawater, which helped in carbonic acid ionization, and thereby increased the concentration of ionized CO_2 species such as CO_3^{2-} and HCO_3^-. However, Ca^{2+} and Mg^{2+} in seawater led to the precipitation of metal carbonates like $CaCO_3$ and $MgCO_3$. The CO_2 removal efficiency was more than 95%. But, the low concentration of CO_2 in the feed gas (15 %v/v) was the limitation of this application. Therefore, we improved this technology further in our work to tackle the high concentration of CO_2 in the feed stream using microreactor technology [51,52].

Effective removal of CO_2 through gas absorption largely relies on the interfacial areas between gas and liquid phases. Different types of equipment such as spray columns, packed columns, and wetted wall columns, each with a unique design, are available for CO_2 absorption. Among these, spray columns and packed columns can produce a relatively large gas—liquid interfacial area. Microchannels can also be used as microcontactors to increase the efficiency of CO_2 absorption because of their high surface-to-volume ratios. Kittiampon et al. [53] conducted a study using aqueous ammonia for CO_2 absorption in a microchannel and achieved a high overall mass transfer coefficient and removal efficiency of 96.6%. The system can be operated continuously, and the production capacity can be adjusted using the numbering up technique. Microchannel has also been utilized as a gas—liquid contactor for water scrubber applications, demonstrating high CO_2 removal efficiency and mass transfer coefficient [54—58].

This chapter discusses the absorption processes which are commonly used for the removal of CO_2 from CO_2-rich natural gas streams. These processes involve the use of solvents such as MEA, DEA, and MDEA to selectively absorb CO_2 from the gas stream. The choice of solvent depends on various factors such as the nature of the gas stream, the required purity level, and the operating conditions. There has been significant progress and development of new technologies and solvents for CO_2 absorption and removal from natural gas streams rich in CO_2. These include new solvents with higher CO_2 absorption capacity and selectivity, faster kinetics, and lower regeneration energy needs; mixed solvents; aqueous solvents based on alkali or alkaline earth hydroxides; membrane-based approaches; and new reactor and process configurations. The collected CO_2 can be used in a variety of applications, including enhanced oil recovery, CCS, chemical production, the food and beverage industry, medical applications, agriculture, and renewable energy.

2. Amine absorption

Fig. 10.1 is a graphical view of a CO_2 removal plant using amine absorption.

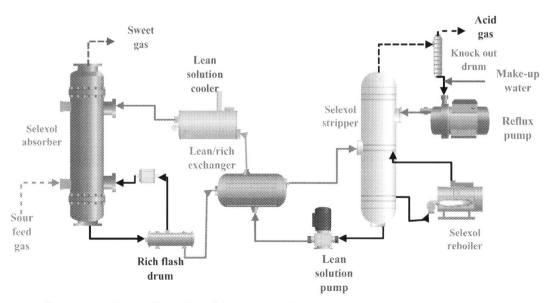

Figure 10.1 A diagram illustrating a CO_2 removal facility that utilizes amine absorption technology.

2.1 Plant details

Selexol is a registered trademark for a solvent that is used in the Selexol process, which is a method for removing CO_2 from gas streams. The Selexol process is a type of amine absorption process, which means that it uses an amine solvent to absorb the CO_2 from the gas stream [59]. The Selexol absorber is the heart of the Selexol process. The absorber is a tall column that is filled with packing material. The gas stream enters the bottom of the column and flows upward, while the lean solvent enters the top of the column and flows downward. The CO_2 in the gas stream is absorbed by the lean solvent, and the rich solvent leaves the bottom of the column. The rich solvent is then heated to regenerate it, and the CO_2 is released. The CO_2 is then compressed and stored. The Selexol process is a very effective method for removing CO_2 from gas streams. It is also a very reliable process, and it can be used to remove CO_2 from a wide variety of gas streams. The Selexol process is a mature technology, and it has been used in commercial applications for many years [60].

Here are some of the advantages of using Selexol in a CO_2 removal plant:
- Selexol is a very effective solvent for CO_2 absorption.
- Selexol is a very reliable solvent, and it can be used to remove CO_2 from a wide variety of gas streams.
- Selexol is a mature technology, and it has been used in commercial applications for many years.

Here are some of the disadvantages of using Selexol in a CO_2 removal plant:
- Selexol is a relatively expensive solvent.
- Selexol can be corrosive, so it is important to use the correct materials for construction in a Selexol plant.
- Selexol can be flammable, so it is important to take safety precautions when operating a Selexol plant.

All in all, the Selexol process is a very effective and reliable method for removing CO_2 from gas streams. It is a mature technology that has been used in commercial applications for many years [61].

The knock-out drum is typically located between the gas pretreatment unit and the amine absorber. The gas stream enters the knock-out drum from the gas pretreatment unit and flows upward. Any liquids that are present in the gas stream will collect at the bottom of the knock-out drum. The gas stream will then flow out of the top of the knock-out drum and into the amine absorber. The knock-out drum is an important part of a CO_2 removal plant using amine absorption. It helps to protect the amine solvent and improve the efficiency of the CO_2 removal process. A knock-out drum is a type of vessel that is used to separate liquids from gases. In a CO_2 removal plant using amine absorption, the knock-out drum is used to separate any liquids that may be present in the gas stream before it enters the amine absorber. This is important because liquids can damage the amine solvent and reduce the efficiency of the CO_2 removal process [62].

Here are some of the advantages of using a knock-out drum in a CO_2 removal plant:
- It protects the amine solvent from damage by liquids.
- It improves the efficiency of the CO_2 removal process.
- It is a relatively simple and inexpensive device.

Here are some of the disadvantages of using a knock-out drum in a CO_2 removal plant:
- It can add to the complexity of the plant.
- It can require additional maintenance.
- It can add to the cost of the plant.

On the whole, the use of a knock-out drum in a CO_2 removal plant is a good way to protect the amine solvent and improve the efficiency of the CO_2 removal process.

A rich flash drum is a vessel used in amine absorption processes to separate hydrocarbons from the rich amine solution. The rich amine solution is pumped from the absorber to the rich flash drum, where it is partially vaporized. The hydrocarbons in the rich amine solution vaporize and are drawn off from the top of the drum, while the lean amine solution is drawn off from the bottom of the drum and recycled back to the absorber. The rich flash drum is an important part of the amine absorption process because it helps to remove hydrocarbons from the amine solution. Hydrocarbons can foul the regenerator and reduce the efficiency of the process. The rich flash drum is typically a vertical cylindrical vessel with a conical bottom. The vessel is equipped with a feed inlet, a draw-off outlet for the lean amine solution, and a vent outlet for the hydrocarbons. The rich amine solution from the absorber enters the bottom of the vessel and flows upward. As the solvent flows upward, it is partially vaporized. The hydrocarbons in the solvent vaporize and are drawn off from the top of the vessel. The lean amine solution is then drawn off from the bottom of the vessel and recycled back to the absorber. The rich flash drum is an important part of the amine absorption process. It helps to remove hydrocarbons from the amine solution and improve the efficiency of the process.

A lean solution cooler is a type of heat exchanger that is used to cool the lean amine solution in a CO_2 removal plant using amine absorption. The lean amine solution is heated up in the absorber as it absorbs CO_2 from the gas stream. The lean solution cooler helps to cool the lean amine solution back down to its original temperature before it is recycled back to the absorber. This helps to improve the efficiency of the absorption process by ensuring that the lean amine solution can absorb more CO_2 from the gas stream.

The lean solution cooler is typically a shell and tube heat exchanger. The lean amine solution flows through the tubes of the heat exchanger, while a cooling fluid flows through the shell of the heat exchanger. The cooling fluid can be water, air, or another type of coolant. The lean solution cooler is an important part of a CO_2 removal plant using amine absorption. It helps to improve the efficiency of the absorption process and to reduce the amount of energy that is required to operate the plant. Here are some of the advantages of using a lean solution cooler in a CO_2 removal plant:
- It helps to improve the efficiency of the absorption process.
- It reduces the amount of energy that is required to operate the plant.
- It is a relatively simple and inexpensive device.

Here are some of the disadvantages of using a lean solution cooler in a CO_2 removal plant:
- It can add to the complexity of the plant.
- It can require additional maintenance.
- It can add to the cost of the plant.

Overall, the use of a lean solution cooler in a CO_2 removal plant is a good way to improve the efficiency of the process and reduce the amount of energy that is required to operate the plant.

A Selexol stripper is a type of distillation column used to recover CO_2 from a Selexol solvent. The Selexol solvent is a physical solvent that is used to absorb CO_2 from gas streams. The Selexol stripper is used to recover the CO_2 from the Selexol solvent so that it can be reused.

The Selexol stripper is typically a vertical cylindrical column with a conical bottom. The column is filled with a packing material that provides a large surface area for the CO_2 to come into contact with the Selexol solvent. The Selexol solvent is pumped from the absorber to the top of the stripper and flows downward. As the solvent flows downward, it is heated up by a reboiler. The heat causes the CO_2 to vaporize and be drawn off from the top of the column. The lean Selexol solvent is then drawn off from the bottom of the column and recycled back to the absorber.

The Selexol stripper is an important part of a CO_2 removal plant using Selexol absorption. It helps to recover the CO_2 from the Selexol solvent so that it can be reused. This helps to reduce the amount of Selexol solvent that is required, which can save money.

Here are some of the advantages of using a Selexol stripper in a CO_2 removal plant:
- It helps to recover the CO_2 from the Selexol solvent so that it can be reused.
- It reduces the amount of Selexol solvent that is required, which can save money.
- It is a relatively simple and inexpensive device.

Here are some of the disadvantages of using a Selexol stripper in a CO_2 removal plant:
- It can add to the complexity of the plant.
- It can require additional maintenance.
- It can add to the cost of the plant.

Overall, the use of a Selexol stripper in a CO_2 removal plant is a good way to recover the CO_2 from the Selexol solvent and save money.

The gas enters the plant and is first pretreated to remove any contaminants. The gas is then heated and mixed with an amine solvent. The CO_2 in the gas is absorbed by the amine solvent. The amine solvent is then heated to regenerate it, and the CO_2 is released. The CO_2 is then compressed and stored. The gas is then cooled and dehydrated, and it is then ready to be transported or used. This is just one example of a CO_2 removal plant. Many other technologies can be used to remove CO_2 from natural gas. The choice of technology will depend on the specific application and the desired level of CO_2 removal. A full flowsheet for recovering CO_2 from natural gas typically involves several processing steps. Here is a general overview of the steps involved [19,63–67].

Pretreatment: The natural gas feed stream is first pretreated to remove impurities such as water, hydrogen sulfide, and other contaminants that could potentially interfere with the CO_2 capture process. This may involve the use of filters, scrubbers, or other types of pretreatment equipment.

CO_2 absorption: The pretreated gas stream is then directed to an absorption column where a solvent, such as MEA, DEA, or MDEA, is used to capture the CO_2 from the gas. The gas is typically cooled and compressed before entering the column to increase the solubility of the CO_2 in the solvent. The solvent captures the CO_2 from the gas stream and forms a CO_2-rich solution.

CO_2 regeneration: The CO_2-rich solvent solution is then sent to a regeneration column, where the CO_2 is stripped from the solvent using steam or other means of heat. The CO_2 is then collected and separated from the solvent, while the regenerated solvent is recycled back to the absorption column to capture more CO_2.

CO_2 purification: The collected CO_2 may undergo further purification to remove any remaining impurities, such as water or nitrogen, using distillation, absorption, or membrane separation techniques.

CO_2 storage or utilization: The purified CO_2 can then be compressed and transported for storage or utilized in various industrial applications, such as enhanced oil recovery, chemical production, or food and beverage processing.

It is worth noting that the flowsheet for recovering CO_2 from natural gas involves multiple steps of gas pretreatment, CO_2 absorption, CO_2 regeneration, CO_2 purification, and CO_2 storage or utilization. The specific details of each step may vary depending on the specific technology and equipment used for CO_2 capture and processing.

2.2 CO_2 separation processes

Both chemical and physical absorptive processes are used to eliminate CO_2 from natural gas. However, the required flow rate of solvent, as well as the required energy for regeneration, is determinative in the main cost criteria for absorptive processes [68]. CO_2 gas causes the accumulation of greenhouse gases, warmer earth climate, rising sea levels, droughts, and acid rains so that the most important loss of CO_2 in the atmosphere is the greenhouse phenomenon caused by the increase in the number of greenhouse gases (SF_6, CH_4, N_2O, and CO_2). These gases in the Earth's atmosphere cause the atmosphere to hold more energy and heat and cause great damage to agriculture, groundwater, and animals. CO_2 in combination with water in favorable weather conditions causes acid rains that cause soil erosion and construction materials. Separation of sour gases, that is, carbon dioxide and hydrogen sulfide, is also important in the refining and petrochemical industries because they cause corrosion issues in the equipment and pipelines. They are known as catalyst toxins. As a result, the study of the thermodynamics of electrolyte solutions has had significant growth in the chemical, biochemistry, and environmental industries [69]. Weak electrolytes such as CO_2 and SO_2 separation of these materials require equilibrium information of vapor and liquid phases, which are used to design absorption or separation units of gases. Further, accurate equilibrium information is necessary for the optimal use of industrial solvents in petrochemical, refinery, and industrial units. In the process of sweetening natural gas by alkanol amines (contain both hydroxyl (—OH) and amino (—NH$_2$, —NHR, and —NR$_2$) functional groups on an alkane backbone [70]), first acidic gases must be dissolved in the liquid phase and then react with an amine solution. To increase the solubility of these gases in the liquid phase and the absorption towers of high pressures, there are different processes for absorbing acidic gases with ethanol amines.

2.2.1 CO$_2$ gas separation

Different methods can be used to separate carbon dioxide. Despite that, the adsorption method in liquid solvents is usually the most economical one, according to economic considerations, the absorbent solvent must have a high capacity for CO$_2$ absorption. In addition, it can be recovered and has a specific absorption rate; liquids used to absorb carbon dioxide may only physically dissolve it or may contain a substance that reacts chemically with dissolved gas. Usually, the specific absorption rate in physical solvents is much less than the specific absorption rate in chemical solvents. Therefore, economically, the use of chemical solvents is preferred.

2.2.2 Chemical absorbent

Gas is absorbed by chemical solvents, which itself is classified into four categories.

2.2.2.1 Alkanol amines

These solvents are a mixture of amine and water compounds that are used in most absorption units, while alkanol amine can also be obtained by heating chlorohydrin and ammonia solution [71,72]. Ethanol amines are also obtained from the combination of ammonia and ethylene oxide. Alkanolamines have at least one hydroxyl group and one amine group. It can be said that the hydroxyl group reduces vapor pressure and increases solubility in water. However, the amine group creates the necessary alkalinity in the aqueous solution so that the resulting solution can absorb acidic gases. One of the important points in amine units is solution storage. The cleaner the amine, the better it will work. Contaminated amine is the main cause of corrosion and other major problems that may arise in the unit.

The amines themselves are divided into four categories. Amines of the first type have an alcoholic basis and have a strong basic property and create precipitation such as MEA, which has a high absorption rate and causes corrosion. As a result, it should be low concentration. Diglycolamine (DGA) has a high absorption rate and does not have a corrosion problem at high concentrations. Therefore, it is an economical option.

Type II amines: DEA ethanolamine does not have a high absorption rate as high as amine type I. However, it has fewer corrosion problems and high viscosity. It is also selective toward H_2S. As such, it is suitable for natural gas treatment.

Type III amines: TEA ethanolamine, absorption rate is lower than amine type I and II and at high temperatures. It breaks down. Thus, it is not a good amine.

DIPA: Diisopropanolamine has a low absorption rate but does not have a corrosive problem. However, because the high freezing point causes blockages in pipes and fittings, its tendency to absorb H_2S is more. MDEA DEA works like DIPA and only absorbs H_2S

2.2.2.2 Sterically hindered amines

Amines with impeded spatial are a group of chemicals that in recent years their use in various industries as reactants in aqueous and nonaqueous solutions for CO_2 absorption has been considered. According to the definition, these compounds are amines in which a primary amine group is connected to a type III carbon or a secondary amine group is connected to a type II carbon or type III. Due to being partially shielded by neighboring groups of the nitrogen atom in the amine molecule, the sterically hindered amines (as the main solvent) can be more efficient in selective reactions with small CO_2 than larger hydrocarbon molecules. In detail, the intermediate reactions (side reactions) like the carbamate formation are significantly reduced owing to the steric hindrance to the reacting CO_2. Consequently, the reduced side reactions result in enhanced capacity up to theoretically double, as well as increased reaction rates of the acid gases with different amine molecules [73–75].

2-Amino-2-methyl-1-propanol (AMP) is a group of amines with spatial inhibition that high loading capacity for CO_2, very low corrosive effects, very limited degradation and decay of the solvent, low recycling solvent rate, low cost of equipment, and low maintenance and operational costs for this solvent can be counted [76,77]. Furthermore, AMP has sterically hindered MEA with high availability. It is obtained by replacing two methyl groups instead of two hydrogen atoms attached to carbon [78,79] (Fig. 10.2). Recently, many studies have been carried out on its reaction with acidic gases, especially CO_2 [78,80,81]. If an aqueous AMP solution is used to absorb CO_2, acidic gas can react with both the alcoholic and amino functional groups since AMP is

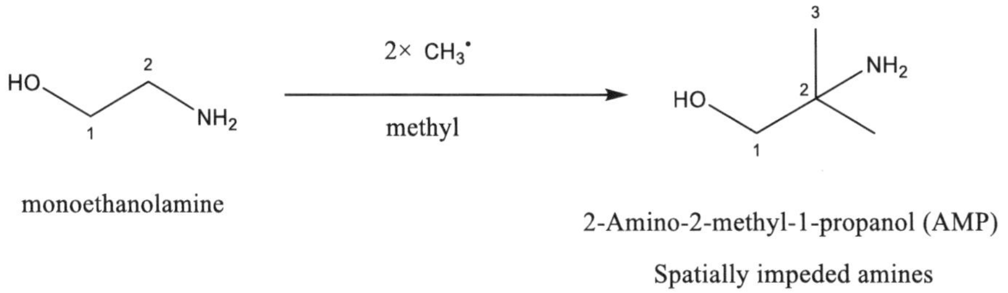

Figure 10.2 Synthesis of sterically Hindered amines 2-amino-2-methyl-1-propanol (AMP) from monoethanolamine.

an amino alcohol [82,83]. CO_2 reaction with AMP's amino group will cause the production of three types of carbamates, bicarbonate, and carbonate ions [82]. The AMP's alcoholic group also produces alkyl carbonate in reaction with CO_2 [84]. However, under reaction conditions, the pH of the solution will never exceed 12 (usually pH of the solution will be in the range of 7.5–9). Thus, the second reaction (AMP's alcoholic group) can be ignored (Fig. 10.2) [78].

Several absorption processes can be used for CO_2 removal from CO_2-rich natural gas.

Chemical absorption: This process involves using a chemical solvent, such as amine, to absorb CO_2 from the natural gas stream. The solvent reacts with the CO_2 to form a chemical compound that can be easily separated from the natural gas. This method is widely used in the industry, but it requires significant energy for solvent regeneration and disposal [85].

The most widely used solvents for CO_2 absorption in natural gas processing are aqueous solutions of amines [86]. Amines are organic compounds that contain nitrogen atoms and have a high affinity for CO_2 [87]. Aqueous solutions of amines are commonly used because they are effective at removing CO_2 from natural gas and are relatively easy to regenerate and reuse [88]. Some of the commonly used amines for CO_2 absorption in natural gas processing are (1) MEA, (2) DEA, (3) MDEA, (4) diisopropanolamine (DIPA), (5) triethanolamine (TEA) [89–91]. MEA is the most widely used amine for CO_2 absorption in natural gas processing, followed by MDEA and DEA [92]. These amines have different properties and performance characteristics, and the choice of which amine to use will depend on several factors, including the specific CO_2 concentration and other components in the gas stream, the required purity of the recovered CO_2, and the desired overall efficiency and economics of the absorption process. In addition to amines, other solvents, such as methanol and ethanol, have also been used for CO_2 absorption in natural gas processing, but they are typically less efficient and more expensive than amines [93,94]. Table 10.2 summarizes the various solvents commonly used in CO_2 removal from natural acidic gas.

It is important to note that the advantages and disadvantages of each solvent will vary depending on the specific application. For example, MEA may be a more effective solvent for removing CO_2 from natural gas than MDEA, but MDEA may be a more energy-efficient solvent. Ultimately, the best solvent for removing CO_2 from natural acidic gas will depend on a variety of factors, including the CO_2 concentration, the gas stream, and the cost.

Physical absorption: In this process, a physical solvent, such as methanol, is used to absorb CO_2 from the natural gas stream. The solvent is then heated to release the CO_2, which can be separated and stored. Physical absorption is less energy-intensive than chemical absorption, but it is less efficient at low CO_2 concentrations [101].

Membrane separation: This process involves passing the CO_2-rich natural gas through a membrane that selectively separates the CO_2 from the other components. Membrane separation is energy-efficient and can be used at low CO_2 concentrations, but it requires high capital costs [102,103].

Cryogenic separation: This process involves cooling the natural gas stream to very low temperatures, which causes the CO_2 to condense and separate from the other components. Cryogenic separation is highly efficient, but it is also energy-intensive and requires a significant amount of infrastructure [104].

Each of these processes has its advantages and disadvantages, and the choice of which process to use will depend on several factors, including the concentration of CO_2 in the natural gas stream, the required purity of the CO_2, and the available resources and infrastructure. The best and widely used separation method for CO_2 removal from CO_2-rich natural gas depends on various factors, such as the CO_2 concentration in the gas stream, the required purity of CO_2, available infrastructure, and resources [105,106]. However, chemical absorption using an amine solvent is the most common and well-established method for CO_2 removal from CO_2-rich natural gas [13]. It is widely used in industrial applications, and the technology has been developed and optimized over several decades. Additionally, amine-based absorption processes can achieve high levels of CO_2 removal, and the recovered CO_2 can be further processed for storage or utilization [94,107]. Other methods, such as physical absorption, membrane separation, and cryogenic separation, have advantages and disadvantages compared to chemical absorption, and they are typically

Table 10.2 Various solvents commonly used in CO₂ removal from natural acidic gas.

Solvent	Type	Advantages	Disadvantages	References
Monoethanolamine (MEA)	Amine	- Effective for a wide range of CO_2 concentrations. - Can be used with a variety of gas streams. - Relatively well-established technology.	- Energy intensive - Can be corrosive - Can produce wastewater.	[26,95]
Diethanolamine (DEA)	Amine	- More effective than MEA for high CO_2 concentrations - Less corrosive than MEA.	- More expensive than MEA - Can produce wastewater.	[96,97]
Methyldiethanolamine (MDEA)	Amine	- More effective than DEA for high CO_2 concentrations - Less corrosive than DEA. - Produces less wastewater than MEA or DEA.	- More expensive than MEA or DEA.	[98]
Dimethylether of polyethylene glycol (DPEG)	Physical solvent	- Energy-efficient - Low - Maintenance - Produces no wastewater	- Not as effective as amine absorption for high CO_2 concentrations - Can be expensive.	[99]
Methanol	Physical solvent	- Effective for a wide range of CO_2 concentrations. - Can be used with a variety of gas streams. - Relatively inexpensive.	- More energy - Intensive than amine absorption - Can be corrosive - Produces wastewater	[97]
Sulfolane	Physical solvent	- Effective for a wide range of CO_2 concentrations. - Can be used with a variety of gas streams. - Less corrosive than amine solvents.	- More expensive than amine solvents - Produces wastewater	[100]

used in specific applications where their specific benefits are particularly advantageous [26,97]. For example, membrane separation is useful for low CO_2 concentrations and in remote locations [108], while cryogenic separation is often used for high CO_2 concentrations and large-scale projects [109].

3. CO$_2$ removal flowsheet

Carbon dioxide removal from natural gas is a process that is used to reduce the amount of CO$_2$ in natural gas. This is done to meet pipeline specifications or to produce a higher-value product [110]. Various technologies are available for removing CO$_2$ from natural gas, and these are discussed in the following:

3.1 Amine absorption

This is the most common technology for CO$_2$ removal from natural gas. In amine absorption, a solvent is used to absorb the CO$_2$ from the gas. The solvent is then heated to regenerate it, and the CO$_2$ is released [111–113].

3.2 Membrane separation

This technology uses membranes to separate the CO$_2$ from the gas. Membranes are thin sheets of material that allow some molecules to pass through them while blocking others. In the case of CO$_2$ removal, the membranes are designed to allow the gas to pass through while blocking the CO$_2$ [114,115].

3.3 Cryogenic distillation

This technology uses cold temperatures to separate the CO$_2$ from the gas. The gas is cooled until the CO$_2$ condenses and can be removed [116,117].

The flowsheet for a CO$_2$ removal plant will vary depending on the technology that is used. However, all CO$_2$ removal plants will have the following basic steps.

3.4 Gas pretreatment

The gas is first pretreated to remove any contaminants that could damage the equipment or interfere with the CO$_2$ removal process [118].

3.5 CO$_2$ removal

The CO$_2$ is removed from the gas using one of the technologies described above [119,120].

3.6 CO$_2$ compression

The CO$_2$ is compressed to high pressure so that it can be transported or stored [120].

3.7 Gas dehydration

The gas is dehydrated to remove any water vapor that could condense and damage the equipment [121–123].

3.8 Gas conditioning

The gas is conditioned to meet pipeline specifications or to produce a higher-value product [124,125].

The cost of CO$_2$ removal from natural gas varies depending on the technology that is used, the size of the plant, and the location of the plant [126]. However, the cost of CO$_2$ removal is typically a small fraction of the overall cost of producing natural gas [127]. CO$_2$ removal from natural gas is a valuable tool for reducing greenhouse gas emissions. By removing CO$_2$ from natural gas, we can help to mitigate the effects of climate change [128].

Table 10.3 shows the advantages and disadvantages of various methods of CO$_2$ removal from natural acidic gas. Each method has its own unique set of advantages and drawbacks that should be taken into consideration when selecting the best approach for a given application.

It is important to note that the advantages and disadvantages of each method will vary depending on the specific application. For example, amine absorption may be a more effective method for removing CO$_2$ from natural gas than membrane separation, but membrane separation may be a more energy-efficient method [134,135]. Ultimately, the best method for removing CO$_2$ from natural acidic gas will depend on a variety of factors, including the CO$_2$ concentration, the gas stream, and the cost [136–138].

4. CO$_2$ absorption plant: Unit operations and parameters

A CO$_2$ absorption plant typically consists of several unit operations that work together to capture and separate CO$_2$ from a gas stream. The specific unit operations and parameters used can vary depending on the type of solvent, feed gas composition, and other factors [18,25]. However, here is a general overview of

Table 10.3 Outlining the pros and cons of different approaches for removing CO_2 from acidic gas.

Method	Advantages	Disadvantages	References(s)
Amine absorption	- Effective for a wide range of CO_2 concentrations. - Can be used with a variety of gas streams. - Relatively well-established technology.	- Energy-intensive. - Can be corrosive. - Can produce wastewater.	[24]
Membrane separation	- Energy-efficient. - Low-maintenance. - Produces no wastewater.	- Not as effective as amine absorption for high CO_2 concentrations. - Can be expensive.	[129,130]
Cryogenic distillation	- Highly effective for high CO_2 concentrations. - Produces no wastewater.	- Energy-intensive. - Can be expensive.	[130,131]
Carbon capture and storage (CCS)	- Permanently removes CO_2 from the atmosphere. - Can be used with a variety of gas streams.	- Expensive. - Requires a large infrastructure.	[132,133]

the major unit operations and some key parameters involved in a CO_2 absorption plant [139–141].

4.1 Gas pretreatment

Before CO_2 absorption, the gas stream is typically treated to remove impurities that could interfere with the absorption process. This may involve filtration, desulfurization, dehydration, or other pretreatment steps [118,142].

4.2 Absorption column

The pretreated gas stream is then directed to an absorption column, where it is contacted with a solvent that selectively captures the CO_2. The absorption column may consist of multiple trays or packing material to increase the contact between the gas and solvent phases. The gas stream may be cooled and compressed before entering the column to increase the solubility of the CO_2 in the solvent [142].

4.3 Solvent regeneration

Once the solvent is loaded with CO_2, it is directed to a regeneration column, where the CO_2 is stripped from the solvent using heat. The solvent may be regenerated using steam or other heat sources, and the CO_2-rich gas is collected for further processing [142].

4.4 CO_2 purification

The collected CO_2 gas may undergo further purification to remove any remaining impurities, such as water or nitrogen, using distillation, absorption, or membrane separation techniques [143].

4.5 CO_2 compression

The purified CO_2 gas is then compressed to the desired pressure for storage or transport [142].

Some of the key parameters involved in a CO_2 absorption plant.

4.6 Solvent type

Different solvents have different properties, such as selectivity, solubility, and degradation resistance, that can affect their performance in CO_2 absorption [144].

4.7 Gas feed composition

The composition of the gas feed stream, including the concentration of CO_2 and other impurities, can affect the efficiency of CO_2 absorption and the performance of the solvent [145].

4.8 Absorption column operating parameters

The temperature, pressure, and flow rate of the gas and solvent streams in the absorption column can affect the mass transfer and the efficiency of CO_2 capture [146].

4.9 Regeneration column operating parameters

The temperature, pressure, and steam flow rate in the regeneration column can affect the amount of CO_2 that is desorbed from the solvent and the efficiency of solvent regeneration [147].

4.10 Purification parameters

The type of purification technique used, as well as the operating parameters, can affect the purity and yield of the recovered CO_2 gas. In general, a CO_2 absorption plant involves multiple unit operations and parameters that work together to capture and separate CO_2 from a gas stream. The efficiency and effectiveness of the plant depend on the specific design and operation of each unit operation and parameter [146].

5. Current applications and cases

Investigations explore the capacity of different water-based solutions containing potassium lysinate (LysK) to capture CO_2. This choice of solvent is based on its high kinetic constant (kov) and CO_2 absorption speed, making it a viable candidate for capturing CO_2 after combustion in flue gas. The investigation involves three distinct concentrations of LysK (8.9%, 17.4%, and 32.9% by weight) added to a 30% by weight aqueous MEA solution. These tests are carried out in a closed-cycle absorption setup within a bench-scale column (with a diameter of 80 mm and a packed height of 900 mm). The experiments are conducted under conditions mimicking synthetic flue gas (with a 4% CO_2 concentration on a molar basis, the rest being N_2) and a liquid-to-gas ratio of 1.39 mol/mol. These conditions align with previous experiments involving MEA and Potassium prolinate (ProK) solutions and are intended for the application of CO_2 capture in natural gas combined cycles [148].

To evaluate the CO_2 absorption capacities of different water-based solutions containing LysK through empirical examinations, LysK is chosen as a promising candidate for post-combustion CO_2 capture, given its previously documented high overall kinetic constant (kov) and CO_2 absorption rate in the literature. Three different LysK concentrations (8.9%, 17.4%, and 32.9% w/w) are compared to a 30% w/w aqueous MEA solution in absorption tests carried out on a bench-scale column. These tests are executed under simulated flue gas conditions, where the CO_2 concentration is 4% on a molar basis, and nitrogen constitutes the rest, with liquid-to-gas ratios of 1.39 mol/mol. These conditions align with prior experiments conducted on MEA and ProK solutions and are designed to replicate the potential application of these solvents for CO_2 capture in natural gas combined cycles [148].

The traditional approach to capturing CO_2 in the context of reducing emissions from fossil fuel–based power generation involves amine-based chemical absorption. In this method, primary amines like MEA have commonly been employed as the standard absorbent. However, amines come with several drawbacks, including their limited capacity for CO_2, susceptibility to thermal and oxidative breakdown, and high energy consumption. Consequently, there is a growing interest in exploring alternative approaches for CO_2 capture. Recent research has indicated that aqueous solutions containing amino acid salts, which possess qualities such as low toxicity, resistance to thermal and oxidative degradation, minimal volatility, and rapid reaction kinetics, could serve as promising alternatives for reducing CO_2 emissions from postcombustion flue gas processes [149–153]. Amino acid salts can achieve high CO_2 loadings and capacities, which could potentially reduce the need for circulating solvents. The absorption kinetics and CO_2 absorption rate are higher, which could result in a smaller absorber size and lower related CAPEX [153]. Foam formation is averted, thanks to the elevated surface tension, and there is an enhancement in the process of mass transfer. Furthermore, the solvent's minimal evaporation contributes to reduced requirements for replenishing the solvent. Nonetheless, many uncertainties remain concerning the use of aqueous solutions containing amino acid salts for CO_2 absorption, especially regarding their real and repeated CO_2 capture capacity, resistance to high temperatures, the chemical composition and potential harmfulness of degradation byproducts, the energy needed for solvent regeneration, and the associated costs in the event of implementing this technology on a large scale in a power plant or industrial facility. Some research also indicates that amino acid salt solutions exhibit a higher specific heat absorption compared to aqueous MEA [154]. The feature discussed in the original text has both advantages and disadvantages. It is beneficial as it reduces the heat of vaporization due to the higher equilibrium CO_2 partial pressure in the reboiler [155]. Nevertheless, this characteristic has both advantages and disadvantages. On the positive side, it reduces the heat needed for vaporization due to the higher equilibrium CO_2 partial pressure in the reboiler. However, it can also result in an overall increase in the energy required for solvent regeneration due to its substantial heat absorption. Moreover, the use of amino acid salts for CO_2 absorption may trigger solid precipitation at high absorption levels. This precipitation can augment the solvent's absorption capacity by pushing the system's equilibrium toward the product side of the absorption process. The complete heat of dissolution in amino acid salt systems can exhibit either an endothermic or exothermic nature. If it is exothermic, it can provide cooling during the absorption phase and heating during solvent regeneration, which proves advantageous from an energy perspective. On the flip side, if it is endothermic, it may result in even higher energy demands for the solvent regeneration process [156].

Existing literature includes prior investigations into methods employing amino acid salts for the removal of acid gases. Siemens has already developed and tested a carbon capture technique known as "POSTCAP" at a pilot scale. They utilized a validated process model to anticipate the precise thermal energy requirement for solvent regeneration in a full-scale application of an amino acid–based capture system within a coal-fired reference power plant. The findings demonstrated that the thermal energy demand amounted to 2.7 MJ/kgCO$_2$ captured, which is lower than the established benchmark of 3.7 MJ/kgCO$_2$ for absorption using MEA solutions. These promising outcomes strongly suggest the need for further exploration of amino acid-based technologies in the context of flue gas decarbonization [157,158].

Earlier research has explored the application of amino acid salts for removing acid gases, and Siemens has conducted pilot-scale development and testing of a carbon capture process named "POSTCAP" [159]. By employing a validated process model, scientists successfully anticipated the precise thermal energy requirement for regenerating the solvent in a full-scale capture unit based on amino acids within a coal-fired power plant. The outcomes revealed that this demand stood at 2.7 MJ/kg CO$_2$ captured, notably lower than the established benchmark of 3.7 MJ/kg CO$_2$ for MEA-based absorption solutions. These encouraging results underscore the necessity for additional research in the realm of amino acid–based technologies aimed at mitigating flue gas emissions [160,161].

LysK is considered a highly promising amino acid molecule for CO$_2$ postcombustion capture when compared to other salts such as potassium prolinate, sarcosinate, histidinate, and taurinate (referred to as ProK, SarK, HisK, and TauK respectively), as per research conducted by Mai Lerche [162]. Its CO$_2$ solubility in related aqueous solutions is good, as evidenced by studies conducted by Shen et al. [163,164] and Zhao et al. [165]. As a result, LysK has been designated as a solute with potential for use in solvent-based CO$_2$ capture.

Bench-scale experimental tests were conducted to study the effectiveness of LysK solutions in capturing CO$_2$ in natural gas combined cycle decarbonization [148,166]. The study builds upon prior research on ProK versus MEA solutions and provides a quantitative evaluation of the absorption performance of the selected solvents in terms of CO$_2$ absorption, solvent capacity, and loading increase over time [148,167]. The research methodology and data analysis presented in this study offer a reliable and cost-effective approach for screening solvents for CCS applications. The literature review has identified LysK as a promising option among amino acid salts solutions [163], and therefore, the study tested LysK solutions at different concentrations of 8.9 %w/w, 17.4 %w/w, and 32.9 %w/w, as per the previously proposed methodology [148,168]. In a controlled experimental setup, tests were carried out using a closed-cycle system that received a continuous supply of synthetic gas mimicking the flue gases generated by NGCC power plants. This synthetic gas comprised approximately 4% carbon dioxide, with the remainder being nitrogen. A specific volume of liquid solvent was continuously circulated until it became fully saturated. To establish a comparison with the current industry standard, a 30% MEA solution was selected as the reference point for this research. Furthermore, the study investigated the impact of increased loading by utilizing LysK solutions with a concentration of 32.9% w/w. This was achieved by adjusting the amount of circulating solvent within the closed-cycle system while maintaining the same quantity of solute as in the reference MEA solution [169].

LysK achieved a higher loading compared to MEA solutions, with a maximum value of 0.98 molCO$_2$/molAlk for LysK and 0.52 mol CO$_2$/molAlk for MEA, resulting in an approximately 88% increase [165,170]. The increase in loading over time was also higher for LysK, with a 32.9% w/w increase compared to the reference solvent after reaching a threshold loading of about 0.47 molCO$_2$/molAlk [148]. Based on the calculated capacities, a concentration range of 4.0—4.4 m with an average value of 4.2 m (about 43.7% w/w) of LysK in water was identified as a promising option for further investigation and study for up-scaling purposes to ensure the same capacity level as the 30% MEA solution [148].

In upcoming experiments, a 43.7% w/w LysK solution will be tested to confirm its effectiveness as a solvent. If the results are positive, additional tests will be conducted to determine vapor—liquid equilibrium curves at various temperatures. This information is necessary for determining the appropriate size of absorbers and evaluating the economic feasibility of using nonprecipitating amino acid salt solutions for large-scale CO$_2$ capture. Furthermore, the study will explore other relevant amino acid salts recommended in the literature using the same testing approach. No information from the original text has been omitted in the paraphrased version [148].

Despite the high importance of natural gas for extensive industrial and urban applications, many resources cannot be economically extracted due to the high content of CO$_2$ (wide range of 40 mol-%–80 mol-%). However, the growing global demand for this valuable resource leads to the development of technical methods for its economic extraction. The main basis of these methods is using combined technologies to efficiently remove CO_2 with minimum energy consumption. Some techniques are cryogenic distillation, membrane separation, and physical absorption with methanol. These techniques along with their hybrid ones are optimized to meet the standard CO_2 contents of pipeline transport (2–3 mol-%) to LNG specification (50 ppm). In this regard, a framework for systematic optimization-based process design is essential. This framework is made up of robust and computationally efficient reduced models for unit operations. As such, it warrants the automated performance of mass optimization calculations [171]. Since natural gas has fewer CO_2 emissions for power production or heating purposes than coal or oil, it has attracted a lot of attention. However, this valuable source has a wide range of variations in terms of composition so that methane exists as the main material plus a range of higher alkanes and other hydrocarbons. Besides, nitrogen, carbon dioxide, hydrogen sulfide, and water are common impurities [15,172]. In some sources, one of the impurities, especially CO_2 is significantly increased. One of the main drawbacks of such intensive CO_2 is its corrosive effect in combination with water. In addition, it is necessary to remove CO_2 to produce the final product by heating standard value. Alongside the costs of gas processing, due to the increasing sensitivities of increasing greenhouse gases [4,5,7,173], the environmental costs of CO_2 capture and sequestration should be considered [13]. Consequently, many natural gas sources are abandoned because of economic viability owing to elevated CO_2 contents. Even so, the greedy global demand for natural gas has led to the development of efficient technologies for CO_2 separation. In addition to eliminating CO_2, these new technologies have programs to the CO_2 management to increase oil recovery and CO_2 storage. They have provided an exceptional opportunity for unconventional reserves of natural gas to enter transmission pipelines [13,174].

Despite amine absorption as the well-established technology for CO_2 removal in commercial operations, there are additional updates of gas processing techniques developing based on flash separation, distillation, absorption, membrane separation, adsorption, or hybrid combinations aforesaid [13,175]. Distillation-based CO_2 removal from CH_4-rich natural gas is now developed to technologies of the Ryan Holmes process [176,177], the Controlled Freeze Zone technology [178,179], and dual pressure distillation [180,181]. Furthermore, there are developed methods based on flash separation: (i) Twister technology [182], (ii) the CryoCell [104], and (iii) condensed rotational speed separation [183,184]. The other group of developed techniques is based on absorptive processes including chemical and physical absorption. In this way, at low temperatures and high CO_2 partial pressures in the raw natural gas, physical solvents like Rectisol (methanol) [171,185], Flour Solvent (propylene carbonate) [186,187], Purisol (N-methyl-2-pyrrolidone) [188], and Selexol (mixture of the homologs of the dimethyl ether of polyethylene glycols) [13,189] are now well known [13]. Various membrane modules and membrane materials are being rapidly developed. For example, although cellulose acetate (supplied by, e.g., W.R. Grace, Honeywell UOP Separex, NATCO Cynara) is an established material for separating CO_2 from natural gas, ultramodern materials like polyimide (supplied by Air Liquide Medal) and perfluoro polymer membranes (supplied by ABB/MTR) may challenge such old-fashioned ones [15,190]. HISELECT membrane package units have been developed to use in industrial natural gas processing plants. In these units, to approach efficient gas separation, mixed matrix membranes containing blends of polymers with zeolites are employed [191]. The most widely used solid porous materials for adsorption are Metal-organic frameworks, activated carbons, and NaX zeolites [13,192]. Amine absorption with membrane processes can effectively use in CO_2 absorption from a natural gas containing about 5 mol % of CO_2. In comparison with the amine absorption, although the membrane-based sweetening yields natural gas with higher CO_2 impurities, its main benefit is its inexpensive concerning capital investment. However, for the gas feed with high content of CO_2 (13 mol-% −40 mol-%), chemical absorption especially with ionic liquids may be of high efficiency [193].

As-captured CO_2 has also special applications such as a refrigerant in a cryogenic process and an enhancer booster in oil recovery. In addition, the newer application is possible through innovative photocatalysts for CO_2 transformation into invaluable organic products. Moreover, CO_2 feedstock can be reduced to formate by using iron-based metal-organic frameworks [194]. The other valuable tested ideas are CO_2 conversion to methane as the main component of the natural gas with novel catalysts like spatially separated Au and CoO dual cocatalysts on hollow TiO_2 [195] and advanced metal-free catalysts [196].

There is a challenging problem in oil production from offshore huge reservoirs in deep waters. Since there are high gas-oil ratios and high CO_2 content, the extraction must process a huge raw CO_2-rich natural gas flow rate. Then, the as-concentrated CO_2

must be separated and transferred to the appropriate stations. Gas processing in offshore oil fields includes three phases: (i) water dew-point adjustment (WDPA), (ii) hydrocarbon dew-point adjustment (HCDPA), and (iii) CO_2 removal. This processing by-product is CO_2 with the benefits of avoiding inert transportation and boosting oil production by injecting it into the reservoir. Typically, TEG absorption for WDPA, JouleThomson expansion for HCDPA and membrane permeation for CO_2 removal are the main options for treating gas in offshore rigs. But there are innovative methods for CO_2 removal. The ionic-liquid [Bmim][NTf$_2$] for simultaneous WDPA and HCDPA is efficient in CO_2-rich, natural gas processing. Rigorous simulations, in conjunction with the ionic-liquid implications in vapor–liquid equilibrium and heat effects, show high-pressure selective stripping of CO_2. It leads to lower compression power for enhanced oil recovery utilization. From an economic viewpoint, the new absorption technology shows 19.3% higher revenues, 23.6% less manufacturing costs, and 18% less investment. All in all, a 37% higher net present value (NPV) is obtained in comparison with conventional gas processing [103]. Chemical absorption with amines demonstrates an excellent efficiency in the selective separation of CO_2/CH_4 mixture with minimum hydrocarbon losses. However, some flaws such as high footprint, solvent losses, and high heat ratio for CO_2 stripping and corrosion are incentives to further improve this process [197,198]. In this regard, ILs have been introduced as an alternative. Because of the flexible properties such as their low volatility, high thermal stability, high CO_2 solubility, and low heat consumption for CO_2 stripping at low pressure, these ILs have the main advantages for this replacement. A wide range of ILs with specific properties for given applications can be synthesized due to their specific characteristics including the low-temperature melting salts with cation–anion combinations [199,200]. However, it should be emphasized that these series of solvents have higher prices and viscosities relative to other solvents like amines [199].

IL systems may be utilized in post and precombustion CO_2 capture. In comparison with the conventional MEA process, the IL Bmim][Ac] decreases heat demand by 16% with a higher CO_2 purity in a developed CO_2 capture process from flue-gas [201]. From an economic point of view, the IL-base process indicates an 11% reduction in total fixed capital investment (FCI) and a 12% in equipment footprint. A physical IL solvent [Hmim][Tf$_2$N] can effectively capture CO_2 (95.12% CO_2 recovery, and 1.23% mol of solvent loss) from fuel-gas in a conceptual process with four adiabatic absorbers and three low-pressure flashes for IL regeneration [202]. Some ILs are not only efficient absorbents for CO_2 but also, they are highly hygroscopic to dehydrate raw gas. The IL [C$_{10}$mim][TfO] is used for postcombustion capture, the overall cost of which is 26% less expensive than the conventional MEA process. However, the dynamic analysis shows highly nonlinear behavior of the IL plant [203]. The IL [Bmim][NTf$_2$] may save 30% of heat consumption in CO_2 capture from power plants [204]. Two alternatives for decarbonated shale gas using [Bmim][NTf$_2$] under single and multistage IL regenerations are assessed. They show reductions in total energy consumption of 42.8% and 66.04%, respectively [103].

6. Specific characterizations and properties of CO_2 absorption

The specific characterizations and properties of CO_2 absorption depend on the method used for the absorption process. However, in general, some of the key characteristics and properties of CO_2 absorption include:

6.1 Selectivity

CO_2 absorption is selective, meaning that the solvent used for absorption will selectively remove CO_2 from the gas stream while leaving other components, such as methane and nitrogen, largely untouched [205].

6.2 Solubility

CO_2 is highly soluble in certain solvents, such as aqueous solutions of amines, which makes it an excellent candidate for absorption [206].

6.3 Reversibility

CO_2 absorption is a reversible process, meaning that the solvent can be regenerated and reused to absorb more CO_2. The regenerated solvent can then be recycled back into the absorption process, reducing waste and overall costs [207,208].

6.4 Temperature dependence

The rate and extent of CO_2 absorption depend on temperature, with higher temperatures generally favoring lower levels of CO_2 absorption [207].

6.5 Kinetics

The kinetics of CO_2 absorption can be affected by factors such as solvent flow rate, CO_2 concentration, and gas flow rate [209].

6.6 Reaction byproducts

Depending on the type of solvent used for CO_2 absorption, reaction byproducts may be generated during the absorption process. These byproducts can affect the overall efficiency and economics of the absorption process and may require additional treatment or disposal [210].

All-inclusive, CO_2 absorption is a complex process that requires careful consideration of the properties and characteristics of both the gas stream and the solvent used for absorption. Proper design and optimization of the absorption process can lead to effective CO_2 removal while minimizing costs and environmental impacts.

7. Novel methods and solvents

There are new and novel methods and solvents with specific properties for CO_2 absorption that are currently being developed and researched. Some of these methods and solvents include.

7.1 Ionic liquids

Ionic liquids are salts that are liquid at room temperature and have been investigated as solvents for CO_2 absorption due to their low volatility, high thermal stability, and tunable properties. They can be tailored to have specific CO_2 solubility and selectivity, making them a promising option for CO_2 absorption (Table 10.4) [217].

7.2 Hybrid solvents

Hybrid solvents are a combination of two or more solvents, such as amines and ionic liquids, that are used together to improve the properties and performance of the absorption process. These solvents can have improved selectivity, stability, and efficiency compared to individual solvents [218].

7.3 Nanoparticles

Nanoparticles, such as silica and metal-organic frameworks (MOFs), have been investigated for CO_2 absorption due to their high surface area and tunable properties. They can be

Table 10.4 Some typical ionic liquids in CO$_2$ absorption from the natural gas feed.

Ionic liquid	Advantages	Disadvantages	References
1-Butyl-3-methylimidazolium chloride ([bmim][Cl])	- High CO$_2$ absorption capacity. - Low volatility. - Thermally stable. - Noncorrosive.	- Expensive. - Can be difficult to recycle and regenerate.	[211]
1-Butyl-3-methylimidazolium bis(trifluoromethylsulfonyl)imide ([bmim][Tf$_2$N])	- High CO$_2$ absorption capacity. - Low volatility. - Thermally stable. - Noncorrosive.	- Expensive. - Can be difficult to recycle and regenerate.	[212,213]
1-Ethyl-3-methylimidazolium bis(trifluoromethylsulfonyl)imide ([emim][Tf$_2$N])	- High CO$_2$ absorption capacity. - Low volatility. - Thermally stable. - Noncorrosive.	- Expensive. - Can be difficult to recycle and regenerate.	[214]
1-Butyl-3-methylimidazolium tetrafluoroborate ([bmim][BF$_4$])	- High CO$_2$ absorption capacity. - Low volatility. - Thermally stable. - Noncorrosive.	- Expensive. - Can be difficult to recycle and regenerate.	[215]
1-Ethyl-3-methylimidazolium tetrafluoroborate ([emim][BF$_4$])	- High CO$_2$ absorption capacity. - Low volatility. - Thermally stable. - Noncorrosive.	- Expensive. - Can be difficult to recycle and regenerate.	[216]

functionalized with specific chemical groups to enhance CO$_2$ solubility and selectivity [219].

7.4 Membrane-based absorption

Membrane-based absorption combines membrane separation with absorption, allowing for the separation of CO$_2$ from natural gas without the need for a liquid solvent. This method can have lower energy requirements and reduced environmental impacts compared to traditional absorption processes [220].

These new and novel methods and solvents offer potential advantages over traditional absorption methods, including improved selectivity, efficiency, and environmental performance. However, further research and development is needed to fully understand and optimize their performance for industrial applications.

The advantages and disadvantages of each IL will vary depending on the specific application. For example, [bmim][Cl] may be a more effective IL for removing CO_2 from natural gas than [emim][BF_4], but [emim][BF_4] may be more thermally stable [221–223]. Ultimately, the best IL for removing CO_2 from natural gas will depend on a variety of factors, including the CO_2 concentration, the gas stream, and the cost [224,225]. ILs are a promising new class of solvents for CO_2 absorption from natural gas. They offer several advantages over traditional solvents, such as high CO_2 absorption capacity, low volatility, thermal stability, and noncorrosiveness. However, they are also more expensive than traditional solvents and can be difficult to recycle. Table 10.5 shows various ILs that have been studied for their ability to absorb CO_2 from natural gas feed [241,242].

Noting that CO_2 absorption capacity of an ionic liquid can vary depending on the operating conditions and impurities in the natural gas feed. Therefore, the values listed in the table should be considered as general estimates.

8. Scales up of the ionic liquid–based technologies

There have been several pilot-scale and demonstration-scale applications of the new technologies for CO_2 absorption that I mentioned earlier. Here are summaries of some examples [203,229,243–246]:
- Ionic liquids: Several pilot-scale studies have been conducted to investigate the use of ionic liquids for CO_2 absorption. For example, a pilot-scale plant using an ionic liquid was built in Germany to capture CO_2 from flue gas from a waste incineration plant. The plant achieved a CO_2 capture rate of up to 95% and demonstrated the feasibility of using ionic liquids for large-scale CO_2 capture [247,248].
- Hybrid solvents: A demonstration plant using a hybrid solvent composed of amines and ionic liquids was built in the

Table 10.5 Several ionic liquids used for CO_2 absorption from natural gas streams.

Ionic liquid	CO_2 absorption capacity (mol CO_2/mol IL)	Operating conditions	References
[BMI][Tf$_2$N] (1-butyl-3-methylimidazolium bis(trifluoromethylsulfonyl)imide)	0.56	30°C, 1 atm	[226—228]
[BMP][Br] (3-butyl-1-methylpyridinium bromide)	0.47	40°C, 1 atm	[229,230]
[C4MIM][Tf$_2$N] (1-butyl-3-methylimidazolium bis(trifluoromethylsulfonyl)imide)	0.59	25°C, 1 atm	[231—233]
[C4MIM][HSO$_4$] (1-butyl-3-methylimidazolium hydrogen sulfate)	0.58	40°C, 2.5 atm	[234—236]
[EMIM][Tf$_2$N] (1-ethyl-3-methylimidazolium bis(trifluoromethylsulfonyl)imide)	0.60	40°C, 1 atm	[214,237]
[EMIM][HSO$_4$] (1-ethyl-3-methylimidazolium hydrogen sulfate)	0.45	30°C, 1 atm	[238]
[MMIM][Tf$_2$N] (1-methyl-3-octylimidazolium bis(trifluoromethylsulfonyl)imide)	0.60	25°C, 1 atm	[239]
[OMIM][Tf$_2$N] (1-octyl-3-methylimidazolium bis(trifluoromethylsulfonyl)imide)	0.68	25°C, 1 atm	[240]

Note: Tf_2N, bis(trifluoromethylsulfonyl)imide; HSO_4, hydrogen sulfate.

Netherlands to capture CO_2 from flue gas from a coal-fired power plant. The plant achieved a CO_2 capture rate of 90%, and the use of the hybrid solvent allowed for a smaller and more efficient absorption unit compared to traditional amine-based processes [249,250].

- Nanoparticles: Several pilot-scale studies have been conducted to investigate the use of nanoparticles for CO_2 absorption. For example, a pilot-scale plant using an MOF-based adsorbent was built in South Korea to capture CO_2 from flue gas from a cement plant. The plant achieved a CO_2 capture rate of 90%, and the MOF-based adsorbent showed promising selectivity and stability for CO_2 absorption [251,252].
- Membrane-based absorption: Several demonstration-scale plants using membrane-based absorption have been built for CO_2 capture from natural gas. For example, a demonstration plant using a membrane contactor was built in Australia to capture CO_2 from natural gas produced from coal seam gas. The plant achieved a CO_2 capture rate of up to 90%, and the membrane contactor showed promise as a low-energy and low-environmental-impact option for CO_2 capture [249].

These pilot-scale and demonstration-scale applications demonstrate the potential of the new technologies for CO_2 absorption and provide valuable information for the development and optimization of these technologies for industrial-scale applications.

9. Conclusion and future outlooks

Absorption processes are commonly used for the removal of CO_2 from CO_2-rich natural gas streams. In this process, a liquid solvent is used to selectively absorb CO_2 from the gas stream. The absorbed CO_2 is then separated from the solvent, and the cleaned gas is released back into the atmosphere. There are different types of absorption processes, including physical absorption and chemical absorption. Physical absorption involves the use of solvents that dissolve CO_2 in their liquid phase without any chemical reaction. Chemical absorption, on the other hand, involves a chemical reaction between the CO_2 and the solvent. The choice of solvent depends on various factors such as the nature of the gas stream, the required purity level, and the operating conditions. Some commonly used solvents include (MEA, DEA, and MDEA). The absorption process can be operated at different pressure and temperature conditions, with higher pressure and lower temperature leading to better CO_2 capture efficiency. However, this also results in higher energy requirements for the process. Overall, absorption processes are an effective and widely adopted method for removing CO_2 from CO_2-rich natural gas streams.

There has been significant progress and development of new technologies and solvents for CO_2 absorption and removal from natural gas streams rich in CO_2. Some of the key trends and future directions include (i) the use of new solvents with higher CO_2 absorption capacity and selectivity, faster kinetics, and lower regeneration energy needs. Promising candidates include amino acid-based solvents, ILs, branch-chained amines, etc. These can significantly improve the efficiency and reduce the costs of CO_2 capture; (ii) Use of mixed solvents, for example, combining amines with ionic liquids, to leverage the benefits of multiple solvents and overcome their limitations. Mixed solvents often show enhanced performance; (iii) development of aqueous solvents based on alkali or alkaline earth hydroxides that have lower regeneration energy needs. However, their CO_2 capacity is typically lower than amines; (iv) use of membrane-based approaches, for example, polymer membranes, carbon membranes, and zeolite membranes. These can lower the solvent regeneration needs but often at a higher capital cost. Membrane-solvent hybrids can also be promising; (v) Exploration of new reactor and process configurations to improve kinetics, increase contact between phases, reduce internal recycle needs, etc. For example, the use of structured packing, flow accelerators, membranes, etc.; (vi) Development of solvents and processes suitable for CO_2 capture at lower partial pressures, e.g., pre-combustion capture. This is important but more challenging; (vii) Integrating CO_2 capture with CO_2 utilization, e.g., for enhanced oil recovery. This can improve the overall sustainability and economics of CO_2 management technologies; (viii) Use of hybrid technologies, e.g., combining absorption with adsorption, or membranes with chemical solvents. Such integrated systems could lead to improved performance; (ix) progress in these areas should enable significant improvements in CO_2 capture from natural gas and more broadly for mitigation of CO_2 emissions. But further R&D is still needed to develop more economical, scalable, and sustainable solutions.

The collected CO_2 from natural gas absorption unit operations can be used in a variety of applications, including (a) enhanced oil recovery: CO_2 can be injected into oil reservoirs to increase the pressure and improve oil recovery rates; (b) CCS: CO_2 can be stored underground in geological formations to reduce greenhouse gas emissions; (c) chemical production: CO_2 can be used as a feedstock for the production of chemicals such as methanol and urea; (d) Food and beverage industry: CO_2 is used in the production of carbonated beverages and in the packaging of food products to extend their shelf life; (e) medical applications: CO_2 is used in medical applications such as laparoscopic surgery and respiratory therapy; (f) agriculture: CO_2 can be used to enhance plant growth in greenhouses and other controlled environments; (g) renewable energy: CO_2 can be converted into fuels such as methane and hydrogen through a process called carbon capture and utilization.

Abbreviations and symbols

[BMI][Tf$_2$N]	1-butyl-3-methylimidazolium bis(trifluoromethylsulfonyl)imide
[bmim][BF$_4$]	1-butyl-3-methylimidazolium tetrafluoroborate
[bmim][Cl]	1-butyl-3-methylimidazolium chloride
[bmim][Tf$_2$N]	1-butyl-3-methylimidazolium bis(trifluoromethylsulfonyl)imide
[BMP][Br]	3-butyl-1-methylpyridinium bromide
[C4MIM][HSO$_4$]	1-butyl-3-methylimidazolium hydrogen sulfate
[C4MIM][Tf$_2$N]	1-butyl-3-methylimidazolium bis(trifluoromethylsulfonyl)imide

[emim][BF$_4$]	1-ethyl-3-methylimidazolium tetrafluoroborate
[EMIM][HSO$_4$]	1-ethyl-3-methylimidazolium hydrogen sulfate
[emim][Tf$_2$N]	1-ethyl-3-methylimidazolium bis(trifluoromethylsulfonyl)imide
[EMIM][Tf$_2$N]	1-ethyl-3-methylimidazolium bis(trifluoromethylsulfonyl)imide
[MMIM][Tf$_2$N]	1-methyl-3-octylimidazolium bis(trifluoromethylsulfonyl)imide
[OMIM][Tf$_2$N]	1-octyl-3-methylimidazolium bis(trifluoromethylsulfonyl)imide
AMP	2-amino-2-methyl-1-propanol
CAPEX	capital expenditure
CCS	carbon capture and storage
CO$_2$	carbon dioxide
DEA	diethanolamine
DGA	diglycolamine
DIPA	diisopropanolamine
DPEG	dimethylether of polyethylene glycol
FCI	fixed capital investment
H$_2$S	hydrogen sulfide
HCDPA	hydrocarbon dew-point adjustment
HisK	potassium histidinate
HSO$_4$	hydrogen sulfate
ILs	ionic liquids
kov	kinetic constant
LNG	liquefied natural gas
LysK	potassium lysinate
MDEA	methyl diethanolamine
MEA	Monoethanolamine
MOFs	metal-organic frameworks
NaOH	sodium hydroxide
NH$_3$	ammonia
NPV	net present value
POSTCAP	carbon capture process called
ProK	potassium prolinate
PZ	piperazine
SarK	potassium sarcosinate
Selexol	a registered trademark for a solvent
TauK	potassium taurinate
TEA	triethanolamine
Tf$_2$N	bis(trifluoromethylsulfonyl)imide,
WDPA	water dew-point adjustment

References

[1] Behrad Vakylabad A. Chapter 11 - Purification of syngas with nanofluid from mathematical modeling viewpoints. In: Rahimpour MR, Makarem MA, Meshksar M, editors. Advances in synthesis gas: methods, technologies and applications, vol. 4. Elsevier; 2023. p. 305—42.

[2] Behrad Vakylabad A. Chapter 4 - Syngas purification by modified solvents with nanoparticles. In: Rahimpour MR, Makarem MA, Meshksar M, editors. Advances in synthesis gas: methods, technologies and applications, vol. 2. Elsevier; 2023. p. 101—30.

[3] Vakylabad AB. Chapter 18 - Mass transfer enhancement in solar stills by nanofluids. In: Rahimpour MR, Makarem MA, Kiani MR, Sedghamiz MA, editors. Nanofluids and mass transfer. Elsevier; 2022. p. 431—47.

[4] Makarem MA, Moravvej Z, Rahimpour MR, Vakylabad AB. 8 - Biofuel production from microalgae and process enhancement by metabolic engineering and ultrasound. In: Rahimpour MR, Kamali R, Amin Makarem M, Manshadi MKD, editors. Advances in bioenergy and microfluidic applications. Elsevier; 2021. p. 209–30.

[5] Hassanzadeh A, Vakylabad AB. 11 - Fuel cells based on biomass. In: Rahimpour MR, Kamali R, Amin Makarem M, Manshadi MKD, editors. Advances in bioenergy and microfluidic applications. Elsevier; 2021. p. 275–301.

[6] Herzog H, Golomb D. Carbon capture and storage from fossil fuel use. Encyclopedia of Energy 2004;1(6562):277–87.

[7] Moravvej Z, Makarem MA, Rahimpour MR. The fourth generation of biofuel. In: Second and third generation of feedstocks. Elsevier; 2019. p. 557–97.

[8] Pini R, Storti G, Mazzotti M. A model for enhanced coal bed methane recovery aimed at carbon dioxide storage: the role of sorption, swelling and composition of injected gas. Adsorption 2011;17:889–900.

[9] Peters L, Hussain A, Follmann M, Melin T, Hägg M-B. CO2 removal from natural gas by employing amine absorption and membrane technology—a technical and economical analysis. Chemical Engineering Journal 2011;172(2–3):952–60.

[10] Tsubaki S, Furusawa K, Yamada H, Kato T, Higashii T, Fujii S, et al. Insights into the dielectric-heating-enhanced regeneration of CO2-rich aqueous amine solutions. ACS Sustainable Chemistry & Engineering 2020; 8(36):13593–9.

[11] Behrad Vakylabad A. Chapter 4 - Mass transfer mechanisms in nanofluids. In: Rahimpour MR, Makarem MA, Kiani MR, Sedghamiz MA, editors. Nanofluids and mass transfer. Elsevier; 2022. p. 97–113.

[12] Mofarahi M, Khojasteh Y, Khaledi H, Farahnak A. Design of CO2 absorption plant for recovery of CO2 from flue gases of gas turbine. Energy 2008;33(8):1311–9.

[13] Stewart M, Arnold K. Gas sweetening and processing field manual. Gulf Professional Publishing; 2011.

[14] Rufford TE, Smart S, Watson GC, Graham B, Boxall J, Da Costa JD, et al. The removal of CO2 and N2 from natural gas: a review of conventional and emerging process technologies. Journal of Petroleum Science and Engineering 2012;94:123–54.

[15] Wall TF. Combustion processes for carbon capture. Proceedings of the Combustion Institute 2007;31(1):31–47.

[16] Baker RW, Lokhandwala K. Natural gas processing with membranes: an overview. Industrial & Engineering Chemistry Research 2008;47(7): 2109–21.

[17] Mumford KA, Smith KH, Anderson CJ, Shen S, Tao W, Suryaputradinata YA, et al. Post-combustion capture of CO2: results from the solvent absorption capture plant at Hazelwood power station using potassium carbonate solvent. Energy & Fuels 2012;26(1):138–46.

[18] Mudhasakul S, Ku H-m, Douglas PL. A simulation model of a CO2 absorption process with methyldiethanolamine solvent and piperazine as an activator. International Journal of Greenhouse Gas Control 2013;15: 134–41.

[19] Idem R, Wilson M, Tontiwachwuthikul P, Chakma A, Veawab A, Aroonwilas A, et al. Pilot plant studies of the CO2 capture performance of aqueous MEA and mixed MEA/MDEA solvents at the University of Regina

CO2 capture technology development plant and the boundary dam CO2 capture demonstration plant. Industrial & Engineering Chemistry Research 2006;45(8):2414−20.

[20] Cousins A, Wardhaugh L, Feron P. A survey of process flow sheet modifications for energy efficient CO2 capture from flue gases using chemical absorption. International Journal of Greenhouse Gas Control 2011;5(4):605−19.

[21] Wang M, Lawal A, Stephenson P, Sidders J, Ramshaw C. Post-combustion CO2 capture with chemical absorption: a state-of-the-art review. Chemical Engineering Research and Design 2011;89(9):1609−24.

[22] Ebenezer SA, Gudmunsson J. Removal of carbon dioxide from natural gas for LNG production. Semester project work; 2005.

[23] Abotaleb A, Gladich I, Alkhateeb A, Mardini N, Bicer Y, Sinopoli A. Chemical and physical systems for sour gas removal: an overview from reaction mechanisms to industrial implications. Journal of Natural Gas Science and Engineering 2022:104755.

[24] Baxter L, Baxter A, Burt S, editors. Cryogenic CO2 capture as a cost-effective CO2 capture process. International Pittsburgh Coal Conference; 2009.

[25] Figueroa JD, Fout T, Plasynski S, McIlvried H, Srivastava RD. Advances in CO2 capture technology—the US Department of Energy's Carbon Sequestration Program. International Journal of Greenhouse Gas Control 2008;2(1):9−20.

[26] Aaron D, Tsouris C. Separation of CO2 from flue gas: a review. Separation Science and Technology 2005;40(1−3):321−48.

[27] Veawab A, Tontiwachwuthikul P, Aroonwilas A, Chakma A, editors. Performance and cost analysis for CO2 capture from flue gas streams: absorption and regeneration aspects. Greenhouse Gas Control Technologies-6th International Conference. Elsevier; 2003.

[28] Ntiamoah A, Ling J, Xiao P, Webley PA, Zhai Y. CO2 capture by temperature swing adsorption: use of hot CO2-rich gas for regeneration. Industrial & Engineering Chemistry Research 2016;55(3):703−13.

[29] Wang L, Liu S, Wang R, Li Q, Zhang S. Regulating phase separation behavior of a DEEA−TETA biphasic solvent using sulfolane for energy-saving CO2 capture. Environmental Science & Technology 2019;53(21):12873−81.

[30] Yeh AC, Bai H. Comparison of ammonia and monoethanolamine solvents to reduce CO2 greenhouse gas emissions. Science of the Total Environment 1999;228(2-3):121−33.

[31] López-Pacheco IY, Rodas-Zuluaga LI, Fuentes-Tristan S, Castillo-Zacarías C, Sosa-Hernández JE, Barceló D, et al. Phycocapture of CO2 as an option to reduce greenhouse gases in cities: carbon sinks in urban spaces. Journal of CO2 Utilization 2021;53:101704.

[32] Anwar M, Fayyaz A, Sohail N, Khokhar M, Baqar M, Yasar A, et al. CO2 utilization: turning greenhouse gas into fuels and valuable products. Journal of Environmental Management 2020;260:110059.

[33] Araújo OdQF, Interlenghi SF, de Medeiros JL. Carbon management in the CO2-rich natural gas to energy supply-chain. In: Advances in carbon management technologies. CRC Press; 2020. p. 286−322.

[34] de Medeiros JL, de Oliveira Arinelli L, Teixeira AM, Araújo OdQF. Offshore processing of CO2-rich natural gas with supersonic separator. Springer; 2019.

[35] Bakonyi P, Peter J, Koter S, Mateos R, Kumar G, Koók L, et al. Possibilities for the biologically-assisted utilization of CO2-rich gaseous waste streams generated during membrane technological separation of biohydrogen. Journal of CO2 Utilization 2020;36:231–43.

[36] da Cunha GP, de Medeiros JL, Araújo OdQF. Carbon capture from CO2-rich natural gas via gas-liquid membrane contactors with aqueous-amine solvents: a review. Gases 2022;2(3):98–133.

[37] Li H, Guo H, Shen S. Low-energy-consumption CO2 capture by liquid–solid phase change absorption using water-lean blends of amino acid salts and 2-alkoxyethanols. ACS Sustainable Chemistry & Engineering 2020;8(34):12956–67.

[38] Vega F, Baena-Moreno F, Fernández LMG, Portillo E, Navarrete B, Zhang Z. Current status of CO2 chemical absorption research applied to CCS: towards full deployment at industrial scale. Applied Energy 2020;260: 114313.

[39] Sifat NS, Haseli Y. A critical review of CO2 capture technologies and prospects for clean power generation. Energies 2019;12(21):4143.

[40] Hasan S, Abbas AJ, Nasr GG. Improving the carbon capture efficiency for gas power plants through amine-based absorbents. Sustainability 2020; 13(1):72.

[41] Zhang Y, Ma L, Lv Y, Tan T. Facile manufacture of COF-based mixed matrix membranes for efficient CO2 separation. Chemical Engineering Journal 2022;430:133001.

[42] Benrabaa R, Boukhlouf H, Löfberg A, Rubbens A, Vannier R-N, Bordes-Richard E, et al. Nickel ferrite spinel as catalyst precursor in the dry reforming of methane: synthesis, characterization and catalytic properties. Journal of Natural Gas Chemistry 2012;21(5):595–604.

[43] Li H, Zhang R, Wang T, Sun X, Hou C, Xu R, et al. Simulation of H2S and CO2 removal from IGCC syngas by cryogenic distillation. Carbon Capture Science & Technology 2022;3:100012.

[44] Costa EP, Roccamante M, Amorim CC, Oller I, Pérez JAS, Malato S. New trend on open solar photoreactors to treat micropollutants by photo-Fenton at circumneutral pH: increasing optical pathway. Chemical Engineering Journal 2020;385:123982.

[45] Surra E, Ribeiro RP, Santos T, Bernardo M, Mota JP, Lapa N, et al. Evaluation of activated carbons produced from Maize Cob Waste for adsorption-based CO2 separation and biogas upgrading. Journal of Environmental Chemical Engineering 2022;10(1):107065.

[46] Zarei V, Nasiri A. Stabilizing Asmari Formation interlayer shales using water-based mud containing biogenic silica oxide nanoparticles synthesized. Journal of Natural Gas Science and Engineering 2021;91: 103928.

[47] Ochedi FO, Yu J, Yu H, Liu Y, Hussain A. Carbon dioxide capture using liquid absorption methods: a review. Environmental Chemistry Letters 2021;19:77–109.

[48] Niu Z-q, Guo Y-c, Lin W-y. Comparison of capture efficiencies of carbon dioxide by fine spray of aqueous ammonia and MEA solution. Journal of Chemical Engineering of Chinese Universities 2010;24:514–7.

[49] Tan L, Shariff A, Lau K, Bustam M. Factors affecting CO2 absorption efficiency in packed column: a review. Journal of Industrial and Engineering Chemistry 2012;18(6):1874–83.

[50] Akkarawatkhoosith N, Nopcharoenkul W, Kaewchada A, Jaree A. Mass transfer correlation and optimization of carbon dioxide capture in a

microchannel contactor: a case of CO2-rich gas. Energies 2020;13(20): 5465.
[51] Li H, Tang Z, Li N, Cui L, Mao X-z. Mechanism and process study on steel slag enhancement for CO2 capture by seawater. Applied Energy 2020;276: 115515.
[52] Nishikawa N, Morishita M, Uchiyama M, Yamaguchi F, Ohtsubo K, Kimuro H, et al. CO2 clathrate formation and its properties in the simulated deep ocean. Energy Conversion and Management 1992;33(5-8): 651−7.
[53] Kittiampon N, Kaewchada A, Jaree A. Carbon dioxide absorption using ammonia solution in a microchannel. International Journal of Greenhouse Gas Control 2017;63:431−41.
[54] Bergamasco L, Izquierdo S, Pagonabarraga I, Fueyo N. Multi-scale permeability of deformable fibrous porous media. Chemical Engineering Science 2015;126:471−82.
[55] Yin Y, Guo R, Zhu C, Fu T, Ma Y. Enhancement of gas-liquid mass transfer in microchannels by rectangular baffles. Separation and Purification Technology 2020;236:116306.
[56] Liu J, Wang S, Zhao B, Tong H, Chen C. Absorption of carbon dioxide in aqueous ammonia. Energy Procedia 2009;1(1):933−40.
[57] Qing Z, Yincheng G, Zhenqi N. Experimental studies on removal capacity of carbon dioxide by a packed reactor and a spray column using aqueous ammonia. Energy Procedia 2011;4:519−24.
[58] Qi G, Wang S, Yu H, Wardhaugh L, Feron P, Chen C. Development of a rate-based model for CO2 absorption using aqueous NH3 in a packed column. International Journal of Greenhouse Gas Control 2013;17:450−61.
[59] Pennline HW, Luebke DR, Jones KL, Myers CR, Morsi BI, Heintz YJ, et al. Progress in carbon dioxide capture and separation research for gasification-based power generation point sources. Fuel Processing Technology 2008;89(9):897−907.
[60] Padurean A, Cormos C-C, Agachi P-S. Pre-combustion carbon dioxide capture by gas−liquid absorption for Integrated Gasification Combined Cycle power plants. International Journal of Greenhouse Gas Control 2012;7:1−11.
[61] Descamps C, Bouallou C, Kanniche M. Efficiency of an Integrated Gasification Combined Cycle (IGCC) power plant including CO2 removal. Energy 2008;33(6):874−81.
[62] Verma N, Verma A. Amine system problems arising from heat stable salts and solutions to improve system performance. Fuel Processing Technology 2009;90(4):483−9.
[63] Le Moullec Y, Kanniche M. Screening of flowsheet modifications for an efficient monoethanolamine (MEA) based post-combustion CO2 capture. International Journal of Greenhouse Gas Control 2011;5(4):727−40.
[64] Oh S-Y, Yun S, Kim J-K. Process integration and design for maximizing energy efficiency of a coal-fired power plant integrated with amine-based CO2 capture process. Applied Energy 2018;216:311−22.
[65] Aboudheir A, McIntyre G. Industrial design and optimization of CO2 capture, dehydration, and compression facilities. HTC Purenergy; 2008. Regina, SK, Canada and Bryan Research & Engineering, Bryan, Texas, USA/Report, http://www.bre.com.
[66] Kumar S, Cho JH, Moon I. Ionic liquid-amine blends and CO2BOLs: prospective solvents for natural gas sweetening and CO2 capture

technology—a review. International Journal of Greenhouse Gas Control 2014;20:87—116.
[67] Wang R, Li D, Liang D. Modeling of CO2 capture by three typical amine solutions in hollow fiber membrane contactors. Chemical Engineering and Processing: Process Intensification 2004;43(7):849—56.
[68] Kidnay AJ, Parrish WR, McCartney DG. Fundamentals of natural gas processing. CRC Press; 2019.
[69] Vakylabad AB. Treatment of highly concentrated formaldehyde effluent using adsorption and ultrasonic dissociation on mesoporous copper iodide (CuI) nano-powder. Journal of Environmental Management 2021; 285:112085.
[70] Bhushan B. Self-assembled monolayers for controlling hydrophobicity and/or friction and wear. In: Modern tribology handbook. CRC Press; 2000. p. 939—60. Two volume set.
[71] Aroonwilas A, Veawab A. Characterization and comparison of the CO2 absorption performance into single and blended alkanolamines in a packed column. Industrial & Engineering Chemistry Research 2004;43(9): 2228—37.
[72] Frauenkron M, Aktiengesellschaft B, Ruider GN, Ho H. Ethanolamines and propanolamines. Environmental Protection 2012;421(8).
[73] Sartori G, Ho W, Savage D, Chludzinski G, Wlechert S. Sterically-hindered amines for acid-gas absorption. Separation and Purification Methods 1987;16(2):171—200.
[74] Sartori G, Savage DW. Sterically hindered amines for carbon dioxide removal from gases. Industrial & Engineering Chemistry Fundamentals 1983;22(2):239—49.
[75] Bougie F, Iliuta MC. Sterically hindered amine-based absorbents for the removal of CO2 from gas streams. Journal of Chemical & Engineering Data 2012;57(3):635—69.
[76] Nwaoha C, Saiwan C, Supap T, Idem R, Tontiwachwuthikul P, Rongwong W, et al. Carbon dioxide (CO2) capture performance of aqueous tri-solvent blends containing 2-amino-2-methyl-1-propanol (AMP) and methyldiethanolamine (MDEA) promoted by diethylenetriamine (DETA). International Journal of Greenhouse Gas Control 2016;53:292—304.
[77] Conway W, Bruggink S, Beyad Y, Luo W, Melián-Cabrera I, Puxty G, et al. CO2 absorption into aqueous amine blended solutions containing monoethanolamine (MEA), N, N-dimethylethanolamine (DMEA), N, N-diethylethanolamine (DEEA) and 2-amino-2-methyl-1-propanol (AMP) for post-combustion capture processes. Chemical Engineering Science 2015; 126:446—54.
[78] Yih SM, Shen KP. Kinetics of carbon dioxide reaction with sterically hindered 2-amino-2-methyl-1-propanol aqueous solutions. Industrial & Engineering Chemistry Research 1988;27(12):2237—41.
[79] Mandal B, Bandyopadhyay S. Absorption of carbon dioxide into aqueous blends of 2-amino-2-methyl-1-propanol and monoethanolamine. Chemical Engineering Science 2006;61(16):5440—7.
[80] Li M-H, Chang B-C. Solubilities of carbon dioxide in water+ monoethanolamine+ 2-amino-2-methyl-1-propanol. Journal of Chemical and Engineering Data 1994;39(3):448—52.
[81] Mandal BP, Biswas A, Bandyopadhyay S. Selective absorption of H2S from gas streams containing H2S and CO2 into aqueous solutions of N-

methyldiethanolamine and 2-amino-2-methyl-1-propanol. Separation and Purification Technology 2004;35(3):191−202.

[82] Chakraborty A, Astarita G, Bischoff K. CO2 absorption in aqueous solutions of hindered amines. Chemical Engineering Science 1986;41(4): 997−1003.

[83] Yamada H, Chowdhury FA, Matsuzaki Y, Goto K, Higashii T, Kazama S. Effect of alcohol chain length on carbon dioxide absorption into aqueous solutions of alkanolamines. Energy Procedia 2013;37:499−504.

[84] Barzagli F, Mani F, Peruzzini M. Efficient CO2 absorption and low temperature desorption with non-aqueous solvents based on 2-amino-2-methyl-1-propanol (AMP). International Journal of Greenhouse Gas Control 2013;16:217−23.

[85] Koronaki IP, Prentza L, Papaefthimiou V. Modeling of CO2 capture via chemical absorption processes – an extensive literature review. Renewable and Sustainable Energy Reviews 2015;50:547−66.

[86] Dashti A, Raji M, Alivand MS, Mohammadi AH. Estimation of CO2 equilibrium absorption in aqueous solutions of commonly used amines using different computational schemes. Fuel 2020;264:116616.

[87] Low GKC, McEvoy SR, Matthews RW. Formation of nitrate and ammonium ions in titanium dioxide mediated photocatalytic degradation of organic compounds containing nitrogen atoms. Environmental Science & Technology 1991;25(3):460−7.

[88] Deng S, Bai R. Removal of trivalent and hexavalent chromium with aminated polyacrylonitrile fibers: performance and mechanisms. Water Research 2004;38(9):2424−32.

[89] Ko J-J, Tsai T-C, Lin C-Y, Wang H-M, Li M-H. Diffusivity of nitrous oxide in aqueous alkanolamine solutions. Journal of Chemical & Engineering Data 2001;46(1):160−5.

[90] Hamzehie M, Mazinani S, Davardoost F, Mokhtare A, Najibi H, Van der Bruggen B, et al. Developing a feed forward multilayer neural network model for prediction of CO2 solubility in blended aqueous amine solutions. Journal of Natural Gas Science and Engineering 2014;21:19−25.

[91] Penttilä A, Dell'Era C, Uusi-Kyyny P, Alopaeus V. The Henry's law constant of N2O and CO2 in aqueous binary and ternary amine solutions (MEA, DEA, DIPA, MDEA, and AMP). Fluid Phase Equilibria 2011;311: 59−66.

[92] Kittel J, Fleury E, Vuillemin B, Gonzalez S, Ropital F, Oltra R. Corrosion in alkanolamine used for acid gas removal: from natural gas processing to CO2 capture. Materials and Corrosion 2012;63(3):223−30.

[93] Afkhamipour M, Mofarahi M. Review on the mass transfer performance of CO2 absorption by amine-based solvents in low-and high-pressure absorption packed columns. RSC advances 2017;7(29):17857−72.

[94] Rao AB, Rubin ES. A technical, economic, and environmental assessment of amine-based CO2 capture technology for power plant greenhouse gas control. Environmental Science & Technology 2002;36(20):4467−75.

[95] Camper D, Bara JE, Gin DL, Noble RD. Room-temperature ionic liquid–amine solutions: tunable solvents for efficient and reversible capture of CO2. Industrial & Engineering Chemistry Research 2008;47(21):8496−8.

[96] Vega F, Cano M, Camino S, Fernández LMG, Portillo E, Navarrete B. Solvents for carbon dioxide capture. In: Carbon dioxide chemistry, capture and oil recovery; 2018. p. 142−63.

[97] Olajire AA. CO2 capture and separation technologies for end-of-pipe applications−a review. Energy 2010;35(6):2610−28.

[98] Hosseini-Ardali SM, Hazrati-Kalbibaki M, Fattahi M, Lezsovits F. Multi-objective optimization of post combustion CO2 capture using methyldiethanolamine (MDEA) and piperazine (PZ) bi-solvent. Energy 2020;211:119035.
[99] Kenarsari SD, Yang D, Jiang G, Zhang S, Wang J, Russell AG, et al. Review of recent advances in carbon dioxide separation and capture. Rsc Advances 2013;3(45):22739—73.
[100] Ghanbarabadi H, Khoshandam B. Simulation and comparison of Sulfinol solvent performance with Amine solvents in removing sulfur compounds and acid gases from natural sour gas. Journal of Natural Gas Science and Engineering 2015;22:415—20.
[101] Ban ZH, Keong LK, Mohd Shariff A. Physical absorption of CO2 capture: a review. Advanced Materials Research 2014;917:134—43.
[102] Chalermthai P, Akkarawatkhoosith N, Kaewchada A, Jaree A. Carbon dioxide removal via absorption using artificial seawater in a microchannel for the case of CO2-rich gas. Chemical Engineering and Processing-Process Intensification 2022;175:108928.
[103] Barbosa LC, Araújo OdQF, de Medeiros JL. Carbon capture and adjustment of water and hydrocarbon dew-points via absorption with ionic liquid [Bmim][NTf2] in offshore processing of CO2-rich natural gas. Journal of Natural Gas Science and Engineering 2019;66:26—41.
[104] Hart A, Gnanendran N. Cryogenic CO2 capture in natural gas. Energy Procedia 2009;1(1):697—706.
[105] Gupta M, Coyle I, Thambimuthu K, editors. CO2 capture technologies and opportunities in Canada. 1st Canadian CC&S Technology Roadmap Workshop; 2003. Citeseer.
[106] Adams D. Flue gas treatment for CO2 capture. IEA Clean Coal Centre London; 2010.
[107] Yamada H. Amine-based capture of CO2 for utilization and storage. Polymer Journal 2021;53(1):93—102.
[108] Bernardo P, Drioli E. Membrane gas separation progresses for process intensification strategy in the petrochemical industry. Petroleum Chemistry 2010;50:271—82.
[109] Shen M, Hu Z, Kong F, Tong L, Yin S, Liu C, et al. Comprehensive technology and economic evaluation based on the promotion of large-scale carbon capture and storage demonstration projects. Reviews in Environmental Science and Bio/Technology 2023:1—63.
[110] Machado PB, Monteiro JG, Medeiros JL, Epsom HD, Araujo OQ. Supersonic separation in onshore natural gas dew point plant. Journal of Natural Gas Science and Engineering 2012;6:43—9.
[111] Zhang R, Li Y, He X, Niu Y, Li Ce, Amer MW, et al. Investigation of the improvement of the CO2 capture performance of aqueous amine sorbents by switching from dual-amine to trio-amine systems. Separation and Purification Technology 2023;316:123810.
[112] Ellaf A, Taqvi SAA, Zaeem D, Siddiqui FUH, Kazmi B, Idris A, et al. Energy, exergy, economic, environment, exergo-environment based assessment of amine-based hybrid solvents for natural gas sweetening. Chemosphere 2023;313:137426.
[113] He X, He H, Barzagli F, Amer MW, Li Ce, Zhang R. Analysis of the energy consumption in solvent regeneration processes using binary amine blends for CO2 capture. Energy 2023;270:126903.
[114] Joarder MSA, Rashid F, Abir MA, Zakir MG. A prospective approach to separate industrial carbon dioxide and flue gases. Separation Science and Technology 2023;58(10):1795—805.

[115] Wang T, Zeng S, Gu Z. Crown ether nanopores in graphene membranes for highly efficient CO2/CH4 and CO2/CO separation: a theoretical study. ACS Applied Nano Materials 2023;6(13):12372−80.
[116] Ariadji T, Adisasmito S, Mucharam L, Abdassah D. A new approach of CO2 separation by applying rapid expansion of supercritical CO2 rich natural gas. Petroleum Research 2023;8(1):71−6.
[117] Nandakishora Y, Sahoo RK, Murugan S, Gu S. 4E analysis of the cryogenic CO2 separation process integrated with waste heat recovery. Energy 2023; 278:127922.
[118] Weir H, Sanchez-Fernandez E, Charalambous C, Ros J, Monteiro JGM-S, Skylogianni E, et al. Impact of high capture rates and solvent and emission management strategies on the costs of full-scale post-combustion CO2 capture plants using long-term pilot plant data. International Journal of Greenhouse Gas Control 2023;126:103914.
[119] Liu C, Zhou S, Yu D, Etschmann B, Zhang L. Flowsheet modelling and techno-economic analysis of CO2 capture coupled pyro-hydrolysis of CaCl2 waste for HCl acid regeneration. Journal of Cleaner Production 2023;419:138195.
[120] Arshad N, Alhajaj A. Process synthesis for amine-based CO2 capture from combined cycle gas turbine power plant. Energy 2023;274:127391.
[121] Øi LE, Fazlagic M, editors. Glycol dehydration of captured carbon dioxide using Aspen Hysys simulation. In: Proceedings of the 55th Conference on Simulation and Modelling (SIMS 55), Modelling, Simulation and Optimization. Aalborg, Denmark: Linköping University Electronic Press; 2014.
[122] Nedoma M, Netušil M, Hrdlička J. Integration of adsorption based post-combustion carbon dioxide capture for a natural gas-fired combined heat and power plant. Fuel 2023;354:129346.
[123] Nikkhah H, Nikkhah A, Ghalavand Y. Acid gas preparation for enhanced oil recovery: techno-economic analysis of different dehydration processes. Separation Science and Technology 2023:1−13.
[124] Pessoaa FLP, da Silva Calixtoa EE, Ávilaa JS, da Costa Amaralb M. Supersonic separation: natural gas conditioning case study. VI International Symposium on Innovation and Technology (SIINTEC), Salvador–BA, 2022.
[125] Tsoy N, Steubing B, Guinée JB. Ex-ante life cycle assessment of polyols using carbon captured from industrial process gas. Green Chemistry 2023; 25:5526−38.
[126] Gardarsdottir SO, De Lena E, Romano M, Roussanaly S, Voldsund M, Pérez-Calvo J-F, et al. Comparison of technologies for CO2 capture from cement production—Part 2: cost analysis. Energies 2019;12(3):542.
[127] Khan MHA, Daiyan R, Neal P, Haque N, MacGill I, Amal R. A framework for assessing economics of blue hydrogen production from steam methane reforming using carbon capture storage & utilisation. International Journal of Hydrogen Energy 2021;46(44):22685−706.
[128] Gambhir A, Tavoni M. Direct air carbon capture and sequestration: how it works and how it could contribute to climate-change mitigation. One Earth 2019;1(4):405−9.
[129] Shahbaz M, Rashid N, Saleem J, Mackey H, McKay G, Al-Ansari T. A review of waste management approaches to maximise sustainable value of waste from the oil and gas industry and potential for the State of Qatar. Fuel 2023;332:126220.
[130] Rath GK, Pandey G, Singh S, Molokitina N, Kumar A, Joshi S, et al. Carbon dioxide separation technologies: applicable to net zero. Energies 2023; 16(10):4100.

[131] Gayathri R, Ranjitha J, Vijayalakshmi S. Carbon dioxide capture and bioenergy production by utilizing the biological system. In: Sustainable utilization of carbon dioxide: from waste to product. Springer; 2023. p. 159—94.

[132] Satterfield T, Nawaz S, St-Laurent GP. Exploring public acceptability of direct air carbon capture with storage: climate urgency, moral hazards and perceptions of the 'whole versus the parts'. Climatic Change 2023; 176(2):14.

[133] Davoodi S, Al-Shargabi M, Wood DA, Rukavishnikov VS, Minaev KM. Review of technological progress in carbon dioxide capture, storage, and utilization. Gas Science and Engineering 2023:205070.

[134] Gautam A, Mondal MK. Review of recent trends and various techniques for CO2 capture: special emphasis on biphasic amine solvents. Fuel 2023; 334:126616.

[135] Muntaha N, Rain MI, Goni LK, Shaikh MAA, Jamal MS, Hossain M. A review on carbon dioxide minimization in biogas upgradation technology by chemical absorption processes. ACS omega 2022;7(38):33680—98.

[136] Aghel B, Behaein S, Alobaid F. CO2 capture from biogas by biomass-based adsorbents: a review. Fuel 2022;328:125276.

[137] Kianfar E. A review of recent advances in carbon dioxide absorption—stripping by employing a gas—liquid hollow fiber polymeric membrane contactor. Polymer Bulletin 2022:1—37.

[138] Tengku Hassan TNA, Shariff AM, Mohd Pauzi MMi, Khidzir MS, Surmi A. Insights on cryogenic distillation technology for simultaneous CO2 and H2S removal for sour gas fields. Molecules 2022;27(4):1424.

[139] Keith DW, Holmes G, Angelo DS, Heidel K. A process for capturing CO2 from the atmosphere. Joule 2018;2(8):1573—94.

[140] Aroonwilas A, Veawab A. Integration of CO2 capture unit using single- and blended-amines into supercritical coal-fired power plants: Implications for emission and energy management. International Journal of Greenhouse Gas Control 2007;1(2):143—50.

[141] Zhang Y, Chen H, Chen C-C, Plaza JM, Dugas R, Rochelle GT. Rate-based process modeling study of CO2 capture with aqueous monoethanolamine solution. Industrial & Engineering Chemistry Research 2009;48(20): 9233—46.

[142] Einbu A, Pettersen T, Morud J, Tobiesen A, Jayarathna C, Skagestad R, et al. Energy assessments of onboard CO2 capture from ship engines by MEA-based post combustion capture system with flue gas heat integration. International Journal of Greenhouse Gas Control 2022;113: 103526.

[143] Magli F, Spinelli M, Fantini M, Romano MC, Gatti M. Techno-economic optimization and off-design analysis of CO2 purification units for cement plants with oxyfuel-based CO2 capture. International Journal of Greenhouse Gas Control 2022;115:103591.

[144] Sayyah Alborzi Z, Amini Y, Amirabedi P, Raveshiyan S, Hassanvand A. Computational fluid dynamics simulation of a membrane contactor for CO2 separation: two types of membrane evaluation. Chemical Engineering & Technology 2023;46:2034—45.

[145] Streb A, Mazzotti M. Performance limits of neural networks for optimizing an adsorption process for hydrogen purification and CO2 capture. Computers & Chemical Engineering 2022;166:107974.

[146] Quan H, Dong S, Zhao D, Li H, Geng J, Liu H. Generic AI models for mass transfer coefficient prediction in amine-based CO2 absorber, Part II: RBFNN and RF model. AIChE Journal 2023;69(1):e17904.

[147] Zhang Z, Hong S-H, Lee C-H. Role and impact of wash columns on the performance of chemical absorption-based CO2 capture process for blast furnace gas in iron and steel industries. Energy 2023;271:127020.

[148] Conversano A, Porcu A, Mureddu M, Pettinau A, Gatti M. Bench-scale absorption testing of aqueous potassium lysinate as a new solvent for CO2 capture in natural gas-fired power plants. International Journal of Greenhouse Gas Control 2021;106:103268.

[149] Gatti M, Martelli E, Di Bona D, Gabba M, Scaccabarozzi R, Spinelli M, et al. Preliminary performance and cost evaluation of four alternative technologies for post-combustion CO2 capture in natural gas-fired power plants. Energies 2020;13(3):543.

[150] MacDowell N, Florin N, Buchard A, Hallett J, Galindo A, Jackson G, et al. An overview of CO2 capture technologies. Energy & Environmental Science 2010;3(11):1645–69.

[151] DuPart M, Bacon T, Edwards D. Understanding corrosion in alkanolamine gas treating plants: part 2. Hydrocarbon Processing;(United States) 1993;72(5).

[152] Mumford KA, Wu Y, Smith KH, Stevens GW. Review of solvent based carbon-dioxide capture technologies. Frontiers of Chemical Science and Engineering 2015;9:125–41.

[153] Budzianowski WM. Energy efficient solvents for CO2 capture by gas-liquid absorption: compounds, blends and advanced solvent systems. Springer; 2016.

[154] Majchrowicz ME. Amino acid salt solutions for carbon dioxide capture. University of Twente Netherlands; 2014.

[155] Oexmann J, Kather A. Minimising the regeneration heat duty of post-combustion CO2 capture by wet chemical absorption: the misguided focus on low heat of absorption solvents. International Journal of Greenhouse Gas Control 2010;4(1):36–43.

[156] van der Ham L, Goetheer E, Fernandez ES, Abu-Zahra M, Vlugt T. Precipitating amino acid solutions. Absorption-based post-combustion capture of carbon dioxide. Elsevier; 2016. p. 103–19.

[157] Jockenhövel T, Schneider R. Towards commercial application of a second-generation post-combustion capture technology—pilot plant validation of the siemens capture process and implementation of a first demonstration case. Energy Procedia 2011;4:1451–8.

[158] Fernandez ES, Goetheer EL, Manzolini G, Macchi E, Rezvani S, Vlugt TJ. Thermodynamic assessment of amine based CO2 capture technologies in power plants based on European Benchmarking Task Force methodology. Fuel 2014;129:318–29.

[159] Hu G, Smith KH, Wu Y, Mumford KA, Kentish SE, Stevens GW. Carbon dioxide capture by solvent absorption using amino acids: a review. Chinese Journal of Chemical Engineering 2018;26(11):2229–37.

[160] Gazzani M, Macchi E, Manzolini G. CO2 capture in integrated gasification combined cycle with SEWGS—Part A: thermodynamic performances. Fuel 2013;105:206–19.

[161] Aronu UE, Mejdell T, Hjarbo KW, Chikukwa A, Grimstvedt AM, Lund A. Baseline test of capture from bio flue gas with MEA at Tiller plant. SINTEF Rapport 2021.

[162] Lerche BM. CO2 capture from flue gas using amino acid salt solutions. 2012.

[163] Shen S, Yang Y-n, Bian Y, Zhao Y. Kinetics of CO2 absorption into aqueous basic amino acid salt: potassium salt of lysine solution. Environmental Science & Technology 2016;50(4):2054–63.

[164] Shen S, Zhao Y, Bian Y, Wang Y, Guo H, Li H. CO2 absorption using aqueous potassium lysinate solutions: vapor–liquid equilibrium data and modelling. The Journal of Chemical Thermodynamics 2017;115:209–20.

[165] Zhao Y, Bian Y, Li H, Guo H, Shen S, Han J, et al. A comparative study of aqueous potassium lysinate and aqueous monoethanolamine for postcombustion CO2 capture. Energy & Fuels 2017;31(12):14033–44.

[166] Conversano A, Delgado S, Coquelet C, Consonni S, Gatti M. CO2 solubility modelling in non-precipitating aqueous solutions of potassium lysinate. Separation and Purification Technology 2022;300:121855.

[167] Guo H, Li H, Shen S. CO2 capture by water-lean amino acid salts: absorption performance and mechanism. Energy & Fuels 2018;32(6):6943–54.

[168] Mauch A, Wunderlich S, Zarnkow M, Becker T, Jacob F, Arendt EK. Part II. The use of malt produced with 70% less malting loss for beer production: impact on processability and final quality. Journal of the American Society of Brewing Chemists 2011;69(4):239–54.

[169] Choi GN, Chu R, Degen B, Wen H, Richen PL, Chinn D. CO2 removal from power plant flue gas–cost efficient design and integration study. Carbon Dioxide Capture Storage Deep Geol Form 2005;1:99–116.

[170] Cremona R, Delgado S, Valtz A, Conversano A, Gatti M, Coquelet C. Density and viscosity measurements and modeling of CO2-loaded and unloaded aqueous solutions of potassium lysinate. Journal of Chemical & Engineering Data 2021;66(12):4460–75.

[171] Ecker A-M, Klein H, Peschel A. Systematic and efficient optimisation-based design of a process for CO2 removal from natural gas. Chemical Engineering Journal 2022:136178.

[172] Kidnay AJ, Parrish WR. Fundamentals of natural gas processing. CRC press; 2006.

[173] Moravvej Z, Soroush E, Makarem MA, Rahimpour MR. Thermochemical routes for hydrogen production from biomass. In: Advances in bioenergy and microfluidic applications. Elsevier; 2021. p. 193–208.

[174] Burgers W, Northrop P, Kheshgi H, Valencia J. Worldwide development potential for sour gas. Energy Procedia 2011;4:2178–84.

[175] Berstad D, Nekså P, Anantharaman R. Low-temperature CO2 removal from natural gas. Energy Procedia 2012;26:41–8.

[176] Holmes AS, Ryan JM. Cryogenic distillative separation of acid gases from methane. Google Patents; 1982.

[177] Holmes AS, Ryan JM. Distillative separation of carbon dioxide from light hydrocarbons. Google Patents; 1982.

[178] Valencia JA, Denton RD. Method and apparatus for separating carbon dioxide and other acid gases from methane by the use of distillation and a controlled freezing zone. Google Patents; 1985.

[179] Kelley B, Valencia J, Northrop P, Mart C. Controlled Freeze Zone™ for developing sour gas reserves. Energy Procedia 2011;4:824–9.

[180] Pellegrini LA. Process for the removal of CO2 from acid gas. WO Patent; 2014.

[181] Langè S, Pellegrini LA, Vergani P, Lo Savio M. Energy and economic analysis of a new low-temperature distillation process for the upgrading of high-CO2 content natural gas streams. Industrial & Engineering Chemistry Research 2015;54(40):9770–82.

[182] Van Veen J, Betting M. Removing solids from a fluid. Google Patents; 2001.

[183] Willems G, Golombok M, Tesselaar G, Brouwers J. Condensed rotational separation of CO2 from natural gas. AIChE Journal 2010;56(1):150–9.

[184] Brouwers B. Rotational particle separator: a new method for separating fine particles and mists from gases. Chemical Engineering & Technology: Industrial Chemistry-Plant Equipment-Process Engineering-Biotechnology 1996;19(1):1–10.

[185] Ranke G, Weiss H. Separation of gaseous components from a gaseous mixture by physical scrubbing. Google Patents; 1982.

[186] Kohl AL, Miller FE. Organic carbonate process for carbon dioxide. Google Patents; 1960.

[187] Mak J, Wierenga D, Nielsen D, Graham C, Viejo A, editors. New physical solvent treating configurations for offshore high pressure CO2 removal. Offshore Technology Conference; 2003. OnePetro.

[188] Feron PH, ten Asbroek N. New solvents based on amino-acid salts for CO2 capture from flue gases. In: Greenhouse gas control technologies. vol. 7; 2005. p. 1153–8. Elsevier.

[189] Kohl A, Nielsen R. Gas purification. Houston, Texas: Gulf Publishing Company; 1997.

[190] Maqsood K, Mullick A, Ali A, Kargupta K, Ganguly S. Cryogenic carbon dioxide separation from natural gas: a review based on conventional and novel emerging technologies. Reviews in Chemical Engineering 2014; 30(5):453–77.

[191] Zagho MM, Hassan MK, Khraisheh M, Al-Maadeed MAA, Nazarenko S. A review on recent advances in CO2 separation using zeolite and zeolite-like materials as adsorbents and fillers in mixed matrix membranes (MMMs). Chemical Engineering Journal Advances 2021;6:100091.

[192] Kayal S, Chakraborty A. Activated carbon (type Maxsorb-III) and MIL-101 (Cr) metal organic framework based composite adsorbent for higher CH4 storage and CO2 capture. Chemical Engineering Journal 2018;334:780–8.

[193] Hospital-Benito D, Lemus J, Moya C, Santiago R, Ferro V, Palomar J. Techno-economic feasibility of ionic liquids-based CO2 chemical capture processes. Chemical Engineering Journal 2021;407:127196.

[194] Wang D, Huang R, Liu W, Sun D, Li Z. Fe-based MOFs for photocatalytic CO2 reduction: role of coordination unsaturated sites and dual excitation pathways. Acs Catalysis 2014;4(12):4254–60.

[195] Zhu S, Liao W, Zhang M, Liang S. Design of spatially separated Au and CoO dual cocatalysts on hollow TiO2 for enhanced photocatalytic activity towards the reduction of CO2 to CH4. Chemical Engineering Journal 2019; 361:461–9.

[196] Mondal S, Powar NS, Paul R, Kwon H, Das N, Wong BM, et al. Nanoarchitectonics of metal-free porous polyketone as photocatalytic assemblies for artificial photosynthesis. ACS Applied Materials & Interfaces 2021;14(1):771–83.

[197] Araújo OdQF, de Medeiros JL. Carbon capture and storage technologies: present scenario and drivers of innovation. Current Opinion in Chemical Engineering 2017;17:22–34.

[198] Huang Y, Zhang X, Zhang X, Dong H, Zhang S. Thermodynamic modeling and assessment of ionic liquid-based CO2 capture processes. Industrial & Engineering Chemistry Research 2014;53(29):11805–17.

[199] Zeng S, Zhang X, Bai L, Zhang X, Wang H, Wang J, et al. Ionic-liquid-based CO2 capture systems: structure, interaction and process. Chemical Reviews 2017;117(14):9625–73.

[200] Vega LF, Vilaseca O, Llovell F, Andreu JS. Modeling ionic liquids and the solubility of gases in them: recent advances and perspectives. Fluid Phase Equilibria 2010;294(1-2):15–30.
[201] Shiflett MB, Drew DW, Cantini RA, Yokozeki A. Carbon dioxide capture using ionic liquid 1-butyl-3-methylimidazolium acetate. Energy & Fuels 2010;24(10):5781–9.
[202] Basha OM, Keller MJ, Luebke DR, Resnik KP, Morsi BI. Development of a conceptual process for selective CO2 capture from fuel gas streams using [hmim][Tf2N] ionic liquid as a physical solvent. Energy & fuels 2013;27(7):3905–17.
[203] Valencia-Marquez D, Flores-Tlacuahuac A, Ricardez-Sandoval L. Technoeconomic and dynamical analysis of a CO2 capture pilot-scale plant using ionic liquids. Industrial & Engineering Chemistry Research 2015;54(45):11360–70.
[204] Ma Y, Gao J, Wang Y, Hu J, Cui P. Ionic liquid-based CO2 capture in power plants for low carbon emissions. International Journal of Greenhouse Gas Control 2018;75:134–9.
[205] Li Y, Wang Y, Chen B, Wang L, Yang J, Wang B. Nitrogen-doped hierarchically constructed interconnected porous carbon nanofibers derived from polyaniline (PANI) for highly selective CO2 capture and effective methanol adsorption. Journal of Environmental Chemical Engineering 2022;10(6):108847.
[206] García G, Aparicio S, Ullah R, Atilhan M. Deep eutectic solvents: physicochemical properties and gas separation applications. Energy & Fuels 2015;29(4):2616–44.
[207] Wang G, Hou W, Xiao F, Geng J, Wu Y, Zhang Z. Low-viscosity triethylbutylammonium acetate as a task-specific ionic liquid for reversible CO2 absorption. Journal of Chemical & Engineering Data 2011;56(4):1125–33.
[208] Silaban A, Harrison DP. High temperature capture of carbon dioxide: characteristics of the reversible reaction between CaO(s) and CO2(g). Chemical Engineering Communications 1995;137(1):177–90.
[209] Sevilla M, Valle-Vigón P, Fuertes AB. N-doped polypyrrole-based porous carbons for CO2 capture. Advanced Functional Materials 2011;21(14):2781–7.
[210] Lu Q, Jiao F. Electrochemical CO2 reduction: electrocatalyst, reaction mechanism, and process engineering. Nano Energy 2016;29:439–56.
[211] Huang J, Rüther T. Why are ionic liquids attractive for CO2 absorption? An overview. Australian journal of chemistry 2009;62(4):298–308.
[212] Kumar R, Ahmadi MH, Rajak DK, Nazari MA. A study on CO2 absorption using hybrid solvents in packed columns. International Journal of Low-Carbon Technologies 2019;14(4):561–7.
[213] Lin K-YA, Park A-HA. Effects of bonding types and functional groups on CO2 capture using novel multiphase systems of liquid-like nanoparticle organic hybrid materials. Environmental Science & Technology 2011;45(15):6633–9.
[214] Luis P, Van Gerven T, Van der Bruggen B. Recent developments in membrane-based technologies for CO2 capture. Progress in Energy and Combustion Science 2012;38(3):419–48.
[215] Huang S, Wang Y, Hou K, Wang P, He M, Liu X. Thermodynamic analysis of an efficient pressure-swing CO2 capture system based on ionic liquid with residual pressure energy recovery. Journal of Cleaner Production 2023:137665.

[216] Ozkutlu M, Orhan OY, Ersan HY, Alper E. Kinetics of CO2 capture by ionic liquid—CO2 binding organic liquid dual systems. Chemical Engineering and Processing: Process Intensification 2016;101:50—5.
[217] Mulk WU, Ali SA, Shah SN, Shah MUH, Zhang Q-J, Younas M, et al. Breaking boundaries in CO2 capture: Ionic liquid-based membrane separation for post-combustion applications. Journal of CO2 Utilization 2023;75:102555.
[218] Shi W, Maginn EJ. Molecular simulation of ammonia absorption in the ionic liquid 1-ethyl-3-methylimidazolium bis (trifluoromethylsulfonyl) imide ([emim][Tf2N]). AIChE Journal 2009;55(9):2414—21.
[219] Ahmady A, Hashim MA, Aroua MK. Experimental investigation on the solubility and initial rate of absorption of CO2 in aqueous mixtures of methyldiethanolamine with the ionic liquid 1-butyl-3-methylimidazolium tetrafluoroborate. Journal of Chemical & Engineering Data 2010;55(12):5733—8.
[220] Althuluth M, Mota-Martinez MT, Kroon MC, Peters CJ. Solubility of carbon dioxide in the ionic liquid 1-ethyl-3-methylimidazolium tris (pentafluoroethyl) trifluorophosphate. Journal of Chemical & Engineering Data 2012;57(12):3422—5.
[221] Jiang Y, Wu Y, Wang W, Li L, Zhou Z, Zhang Z. Permeability and selectivity of sulfur dioxide and carbon dioxide in supported ionic liquid membranes. Chinese Journal of Chemical Engineering 2009;17(4):594—601.
[222] Zare M, Haghtalab A, Ahmadi AN, Nazari K. Experiment and thermodynamic modeling of methane hydrate equilibria in the presence of aqueous imidazolium-based ionic liquid solutions using electrolyte cubic square well equation of state. Fluid Phase Equilibria 2013;341:61—9.
[223] Acidi A, Hasib-ur-Rahman M, Larachi F, Abbaci A. Ionic liquids [EMIM][BF4], [EMIM][Otf] and [BMIM][Otf] as corrosion inhibitors for CO2 capture applications. Korean Journal of Chemical Engineering 2014;31(6):1043—8.
[224] Taheri M, Zhu R, Yu G, Lei Z. Ionic liquid screening for CO2 capture and H2S removal from gases: the syngas purification case. Chemical Engineering Science 2021;230:116199.
[225] Odunlami OA, Vershima DA, Oladimeji TE, Nkongho S, Ogunlade SK, Fakinle BS. Advanced techniques for the capturing and separation of CO2 — a review. Results in Engineering 2022;15:100512.
[226] Lian S, Song C, Liu Q, Duan E, Ren H, Kitamura Y. Recent advances in ionic liquids-based hybrid processes for CO2 capture and utilization. Journal of Environmental Sciences 2021;99:281—95.
[227] Zunita M, Hastuti R, Alamsyah A, Khoiruddin K, Wenten IG. Ionic liquid membrane for carbon capture and separation. Separation & Purification Reviews 2022;51(2):261—80.
[228] Liu H, Maginn E, Visser AE, Bridges NJ, Fox EB. Thermal and transport properties of six ionic liquids: an experimental and molecular dynamics study. Industrial & Engineering Chemistry Research 2012;51(21):7242—54.
[229] Zhou G, Jiang K, Wang Z, Liu X. Insight into the behavior at the hygroscopicity and interface of the hydrophobic imidazolium-based ionic liquids. Chinese Journal of Chemical Engineering 2021;31:42—55.
[230] Moazezbarabadi A, Wei D, Junge H, Beller M. Improved CO2 capture and catalytic hydrogenation using amino acid based ionic liquids. ChemSusChem 2022:e202201502.

[231] Wang Y, Lu Y, Wang C, Zhang Y, Huo F, He H, et al. Two-dimensional ionic liquids with an anomalous stepwise melting process and ultrahigh CO2 adsorption capacity. Cell Reports Physical Science 2022;3(7):100979.

[232] Carvalho PJ, Álvarez VH, Marrucho IM, Aznar M, Coutinho JAP. High pressure phase behavior of carbon dioxide in 1-butyl-3-methylimidazolium bis(trifluoromethylsulfonyl)imide and 1-butyl-3-methylimidazolium dicyanamide ionic liquids. The Journal of Supercritical Fluids 2009;50(2):105–11.

[233] Busato M, D'Angelo P, Lapi A, Tolazzi M, Melchior A. Solvation of Co2+ ion in 1-butyl-3-methylimidazolium bis(trifluoromethylsulfonyl)imide ionic liquid: a molecular dynamics and X-ray absorption study. Journal of Molecular Liquids 2020;299:112120.

[234] Kanehashi S, Kishida M, Kidesaki T, Shindo R, Sato S, Miyakoshi T, et al. CO2 separation properties of a glassy aromatic polyimide composite membranes containing high-content 1-butyl-3-methylimidazolium bis(trifluoromethylsulfonyl)imide ionic liquid. Journal of Membrane Science 2013;430:211–22.

[235] Zeng Q, Zhang J, Cheng H, Chen L, Qi Z. Corrosion properties of steel in 1-butyl-3-methylimidazolium hydrogen sulfate ionic liquid systems for desulfurization application. RSC Advances 2017;7(77):48526–36.

[236] Grishina E, Ramenskaya L, Gruzdev M, Kraeva O. Water effect on physicochemical properties of 1-butyl-3-methylimidazolium based ionic liquids with inorganic anions. Journal of Molecular Liquids 2013;177:267–72.

[237] Pan J, Muppaneni T, Sun Y, Reddy HK, Fu J, Lu X, et al. Microwave-assisted extraction of lipids from microalgae using an ionic liquid solvent [BMIM][HSO4]. Fuel 2016;178:49–55.

[238] Karousos D, Vangeli O, Athanasekou C, Sapalidis A, Kouvelos E, Romanos GE, et al. Physically bound and chemically grafted activated carbon supported 1-hexyl-3-methylimidazolium bis (trifluoromethylsulfonyl) imide and 1-ethyl-3-methylimidazolium acetate ionic liquid absorbents for SO2/CO2 gas separation. Chemical Engineering Journal 2016;306:146–54.

[239] Erkoç T, Sevgili LM, Çavus S. Liquid–liquid extraction of linalool from methyl eugenol with 1-ethyl-3-methylimidazolium hydrogen sulfate [EMIM][HSO4] ionic liquid. Open Chemistry 2019;17(1):564–70.

[240] Karousos DS, Labropoulos AI, Tzialla O, Papadokostaki K, Gjoka M, Stefanopoulos KL, et al. Effect of a cyclic heating process on the CO2/N2 separation performance and structure of a ceramic nanoporous membrane supporting the ionic liquid 1-methyl-3-octylimidazolium tricyanomethanide. Separation and Purification Technology 2018;200:11–22.

[241] Lei Z, Zhang B, Zhu J, Gong W, Lü J, Li Y. Solubility of CO2 in methanol, 1-octyl-3-methylimidazolium bis(trifluoromethylsulfonyl)imide, and their mixtures. Chinese Journal of Chemical Engineering 2013;21(3):310–7.

[242] Valencia-Marquez D, Flores-Tlacuahuac A, Ricardez-Sandoval L. A controllability analysis of a pilot-scale CO2 capture plant using ionic liquids. AIChE Journal 2016;62(9):3298–309.

[243] Kalb RS. Toward industrialization of ionic liquids. In: Commercial applications of ionic liquids. Springer; 2020. p. 261–82.

[244] Ramdin M, de Loos TW, Vlugt TJ. State-of-the-art of CO2 capture with ionic liquids. Industrial & Engineering Chemistry Research 2012;51(24):8149–77.

[245] Bara JE, Camper DE, Gin DL, Noble RD. Room-temperature ionic liquids and composite materials: platform technologies for CO2 capture. Accounts of Chemical Research 2010;43(1):152−9.

[246] Scholes CA, Kentish SE, Qader A. Membrane gas-solvent contactor pilot plant trials for post-combustion CO2 capture. Separation and Purification Technology 2020;237:116470.

[247] Zhang X, Zhang X, Dong H, Zhao Z, Zhang S, Huang Y. Carbon capture with ionic liquids: overview and progress. Energy & Environmental Science 2012;5(5):6668−81.

[248] Isosaari P, Srivastava V, Sillanpää M. Ionic liquid-based water treatment technologies for organic pollutants: current status and future prospects of ionic liquid mediated technologies. Science of the Total Environment 2019;690:604−19.

[249] Haider J, Qyyum MA, Riaz A, Naquash A, Kazmi B, Yasin M, et al. State-of-the-art process simulations and techno-economic assessments of ionic liquid-based biogas upgrading techniques: challenges and prospects. Fuel 2022;314:123064.

[250] Chao L, Niu T, Xia Y, Chen Y, Huang W. Ionic liquid for perovskite solar cells: an emerging solvent engineering technology. Accounts of Materials Research 2021;2(11):1059−70.

[251] Ostadjoo S, Berton P, Shamshina JL, Rogers RD. Scaling-up ionic liquid-based technologies: how much do we care about their toxicity? Prima facie information on 1-ethyl-3-methylimidazolium acetate. Toxicological Sciences 2018;161(2):249−65.

[252] Ahmad NNR, Leo CP, Mohammad AW, Shaari N, Ang WL. Recent progress in the development of ionic liquid-based mixed matrix membrane for CO2 separation: a review. International Journal of Energy Research 2021;45(7):9800−30.

Adsorption techniques for natural gas sweetening

11

Swing technologies for natural gas sweetening: Pressure, temperature, vacuum, electric, and mixed swing processes

Meisam Ansarpour[1] and Masoud Mofarahi[1,2]
[1]*Department of Chemical Engineering, Faculty of Petroleum, Gas and Petrochemical Engineering, Persian Gulf University, Bushehr, Iran;*
[2]*Department of Chemical and Biomolecular Engineering, Yonsei University, Seoul, South Korea*

1. Introduction

1.1 Natural gas

There are various geographical sources for producing natural gas, and consequently, these sources are different from each other [1]. Some of these compositions are listed in Table 11.1. Moreover, because of the growing demand for natural gas, H_2S and CO_2-rich sources (concentration higher than 20.0%) are also utilized [2].

The raw natural gas should be refined by different sweetening processes to eliminate the acidic part due to some reasons, such as:

❖ The H_2S present in natural gas is so disastrous.
❖ To reach the commercially sales gas composition ($H_2S < 4.0$ ppm, CO_2: meet the gross calorific value).
❖ To fit with the downstream process specification. For instance, the CO_2 content of natural gas can freeze at $-70°C$ in a cryogenic process.
  Absorbed CO_2 can be used in enhanced oil recovery (EOR) process.
  With regard to the number of carbon atoms content (more than C_2^+), natural gas divide to three types: less than 10.0 vol.% (lean gas), more than 10.0 vol.% (heavy gas), and the gas with

Table 11.1 Different compositions from various sources for natural gas [3–6].

Component	Ria Arriba Country (New Mexico) (dry and sweet)	Laeq (France) (sour)	Dalan (Persian Gulf) (wet and sweet)	Kansas (USA) (sweet)	Egypt (sour)
CH_4	96.91	69	80.6	72.89	62.73
C_2H_6	1.33	3	2.58	6.27	11.88
C_3H_8	0.19	0.9	0.87	3.74	7.58
C_4H_{10}	0.05	0.5	0.52	1.38	2.43
C_5^+	0.02	0.5	2.33	0.62	2.8
N_2	0.68	1.5	11.12	14.65	0.37
H_2S	–	15.3	–	–	8.4
CO_2	0.82	9.3	1.91	–	2.56
H_2O	–	–	0.07	–	1.25

very high concentration of C_2^+ that causes phase change in production (condensate) [7–9]. Besides, H_2S and CO_2 are significant components to classify the exploited gas into sour gas ($H_2S > 1.0$ vol.% and $CO_2 > 2.0$ vol.%) and sweet gas (below the mentioned concentrations) [8,9]. It should be noted that further impurities such as Hg, He, and As exist in natural gas [10,11].

There are two industrially important uses for natural gas: liquid natural gas (LNG) and pipeline gas. The gas specifications for each one of them are listed in Table 11.2. Higher levels of impurities are endurable in pipeline transport of natural gas than transport in LNG form [12]. Specially, components like CO_2, H_2O, or aromatic hydrocarbons cause issues in cryogenic heat exchanger units for LNG preparation [10].

1.2 Adsorption

The adsorption process is utilized for purification and separation of different streams. This process significantly depends on two parameters: (i) adsorbent and (ii) engineering design of the adsorption process [15,16]. Moreover, there are two main steps in the adsorption process that are: (i) adsorption of impurities or adsorbate from the feed stream and (ii) regeneration, which desorbs the adsorbent species from the adsorbent. The concept of these two steps is shown in Fig. 11.1. It is clear that useful outputs can be obtained from both of these steps. The effluent from the adsorption step is purified from adsorbed components. Next,

Table 11.2 Specifications for liquid natural gas (LNG) and pipeline gas [8,9,13,14].

Impurity	Feed to LNG plant	Pipeline gas
H_2O	<0.1 ppm$_v$	<120.0 ppm$_v$
H_2S	<4.0 ppm$_v$	2.7–22.9 mg Sm^{-3}
CO_2	<50.0 ppm$_v$	<2.0%
Total sulfur	<20.0 ppm$_v$	5.7–22.9 mg Sm^{-3}
N_2	<1.0 vol%	3.0 vol%
Hg	<0.01 mg Nm^{-3}	—
C_4	<2.0 vol%	C_3^+: dew point > −10°C cricondentherm
C_5^+	<0.1 vol%	—
Aromatics	<2.0 ppm$_v$	—

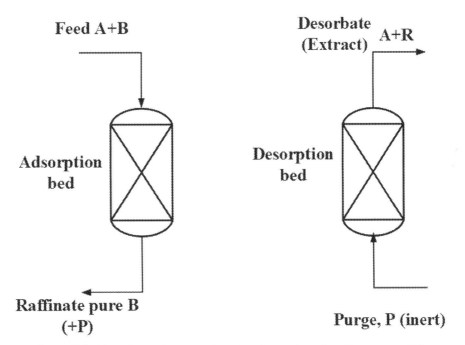

Figure 11.1 Adsorption and regeneration steps in a swing adsorption process [17].

the effluent from the regeneration step contains a more pure and durable adsorbed species that is called the "extract" product [17].

The term "swing" links with various parameters, like temperature, pressure, electricity supply, and vacuum, that are employed in the regeneration step [18–20]. The adsorption process depends on the gas phase principles, which can be adsorbed by the solid phase. Generally, adsorption is processed in an exothermic procedure in which sorbent species bind to the porous solid but do not diffuse [21]. In the following, the adsorbent will be regenerated through a cyclic procedure, either decreasing the pressure, vacuuming the column, fluctuating between low and high temperatures, or using an electric supply, which removes the adsorbent from the sorbent. Consequently, swing processes are called according to their regeneration method, pressure swing adsorption (PSA), vacuum swing adsorption (VSA), temperature swing adsorption (TSA), and electric swing adsorption (ESA). In general, between the mentioned swing technologies, PSA is the most used technology in industries because of its rapid cycle time [17].

In the following, various adsorption processes used for natural gas sweetening are explained.

2. Swing adsorption processes

As mentioned before, there are various types of swing adsorption processes, such as PSA, TSA, VSA, and ESA, in regard to their regeneration method. These different technologies are discussed in this section.

2.1 Pressure swing adsorption

PSA is a technology where regeneration is completed by decreasing the column's pressure since the capacity of adsorption reduces with pressure. For a further reduction in the partial pressure of the adsorbate component, an inert gas is passed through the adsorb column (purge step) after depressurization. This process was initially presented in the 1930s [22–24] and commercially utilized in the 1960s [25,26]. Further information on PSA progression is listed in Table 11.3.

Pressure-based swing adsorption is progressively being noted for commercial applications like separation and purification of different gaseous blends (CO_2, NOx, SO_2, and volatile organic compound [VOC] removal) [28]. Furthermore, it has become one of the most appropriate and commonly applied processes for purification, separation of hydrocarbons and petrochemicals, air drying, and also biogas usage [29]. The PSA process for CH_4/CO_2 separation from biogas is more attractive than membrane separation technology due to some benefits such as easy

Table 11.3 Milestones of the pressure swing adsorption (PSA) process [17,27].

Date	Progress
1930–33	First patent on PSA
1953–54	First research paper on principle of PSA
1955–56	Synthetic zeolites produced in commercial scale
1957–58	Development of VSA
1960–70	Commercialization of PSA
1970–73	Air separation using PSA
1972–73	O_2 selective carbon molecular sieve (CMS) produced commercially
1976	PSA N_2 process using CMS
1976–80	Small-scale medical O_2 units
1982–88	Large-scale VSA process for air separation
2000–13	Application of PSA in biogas purification
2013–17	Development of multibed PSA process

maintenance, low operational cost, low capital, flexibility, and up to 16 beds for a 100,000 Nm^3/h capacity [30,31].

2.1.1 Procedure

A 4-step PSA cycle in a single column is shown in Fig. 11.2, which is a simple but frequent and standard case. The following is a brief description of N_2 separation using helium as a purge gas and activated carbon (AC) and zeolite 5A as adsorbents.

- Compression: After emptying the column bed of N_2 at low pressure (P_L), the column is pressurized by helium or feed gas at high pressure (P_H). In this step, N_2 is adsorbed in the first adsorbent layers.
- High-pressure production step: Under high-pressure operating conditions, the end of the column is opened to exit the pure helium as a product. This process is continuous till the N_2 front starts breaking through at the product end.
- Decompression step: At N_2 breakthrough, production and feed are interrupted, and the adsorption bed is emptied down to P_L, usually in a countercurrent direction. N_2 is then desorbed, and a waste enriched in N_2 is obtained.
- Low-pressure purge: The residual N_2 is rinsed or purged out, by pure He produced from the second stage and the column remaining at P_L. Desorption happens through a reduction of the partial pressure of N_2 [32].

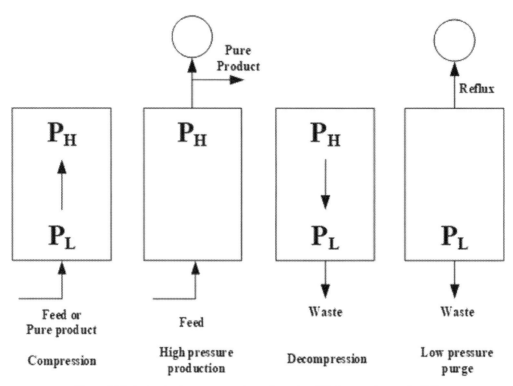

Figure 11.2 A standard pressure swing adsorption (PSA) operating cycle [32].

2.1.2 Various parameters affect the PSA process

2.1.2.1 Effects of the cycle time

One of the greatest impressive factors in the PSA technology is cycle time since it is essential to offer enough residence time to the sorbent to reach the desired output [33]. Cycle time points to the one adsorption–desorption cycle's duration that shows the efficiency of the process [34]. Thus, shorter cycle time leads to higher productivity. However, the cycle time is evaluated based on the breakthrough curve, which principally states the adsorption duration [35]. In overall, breakthrough determines the saturation of the adsorption bed and presents the outlet concentration against the initial concentration in the feed stream against time, as can be seen from Fig. 11.3. Breakthrough happens when the adsorbate saturates the adsorbent. In particular, a breakthrough is achieved when the initial and final concentrations of the adsorbent become equal. At the saturation time and beyond, adsorption stops to occur in the bed, which leads to an accurate estimation of the adsorption time (or residence time). Briefly,

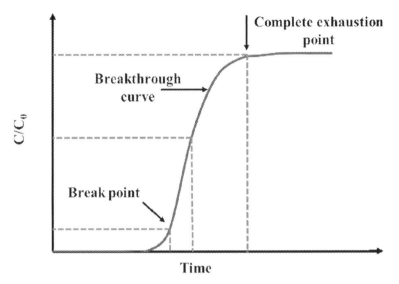

Figure 11.3 Breakthrough curve, C = final concentration, C_0 = initial concentration [27].

the average time that an adsorbent is retained in an adsorption column before becoming fully saturated is called residence time [27].

Notably, enhancing the residence time increases the product recovery and reduces the product purity as a result of the adsorbent's saturation. In their research, Magomnang et al. [36] considered zeolite 13X as an adsorbent to estimate the performance of PSA with a breakthrough time of 25–35 min; however, it also depends on the regeneration step [36]. Similarly, Canevesi et al. [37] investigated the PSA process using CMS adsorbent and reported that the cyclic steady state approaches after 100 cycles [37]. Moreover, Khunpolgrang et al. [38] optimized the partial pressure and vacuum pressure to 400 kPa and to 8.0 kPa, respectively. To avoid a significant adsorption heat demand, they recommended that the residence duration be between 40% and 50% of the breakthrough time. It should be noted that breakthrough time increases with adsorption pressure since more adsorbate is adsorbed at greater pressure [38].

2.1.2.2 The effect of the flow rate

In addition to the cycle time, flow rate is another significant parameter in defining the process's efficiency and purity. It is reported that by increasing feed flow rate, purity will reduce, which is obvious if the adsorption time does not change, fewer adsorbents will be available to adsorb the gas molecules, as can be seen in Fig. 11.4. So, raising the feed flow rate leads to quick saturation of adsorbents and an increment of the adsorption zone, which causes a reduction in product purity. Besides, the product recovery in the exhaust stream reduces because of the transfer of unabsorbed species at higher flow rates of feed stream [27]. For instance, Magomnang et al. [36] investigated the effect of the inlet and outlet gas flow rates in the range of 0–15 LPM and 0.5–1 LPM, respectively. They also observed that a reduction in feed flow rate resulted in an improvement in CO_2 separation due to the presence of a higher number of adsorption sites [36]. In another study, Shen et al. [39] reported the same results to purify the biogas via PSA process with gas flow rates varying from 5–15 SLPM [39].

2.1.2.3 Effect of the pressure

On the other hand, the pressure and flow rate can be correlated with each other in the adsorption process. It is evident that pressure is the core of the PSA process and controlling that can lead to higher process efficiency [40]. The equilibrium relationship between the adsorbent and adsorbate is the basis of the selection of pressure. In particular, for the adsorption technology, an isotherm displays the equilibrium loading of the components, which directly relates to their partial pressure. Besides, the purest product is probable only at high pressure in a PSA process. In contrast, using high-pressure PSA systems leads to more energy loss and compression costs Furthermore, an increment

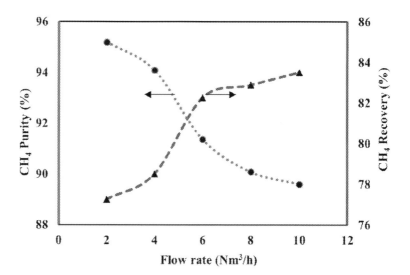

Figure 11.4 Effect of the flow rate on methane recovery and purity [27].

in the adsorption pressure results in a higher product retention in the adsorption bed, causing reduction in product recovery [41]. As can be seen from Figs. 11.5 and 11.6, for a methane PSA process, the recovery ratio reduces more quickly than the purity by increasing the adsorption pressure, demonstrating a direct correlation between the product recovery ratio and the adsorption pressure. Besides, the pressure of the desorption step is another impressive parameter that influences the methane recovery and purity in an adsorption system. It is reported that a higher amount of methane is produced at lower desorption pressures, which leads to a reduction in the CH_4 recovery ratio [27]. In another study, Khunpolgrang et al. [38] used zeolite 13X as an adsorbent to study the behavior of combined vacuum regeneration and N_2 as a purging gas for CH_4/CO_2 separation. The adsorption is performed at 8 kPa vacuum pressure and 400 kPa adsorption pressure to reach 90.0% purity of methane. Moreover, it is feasible to change the cycle time of the PSA process by varying the adsorption pressure [38].

2.1.2.4 Effect of purge/feed (P/F) ratio

Purging is another influential step in the adsorption process since using a pure component in the regeneration step results in a strongly adsorbed species to be flushed back toward the exhaust, so, an adsorbent column free of adsorbate and ready to use in a new cycle will be available. Therefore, the amount of purge gas affects the recovery and purity of the product as

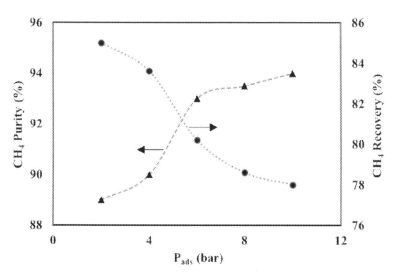

Figure 11.5 Effect of the adsorption pressure on methane recovery and purity [27].

Figure 11.6 Effect of the desorption pressure on methane recovery and purity [27].

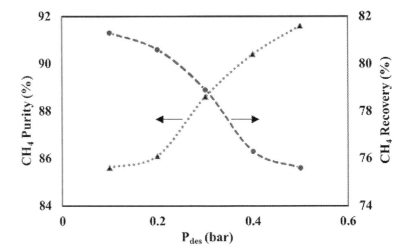

shown in Fig. 11.7 [42]. Since the saturated adsorbents are usually regenerated by decreasing the desorption pressure in a PSA process; so, purge step with low-pressure is required [43]. Generally, increasing purge volume causes improvement in purity and reduction in product recovery. This is because purge gas provides a regenerated column with adsorbents of less loading, and in contrast, it has a negative effect of recovery due to it being done by utilizing the product [44,45]. Shen et al. [46] observed the

Figure 11.7 Effect of the purge/feed ratio on methane recovery and purity [27].

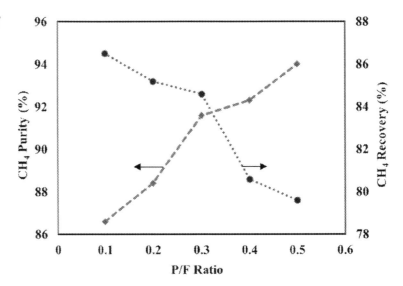

purge/feed ratio between 0.08 and 0.1 was an optimum value to reach desirable product recovery and purity and also, minimum energy consumption after a detailed investigation on the influences of a P/F to all process factors such as adsorption and desorption pressure and cycle time [46].

2.1.2.5 Effects of the adsorbents

The most common adsorbents purified via PSA system are zeolites, CMS, AC, metal organic-framework (MOF), silica gels, and alumina [47]. Pore diameter of adsorbents is an impressive factor in choosing appropriate adsorbents to reach high purity of product. In general, the pore diameter of adsorbent should be less than the molecular diameter of the objective gas. For example, for a CO_2/CH_4 separation via PSA process, the pore diameter of adsorbent should be smaller than CH_4 diameter (3.8Å) and bigger than CO_2 diameter (3.2Å); thus the CO_2 molecules get adsorbed inside the adsorbent's pores [27]. Fig. 11.8 shows an imagination for an adsorption process in regard with the adsorbent and adsorbate size.

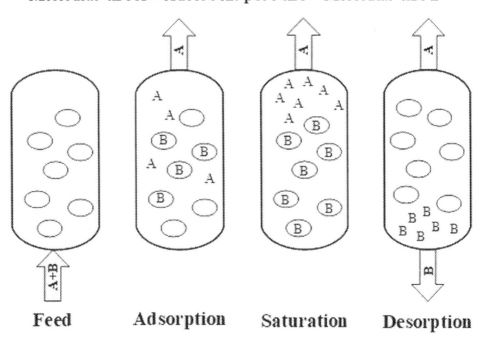

Figure 11.8 Adsorbent and adsorbate size in adsorption process.

2.1.2.5.1 Zeolite Zeolites are natural or synthetic crystalline aluminosilicates, which are mostly used to adsorb H_2S, CO_2, H_2O, and mercaptans according to their polarity. Various types of zeolites exist with different pore diameter and chemical compositions which determine their selectivity affinity. The synthetic zeolites like zeolite 13X are common adsorbents for biogas purification. The pore diameter of zeolite is about 8Å, which is suitable to adsorb H_2S, CO_2, and H_2O that have molecular diameter less than 8Å [48]. It is also reported that zeolite 13X showed better performance for CO_2 adsorption than zeolite 4A because of larger pore diameter and consequently, higher adsorption capacity [49]. In contrast, regeneration of zeolite 4A is higher than zeolite 13X by purging perhaps because of smaller pore size, which selectively adsorbs CO_2. Besides, natural zeolites have shown good performance in swing adsorption technologies with no loss of activity after five cycles [47]. In a research [29], a two-unit PSA system was used to purify the biogas in the presence of zeolite 5A as an adsorbent (Fig. 11.9). They reached biomethane purity of 97% and almost pure CO_2. Furthermore, they observed 99% recovery of CH_4 due to reusing CO_2 effluent from the second PSA unit in a first adsorption column [29].

In a similar research, Santos et al. [30] also studied the influence of recycling streams using zeolite 13X with a two-column PSA system and six cycle steps on biomethane upgrade. They observed more than 99% of purity and 85% of recovery and also 0.12 kW/mol energy consumption [30]. Santos et al. [50] evaluated the use of zeolite 13X as an adsorbent in a single column PSA unit for CH_4/CO_2 separation. They also claimed the ratio of adsorption pressure to the purge gas plays a significant role in the recovery and purity of the product. For instance, a reduction in desorption pressure results in a reduction in methane CH_4 and an increment in CH_4 purity [50]. In another investigation, Tagliabue et al. [8] compared 13X and 5A zeolites with natural zeolite clinoptilolite as adsorbents in a PSA process and reported natural zeolite had higher efficiency and lower cost than the other ones [8]. Moreover, the Heck group [51] studied the separation of $CO_2/CH_4/H_2S$ gas mixtures using various types of zeolites. They observed zeolite 5A and 13X outperform zeolite 4A with 98% methane purity and also 99.6% CO_2 purity in exhaust stream [51].

2.1.2.5.2 Carbon molecular sieve CMSs are a crystalline material with uniform and accurate pore size. CMSs are a unique kind of activated carbon with almost molecular pore size and a few small pores according to its structure from AC. In general, activated carbon (AC) separates components based on the

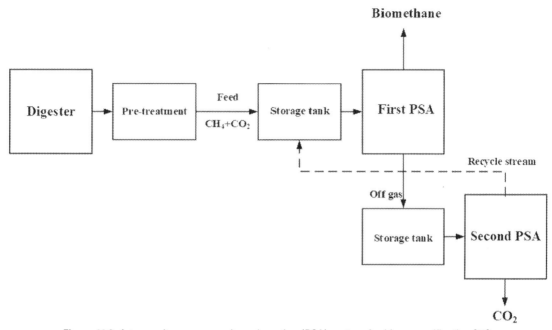

Figure 11.9 A two-unit pressure swing adsorption (PSA) system for biogas purification [29].

adsorption equilibrium constant, while separation via CMS is done based on the adsorption rate. For example, the CMS 3K has the kinetic selectivity and is not influenced by its equilibrium selectivity [52]. So, the adsorption capacity for carbon dioxide is much faster than methane, especially at low temperatures where CO_2 regenerates more easily than high temperatures. Grande and Rodrigues claimed that CMS 3K had 0.27 kW/mol unit consumption against zeolite 13X with 0.41 kW/mol product consumption [53]. The adsorption rate for various gases in the presence of CMS 3K as an adsorbent was studied by Reid and Thomas, and they reported the order of adsorption rates are as follows: carbon dioxide > oxygen > nitrogen > methane [54].

Cavenati et al. [55] studied the separation process in a CH_4/CO_2 system using CMS 3K and observed more than 96% purity and 75% recovery of methane [55]. In another work, Canevesi et al. [56] investigated biogas upgrading and calculated the breakthrough curve at various pressures (0.25, 0.5, 1, and 5 bar) using CMS. They concluded biomethane purity was 97.5% with over 90% recovery at 5 bar (adsorption step) and 0.5 bar (desorption step) pressures [56].

2.1.2.5.3 Metal-organic framework MOF is the 2D or 3D crystalline porous structure that is formed by the strong coordinating bonds between the multidentate organic linkers and metal cation. MOF is the famous adsorbent in CO_2 capture processes. So, several MOFs adsorbents have been studied for CO_2 separation CH_4/CO_2 systems, and among various types of them, Cu-BTC introduced as the most promising adsorbent [8]. Zhou et al. [57] investigated the performance and adsorption capacity of the MOF at various temperatures (303−373 K) for CO_2 capture systems. They reported a significant CO_2 adsorption capacity (6.6 mol/kg) under optimum conditions (pressure of 2.5 bar and temperature of 303 K) [57]. The benefit of employing MOF as adsorbent is lower power consumption and smaller adsorbent bed size.

The presence of some components like H_2S and H_2O due to their harmful properties for adsorbent is a notable factor in choosing the suitable adsorbent [58]. Notably, the H_2S adsorption is irreversible, so it is crucial to separate H_2S from natural gas before feeding in the PSA process. When carbonaceous materials (like AC) are used, H_2O can be removed readily in the same vessel as CO_2. But it is impossible to simultaneously remove H_2O via zeolite since H_2O adsorption is very steep, consequently very difficult desorption. Wu et al. [41] studied the novel adsorbent MOFs and also conventional adsorbents (CMS 3K and zeolite 13X) for CH_4 purification using PSA technology. The results of the simulation showed that MOFs have 56% energy consumption lower than others and also observed that MOFs revealed a linear isotherm curve at the same partial pressure with others, which facilitate the CO_2 desorption [41].

Out of all, the choosing of an adsorbent is vital for the efficient PSA process. However, it is crucial to determine the behavior of isotherms in PSA system designing, while linear or mild isotherms are more ideal for PSA process [48]. It can be possibly due to the difficulties that exist in the regeneration stage, which increase the energy consumption associated with steep isotherms to desorb adsorbate (i.e., CO_2). So, the selection of adsorbent is a main point in adsorption system designing. Among several commercial adsorbents, the process will not reach the goals if inappropriate adsorbent is chosen. Besides, it is important to progress a systematic strategy to evaluate the adsorbent performance during the initial operation of the adsorption process. Accordingly, a simple method that expedites the decision-making process was developed by Gutierrez et al. [59]. The most simplified method to evaluate adsorbent in different PSA applications is to obtain equilibrium of adsorption data through tests on adsorption isotherm [59].

In addition, the chosen adsorbent should have strong performance (small pore size, high thermal stability, and high abrasion resistance) in the presence of humidity and other harmful contaminants that may be in the feed stream. The adsorbent's cost as well as the bed packing density, which determines the size of the adsorbent bed, must be acceptable and economical. The porous texture mostly controls the adsorption mechanism and therefore, purification performances. Also, the pores usually are categorized according to their sizes to five groups: Macropores (d > 500 Å), mesopores (20 < d ≤ 500 Å), and micropores (d ≤ 20 Å). There are two further division which are supermicropores (7Å < d ≤ 20 Å) and ultramicropores (d ≤ Å) [8].

2.1.3 Literature

Izumi [60] investigated CO_2 purification in a two-stage PSA system. The author used a dual bed process in the first stage and a four-bed process in the second one and observed the CO_2 purity up to 99% [60]. Furthermore, Kim et al. [61] investigated the CO_2 and N_2 adsorption using X-type zeolite as an adsorbent, experimentally and numerically [61]. Hwang and Lee [62] explored the breakthrough curves for CO_2 and CO adsorption and desorption using AC in an experiment and modeling study [62].

A novel PSA method with an intermediate feed inlet position that is controlled by dual refluxes was developed by Diagne et al. [63]. They investigated the effect of various CO_2 feed concentration and feed inlet position on CO_2 purity. Their experimental results revealed that this new PSA process could enrich CO_2 from 20% to 90%. Furthermore, feed pressure and desorption pressure were 1.0 and 0.12 atm, respectively [63]. After that, they also developed a new PSA process and reported that the optimal feed inlet position to be located between 0.4 and 0.6 of the column length, and the best reflux ratio was between 0.3 and 0.6 [64].

In the literature [65,66], the selected adsorbent is zeolite 13X. In spite of having a high adsorption heat, zeolite 13X also has a minimal need for purging, a large working capacity, and a high level of equilibrium selectivity. Thus, it brings high product recovery and purity from the PSA cycle. Cho et al. [67] explored a two-stage PSA system using zeolite 13X as adsorbent in a CO_2 separation process. They employed various stages such as adsorption, feed pressurization, blowdown, pressure equalization, and purging in low pressure. Their first PSA stage showed 63.2% purity of CO_2 with 92.4% recovery. In the next stage, the effluent from the first stage purified to 99% with a recovery of 88%. Besides, the process power consumption was 641.5–770 kWh tonne^{-1} CO_2 captured, and also the overall recovery of the process was 80% [67].

It is probable that traditional PSA systems will be appropriate for the CO_2 separation from large sources. Moreover, Chaffee et al. [68] showed that the PSA process is not cost-effective [68]. This may be the result of technical challenges with the PSA process' pressurization phases and the integration of the massive flow in a typical power plant's flue gas stack. However, there are advantages to burning fuel at a little higher pressure [69], and for such power plants, quick PSA application may be simpler. Rapid TSA and rapid VSA are more suitable for the CO_2 separation. Notably, the CO_2 capture step in TSA and VSA should be integrated with the complex thermodynamics of the power generation plant itself. Due to the numerous components present in the flue gases and the variations in the TSA/VSA setup with the power plant, the separation performance might be difficult. Recent research studies have studied the integration of carbon capture and sequestration (CCS) with the different components in a power generation system [70–78]. The TSA process is the next discussed process.

2.2 Temperature swing adsorption

In addition to the CH_4, natural gas consists of compounds like H_2O, H_2S, CO_2, and heavy hydrocarbons, which should be eliminated before any commercial and technical use [79]. The composition of natural gas can contain various components, which should be purified based on the technical use. Particularly, TSA is used to separate low hydrocarbon concentrations [6]. However, the adsorption process will be different according to the feed concentration and components [9,80].

Firstly, in the 1950s, TSA was patented and investigated for natural gas purification [81–83]. Subsequently, TSA technology faced rapid development and progression. For instance, a wide range of industrial applications can be found with multifixed beds, indirect solutions, or coupling with other adsorption technologies such as PSA and VSA [84–89]. Besides, the TSA technology is one of the best technologies for dehydration of natural gas.

2.2.1 Procedure

Fig. 11.10 indicates a schematic with double fixed beds [9,90]. The simplest design for the TSA system is a two-bed system to

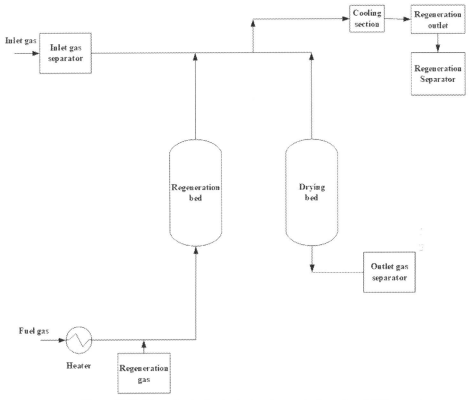

Figure 11.10 Flow sheet of a basic two-tower desiccant unit [90].

attain a continuous flow. Nevertheless, in the current TSA designs, multifixed beds are employed to obtain a cost-efficient process [90]. A fixed bed is provided with the moist feed gas, and mostly, the flow direction is from top to bottom because of the high gas flow rates that can be achieved [90]. Then, a dehumidified natural gas as product is exited at the outlet. Subsequently, in the second column, the regeneration process is done by a hot product gas in a diverse flow direction compared to the previous column (from bottom to top). Heat exchanger is used to heat the regeneration gas to the appropriate temperature, and also, the regeneration cold gas is employed to cool the adsorption bed.

Hydrocarbon and H_2O removal is a major issue in natural gas sweetening. A flow diagram of the TSA system for hydrocarbon recovery and dehydration is illustrated in Fig. 11.11 [91]. This is a three-fixed beds process with continuous product flow. During the first bed, the adsorption of objective components according to their affinity with adsorbent is done, and in the second column,

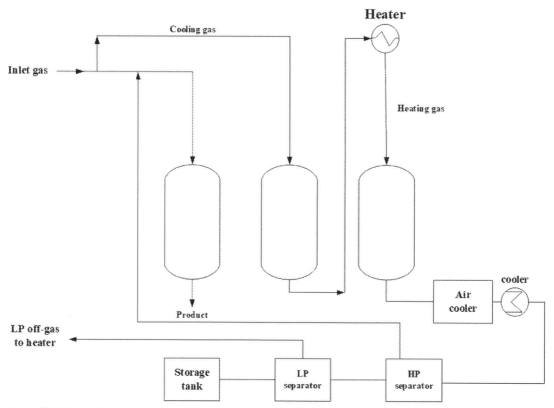

Figure 11.11 Flow diagram of a temperature swing adsorption (TSA) plant for hydrocarbon dew point adjustment [91].

the adsorbent is regenerated using hot purge gas. Then, a concentrated stream of adsorbed gas leaves the regeneration column and after depleting from heavy components and H_2O by multisteps condensation (high- and low-pressure steps) and returns to the feed stream. After the third adsorber was desorbed in the preceding cycle, feed gas was used to cool the regenerating fixed bed to an appropriate temperature for the adsorption process [91]. For natural gas sweetening, the cycle time for TSA process is varied from 0.5 to 2 h (quick cycle TSA in the presence of silica gel as adsorbent) to over 16 h and depends on the various factors such as feed and desirable product concentrations, type of adsorbent, and application [80,92].

The TSA procedure typically contains four steps: (i) utilizing adsorbents to absorb objective components. After that, (ii) hot fluid is sent through the adsorbent bed to desorb the CO_2 saturated bed, which also raises the temperature of the adsorbent bed. Thus, (iii) the cold liquid is used to cool the adsorbent bed. Finally, (iv) the liquid is displaced through the adsorbent and dried using purge gas [27].

The regeneration step in the TSA process [93–97] involves heating the adsorbent component, which lowers its adsorption capability. The bed is heated by commonly circulating a hot gas through it. Since steam is often used to heat materials, it is possible that following desorption a condensation step will be necessary to remove water from the recovered adsorbate. If the desorbed species can be hydrolyzed or forms an azeotrope with water, this is not always easy to do [98]. A hot gas with little adsorption capability may be used when using steam is not ideal. Nitrogen may be effectively employed in certain separations (CO_2 capture, for instance) but because of its low heat capacity (about half that of steam), a higher amount of gas is needed. Thus, a substantially lower concentration of the recovered adsorbate will be obtained. Before beginning a new cycle, the adsorbent must be cooled after heating. Air drying [99], nitrogen oxide (NOx) and low-molecular-weight hydrocarbon removal from gas streams [100], and VOC abatement [89,101] are some of the principal uses of TSA. As an alternative to CO_2 capture, TSA has also been proposed [102]. For CO_2 capture from flue gases and VOC recovery, the combination of PSA and TSA has been investigated [103,104]. As a result of the integration of heat losses from the capture plant, results indicated the process as an economically superior system [103].

2.2.2 Adsorbents in TSA process

Physical adsorption methods such as TSA may be used in the natural gas treatment process chain to remove acidic components like H_2S and CO_2. Since adsorption is only financially and technically reasonable for the low concentrations of H_2S and CO_2 in the inlet gas, it is often employed as an appropriate purification step in coupling with a gas scrubber (such as an alkanolamine scrubber) [11,105]. For the physical adsorption of CO_2 and H_2S, zeolitic adsorbents of types 13X, 4A, and 5A are often used [106,107]. Common dehydration systems use a 4A zeolite to remove CO_2 without the requirement for significant hydrocarbon separation. Zeolites with smaller pores, such 3A zeolite, cannot be used in swing adsorption technologies because of steric hindrance. Mesoporous adsorbents like silica gels have a much lower capacity for CO_2 and H_2S than microporous zeolites [108].

Carbon dioxide adsorbs on zeolites less strongly than hydrogen sulfide because of its lower polarity. Due to the displacement of adsorbed CO_2 molecules by H_2S molecules, there is an accumulation of CO_2 in the gas phase, which results in the formation of a CO_2 peak near the exit of the fixed bed during the concentration process [6]. A maximal concentration of H_2S will be present at the adsorber's output because water is a strong adsorbent and may displace molecules of H_2S that have been adsorbed. The design of the adsorber must take into account the high peak CO_2 and H_2S concentrations of the regeneration cycle, which may be up to 30 times greater than input gas concentrations [6]. With the exception of the notable concentration peaks, the process management of TSA for H_2S or CO_2 removal, which calls for lower regeneration temperatures of 150–200°C, is analogous to that of TSA employed for dehydration [105]. Fig. 11.12 shows a typical distribution within the adsorber.

Polar oxidic adsorbents such as silica gels, zeolites, aluminum oxides, and silica-alumina gels are often employed to clean natural gas because of their excellent regenerability [79,109]. In Table 11.4, an overview of adsorbents utilized in commerce is tabulated.

Zeolites are an example of an oxidic adsorbent that is extensively utilized in the natural gas processing industry. SiO_4 and AlO_4 tetrahedra make up the crystalline, hydrated aluminosilicates known as zeolites. In the natural gas industry, type A zeolites 3A, 4A, and (rarely) 5A are often employed for drying. When carbonyl sulfide (COS) minimization is necessary, 3A zeolite is often employed. Along with dehydration, 4A zeolite is ideal for polishing H_2S and CO_2. The further removal of hydrocarbons often involves the use of 5A zeolite. Faujasite zeolite type X is another frequently used zeolite with larger pore diameter than the type A zeolites [110,111]. Heavy mercaptans and heavy hydrocarbons are routinely separated using this zeolite in the natural gas sector [11,113]. To create cylindrical or spherical pellets with various diameters, the crystalline zeolite particles are often incorporated in macroporous binder material [107,111]. In the microporous region, zeolites typically have a narrow, well-defined pore structure. Undesirable side effects might arise during adsorption in the natural gas processing because the binder material in zeolite adsorbents may have a variable porosity and include contaminants.

In natural gas TSA plants, silica gels can withstand a far larger number of adsorption regeneration cycles because of their superior mechanical strength and increased heat stability compared to zeolites. As a result, silica gels used in short cycle procedures

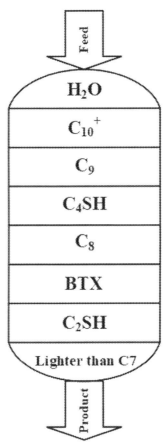

Figure 11.12 Sequence of adsorbed species on a silica gel column [91,92].

Table 11.4 Physical properties of various adsorbents in temperature swing adsorption (TSA) processes in natural gas purification [6,110–112].

Properties	Silica (alumina) gel	Zeolite type A	Aluminum oxide
Specific surface [$m^2\ g^{-1}$]	750–830	350–1100	200–350
Pore volume [$cm^3\ g^{-1}$]	0.4–0.55	0.2–0.4	0.3
Mean pore diameter [Å]	5–22	3–5 8–10 (type X)	26
Pore size distribution	Wide	Narrow	Wide
Structure	Amorphous	Crystalline	Amorphous
Density [$kg\ m^{-3}$]	700–750	690–720	800–880
Heat capacity [$J\ kg^{-1}\ K^{-1}$]	920–1000	200	240
Regeneration temperature [°C]	230	290	240

and zeolitic materials used in long cycle TSA processes both achieve a comparable service life [92]. The appearance of a regeneration reflux is not to be probable when utilizing silica gels since they are created without a binder [92,114].

2.2.3 The adsorption of various components using TSA technology

2.2.3.1 Adsorption of water

Natural gas streams are desiccated using oxidic adsorbents. In contrast to silica gel and alumina, which have much reduced capacities in the low humidity range, zeolites already have a high equilibrium capacity for water at low relative humidity levels (20%) [111]. Zeolites exhibit a saturation influence with a loading plateau after the first rapid rise in capacity because of the limited total pore volume. In theory, high relative humidity might cause capillary condensation in the binder, which would sharply increase loading [115]. Zeolites of type A and type X exhibit different adsorption behaviors. Because there is more room available within the cages of the 13X zeolite, it has the best ability to absorb water [116].

The water's adsorption enthalpy, and therefore the energy input needed for desorption, is a significant consideration for cyclic TSA systems in addition to the equilibrium capacity [2]. Zeolites have excessive heats of adsorption in comparison to other adsorbents, particularly in the range of low loadings. Due to the zeolites' significant heterogeneity, the numbers drastically decline as loadings rise. On the other hand, aluminum oxides and silica gels create bonds with the water that are energetically weaker, which results in a lower heat of adsorption. Because there is less energy heterogeneity, the numbers are more stable. The cations already mentioned provide the water molecule with an adsorption site of high energetic quality, which contributes to the water molecules' strong interaction with the zeolite surface [92]. A substantial energy supply is needed for the fixed bed's desorption and regeneration due to the molecular sieves' strong binding. Consequently, fixed-bed adsorbers that use silica gels and aluminum oxides are often thought to be simpler to desorb [6,110].

The dynamics, which are determined by the shape of the isotherm and the mass transport rate (kinetics), as well as thermodynamic factors like equilibrium capacity and adsorption enthalpy, are crucial for the adsorption process [16,117]. While aluminum oxides and silica gels, which have a mesoporous pore system, have considerably larger mass transfer zones than zeolites, which have extremely acute mass transfer zones due to their precisely defined microporous pore structure contained in a macroporous binder material. This suggests that adding molecular sieves to the fixed bed is substantially more effective than using silica gel or active aluminum oxide [16,80].

Silica gels are often used for the dehydration process with lower purity criteria because of their high absorption capacity at high relative humidity and simple desorbability [6]. Due to their excellent stability in the cyclic process and strong water absorption capacity, aluminum oxides are also employed for predrying [2,112]. Zeolites are often utilized in situations where severe purity standards for natural gases are needed, such as in the subsequent processing of LNG, since they already have extremely high water loads and effectively utilize their bed at low water concentrations in the input gas stream. It is also feasible to use both a zeolite and a silica gel in a combined procedure. A cascade comprising two fixed beds, one containing silica gel and the other zeolite, is utilized since a molecular sieve has a shorter cycle stability [92]. Despite having a great ability to absorb water, type X zeolites are seldom used to dry natural gases because of issues with the co-adsorption of sulfur components and heavy hydrocarbons because of the X zeolite's wide pore aperture [118].

When choosing adsorbents for fine drying of natural gas that has previously been delivered via a pipeline, another factor must be considered. Methanol is usually injected into the pipeline to reduce the production of methane hydrates during the transport process. Methanol and water would be co-adsorbed when using a 4A zeolite [80]. Co-adsorbed methanol, for instance, may cause issues with the regeneration cycle's process technology [80]. As a result, this application mostly uses 3A zeolite.

2.2.3.2 Adsorption of heavy hydrocarbons

For the simultaneous separation of heavy hydrocarbons, TSA is a widely used method. The sequence is the consequence of higher molecular weight hydrocarbons typically adsorbing better than hydrocarbons with shorter chain length [91]. For this reason, heavy hydrocarbon recovery is often classified as aromatic compounds like benzene or C_5^+/C_6^+ hydrocarbons like cyclohexane and neopentane. From a scientific perspective, chain-shaped alkanes, cyclic aliphatic, and aromatic molecules all have highly diverse chemical structures that are predicted to have an impact on the adsorption of C6 + hydrocarbons [119]. In an industrial setting, silica gels and zeolites may be employed for hydrocarbon adsorption. For hydrocarbon dew point modification, type X zeolites and silica gels are mostly employed since LTA 3 and 4A zeolites have tiny pore structures [91].

N-hexane has a somewhat larger capacity on silica gel when compared to cyclohexane, according to a study of their respective capacities. The unbranched alkane's substantially more space-efficient molecular geometry in comparison to the chair-shaped cyclohexane [119] may be one reason for this behavior. The cyclic hydrocarbon has a greater capacity on the 13X zeolites than the n-alkane [120].

2.2.3.3 Adsorption of mercaptans

In a TSA plant, sulfur components known as mercaptans are typically extracted from natural gas during the same process step as the heavy hydrocarbons. The purification of the desorption stream, which requires substantially more work, is the fundamental problem in handling mercaptans.

Due to the mercaptan molecule has a diameter in the range of 0.38–0.51 nm, the faujasite zeolite 13X (NaX) or the zeolites LTA 5A as well as silica-alumina gels and silica gels are primarily utilized for mercaptan adsorption [92,118].

The silica alumina gel has a much lower capacity for mercaptan adsorption than zeolites, and in the loading range studied by Chowanietz et al. [121], no saturation could be shown on the mostly mesoporous silica alumina gel [121]. Chowanietz et al. [122] demonstrated that the typical temperature plateaus during mercaptan desorption in a TSA system also greatly rely on the dipole moments in addition to the adsorption affinity. But, water, which is even more polar, dominates the desorption dynamics in multicomponent systems [122]. Even at extremely low concentrations, the equilibrium load steeply increases, according to the mercaptan isotherms on 5A zeolite. A loading plateau then develops at high gas phase concentrations. Compared to silica-alumina gel, the capacity for both mercaptans is much larger. As anticipated, in the region of low concentrations (500 mol-ppm), the polar ethyl mercaptan obtains greater adsorption capabilities than the methyl mercaptan. The isotherms meet at a concentration in the 700 mol-ppm range, and a lower loading plateau for ethyl mercaptan is discovered [108].

Faujasite NaX (13X) has the greatest capacity for mercaptans. The maximum loads on type X zeolites are greater because the cages of the zeolite are substantially bigger than those of the

LTA zeolite. However, this may result in the coadsorption of heavier hydrocarbons (for instance) [118]. When employing type X zeolites, there is also a difficulty with the COS formation that is discussed in the next section [118].

2.2.4 The formation of COS during TSA process

The production of COS during the adsorption process is a problem with TSA methods for separation of heavy hydrocarbons or dehydration. One of the most significant limitations of adsorption procedures utilizing molecular sieves occurs when H_2S and CO_2 are present in natural gas. The following reaction results in the formation of COS.

$$H_2S + CO_2 \rightleftharpoons COS + H_2O \tag{11.1}$$

COS is a poisonous gas that may cause catalyst poisoning, for instance, if natural gas is utilized as a source of hydrogen for the synthesis of ammonia [6]. COS may affect how well natural gas liquids are recovered during natural gas processing. Although COS by itself is not corrosive, its reaction with CO_2 and H_2S in the presence of water may cause stress corrosion cracking. Molecular sieves 4A, which are often avoided, have very high production rates. When zeolite 5A or other molecular sieves are employed to dehydrate natural gases, COS is also shown to develop. In a series of dynamic adsorption studies, Cines et al. [123] studied the desulfurization of natural gas using several zeolites. The sequence of the increase in catalytic activity for COS generation was 5A, 4A, and then 13X [123]. The mechanism was validated by systematic research by Bulow et al. using zeolites with various cation types and exchange degrees. The quantity of weakly coordinated cations corresponded with the catalytic activity [123]. According to Bulow et al. [124], the proton binds with the oxygen in the zeolite's lattice, whereas the SH group of the H_2S molecule interacts with weakly coordinated cations. The production of COS should be accelerated by this dissociative adsorption, and H_2S should become highly reactive [124]. The zeolites' catalytic activity quickens the reaction rate but has no effect on the equilibrium state. High conversion rates are seen at extremely long contact durations in static studies on a 5A zeolite with limited catalytic activity and on a 3A zeolite whose catalytically active centers are inaccessible to CO_2 and H_2S owing to steric hindrance. According to Lutz et al. [125,126], this phenomenon results from the preferential adsorption of H_2O in the zeolites' smallest cages. The only H_2O molecule that can diffuse into the minuscule sodalite cages is a little one, and as a result, it is cut off from the reaction. In the bigger cages, CO_2, H_2S, and COS adsorb. Due to the separation of water from the COS, the backward reaction, which would change the reaction's equilibrium and remove COS, is prevented [125,126].

2.2.5 Literature

Sonnleitner et al. [127] used Lewatit VP OC 1065 and zeolite 13X as adsorbents in the TSA technology for upgrading biogas, and they found that both materials had better CO_2 adsorption (2.5 mol/kg and 3.6 mol/kg, respectively). Zeolites are the most important adsorbent for TSA; according to the authors' observations, that they are stable up to 463 K temperature [127].

TSA has shortcomings which may prevent its use despite being a separation method for applications like gas drying and VOC capture [128,129]. Because cooling and heating the adsorbent take time, TSA cycles are sometimes quite lengthy (several hours). Long cycles need using a considerable amount of adsorbent inventory, which raises the cost of the adsorbent. Repeated TSA cycling might cause a reduction in the adsorption capacity [130]. TSA is unquestionably a popular method for postcombustion CO_2 separation. However, scaling up for large sources purification presents challenges in terms of cycle order and process characteristics, such as energy consumption, recovery, and purity. Additionally, prolonged severe heating or cooling of the adsorbent is required. As a result, the scientific community was interested in the ESA as a substitute for TSA technology [131].

2.3 Electric swing adsorption

New technologies for raising the adsorbent temperature have been proposed to improve the performance of the TSA process. Adsorbent heating solutions include direct electrical heating [132,133], induced electrical heating [134], and microwave heating [135,136]. There are various transmission bottlenecks for energy from the power source to the adsorbent material with inductive and microwave regeneration [137]. Since the induced current strength diminishes from the exterior surface toward the core of the column, electromagnetic induction heating has significant limits when it comes to heating the middle of common cylindrical columns. Alternative arrangements, such as using a magnetic component in the column's core to reflect the electromagnetic waves, may be an answer. It was suggested to use sleeve-shaped columns, but this requires cooling the inductor that is put around the column with a circulating bath [131]. Besides, microwave regeneration has the ability to heat molecules of various adsorbates in a selective manner depending on their dielectric characteristics. For the separation of nonpolar and polar species, microwave regeneration may be an intriguing method [135]. Devices based on this concept normally utilize moving adsorbents creating considerable attrition effects, while novel experimental setups based in fixed-bed have been developed [135]. Direct electro-thermal regeneration might be regarded as the most promising of these novel regeneration techniques due to the drawbacks of inductive and microwave heating [137]. Despite the three techniques' use of electricity as a drawback, they might be seen as feasible alternatives, particularly where waste heat (used in classic TSA) is either unavailable or insufficient.

One of the separation techniques for purifying and cleaning gases is called ESA. It has already undergone commercial testing for the reduction of VOCs, and further markets, particularly the CO_2 capture from anthropogenic points, are anticipated in the near future [131]. ESA is the specific TSA method in which the heat is produced in situ sending an electric current through a conductor, that is, using the Joule Effect [98,132,133,138,139]. When compared to conventional TSA procedures, the ESA procedure has a number of benefits [132,133,137]:

- The adsorbent receives energy directly, increasing heating efficiency.
- Faster heating permits the design of the smaller systems.
- Higher recovery can be achieved by purge gas flow which is controlled autonomously of the heating rate.
- The heating rate is dependent on neither heat transfer between adsorbent and heating source nor the heat capacity of the heat source.
- Heat and mass fluxes are in the same direction, which should improve regeneration efficiency because of the thermal and gas diffusion effects.

Since it was developed by Fabuss and Dubois [133], this heating method, also known as electrothermal heating, has been around for more than 40 years. The patent described a unique method for removing pollutants from fluid streams that included a particulate adsorbent that was said to be sufficiently electrically conductive to heat up when an electric current was passed through it. It is also possible to use a combination of adsorbents that conduct electricity and those that do not. According to Fabuss and Dubois [133], the ESA technique may be used to remove organic pollutants from water streams as well as organic vapors from air streams. After that, Economy and Lin [140] introduced the in-situ reactivation of AC fibers using the same technique, and Rintoul [141] revealed a system made up of three or more thermally insulated heating beds where electrically conducting carbonaceous particulate materials and a pair of electrodes are placed per column.

Since the beginning of ESA, the significance of the adsorbent has been acknowledged, and research into novel adsorbent materials has continued. Gadkaree and Tyndell [142] reported an electrically heatable AC body for ESA use. The substance is either entirely formed of carbon or is made up of a nonmetallic monolithic structure and activated carbon to a limited extent [142]. For the purpose of CO_2 capture, Delaney et al. [143] pioneered efforts in amine modification of mesoporous carbons by electrothermal regeneration [143]. The regeneration of carbon-coated alumina-silicate ceramic fibers by electrothermal heating was effectively demonstrated by Park and colleagues [144]. The high expense of CO_2 separation and capture is the present obstacle to CO_2 capture and storage. The cost of CO_2 capture is decreased because electrothermal swing adsorption has the potential to be more energy-efficient than traditional TSA and PSA processes. Due to their proven ability to adsorb CO_2 and their high electrical conductivity, activated carbon fiber materials have been used as the adsorbents [145].

2.3.1 Procedure

The process design must be created when the best adsorbent has been chosen to execute an ESA separation. Since the main difference in the process is how heat is supplied, ESA operation is comparable to TSA. The principles of a straightforward ESA cycle are covered in this section. The most straightforward case of an ESA cycle would be an appropriate example. Fig. 11.13 depicts the cycle, which consists of these three steps [131].

I. Feed: in this stage, at the lower temperature of the cycle, the feed mixture comprising the components to be separated is supplied to the adsorption column. The species having a higher affinity for the adsorbent are kept at its surface throughout the feed stage (heavier compound), while those with a lower affinity (lighter compound) are recovered at the bed outlet. Feed stage ends before the breakthrough of the more retained species. The equilibrium is shown by point A in Fig. 11.13B at the end of the feed stage.

II. Electrification: After the feed process, the adsorbent bed is given an electric current, which raises its temperature (Joule effect). During the electrification process, a purge gas flow is delivered to the adsorption bed, as shown in Fig. 11.13B. By

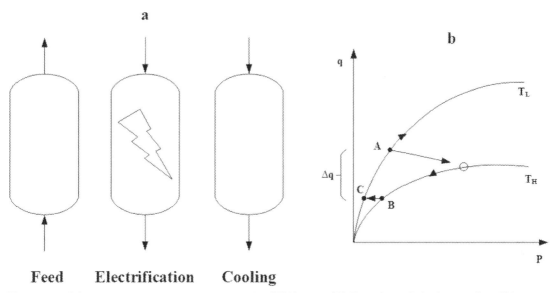

Figure 11.13 (A) A three-step electric swing adsorption (ESA) cycle. (B) The schematic isotherms of an ESA operation [131].

lowering the partial pressure of the adsorbate in the gas phase, the purge flow contributes to the removal of desorbed molecules from the adsorption bed as well as the desorption of the adsorbate. The adsorbent should be (partially) regenerated when the maximum cycle temperature is achieved at the end of the electrification stage, as shown in point B (higher temperature isotherm) in Fig. 11.13B. The adsorbate is recovered at the feed-end during this step. In this situation, the column may be closed while the electrification step is being carried out. The empty point in the higher temperature isotherm happens at a high pressure because there would not be a gas stream leaving the column. The empty point reveals the maximum partial pressure of the adsorbate within the adsorbent. The maximum concentration factor is determined by the ratio of that value to the pressure at the A point.

III. Cooling: To prepare the adsorbent for the beginning of a new ESA cycle, the temperature of the adsorbent must be lowered once the adsorbate is removed from the adsorption bed. In the cooling process, the column is given a purge gas flow to lower its temperature. The system is prepared to begin a new cycle at the conclusion of the cooling stage, which is indicated by point C in Fig. 11.13B. Alternative methods to improve the cooling stage must be researched since it is generally a lengthy procedure (like in the TSA). Bonjour et al. successfully tested the use of a heat-exchanger within the bed [89], and similar tactics may be proposed for the ESA system.

2.3.2 ESA electrification step

The electrification stage distinguishes ESA from traditional TSA. Since the heat is applied directly to the adsorbent in ESA, a high heating efficiency is anticipated. However, other parameters could have an impact on effectiveness. By ignoring the required energy for desorption and energy losses, the energy required to heat the adsorbent can be calculated using [131]:

$$E_{ads} = m_{ads} \int_{T_i}^{T_f} C_{pads} dT \quad (11.2)$$

where E_{ads} is the energy required to heat the adsorbent from the beginning temperature (T_i) to the final (T_f), m_{ads} is the mass of the adsorbent, and c_{pads} is the heat capacity of the adsorbent.

The energy delivered to the system during an ESA electrification step depends on the amount of electricity used and how long the step lasts. This relationship is described by Joule's first law.

$$E_{total} = \int_{0}^{t_{el}} P dt = \int_{0}^{t_{el}} VI dt \quad (11.3)$$

where P is the amount of electricity used, V is the electric voltage applied, t_{el} is the electrification time, and I is the electric current flowing through the conductor. Eq. (11.3) demonstrates that the electrification time and the electric power are operational factors that may be altered to provide the required quantity of energy. Shorter electrification times result from using more electric power, whereas the converse is true when using less electricity.

Since the relation between voltage, electric resistance (R), and current intensity must obey Ohm's law ($V=IR$), the total energy delivered provided by:

$$E_{total} = I^2 \int_{0}^{t_{el}} R dt = V^2 \int_{0}^{t_{el}} \frac{1}{R} dt \quad (11.4)$$

In Eq. (11.4), R points to the overall electric resistance of the adsorption column. Notably, the adsorbent's local electric resistivity is not constant along the adsorption column, although the represented electric resistance in Eq. (11.3) is the total value measured across the adsorption bed.

The fact that the electric power given in an electrification step is not constant because the adsorbent's electric resistance lowers with temperature is a crucial consideration. Therefore, if the voltage is maintained at a constant level, the power will rise; conversely, if the current intensity is maintained at a certain level, the power will fall (Eq. 11.4). Because of this, the temperature rise is never a linear function of time.

2.3.3 Various parameters affect the ESA process
2.3.3.1 Adsorbent

The adsorbate loading also affects the electrical resistance of the adsorbent. According to reports, the resistivity falls as the quantity adsorbed rises [89,146–148]. The power that is provided will also be affected by this impact; it may go up or down depending on whether constant voltage or constant current intensity is being used (Eq. 11.4). As a result, it is difficult to achieve a temperature that is exactly predefined since the electric power provided is never consistent.

There have been some studies about the creation of new adsorbents for ESA applications. Masala et al. [149] provided a research study on a zeolite/electrically conductive carbon monolith appropriate for the ESA procedure. The behavior of zeolite (H-ZSM-5) in the CO_2 capture process was examined in both monolith and powder forms. The monolith was created using 78 weight percent H-ZSM-5 and 22 weight percent phenolic resin, which is created at a high temperature (1073 K) and then transformed into conductive carbon. A piece of adsorbent that measured $0.6 \times 0.6 \times 0.6$ cm was put to the test. The authors provide a resistance value of 29, as determined by a multimeter (over a 20 cm of monolith length).

In a prior study by Regufe et al. [150], the extrusion procedure was used to create a hybrid honeycomb monolith made of zeolite 13X and activated carbon. However, because of the monolith's high electric resistivity, this material's performance in the electrification tests was unsuccessful [150].

2.3.3.2 Voltage

Moon and Shim carried out some electrification steps at various voltages (5–30 V) and discovered that, as predicted by Joule's law, faster heating steps were produced by higher voltage electrification steps [151]. Additionally, when using lower voltages, it takes longer to attain a given temperature than what Eqs. (11.9) and (11.10) indicate. This is brought on by heat loss to the environment [151]. The findings demonstrated that energy losses to the environment are influenced by electric power.

2.3.3.3 Electric power

The amount of electric power used affects the electrothermal desorption efficiency. According to experimental findings, quicker heating causes greater adsorbate concentration peaks and shorter desorption durations [132,139,147,148,151]. For improved ESA performance, increased electric power use for shorter periods of time is advised [132,151]. Studies using a monolith of activated carbon honeycomb revealed that the electric power used has a significant effect on the effectiveness of the Joule effect heating. The electrification step's heating efficiency is 52% when 293 W of electricity is used on average, compared to 23% when 42 W are used. The material, size, and form of the electrodes as well as their other characteristics have a significant impact on the examined system's energy losses. According to the most effective electrodes arrangement examined, 32% of the energy input was lost at the electrodes [152].

2.3.3.4 Purge gas

In the electrification process, Petkovska et al. [132] investigated the effect of the purge gas flow rate. The temperature that was achieved and the regeneration's concentration peak were both impacted by the purge gas flow rate. Lower temperatures and also lower concentration maxima are attained with larger flow rates. Instead of dilution effects, the lower concentration peak can be attributed to the lower temperature reached [132]. Large flow rates shorten the time needed for adsorbent regeneration and reduce the concentration of the recovered adsorbate [147,148].

2.3.4 Literature

Concerns about the atmospheric quantities of these chemicals and their effects on human health and the environment have made the removal of VOCs from various gas streams a topic of utmost relevance [153,154].

For the regeneration of a loaded adsorbent, electrothermal desorption and vacuum may be coupled. The desorption of CO_2 from a carbonaceous adsorbent was studied using this strategy. The regeneration efficiency when using both approaches is 20% greater than when using vacuum alone, according to the authors, and the firmly adsorbed CO_2 may be desorbed [151]. Farant et al. [155] were granted a patent for a device that uses a similar principle for the regeneration of ACFCs. In comparison to previous art, the authors assert that the disclosed approach enables a reduction in the energy required to heat the adsorbent [155].

The effectiveness of ESA as a method to carry out this separation while using various forms of AC as an adsorbent has been shown [156–158], and the results have demonstrated that the adsorbates may be recovered at high quantities [146,159]. With the potential for further cost reduction if improved ($ 0.014/kg of adsorbent to boost its temperature to 160 K), the procedure has been hailed as promising. The cost was projected to be 50% of the average cost when employing steam regeneration [160]. Snyder and Leesch [156] further looked into the adsorbent's cyclic behavior and found that, after 12 cycles, it still preserved 97% −100% of its adsorption capacity [156].

Condensation equipment is typically used in conjunction with ESA systems for VOC recovery in order to recover the adsorbed chemicals [159,161]. A brand-new bench scale system that uses ACFCs to recover harmful VOCs was described by Sullivan and colleagues. The device uses a cryogenic condenser to combine electrothermal adsorbent regeneration with sorbate recovery [162]. Recovery of methyl ethyl ketone (MEK) was found to have removal efficiency over 99.9%. Chmiel et al. [163] described a process in which the regeneration of the adsorbent was carried out electrothermally without the use of purge gas. In order to flush the adsorbate and cool the carbon fibers, the heating is then turned off and a purge gas is provided to the adsorbent [163]. Experimental results from the aforementioned equipment were reported by Sullivan et al. [164]. Without the need of a cryogenic condenser, this enhanced technique enables condensation in the vessel walls at ambient temperature [164].

Activated carbon monoliths (ACMs) were used as the adsorbent in the ESA technique and apparatus that Place et al. reported [165] for the removal of VOCs from air. The monoliths are manufactured by extrusion and sintering of phenolic resin powders that were made by curing novolak resins. The ACM had channels with an open area of 30%–60% and also channel walls and channel size between 0.5 and 1 mm. The disclosed adsorber bed is made up of numerous monoliths that may be electrically coupled either in series or parallel. The adsorber arrangements revealed the same electric resistance. Four monoliths are linked in series to create three groups, which are connected in parallel between each other in both arrangement schemes. Although both arrangements have the same electric resistance, they lead to distinct gas flow routes (6 or 3 parallel channels). Due to the flexibility of the arrangement, the gas velocity may be controlled independently of the electrical characteristics of the column. The VOC recovery is carried out counter-currently to feed (while heating the adsorbent by Joule effect). To ensure that the VOCs to be captured do not penetrate the column during the feed process, a second granular bed may also be used [165].

2.4 Vacuum swing adsorption

VSA is similar to PSA, and is another noncryogenic gas separation technology. One main difference between the VSA and PSA systems is the difference in the operating pressure. The adsorption is conducted at atmospheric or near-atmospheric pressures and the regeneration (or desorption) is performed at reduced pressure. Cyclic adsorption systems are classified based on their regeneration or desorption stage. If regeneration step is done lower than atmospheric pressure, the adsorption process is termed as VSA [131].

The procedure of the VSA process is explained in the following subsection.

2.4.1 Procedure

Fig. 11.14 shows a three-step VSA process from Li et al. study [166]. These three steps comprise.
(I) Pressurization with feed gaseous mixture. The feed gas enters the adsorption column from bottom to top. This step's duration was 45 s. This is predicated on the requirement to use the longest possible bed without breakthrough for CO_2 capture.
(II) Countercurrent evacuation. Based on a balance between cycle scheduling constraints, appropriate desorption kinetics, and vacuum pump capacity, the adsorbed species are sucked into the vacuum line counter-currently for 112 s.
(III) Repressurization step with waste stream. The column is repressurized to ambient pressure with waste gas from the top of the bed for only 3 s [166].

2.4.2 Various parameters affect the VSA process
2.4.2.1 Rinse time

Rinse time is a step that is employed in the VSA process and affects the purity and recovery of the product. The previous research studies showed that increasing the rinse time results in an increment in product's purity and reduction in recovery.

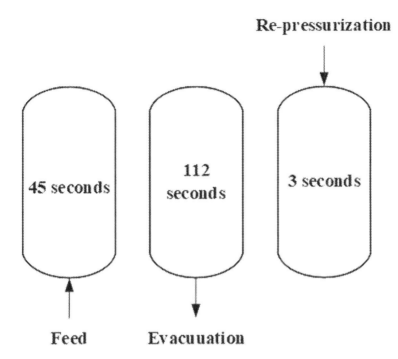

Figure 11.14 Schematic diagram of a three-step vacuum swing adsorption (VSA) cycle [166].

For example, Huang and Eić et al. [167] studied the effect of rinse time in their work and observed as the rinse time of 50 s, high purified CO_2 (>99%) was achieved while the recovery was decreased to less than half. This is due to the more CO_2 requirement to rinse the column with the rise of time, so forcing the CO_2/N_2 mass transfer zone to move further into adsorption bed, for instance, by replacing the weaker adsorbed nitrogen by stronger adsorbed carbon dioxide, and therefore improving the purity of CO_2 through the evacuation. They believed increasing the rinse time more than 50 s will enhance the purity or product up to 100% and recovery can be reached down to zero [167].

2.4.2.2 Evacuation time

Along with the negative effects of greater vacuums, the rate and duration of evacuation can also have an influence on performance. Rapid evacuation causes a large pressure drop, which causes insufficient desorption of the product that has been adsorbed. Due to mass transfer restrictions, high desorption flow rates also restrict the product's ability to transition from the solid phase to the gas phase [168]. The effect of this parameter was investigated in the literature [167], and the results showed a slight enhancement in the purity of the product by increasing the evacuation time. Interestingly, they observed that recovery increased sharply with the rising of evacuation time from 40 to 80 s and after that, increased with lower slope. It should be noted that various evacuation times lead to different evacuation pressures. For instance, the evacuation times of 40, 80, and 120 s, resulted in 35, 24, and 23 kPa final evacuation pressures. Also, the reduction in evacuation pressure causes an improvement in the objective gas purity and recovery and also the selectivity of CO_2/N_2. Notably, the evacuation time should be optimized because it can reduce productivity too [167].

2.4.2.3 Adsorption time

Adsorption time is an influenceable parameter in the swing adsorption system. The purity of the product was reported by Huang and Eić et al. [167] slightly increased by raising the adsorption time while the recovery reduced. The increasing of adsorption time provides a good opportunity for adsorbent to capture more adsorbate, which leads to enhancement in product purity but simultaneously decreases the recovery because more adsorptive break through the column during the adsorption and rinse steps [167]. In contrast, they observed opposite results in another research, which means purity decreased and recovery increased as a result of rising in adsorption time. They believed that this is due to the difference in the PSA configuration. In the PSA process, the broken through gas mixes with the raffinate stream, which leads to a reduction in purity and enhancement in recovery [169].

2.4.2.4 Feed and rinse flow rates

Feed and rinse flow rates are impressive parameters in adsorption performance, which show almost the same behavior. In a CO_2 absorption investigation through a VSA process [167], the results showed rising the rinse and feed flow rates will lead to an enhancement in purity and reduction in recovery, as the observations for adsorption time effect. As mentioned in the previous subsection, more CO_2 will break through the column and consequently, recovery of product reduces, which is predictable in regard with Eq. (11.5). By increasing the feed flow rate, more CO_2 than N_2 was adsorbed thus, the purity of CO_2 increased. On the other hand, by increasing the rinse time, more N_2 was removed, which also enhanced the purity of CO_2. It is clear that the amount of CO_2 used in the rinse step influences the recovery.

$$\text{Recovery}_{CO_2} = \frac{\int_0^{t_V}(|u|C_{CO_2})|_{z=0}dt+\int_0^{t_{VI}}(|u|C_{CO_2})|_{z=0}dt-\int_0^{t_{III}}(uC_{CO_2})|_{z=0}dt}{\int_0^{t_I}(uC_{CO_2})|_{z=0}dt+\int_0^{t_{II}}(uC_{CO_2})|_{z=0}dt} \quad (11.5)$$

2.4.2.5 Operating temperature

Operating temperature is another parameter, which plays a significant role in adsorption processes. The vacuum pressure, operating temperature, and feed concentration have a significant effect on the product's purity and recovery [168]. The result of study for a VSA technology revealed that at temperatures above room temperatures, the recovery of the product reduced, and no obvious change was observed for purity of that. This reduction is due to the decreasing in adsorption capacity of the column when the temperature rises and thus, by reducing the amount of adsorbed product during the evacuation and blowdown steps [167].

2.4.2.6 Feed concentration

The effects of feed concentration on the VSA adsorption process were investigated experimentally and theoretically by Huang and Eić et al. [167]. They studied various feed composition containing: 0.8 CO_2/0.2 N_2, 0.5 CO_2/0.5 N_2, and 0.2 CO_2/0.8 N_2, which are the typical concentrations of CO_2 in the membrane process, precombustion flue gas, and postcombustion flue gas, respectively. They concluded that product purity increased significantly by rising the CO_2 concentration in the feed stream and high purity of CO_2 was achieved (99.8%) at 80% CO_2 inlet concentration and 5.57 cm/s rinse flow rate [167]. The purity of CO_2 at the lower studied CO_2 inlet concentrations was not high due to the presence of large amount of nitrogen in the adsorption bed. Moreover, the recovery of CO_2 reduced significantly from 89.9% to 37.9% by increasing CO_2 concentration from 20% to 80% and also increasing in rinse flow rate [170].

2.4.2.7 Feed pressure

The last but not the least influenceable parameter in the swing adsorption process is feed pressure. The rising in feed pressure will increase the purity and reduce the recovery [167]. There is an optimum pressure for each process, which increases further than that value, and does not affect the purity of the product. For in instance, in a CO_2/N_2 separation system [167], at low feed pressures, large breakthrough time for both CO_2 and N_2 was reported, and low amounts of CO_2 and N_2 were exited during the rinse and adsorption from the column, which leads to low purity and high recover of CO_2. On the contrary, at high pressure, breakthrough times become smaller, causing more purity and less recovery. These results are because of the purity dependency to the extent of the weaker adsorbed N_2, which is being displaced from the column during the rinse step. The purity of CO_2 in the pressure of 211 kPa was found near 99.5% [167].

2.4.3 Literature

Carbon dioxide is a familiar gas for vacuum recovery and can be readily desorbed from adsorbent in a dual-bed configuration [63,64,171–175]. Ho et al. [176] compared CO_2 separation in VSA process and MEA absorption. They concluded that the selectivity of zeolite X is not high enough to reach high purity of CO_2 but, by improving the CO_2/N_2 selectivity could, the purity could be enhanced too, and also the cost of the adsorption will be decreased. Aaron and Tsouris [177] claimed that CO_2 capture cannot be performed in a single PSA and VSA system; however, they proposed that they could be employed as a prestep in the adsorption process. In another study, Krishnamurthy et al. studied the CO_2 separation [178] from a dry gas by VSA technology in a pilot plant with two coupled configurations using zeolite 13X as an adsorbent. Breakthrough experiments were first performed by perturbing a N_2 saturated bed in the presence of 85% N_2 and 15% CO_2. They used a four-step VSA process: pressurization, adsorption, blowdown, and evacuation. The authors concluded that CO_2 was purified up to 95.9% and 86.5% recovery. A four-step cycle with light product pressurization (LPP) employing two beds was taken into consideration to improve the process' efficiency. The new results showed 94.9% purity with a recovery of 89.8% [178].

Chou and Chen [179] studied CO_2 capture in a VSA process using zeolite 13X. They employed the VSA technology with two- and three-bed processes in a simulation investigation. Pressure drop in the adsorptive bed is not performed due to the large size of adsorbent. Instantaneous equilibrium is supposed between the gas and solid phases, and a nonisothermal process is then considered. To perform the simulation, a method of lines with adaptive grid points is applied as a numerical method. The approximation of the spatial derivatives is made from the upwind difference, and the cubic spline estimation is used to calculate the flow rates in the adsorption column. They concluded that CO_2 purity was 63% with a recovery of 67% for three bed configuration [179]. Huang and Eić et al. [167] investigated a CO_2/N_2 separation under various operating conditions such as different feed compositions and different adsorbents (zeolite 13X, AC, CMS), theoretically and experimentally. They reported that the equilibrium selectivity of CO_2 over N_2 for the used adsorbents was found to be in the order as follows: zeolite 13X > CMS > AC. Besides, zeolite 13X showed higher adsorption affinity in CO_2/N_2 systems. It can be as a result of the CO_2/N_2 kinetic selectivity is not high enough, and therefore, the equilibrium selectivity dominates the separation. Briefly, after investigating various parameters, zeolite 13X displayed higher adsorption performance than CMS, and the maximum CO_2 purity and recovery were found to be 99% and 76%, respectively [167].

Ling et al. [168] studied the capture of CO_2 in single- and dual-bed VSA processes with 25 kPa vacuum pressure and using zeolite 13X-APIII. They performed experiments under various temperatures (20–120 °C) and different CO_2 feeding concentrations (15%, 30%, and 50%). They found that pressure equalization between two adsorption beds can strangely enhance the CO_2 purity in comparison with the single-bed cycle. The results showed both recovery and purity of CO_2 were over 90% for CO_2 inlet concentration of 50%. Moreover, the same results are reachable even for 15% CO_2 inlet concentration at very low pressures (1 kPa) [168]. Zhang et al. [174] studied the effect of various operation factors in a CO_2 capture process through a six and nine-step VSA configuration. They observed 82%–95% purity, 60%–80% recovery, and also 4–10 (kW/TPDc) energy consumption by using zeolite 13X in a CO_2 adsorption process [174]. Li et al. [166] experimentally studied the CO_2 separation from dry and wet gas (95% humidity) through a VSA technology and used zeolite 13X as an adsorbent. They reported significant influence of humidity on CO_2 capture after binary breakthrough of CO_2/water vapor was done. They observed that the water zone through the adsorption process migrates a quarter of the way into the adsorption column and stabilizes its position; thus, CO_2 adsorption is reduced. The results showed the purity and recovery of CO_2 were dropped by 18.5% and 22% compared to the dry flue gas. Notably, the temperature rising for the feed wet gas should be considered thereof humidity would be increased [166].

The literature claimed that although very deep vacuum levels (≤ 5 kPa) improve the purity and recovery of the product over 90%, but the requirements for low vacuum level are not cost-effective. Besides, the operating valves, which can endure very low pressures, are too expensive particularly for the rapid VSA system [180].

2.5 Mixed swing adsorption processes

2.5.1 Temperature–Pressure swing adsorption (TPSA)

In overall, there are two types of TPSA process based on the operating temperature: normal temperature-PSA (NT-PSA) and elevated temperature-PSA (ET-PSA). In order to create hydrogen from gases containing 60%–90% H_2, N-TPSA, another physical purification method has been extensively used in the USA, France, Spain, China, Argentina, Brazil, etc. [181]. The NT-PSA has issues such as low HRR (90%), with a high cost and complicated system, when a high HP over 99.999% is necessary [181].

2.5.1.1 Procedure

2.5.1.1.1 NT-PSA process In order to produce fuel cell grade hydrogen from carbon and hydrogen-rich gases like refinery fuel gas, ethylene plant effluent gas, coke oven gas, and reformed gas, it is frequently necessary to use NT-PSAs based on physical adsorbents like silica gel, zeolite, and AC as the final step [17,182]. Due to the prevention of heat regeneration in TSA, the use of NT-PSA lowers the purifying energy consumption. Additionally, NT-PSA's cyclic operation was significantly quicker than TSA's. The NT-PSA procedure is a cyclic one that entails many beds and a number of subsequent steps [183–185]. The normal Skarstrom cycle has the following steps [26,55]. Gas impurities are physically adsorbents' surface during the feed process. In a series of pressure equalization depressurization stages once the adsorbents are saturated, the pressure of the adsorption bed is decreased, and the released gases are supplied to other beds to recover the pressure energy. In the blowdown process, the bed pressure is further decreased to about 1 bar, and the saturated adsorbents are then renewed using a counter-current product purging step. After that, several pressure equalization, pressurization, and re-pressurization procedures are used to restore the bed pressure to its original value. HP, HRR, and productivity are often used to assess NT-PSA performance as they are in Eqs. (11.6–11.8).

$$\text{HP} = \frac{\int_0^{t_{total}} x_{product,H_2} Q_{product,out} dt}{\int_0^{t_{total}} \left(x_{product,H_2} + x_{product,CO_2} + x_{product,CO}\right) Q_{product,out} dt} \quad (11.6)$$

$$\text{HRR} = \frac{\int_0^{t_{total}} x_{product,H_2} Q_{product,out} dt}{\int_0^{t_{total}} \left(x_{feed,CO} + x_{feed,H_2}\right) Q_{feed} dt} \quad (11.7)$$

$$\text{Productivity} = \frac{\int_0^{total}(x_{product,H_2} + x_{product,CO_2} + x_{product,CO})Q_{product,out}dt}{m_{adsorbents,total}t_{total}}$$

(11.8)

NT-PSAs may also be utilized to adsorb CO_2 prior to combustion [186,187] or to produce both H_2 and CO_2 [188–190]. One advantage of NT-PSA is that it makes it possible to get a very high HP. Following NT-PSA treatment, the residual CO and CO_2 concentration may be lowered from a percentage level to below 10 ppm [191]. But even though the reformed gas produced 98% –99.99% HP, NT–PSA had a high complexity and poor HRR.

2.5.1.1.2 ET-PSA process In order to eliminate the trade-off between HP and HRR in NT-PSA, ET-PSA offers a workable alternative that involves the use of K_2CO_3/LDOs or molten salt-modified MgO-based chemisorbents. First, the H_2 loss resulting from the low CO_2/H_2 selectivity may be prevented by using chemisorbents with high capacity and quick kinetics. Second, by raising the adsorption temperature to 200–450°C, steam purging and rinsing may be used. After the adsorption process, the residual H_2 is driven out of the bed by a high-pressure steam rinse, and the H_2 purge is replaced with a low-pressure steam purge to prevent H_2 loss. Because steam can be separated by simply condensing, it was selected as the rinse/purge gas. The 4-bed 8-step cycle [192], the 6-bed 8-step cycle [193], the 7-bed 10-step cycle [194], the 8-bed 11-step cycle [195–201], and the 9-bed 11-step cycle [202] are just a few of the ET-PSA processes that have been developed. Allam et al. [194] proposed a 7-bed, 11-step ET-PSA process to create an H_2-rich decarbonizing stream for gas turbines from the reformed gas. This process used K_2CO_3/MG70 and an iron chrome catalyst and operated at 350–450°C and 35 bar (Fig. 11.15). The recommended system used a co-current CO_2 rinse step and a counter-current steam purge step, with the high-pressure CO_2 coming from the compressed dry CO_2 product and the steam coming from the heat recovery steam generator (HRSG) in the steam turbine system. A constant flow of feed and product gas was produced by simultaneously feeding two beds and purging one bed. Co-current feeding, co-current CO_2 rinse, three-stage pressure equalization depressurization, blow down, and lastly counter-current steam purge are performed in turn in each bed.

The pressure recovery of the CO_2 product required a lot of compression energy even if the CO_2 rinse reduced H_2 loss. The size of the reactors was increased as a result of the installation of a CO_2 rinse step since it also resulted in a decrease in the

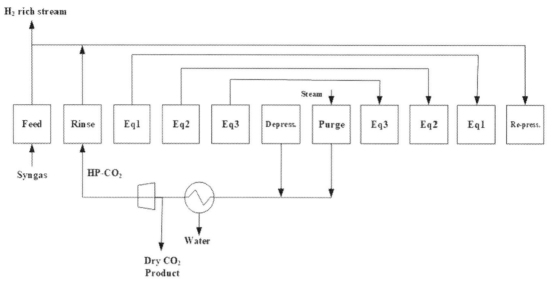

Figure 11.15 7-Bed and 11-step elevated temperature-PSA (ET-PSA) process with co-current CO_2 rinse and counter-current steam purge [194].

CO_2 working capacity. Wright et al. [197] and Van Selow et al. [198,199] substituted a counter-current steam rinse for the CO_2 rinse, mixing the input gas with the effluent gas with a high CO_2 content (Fig. 11.16). The steam rinse cut the volume of both the rinse gas and the purge gas in half as CO_2 was absorbed during the CO_2 rinse stage. Therefore, even though using a steam rinse resulted in a power penalty, it was more than made up for by the compressor's reduced CO_2-product energy consumption. The sequence in each bed is: Co-current feeding, counter-current steam rinse, three-stage pressure equalization depressurization, blow down, and at last, counter-current steam purge.

The primary energy source for ET-PSA is the entire steam consumption for the steam rinse and steam purge steps as opposed to the H_2 loss in NT-PSA. According to Reijer et al. [193], the rinse flow rate had a significant impact on CO_2 purity, whereas the purge flow rate had a significant impact on CO_2 capture ratio. The minimal rinse-to-feed and purge-to-feed ratios for a 6-bed, 8-step ET-PSA were 0.157 and 0.371, respectively, while obtaining a 90% CO_2 capture ratio and a 98% CO_2 purity. Wright et al. [196] showed that adding equalization stages decreased the amount of H_2 that was left in the bed before blow down, allowing the rinse steam to be lowered without appreciably altering the purge steam. The creation of better adsorbents with a beneficial isotherm and adsorption capacity was another potential approach to reducing

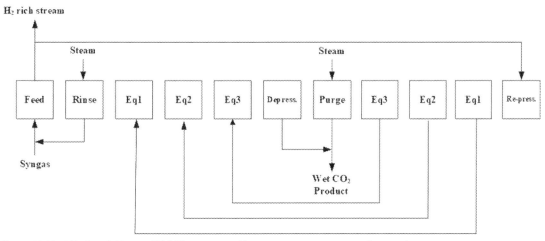

Figure 11.16 7-Bed and 11-step ET-PSA process with counter-current steam rinse and counter-current steam purge [197].

the steam usage in ET-PSA [203,204]. Additionally, Zheng et al. [192] demonstrated that while the purge/feed ratio was fixed, a 20% improvement in adsorption capacity raised HRR by around 5%.

The ET-PSA system's ability to maintain operating continuity may be difficult. Najmi et al. [200] investigated the dynamic behavior of ET-PSA based on the 8-bed, 11-step process proposed by Wright et al. [196]. They found that the H_2 consumption for the repressurization step caused a significant variation in the product flow rate, which may have an impact on the performance of the downstream power unit. In particular, while feeding into the gas turbine with a certain output power, the volatility of the generated H_2-rich stream was not permitted. Setting up a multi-train, multi-bed ET-PSA, where multiple trains operate on different schedules, might smooth out the oscillations in H_2 production [195]. The "pentuple-train" design, which runs two sets of five trains in parallel while allowing a time lag between them, was shown to minimize the variations from 33% to 14% when the number of trains was fixed at 10. A "double-train" system, but with significantly more complicated operation procedures, might further lower the variations to 11%.

Another method to prevent the flow rate changes is to use buffer tanks. In order to create H_2-rich gas with over 95% HP from the shifted gas, Zhu et al. [205] suggested an 8-bed 13-step ET-PSA that includes a counter-current steam purge, a co-current steam rinse, and four pressure equalization stages. The purge duration in this procedure was twice as long as the total of the feeding and rinsing times, which helped to create hydrogen with a greater purity. The R/F and P/F ratios can be modified by adjusting the flow rate of the feeding, rinsing, and purging gases or the proportion of adsorption to rinse time. Changes in the process' feed gas and product gas flow rates were caused by the addition of a rinse stage, the pressurizing stage at the beginning of the feeding phase (for the feed gas), the consumption of steam in the re-pressurization step, and the variable flow rates of feed gas and rinse gas (for the product gas). The discontinuity of the flow rate was compensated with a pressure shift within 0.5 bar [205] by building up buffer tanks five times the size of the adsorption bed both before and after the ET-PSA systems.

2.5.1.2 Literature

A brief literature on NT-PSA and ET-PSA processes is tabulated in Table 11.5.

2.5.2 Vacuum-electric swing adsorption (VESA)

ESA is known as a popular separation technology because of the shorter heating time and higher regeneration efficiency [152,212–214]. In the regeneration step of the ESA process, heat is provided by the Joule Effect reached by passing electrical current through a conductor which is either the adsorbent or is directly adjacent to the adsorbent [131]. In comparison with the TSA process, ESA provides a faster desorption step which reduces the overall process duration [215]. This raises adsorbent regeneration and also the purity and recovery of the product [131]. However, the inert gas performed to purge product from the adsorbent in the ESA desorption process has an adverse effect on the purity [213,214]. The product purity dilution by inert gas in the desorption stage can be avoided with vacuum desorption since the product gas partial pressure gradient is the driving force for product desorption and can be achieved by either total pressure reduction (vacuum) or concentration reduction (a purge). Thus, coupling of VSA and ESA technologies (VESA), that is increasing the column temperature by alternative Joule heat and evacuating product gas using vacuum pump, can enhance the performance efficiency and decrease energy consumption.

2.5.2.1 Procedure

Fig. 11.17 explains a five-step VESA process which are adsorption, electrification, vacuum desorption, cooling, and repressurization. The major differences between this process, and VSA

Table 11.5 Performance of ET-PSA and NT-PSA in a hydrogen production process.

Type of TPSA	Process	Temperature (°C)	Cycle time (s)	H_2 productivity (%)	H_2 recovery (%)	References
NT-PSA	H_2 production	20	510	99.99	65.4	[206]
NT-PSA	H_2 production	21	800	99.999	86	[180]
NT-PSA	H_2 production	30	480	99.9992	62.6	[207]
NT-PSA	H_2 production	25	250	99.99	80	[208]
NT-PSA	H_2 production	60	90.2	99.99	76.2	[209]
NT-PSA	H_2 production	35	670	99.99	80	[210]
NT-PSA	H_2 production	35	800	99.97	79	[211]
ET-PSA	CO_2 adsorption	400	280.2	96.4	98.5	[194]
ET-PSA	CO_2 adsorption	200	960	93.97	96.9	[8]
ET-PSA	CO_2 adsorption	300	1920	89.6	90.5	[135]

and ESA processes are clear. In contrast to the ESA method, vacuum desorption took the place of the purge step in the VESA process. Species that had been adsorbed on the adsorbent or vapourized in the gas phase were then removed from the system under vacuum pressure after the electrification process. N_2 gas was then used to re-pressurize the adsorption column. The cooling procedure mirrored that of the ESA [216].

The following equation [217,218] was used to compute the working capacity of the adsorbent (WC) in accordance with the CO_2 isotherms on the adsorbent.

$$WC = q(T_1, P_1) - q(T_2, P_2) \tag{11.9}$$

The selectivity (S) of component i over component j can also be simply calculated as Eq. (11.10) [65,215]:

$$S = \frac{q_i}{q_j} \tag{11.10}$$

where at the same pressure and temperature, q_i and q_j are the equilibrium adsorption quantities of components i and j, respectively.

The recovery and purity of rich product's may be estimated as follows:

$$Purity = \frac{\int_0^{t_2} C_{i,out} v_{out} dt}{\int_0^{t_2} v_{out} dt} \tag{11.11}$$

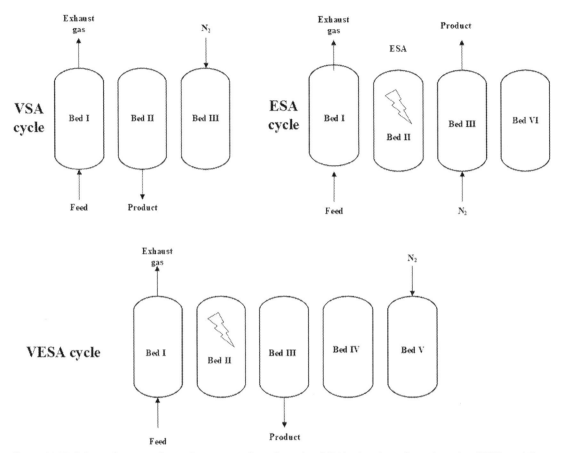

Figure 11.17 Schematic comparison of vacuum swing adsorption (VSA), electric swing adsorption (ESA), and five-step vacuum-electric swing adsorption (VESA) cycles [216].

$$Recovery = \frac{\int_0^{t_2} C_{i,out} v_{out} dt}{\int_0^{t_1} C_{i,in} v_{in} dt} \qquad (11.12)$$

Where, t_1 and t_2 are the adsorption and regeneration times, respectively. $C_{i,\text{out}}$ is an instantaneous mole fraction of component i in the desorption gas, v_{in} and v_{out} [sl/min] are the flowrates of the feed and desorption streams, respectively, and also, $C_{i,\text{in}}$ is the mole fraction in the feed gas.

The energy needed to create joule heat as well as that from the compressor and vacuum pump are all used throughout VESA operations.

$$Energy = \int_0^{t_{feed}} \frac{k}{k-1} \frac{Q_{feed}P_{feed}}{\eta} \left[\left(\frac{P_{feed}}{P_{atm}}\right)^{\frac{k-1}{k}} - 1\right] dt +$$
$$\int_0^{t_{vac}} \frac{k-1}{k} \frac{Q_{vac}P_{vac}}{\eta} \left[\left(\frac{P_{atm}}{P_{vac}}\right)^{\frac{k-1}{k}} - 1\right] dt + \int_0^{t_{elc}} UIdt$$
(11.13)

2.5.2.2 Literature

Zhao et al. used ESA and compared it to traditional VSA for CO_2 capture using a new hybrid zeolite/activated carbon honeycomb. In order to evaluate the benefits of this dual regeneration method for adsorbing CO_2 from a 15% CO_2/N_2 gas stream at low pressure, they next coupled electrical and vacuum swing adsorption (VESA. When the desorption pressures ranged from 30 to 10 kPa, a CO_2 downstream purity of just 17%–23% could be achieved with a straightforward VSA-only cycle. The main cause of this was the adsorbent's weak adsorption properties, which resulted in minimal variation in CO_2 adsorption capacity throughout this pressure range. With ESA, purity of the CO_2 product ranged from 15% to 34%, and recovery ranged from 29% to 78% as the electrification duration was increased from 30 to 180 s. A CO_2 purity of 33% and recovery of 72% were achieved by the combined VESA process with a quick electrification duration of 30 s and a low desorption pressure of 10 kPa. Energy calculations show that, despite VSA's poor purity, the total specific energy for VESA was greater than that of VSA but lower than that of ESA alone [216].

2.5.3 Vacuum temperature swing adsorption (VTSA)

An adsorption mixing TSA and VSA into a so called VTSA process can also be performed. The adsorption capacity of product species reduces as temperature raises [219]. Therefore, vacuum desorption operated coupling with temperature swing (VTSA) should enhance the process efficiency. For instance, Wang et al. [220] considered VTSA process for CO2 capture and observed that the desorption conditions of VTSA system were more feasible compared to the TSA or VSA. Besides, the energy consumption for VTSA process was lower than both TSA and VSA [220].

2.5.3.1 Procedure

Wurzbacher et al. [221] investigated the impact of various operating circumstances on the performance of the VTSA cycle in a CO_2 capture process from air. Fig. 11.18 shows the VTSA cyclic

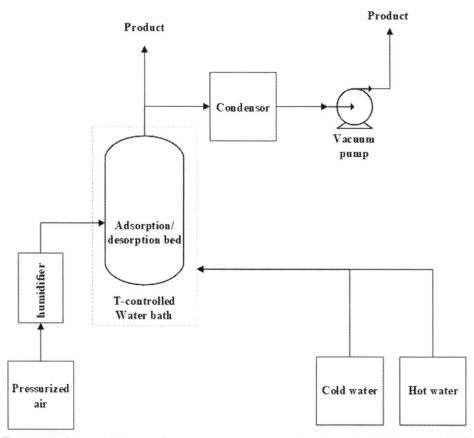

Figure 11.18 Schematic diagram of a vacuum–temperature swing adsorption (VTSA) process [221].

runs were carried out. In order to generate abrupt breakthrough curves and accomplish virtually total removal of the adsorbate from the gas stream, the height of the column is often set to be significantly greater than its diameter in traditional adsorption-based gas separation procedures. In contrast, because of the pressure drop, this need is unnecessary and even undesirable for separating CO_2 from the air. Since a large-scale system with modest pressure drops would have relatively minor pressure dips, the height of the column was determined to be quite short. To quickly heat and cool for desorption and adsorption, the cylinder was maintained in a water bath. One of the air streams was combined with the dry air stream after being bubbled through a humidifier's water bath at a temperature of 25°C. Until the CO_2 concentration in the air exiting the water bath was equal to the

CO_2 concentration in the input air, the water bath was filled with CO_2 before each experiment by bubbling air through it.

By integrating the breakthrough profile, the amount of CO_2 adsorbed in a single cycle, Δq^{ads} (mmolCO_2 g^{-1} sorbent material), was determined:

$$\Delta q^{ads} = \int_{t=0}^{t=t_{ads}} \frac{\dot{n}_{air}(c_0 - c_1)}{m_s} dt \quad (11.14)$$

where $\dot{n}_{air}$ is the air molar flow rate, t_{ads} points to the adsorption time, c_0 and c_1 are the upstream and downstream CO_2 concentrations, respectively. Also, m_s is the mass of the sorbent material in adsorption bed.

The desorption process was started by heating the cylinder to the required desorption temperature after the cylinder had been evacuated to the necessary desorption pressure. By integrating across the desorption process, the amount of CO_2 desorbed in one cycle Δq^{des}_{TVS}, (mmol g^{-1}), was calculated:

$$\Delta q^{des}_{TVS} = \int_{t=0}^{t=t_{des}} \frac{\dot{n}_{CO_2}}{m_s} dt \quad (11.15)$$

where $\dot{n}_{CO_2}$ is the observed molar flow rate of desorbed CO_2 and t_{des} is the desorption time. The cycle was resumed when the cylinder had cooled to below 25 °C after desorption.

Desorption was also carried out using Ar as a purge gas, which was heated to the necessary desorption temperature, as an alternative to desorption under vacuum and for comparison's sake. The term "temperature-concentration swing" refers to this mechanism (TCS). By combining the CO_2 content of the argon stream exiting the cylinder, the CO_2 desorbed Δq^{des}_{TCS}, (mmol g^{-1}) was determined in this instance:

$$\Delta q^{des}_{TCS} = \int_{t=0}^{t=t_{des}} \frac{\dot{n}_{AR} C_1}{m_s} dt \quad (11.16)$$

2.5.3.2 Literature

Experimental analysis is done by Wurzbacher et al. [221] on a VTSA technology that can remove pure CO_2 from both dry and humid ambient air. Under equilibrium and non-equilibrium (short-cycle) circumstances, adsorption/desorption cycles using a packed bed of a sorbent material consisting of commercial silica gel functionalized with diamine are carried out. Thus, a broad variety of operating parameters, including 10–15 mbar desorption pressure, 0%–80% relative humidity during adsorption, and 74–90°C desorption temperature is used to assess the CO_2 capture capability. Per cycle, up to 158 mL of CO_2 with a purity of up to 97.6% may be recovered. Desorption pressures exceeding 100 mbar result in CO_2 capture capacities below 0.03 mmol g^{-1} when the environment is dry. The desorption pressure may be increased to 150 mbar in humid settings with 40% relative humidity during adsorption while capture capacities remain above 0.2 mmol g^{-1}. Over 40 consecutive adsorption/desorption cycles, the sorbent material in the VTSA process has shown stable performance [221].

2.5.4 Vacuum pressure swing adsorption (VPSA)

Adsorption process is a promising technology for gas purification such as CO_2 capture. VSA is the most concentrated adsorption process for carbon dioxide capture from natural gas, since the feed is available at ambient pressure and desorption at subatmospheric pressure is preferable to compression of the feed gas [166,168,174,179]. In the VSA process, the regeneration step often needs a very deep vacuum to reach high purity of CO_2 since the adsorption process is done almost between 0–1 atm [179,222]. Deep vacuum causes low-pressure gas flows, high energy consumption, and pressure drop issues. For further improvement of CO2 capture efficiency, VPSA process was considered [223,224].

2.5.4.1 Procedure

Due to its adaptability in terms of required plant size, operating circumstances, and permitted input gas compositions [225–228], VPSA has received considerable attention. To improve separation performance using VPSA cycles, new techniques [229–231] and materials [232–236] must be developed.

To absorb CH_4 from low concentration sources, nitrogen must be separated from it. VPSA, offers a versatile and scalable approach for separating CH_4 and N_2. Fig. 11.19 shows a pilot-scale setup for a six-bed and twelve-step VPSA system. Without any further pre-treatment, town gas and nitrogen are combined to mix the feed gas. The six steps are: (I) adsorption, (II) equilibrium de-pressurization, (III) desorption, (IV) isolation, (V) equilibrium re-pressurization and finally (VI) repressurization with the light product (N_2-rich) gas stream [237].

The energy analysis for this process overlooks expenses related to cooling water and instrument air since the duty of the compressors and vacuum pumps is what drives the majority of the operating costs in a VPSA process. This is also true in this

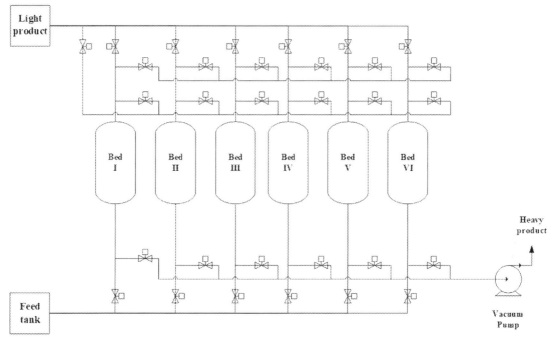

Figure 11.19 A pilot-scale setup for a six-bed and twelve-step vacuum–pressure swing adsorption (VPSA) process [237].

example, where just the vacuum pump and compressor use any substantial amounts of energy. Eq. (11.4), where is the ratio of specific heats, may be used to compute the thermodynamic energy per mole of gas processed needed by pumps (W, kJ/mol). The pumps' efficiency (η) is assumed to be 0.7. As a result, W_{True}, or the real specified work needed, may be expressed as Eq. (11.18) [237].

$$W = \frac{\gamma RT_{\ln}}{\gamma - 1} \left[\left(\frac{P_{out}}{P_{in}} \right)^{\frac{\gamma - 1}{\gamma}} - 1 \right] \quad (11.17)$$

$$W_{True} = \frac{W}{\eta} \quad (11.18)$$

Eqs. (11.19) and (11.20) are used to compute the heating value of gas (Hs) and the Wobbe Index (WI), which stand for the heat produced by burning and interchangeability of a fuel gas, respectively.

$$H_s = \frac{1}{100} \left(H_{sCH_4} f_{CH_4} + H_{sN_2} f_{N_2} \right) \quad (11.19)$$

$$WI = \frac{H_s}{\sqrt{\frac{1}{100}\left(d_{CH_4}f_{CH_4} + d_{N_2}f_{N_2}\right)}} \quad (11.20)$$

It is crucial to recognize that economics are the driving force behind the scaling up of this process. The absence of techno-economic research serves as a deterrent to increased industry participation and interest [237].

2.5.4.2 Literature

Hu et al. [237] study the viability of extracting CH_4 from low concentration sources (4.7%–44.5%) using a new adsorbent (ILZ) in a 112 kg scale VPSA prototype plant. A 3-stage VPSA process was used to produce a product purity of 44.5% CH_4 and an 81% methane recovery from a feed gas with only 4.7% CH_4. Pipelines may then be utilized to transport this product gas, which can subsequently be used in China for either 4T town gases or the production of electricity. The total energy used to capture 1 mole of CH_4 was 133 kJ/mol, which is 85% less than the mole's heating value of 880 kJ/mol. The results showed that reducing GHG emissions is encouraged by the absorption of CH_4 from large but low concentration sources [237].

Grande and Rodrigues [53] used zeolite 13X (equilibrium-based adsorbent) and CMS 3K (kinetic based adsorbent) in a methane production process in the VPSA system. Finally, in terms of purity, both adsorbents reached a purity more than 98%. However, a higher recovery was achieved by CMS 3K that was about 80%. This is possibly due to zeolite 13 X lower adsorption in the pressurization step [53]. In another research by the same group [238], they used a new bed arrangement with a lead trim concept and a combination of both adsorbents in two layers in further research in the same swing technology. The 83.7% recovery and 98% purity of methane was observed [238].

Liu et al. [239] considered zeolite 5A as adsorbent in a three-bed VPSA process to investigate the CO_2 adsorption from a dry gas. They performed adsorption in seven major steps: pressurization, pressure equalization, high-pressure adsorption, heavy product rinse, concurrent depressurization, blowdown, and purging. They concluded the maximum purity and recovery for CO_2 was 79% and 85%, respectively. Moreover, the overall energy consumption was 656 kWh tonne^{-1} CO_2 captured [239].

Lu et al. [240] investigated the CO_2 adsorption using zeolite 13X from a dry gas in a three-bed and single-stage process. The process consisted of the following steps: adsorption, pressure equalization steps, evacuation, blowdown, rinse, and purging. They reported maximum purity of CO_2 was 85%% with a recovery of 80% and 440 kWh tonne^{-1} CO_2 energy consumption. They used a two-stage VPSA system in their next research using the same adsorbent. In the first stage, they observed maximum purity and recovery of CO_2 were 82% and 95%, respectively. However, the second stage was able to concentrate the CO_2 up to 95% with a recovery of 90%. Also, the total energy consumption of the two two-stage VPSA configuration was 675 kWh tonne^{-1} CO_2 [240].

3. Conclusion and future outlooks

The technical challenge is to maintain coherence and physical strength in the scaleup to commercial operations. Future research and development on novel adsorptive methods and materials is anticipated. They will be used to expand markets and enhance the separation performance of currently used applications. A notable trend is the extension of this technology's scope of use through the use of faster cycle processes and innovative adsorber designs. Two interesting areas for new applications of adsorption technology are hybrid gas separation and production ideas like adsorbent membranes and simultaneous sorption-reaction systems. Without a doubt, adsorption research and development will continue to provide difficulties to engineers and scientists for many years to come. It is also critical to establish systematic optimization methodologies for the design of PSA systems given the increased competitive demands for high-performance PSA separations. For example, it may be beneficial to optimize bigger and more intricate PSA systems, such as multibed procedures for the separation of refinery gases. A wide range of novel applications for PSA systems and other periodic separation processes are also provided by effective optimization strategies. In this study, different gas purification technologies (single and mixed processes) by considering their procedures are studied.

Abbreviations and symbols

AC	activated carbon
ACM	activated carbon monolith
C	Concentration
CCS	carbon capture and sequestration
CMS	carbon molecular sieve
COS	carbonyl sulfide
EOR	enhance oil recovery

ESA	electric swing adsorption
ET-PSA	elevated temperature-pressure swing adsorption
I	electric current
LNG	liquid natural gas
LPP	light product pressurization
MEK	methyl ethyl ketone
MOF	metal–organic framework
m_s	mass of the sorbent
NT-PSA	normal temperature–pressure swing adsorption
P/F	purge/feed ratio
PSA	pressure swing adsorption
Q	equilibrium adsorption quantity
R	electric resistance
t_{ads}	adsorption time
TCS	temperature–concentration swing
t_{des}	desorption time
t_{el}	electrification time
TPSA	temperature–pressure swing adsorption
TSA	temperature swing adsorption
V	electric voltage
VESA	vacuum–electric swing adsorption
VOC	volatile organic compound
VPSA	vacuum–pressure swing adsorption
VSA	vacuum swing adsorption
VTSA	vacuum–temperature swing adsorption
WC	working capacity

References

[1] Demirbas A. Methane gas hydrate. In: Green energy and technology. London: Springer; 2010.

[2] Berg F, Pasel C, Eckardt T, Bathen D. Temperature swing adsorption in natural gas processing: a concise overview. ChemBioEng Reviews 2019; 6(3):59–71.

[3] Sayed AE-R, Ashour I, Gadalla M. Integrated process development for an optimum gas processing plant. Chemical Engineering Research and Design 2017;124:114–23.

[4] Saidi M, Parhoudeh M, Rahimpour MR. Mitigation of BTEX emission from gas dehydration unit by application of Drizo process: a case study in Farashband gas processing plant; Iran. Journal of Natural Gas Science and Engineering 2014;19:32–45.

[5] Al-Megren H. Advances in natural gas technology. BoD–Books on Demand; 2012.

[6] Kidnay AJ, Parrish WR. Fundamentals of natural gas processing. CRC Press; 2006.

[7] Faramawy S, Zaki T, Sakr A-E. Natural gas origin, composition, and processing: a review. Journal of Natural Gas Science and Engineering 2016;34:34–54.

[8] Tagliabue M, Farrusseng D, Valencia S, Aguado S, Ravon U, Rizzo C, et al. Natural gas treating by selective adsorption: material science and chemical engineering interplay. Chemical Engineering Journal 2009; 155(3):553–66.

[9] Mokhatab S, Poe WA, Speight JG. Handbook of natural gas transmission and processing. Burlington, MA: Gulf Professional Publishing; 2006.
[10] Mokhatab S, Mak JY, Mokhatab S, Valappil JV. Handbook of liquefied natural gas. Elsevier; 2014.
[11] Mokhatab S, Meyer P. Selecting best technology lineup for designing gas processing units. In: Gas processors association—europe, europe sour gas processing conference; 2009.
[12] Bahadori A. Natural gas processing: technology and engineering design. Gulf Professional Publishing; 2014.
[13] Alcheikhhamdon Y, Hoorfar M. Natural gas quality enhancement: a review of the conventional treatment processes, and the industrial challenges facing emerging technologies. Journal of Natural Gas Science and Engineering 2016;34:689—701.
[14] Baker RW, Lokhandwala K. Natural gas processing with membranes: an overview. Industrial & Engineering Chemistry Research 2008;47(7): 2109—21.
[15] Keller JU, Staudt R. Gas adsorption equilibria: experimental methods and adsorptive isotherms. Springer Science & Business Media; 2005.
[16] Ruthven DM. Principles of adsorption and adsorption processes. John Wiley & Sons; 1984.
[17] Ruthven DM, Farooq S, Knaebel K. Pressure swing adsorption. New York: UCH; 1994. 352pp.
[18] Shen Y, Shi W, Zhang D, Na P, Tang Z, et al. Recovery of light hydrocarbons from natural gas by vacuum pressure swing adsorption process. Journal of Natural Gas Science and Engineering 2019;68:102895.
[19] Riboldi L, Bolland O. Overview on pressure swing adsorption (PSA) as CO2 capture technology: state-of-the-art, limits and potentials. Energy Procedia 2017;114:2390—400.
[20] Ling J, Ntiamoah A, Xiao P, Xu D, Webley P, Zhai Y. Overview of CO2 capture from flue gas streams by vacuum pressure swing adsorption technology. Austin Journal of Chemical Engineering 2014;1(2):1—7.
[21] Dube O, Celik CE, Mcnamara TA. Control of swing adsorption process cycle time with ambient CO2 monitoring. 2019.
[22] Leonard HR, Dargan WH. Separation of gases. Google Patents; 1931.
[23] Perley GA. Method of making commercial hydrogen. Google Patents; 1933.
[24] Finlayson D, Sharp A. Improvements in or relating to the treatment of gaseous mixtures for the purpose of separating them into their components or enriching them with respect to one or more of their components. Patent 1932;365:1932.
[25] De MPG, Daniel D. Process for separating a binary gaseous mixture by adsorption. Google Patents; 1964.
[26] Skarstrom CW. Method and apparatus for fractionating gaseous mixtures by adsorption. Google Patents; 1960.
[27] Shah G, Ahmad E, Pant K, Vijay V. Comprehending the contemporary state of art in biogas enrichment and CO2 capture technologies via swing adsorption. International Journal of Hydrogen Energy 2021;46(9): 6588—612.
[28] Miltner M, Makaruk A, Harasek M. Review on available biogas upgrading technologies and innovations towards advanced solutions. Journal of Cleaner Production 2017;161:1329—37.

[29] Augelletti R, Conti M, Annesini MC. Pressure swing adsorption for biogas upgrading. A new process configuration for the separation of biomethane and carbon dioxide. Journal of Cleaner Production 2017;140:1390–8.

[30] Santos MP, Grande CA, Rodrigues ARE. Pressure swing adsorption for biogas upgrading. Effect of recycling streams in pressure swing adsorption design. Industrial & Engineering Chemistry Research 2011;50(2):974–85.

[31] Abdullah A, Idris I, Shamsudin I, Othman M. Methane enrichment from high carbon dioxide content natural gas by pressure swing adsorption. Journal of Natural Gas Science and Engineering 2019;69:102929.

[32] Daniel T, Wankat P. Gas purification by pressure swing adsorption. Separation & Purification Methods; 2006.

[33] Ebner AD, Mehrotra A, Ritter JA. Graphical approach for complex PSA cycle scheduling. Adsorption 2009;15(4):406–21.

[34] Arvind R, Farooq S, Ruthven D. Analysis of a piston PSA process for air separation. Chemical Engineering Science 2002;57(3):419–33.

[35] Shokroo EJ, Farsani DJ, Meymandi HK, Yadollahi N. Comparative study of zeolite 5A and zeolite 13X in air separation by pressure swing adsorption. Korean Journal of Chemical Engineering 2016;33(4):1391–401.

[36] Magomnang A, Maglinao A, Capareda SC, Villanueva EP. Evaluating the system performance of a pressure swing adsorption (PSA) unit by removing the carbon dioxide from biogas. Indian Journal of Science and Technology 2018;11(17):1–17.

[37] Canevesi RL, Borba CE, da Silva EA, Grande CA. Towards a design of a pressure swing adsorption unit for small scale biogas upgrading at. Energy Procedia 2019;158:848–53.

[38] Khunpolgrang J, Phalakornkule S, Kongnoo A, Phalakornkule C. Alternative PSA process cycle with combined vacuum regeneration and nitrogen purging for CH_4/CO_2 separation. Fuel 2015;140:171–7.

[39] Zhou Y, Shen Y, Fu Q, Zhang D. CO enrichment from low-concentration syngas by a layered-bed VPSA process. Industrial & Engineering Chemistry Research 2017;56(23):6741–54.

[40] Grande CA. Advances in pressure swing adsorption for gas separation. International Scholarly Research Notices; 2012.

[41] Wu B, Zhang X, Xu Y, Bao D, Zhang S. Assessment of the energy consumption of the biogas upgrading process with pressure swing adsorption using novel adsorbents. Journal of Cleaner Production 2015; 101:251–61.

[42] Wiheeb A, Helwani Z, Kim J, Othman M. Pressure swing adsorption technologies for carbon dioxide capture. Separation and Purification Reviews 2016;45(2):108–21.

[43] Jain S, Moharir A, Li P, Wozny G. Heuristic design of pressure swing adsorption: a preliminary study. Separation and Purification Technology 2003;33(1):25–43.

[44] Owens DJ, Ebner AD, Ritter JA. Equilibrium theory analysis of a pressure swing adsorption cycle utilizing a favorable Langmuir isotherm: approach to periodic behavior. Industrial & Engineering Chemistry Research 2012; 51(41):13454–62.

[45] Kim S, Ko D, Moon I. Dynamic optimisation of CH_4/CO_2 separating operation using pressure swing adsorption process with feed composition varies. In: Chemical engineering transactions. Italian Association of Chemical Engineering-AIDIC; 2015. p. 853–8.

[46] Shen Y, Shi W, Zhang D, Na P, Fu B. The removal and capture of CO2 from biogas by vacuum pressure swing process using silica gel. Journal of CO2 Utilization 2018;27:259−71.

[47] Ferella F, Puca A, Taglieri G, Rossi L, Gallucci K. Separation of carbon dioxide for biogas upgrading to biomethane. Journal of Cleaner Production 2017;164:1205−18.

[48] Grande CA. Biogas upgrading by pressure swing adsorption. Biofuel's Engineering Process Technology; 2011. p. 65−84.

[49] Montanari T, Finocchio E, Salvatore E, Garuti G, Giordano A, Pistarino C, et al. CO2 separation and landfill biogas upgrading: a comparison of 4A and 13X zeolite adsorbents. Energy 2011;36(1):314−9.

[50] Santos MNP, Grande CA, Rodrigues AE. Dynamic study of the pressure swing adsorption process for biogas upgrading and its responses to feed disturbances. Industrial & Engineering Chemistry Research 2013;52(15):5445−54.

[51] Heck HH, Hall ML, dos Santos R, Tomadakis MM. Pressure swing adsorption separation of H2S/CO2/CH4 gas mixtures with molecular sieves 4A, 5A, and 13X. Separation Science and Technology 2018;53(10):1490−7.

[52] Sarker AI, Aroonwilas A, Veawab A. Equilibrium and kinetic behaviour of CO2 adsorption onto zeolites, carbon molecular sieve and activated carbons. Energy Procedia 2017;114:2450−9.

[53] Grande CA, Rodrigues AE. Biogas to fuel by vacuum pressure swing adsorption I. Behavior of equilibrium and kinetic-based adsorbents. Industrial & Engineering Chemistry Research 2007;46(13):4595−605.

[54] Reid C, Thomas K. Adsorption of gases on a carbon molecular sieve used for air separation: linear adsorptives as probes for kinetic selectivity. Langmuir 1999;15(9):3206−18.

[55] Cavenati S, Grande CA, Rodrigues AE. Upgrade of methane from landfill gas by pressure swing adsorption. Energy & fuels 2005;19(6):2545−55.

[56] Canevesi RL, Andreassen KA, da Silva EA, Borba CE, Grande CA. Pressure swing adsorption for biogas upgrading with carbon molecular sieve. Industrial & Engineering Chemistry Research 2018;57(23):8057−67.

[57] Zhou K, Chaemchuen S, Verpoort F. Alternative materials in technologies for Biogas upgrading via CO2 capture. Renewable and Sustainable Energy Reviews 2017;79:1414−41.

[58] Kohlheb N, Wluka M, Bezama A, Thrän D, Aurich A, Müller RA. Environmental-economic assessment of the pressure swing adsorption biogas upgrading technology. BioEnergy Research 2021;14(3):901−9.

[59] Álvarez-Gutiérrez N, Gil M, Rubiera F, Pevida C. Simplistic approach for preliminary screening of potential carbon adsorbents for CO2 separation from biogas. Journal of CO2 Utilization 2018;28:207−15.

[60] Izumi J. Process off-gas treatment with pressure swing adsorption. In: Proceedings of symposium on adsorption processes. Taiwan: Chung-Li; 1992.

[61] Kim J-N, Chue K-T, Kim K-I, Cho S-H, Kim J-D. Non-isothermal adsorption of nitrogen-carbon dioxide mixture in a fixed bed of zeolite-X. Journal of Chemical Engineering of Japan 1994;27(1):45−51.

[62] Hwang KS, Lee WK. The adsorption and desorption breakthrough behavior of carbon monoxide and carbon dioxide on activated carbon. Effect of total pressure and pressure-dependent mass transfer coefficients. Separation Science and Technology 1994;29(14):1857−91.

[63] Diagne D, Goto M, Hirose T. New PSA process with intermediate feed inlet position operated with dual refluxes: application to carbon dioxide removal and enrichment. Journal of Chemical Engineering of Japan 1994; 27(1):85–9.

[64] Diagne D, Goto M, Hirose T. Parametric studies on CO2 separation and recovery by a dual reflux PSA process consisting of both rectifying and stripping sections. Industrial & Engineering Chemistry Research 1995; 34(9):3083–9.

[65] Chue K, Kim J, Yoo Y, Cho S, Yang R. Comparison of activated carbon and zeolite 13X for CO2 recovery from flue gas by pressure swing adsorption. Industrial & Engineering Chemistry Research 1995;34(2):591–8.

[66] Kikkinides ES, Yang R, Cho S. Concentration and recovery of carbon dioxide from flue gas by pressure swing adsorption. Industrial & Engineering Chemistry Research 1993;32(11):2714–20.

[67] Cho S-H, Park J-H, Beum H-T, Han S-S, Kim J-N. A 2-stage PSA process for the recovery of CO2 from flue gas and its power consumption. In: Studies in surface science and catalysis. Elsevier; 2004. p. 405–10.

[68] Chaffee AL, Knowles GP, Liang Z, Zhang J, Xiao P, Webley PA. CO2 capture by adsorption: materials and process development. International Journal of Greenhouse Gas Control 2007;1(1):11–8.

[69] Franco A, Diaz AR. The future challenges for "clean coal technologies": joining efficiency increase and pollutant emission control. Energy 2009; 34(3):348–54.

[70] Hedin N, Andersson L, Bergström L, Yan J. Adsorbents for the post-combustion capture of CO2 using rapid temperature swing or vacuum swing adsorption. Applied Energy 2013;104:418–33.

[71] Escosa JM, Romeo LM. Optimizing CO2 avoided cost by means of repowering. Applied Energy 2009;86(11):2351–8.

[72] Steeneveldt R, Berger B, Torp T. CO2 capture and storage: closing the knowing–doing gap. Chemical Engineering Research and Design 2006; 84(9):739–63.

[73] Lucquiaud M, Chalmers H, Gibbins J. Capture-ready supercritical coal-fired power plants and flexible post-combustion CO2 capture. Energy Procedia 2009;1(1):1411–8.

[74] Li H, Yan J. Performance comparison on the evaporative gas turbine cycles combined with different CO2-capture options. International Journal of Green Energy 2009;6(5):512–26.

[75] Li H, Flores S, Hu Y, Yan J. Simulation and optimization of evaporative gas turbine with chemical absorption for carbon dioxide capture. International Journal of Green Energy 2009;6(5):527–39.

[76] Lisbona P, Martinez A, Lara Y, Romeo LM. Integration of carbonate CO2 capture cycle and coal-fired power plants. A comparative study for different sorbents. Energy & Fuels 2010;24(1):728–36.

[77] Fu C, Gundersen T. Heat integration of an oxy-combustion process for coal-fired power plants with CO2 capture by pinch analysis. Chemical Engineering Transactions 2010;21:181–6.

[78] Alabdulkarem A, Hwang Y, Radermacher R. Energy consumption reduction in CO2 capturing and sequestration of an LNG plant through process integration and waste heat utilization. International Journal of Greenhouse Gas Control 2012;10:215–28.

[79] Kohl A, Nielsen R. Gas purification. 5 ed. Houston: Gulf Publishing Company; 1997.

[80] Herold R, Mokhatab S. Optimal design and operation of molecular sieve gas dehydration units—Part 1. Gas Processing & LNG; 2017.
[81] Howard K, Lafferty JL, Montgomery RR. Selective adsorption process. Google Patents; 1958.
[82] Heinrich K. Process for the purification and separation of gas mixtures. Google Patents; 1953.
[83] Berg CH. Adsorption process and apparatus. Google Patents; 1954.
[84] Su F, Lu C, Chung A-J, Liao C-H. CO2 capture with amine-loaded carbon nanotubes via a dual-column temperature/vacuum swing adsorption. Applied Energy 2014;113:706–12.
[85] Su F, Lu C. CO 2 capture from gas stream by zeolite 13X using a dual-column temperature/vacuum swing adsorption. Energy & Environmental Science 2012;5(10):9021–7.
[86] Mulgundmath V, Tezel FH. Optimisation of carbon dioxide recovery from flue gas in a TPSA system. Adsorption 2010;16(6):587–98.
[87] Duarte GS, Schürer B, Voss C, Bathen D. Modeling and simulation of a tube bundle adsorber for the capture of CO2 from flue gases. Chemie Ingenieur Technik 2016;88(3):336–45.
[88] Salazar Duarte G, Schürer B, Voss C, Bathen D. Adsorptive separation of CO2 from flue gas by temperature swing adsorption processes. ChemBioEng Reviews 2017;4(5):277–88.
[89] Bonjour J, Chalfen J-B, Meunier F. Temperature swing adsorption process with indirect cooling and heating. Industrial & Engineering Chemistry Research 2002;41(23):5802–11.
[90] Hubbard RA. Gas conditioning and processing. Oklahoma: John M Campbell and Company; 1994.
[91] Mokhatab S, Northrop S, Mitariten M. Controlling the hydrocarbon dew point of pipeline gas. In: Petroleum technology quarterly; 2017. p. 109–16.
[92] Mitariten M, Lind W. Laurence Reid gas conditioning conference. 2007. Norman, OK.
[93] Chen A-Q, Wankat PC. Analytical scaling of thermal swing adsorption. Separation Science and Technology 1991;26(12):1575–83.
[94] Chen AQ, Wankat PC. Scaling rules and intensification of thermal swing adsorption. AIChE Journal 1991;37(5):785–9.
[95] Kalbassi MA, Allam RJ, Golden TC. Temperature swing adsorption. Google Patents; 1998.
[96] Basmadjian D. On the possibility of omitting the cooling step in thermal gas adsorption cycles. Canadian Journal of Chemical Engineering 1975;53(2):234–8.
[97] Davis MM, LeVan MD. Experiments on optimization of thermal swing adsorption. Industrial & Engineering Chemistry Research 1989;28(6):778–85.
[98] Saysset S, Grévillot G, Lamine A. Adsorption of volatile organic compounds on carbonaceous adsorbent and desorption by direct joule effect. Récents Progrès en Genie des Procédés 1999;68:389–96.
[99] Ahn H, Chang-Ha L. Adsorption dynamics of water in layered bed for air-drying TSA process. AIChE Journal 2003;49(6):1601.
[100] Ojo AF, Fitch FR, Bülow M. Temperature swing adsorption process. Google Patents; 2002.
[101] Yamauchi H, Kodama A, Hirose T, Okano H, Yamada K-I. Performance of VOC abatement by thermal swing honeycomb rotor adsorbers. Industrial & Engineering Chemistry Research 2007;46(12):4316–22.

[102] Mérel J, Clausse M, Meunier F. Carbon dioxide capture by indirect thermal swing adsorption using 13X zeolite. Environmental Progress 2006; 25(4):327−33.

[103] Ishibashi M, Ota H, Akutsu N, Umeda S, Tajika M, Izumi J, et al. Technology for removing carbon dioxide from power plant flue gas by the physical adsorption method. Energy Conversion and Management 1996; 37(6−8):929−33.

[104] Gales L, Mendes A, Costa C. Recovery of acetone, ethyl acetate and ethanol by thermal pressure swing adsorption. Chemical Engineering Science 2003;58(23−24):5279−89.

[105] Stewart M, Arnold K. Gas sweetening and processing field manual. Gulf Professional Publishing; 2011.

[106] Gleichmann K, Unger B, Brandt A. Industrial zeolite molecular sieves. Zeolites-Useful Minerals; 2016.

[107] Gleichmann K, Unger B, Brandt A. Industrielle Herstellung von zeolithischen Molekularsieben. Chemie-Ingenieur-Technik 2017;89: 851−62.

[108] Steuten B, Pasel C, Luckas M, Bathen D. Trace level adsorption of toxic sulfur compounds, carbon dioxide, and water from methane. Journal of Chemical & Engineering Data 2013;58(9):2465−73.

[109] Mokhatab S, Poe WA, Mak JY. Handbook of natural gas transmission and processing: principles and practices. Gulf Professional Publishing; 2018.

[110] Gandhidasan P, Al-Farayedhi AA, Al-Mubarak AA. Dehydration of natural gas using solid desiccants. Energy 2001;26(9):855−68.

[111] Yang RT. Adsorbents: fundamentals and applications. John Wiley & Sons; 2003.

[112] Netusil M, Ditl P. Comparison of three methods for natural gas dehydration. Journal of Natural Gas Chemistry 2011;20(5):471−6.

[113] Schumann K, Unger B, Brandt A. Zeolithe als sorptionsmittel (zeolite as adsorbents). Chemie Ingenieur Technik 2010;82:929−40.

[114] Rastelli H, Shadden JS. Extending molecular sieve life in natural gas dehydration units. In: Gas processors association 86th annual convention proceedings; 2007. San Antonio, Texas.

[115] Do DD. Adsorption analysis: equilibria and kinetics. Imperial College Press; 1998.

[116] Al Ezzi A, Ma H. Equilibrium adsorption isotherm mechanism of water vapor on zeolites 3A, 4A, X, and Y. In: ASME international mechanical engineering congress and exposition. American Society of Mechanical Engineers; 2017.

[117] Moore JD, Serbezov A. Correlation of adsorption equilibrium data for water vapor on F-200 activated alumina. Adsorption 2005;11(1):65−75.

[118] Northrop P, Sundaram N. Modified cycles, adsorbents improve gas treatment, increase mot-sieve life. Oil & Gas Journal 2008;106(29):54−60.

[119] Berg F, Bläker C, Pasel C, Luckas M, Eckardt T, Bathen D. Load-dependent heat of adsorption of C6 hydrocarbons on silica alumina gel. Microporous and Mesoporous Materials 2018;264:208−17.

[120] Bläker C. Experimentelle und theoretische Untersuchungen zur Kombination von Adsorptionsvolumetrie und-kalorimetrie. Shaker Verlag; 2018.

[121] Chowanietz V, Pasel C, Luckas M, Bathen D. Temperature dependent adsorption of sulfur components, water, and carbon dioxide on a silica−alumina gel used in natural gas processing. Journal of Chemical & Engineering Data 2016;61(9):3208−16.

[122] Chowanietz V, Pasel C, Luckas M, Eckardt T, Bathen D. Desorption of mercaptans and water from a silica–alumina gel. Industrial & Engineering Chemistry Research 2017;56(2):614–21.
[123] Fellmuth P, Lutz W, Bülow M. Influence of weakly coordinated cations and basic sites upon the reaction of H2S and CO2 on zeolites. Zeolites 1987;7(4):367–71.
[124] Bülow M, Lutz W, Suckow M. Applications in industry. 1999.
[125] Lutz W, Buhl J-C, Thamm H. A new COS-suppressing zeolite for gas-sweetening. Erdöl Erdgas Kohle; 1999. p. 115.
[126] Lutz W, Seidel A, Boddenberg B. On the formation of COS from H2S and CO2 in the presence of zeolite/salt compounds. Adsorption Science and Technology 1998;16(7):577–81.
[127] Sonnleitner E, Schöny G, Hofbauer H. Assessment of zeolite 13X and Lewatit VP OC 1065 for application in a continuous temperature swing adsorption process for biogas upgrading. Biomass Conversion and Biorefinery 2018;8(2):379–95.
[128] Basmadjian D. Modeling in science and engineering. 1999.
[129] Wankat PC. Separation process engineering. Pearson Education; 2006.
[130] Cavalcante C. Industrial adsorption separation processes: fundamentals, modeling and applications. Latin American Applied Research 2000;30(4):357–64.
[131] Ribeiro R, Grande C, Rodrigues AE. Electric swing adsorption for gas separation and purification: a review. Separation Science and Technology 2014;49(13):1985–2002.
[132] Petkovska M, Tondeur D, Grevillot G, Granger J, Mitrović M. Temperature-swing gas separation with electrothermal desorption step. Separation Science and Technology 1991;26(3):425–44.
[133] Fabuss BM, Du Bois WC. Apparatus and process for desorption of filter beds by electric current. Google Patents; 1971.
[134] Moskal F, Nastaj J. Internal heat source capacity at inductive heating in desorption step of ETSA process. International Communications in Heat and Mass Transfer 2007;34(5):579–86.
[135] Hashisho Z, Rood M, Botich L. Microwave-swing adsorption to capture and recover vapors from air streams with activated carbon fiber cloth. Environmental Science & Technology 2005;39(17):6851–9.
[136] Reuß J, Bathen D, Schmidt-Traub H. Desorption by microwaves: mechanisms of multicomponent mixtures. Chemical Engineering & Technology 2002;25(4):381–4.
[137] Sullivan PD, Rood MJ, Grevillot G, Wander JD, Hay KJ. Activated carbon fiber cloth electrothermal swing adsorption system. Environmental Science & Technology 2004;38(18):4865–77.
[138] Sullivan PD. Organic vapor recovery using activated carbon fiber cloth and electrothermal desorption. University of Illinois at Urbana-Champaign; 2003.
[139] Burchell T, Judkins R, Rogers M, Williams A. A novel process and material for the separation of carbon dioxide and hydrogen sulfide gas mixtures. Carbon 1997;35(9):1279–94.
[140] Economy J, Ry L. Adsorption characteristics of activated carbon fibers. 1976.
[141] Rintoul JC. A heating apparatus for heating solid, particulate material. European Patent Office; 1983.
[142] Gadkaree KP, Tyndell BP. Electrically heatable activated carbon bodies for adsorption and desorption applications. Google Patents; 2000.

[143] Delaney SW, Knowles GP, Chaffee AL. Electrically regenerable mesoporous carbon for CO2 capture. In: ACS national meeting book of abstracts. American Chemical Society; 2006.

[144] Park S, Kwon Y-P, Kwon H-C, Lee J-H, Lee H-W, Lee JC. Electrothermal properties of regenerable carbon contained porous ceramic fiber media. Journal of Electroceramics 2009;22(1):315–8.

[145] An H, Feng B, Su S. CO2 capture by electrothermal swing adsorption with activated carbon fibre materials. International Journal of Greenhouse Gas Control 2011;5(1):16–25.

[146] Baudu M, LeCloirec P, Martin G. Thermal regeneration by Joule effect of activated carbon used for air treatment. Environmental Technology 1992;13(5):423–35.

[147] Yu FD, Luo L, Grevillot G. Electrothermal swing adsorption of toluene on an activated carbon monolith: experiments and parametric theoretical study. Chemical Engineering and Processing: Process Intensification 2007;46(1):70–81.

[148] Yu FD, Luo LA, Grévillot G. Electrothermal desorption using Joule effect on an activated carbon monolith. Journal of Environmental Engineering 2004;130(3):242–8.

[149] Masala A, Vitillo JG, Mondino G, Martra G, Blom R, Grande CA, et al. Conductive ZSM-5-based adsorbent for CO2 capture: active phase vs monolith. Industrial & Engineering Chemistry Research 2017;56(30):8485–98.

[150] Regufe MJ, Ferreira AF, Loureiro JM, Shi Y, Rodrigues A, Ribeiro AM. New hybrid composite honeycomb monolith with 13X zeolite and activated carbon for CO2 capture. Adsorption 2018;24(3):249–65.

[151] Moon S-H, Shim J-W. A novel process for CO2/CH4 gas separation on activated carbon fibers—electric swing adsorption. Journal of Colloid and Interface Science 2006;298(2):523–8.

[152] Ribeiro R, Grande C, Rodrigues A. Electrothermal performance of an activated carbon honeycomb monolith. Chemical Engineering Research and Design 2012;90(11):2013–22.

[153] Europe, U.N.E.C.F.. Protocol to the 1979 convention on long-range transboundary air pollution concerning the control of emissions of volatile organic compounds or their transboundary fluxes, vol. 30. UN; 1991.

[154] Europe, E.C.f., N.U.C.é.p. l'Europe, U.N.E.C.f. Europe. Convention on long-range transboundary air pollution and its protocols, vol. 50. New York: United Nations; 1979.

[155] Farant J-P, Desbiens G. Adsorption of contaminants from gaseous stream and in situ regeneration of sorbent. Google Patents; 2008.

[156] Snyder JD, Leesch JG. Methyl bromide recovery on activated carbon with repeated adsorption and electrothermal regeneration. Industrial & Engineering Chemistry Research 2001;40(13):2925–33.

[157] Petkovska M, Mitrović M. Microscopic modelling of electrothermal desorption. The Chemical Engineering Journal and the Biochemical Engineering Journal 1994;53(3):157–65.

[158] Petkovska M, Mitrovic M. One-dimensional, nonadiabatic, microscopic model of electrothermal desorption process dynamics. Chemical Engineering Research and Design 1994;72(6):713–22.

[159] Lordgooei M, Carmichael KR, Kelly TW, Rood MJ, Larson SM. Activated carbon cloth adsorption-cryogenic system to recover toxic volatile organic compounds. Gas Separation & Purification 1996;10(2):123–30.

[160] Levy R, Hicks R, Gold H. In-place electrically heated regeneration of vapor-phase activated carbon. Foster-Miller INC Waltham MA; 1990.

[161] Martin G, Baudu M, Cloirec PL. Device for treating fluids with an adsorption structure of superposed and spaced sheets and regeneration by joule effect. European Patent Office; 1997.

[162] Sullivan P, Rood MJ, Hay K, Qi S. Adsorption and electrothermal desorption of hazardous organic vapors. Journal of Environmental Engineering 2001;127(3):217−23.

[163] Chmiel H, Schippert E, Möhner C. Method for regenerating electrically conducting adsorbents laden with organic substances. Google Patents; 2002.

[164] Sullivan PD, Rood MJ, Dombrowski KD, Hay KJ. Capture of organic vapors using adsorption and electrothermal regeneration. Air Force Research Lab Tyndall AFB FL Materials And Manufacturing Directorate; 2004.

[165] Place RN, Blackburn AJ, Tennison SR, Rawlinson AP, Crittenden BD. Method and equipment for removing volatile compounds from air. Google Patents; 2005.

[166] Li G, Xiao P, Webley P, Zhang J, Singh R, Marshall M. Capture of CO_2 from high humidity flue gas by vacuum swing adsorption with zeolite 13X. Adsorption 2008;14(2):415−22.

[167] Huang Q, Eić M. Commercial adsorbents as benchmark materials for separation of carbon dioxide and nitrogen by vacuum swing adsorption process. Separation and Purification Technology 2013;103:203−15.

[168] Ling J, Ntiamoah A, Xiao P, Webley PA, Zhai Y. Effects of feed gas concentration, temperature and process parameters on vacuum swing adsorption performance for CO_2 capture. Chemical Engineering Journal 2015;265:47−57.

[169] Huang Q, Malekian A, Eić M. Optimization of PSA process for producing enriched hydrogen from plasma reactor gas. Separation and Purification Technology 2008;62(1):22−31.

[170] Lee KB, Sircar S. Removal and recovery of compressed CO_2 from flue gas by a novel thermal swing chemisorption process. AIChE Journal 2008;54(9):2293−302.

[171] Ebner AD, Ritter JA. Equilibrium theory analysis of dual reflux PSA for separation of a binary mixture. AIChE Journal 2004;50(10):2418−29.

[172] Li X, Hagaman E, Tsouris C, Lee JW. Removal of carbon dioxide from flue gas by ammonia carbonation in the gas phase. Energy & Fuels 2003;17(1):69−74.

[173] Zhang J, Webley PA. Cycle development and design for CO_2 capture from flue gas by vacuum swing adsorption. Environmental Science & Technology 2008;42(2):563−9.

[174] Zhang J, Webley PA, Xiao P. Effect of process parameters on power requirements of vacuum swing adsorption technology for CO_2 capture from flue gas. Energy Conversion and Management 2008;49(2):346−56.

[175] Chen C-Y, Lee K-C, Chou C-T. Concentration and recovery of carbon dioxide from flue gas by vacuum swing adsorption. Journal of the Chinese Institute of Chemical Engineers 2003;34(1):135−42.

[176] Ho MT, Allinson GW, Wiley DE. Reducing the cost of CO_2 capture from flue gases using pressure swing adsorption. Industrial & Engineering Chemistry Research 2008;47(14):4883−90.

[177] Aaron D, Tsouris C. Separation of CO_2 from flue gas: a review. Separation Science and Technology 2005;40(1−3):321−48.

[178] Krishnamurthy S, Rao VR, Guntuka S, Sharratt P, Haghpanah R, Rajendran A, et al. CO2 capture from dry flue gas by vacuum swing adsorption: a pilot plant study. AIChE Journal 2014;60(5):1830–42.
[179] Chou C-T, Chen C-Y. Carbon dioxide recovery by vacuum swing adsorption. Separation and Purification Technology 2004;39(1–2):51–65.
[180] Webley PA. Adsorption technology for CO2 separation and capture: a perspective. Adsorption 2014;20(2):225–31.
[181] Sircar S, Golden T. Purification of hydrogen by pressure swing adsorption. Separation Science and Technology 2000;35(5):667–87.
[182] Yang RT. Gas separation by adsorption processes, vol. 1. World Scientific; 1997.
[183] Batta LB. Selective adsorption process. Google Patents; 1971.
[184] Yamaguchi T, Kobayashi Y. Gas separation process. Toyo Engineering Corporation; 1993.
[185] Fuderer A, Rudelstorfer E. Selective adsorption process. Google Patents; 1976.
[186] Riboldi L, Bolland O. Evaluating Pressure Swing Adsorption as a CO2 separation technique in coal-fired power plants. International Journal of Greenhouse Gas Control 2015;39:1–16.
[187] Casas N, Schell J, Joss L, Mazzotti M. A parametric study of a PSA process for pre-combustion CO2 capture. Separation and Purification Technology 2013;104:183–92.
[188] Sircar S, Kratz W. Simultaneous production of hydrogen and carbon dioxide from steam reformer off-gas by pressure swing adsorption. Separation Science and Technology 1988;23(14–15):2397–415.
[189] Riboldi L, Bolland O. Pressure swing adsorption for coproduction of power and ultrapure H2 in an IGCC plant with CO2 capture. International Journal of Hydrogen Energy 2016;41(25):10646–60.
[190] Shi W, Yang H, Shen Y, Fu Q, Zhang D, Fu B. Two-stage PSA/VSA to produce H2 with CO2 capture via steam methane reforming (SMR). International Journal of Hydrogen Energy 2018;43(41):19057–74.
[191] Majlan EH, Daud WRW, Iyuke SE, Mohamad AB, Kadhum AAH, Mohammad AW, et al. Hydrogen purification using compact pressure swing adsorption system for fuel cell. International Journal of Hydrogen Energy 2009;34(6):2771–7.
[192] Zheng Y, Shi Y, Li S, Yang Y, Cai N. Elevated temperature hydrogen/carbon dioxide separation process simulation by integrating elementary reaction model of hydrotalcite adsorbent. International Journal of Hydrogen Energy 2014;39(8):3771–9.
[193] Reijers R, van Selow E, Cobden P, Boon J, van den Brink R. SEWGS process cycle optimization. Energy Procedia 2011;4:1155–61.
[194] Allam RJ, Chiang R, Hufton JR, Middleton P, Weist EL, White V. Development of the sorption enhanced water gas shift process. Carbon dioxide capture for storage in deep geologic formations. Elsevier; 2005;1: p. 227–56.
[195] Najmi B, Bolland O, Colombo KE. A systematic approach to the modeling and simulation of a sorption enhanced water gas shift (SEWGS) process for CO2 capture. Separation and Purification Technology 2016;157:80–92.
[196] Wright A, White V, Hufton J, Quinn R, Cobden P, van Selow E. CAESAR: development of a SEWGS model for IGCC. Energy Procedia 2011;4:1147–54.

[197] Wright A, White V, Hufton J, van Selow E, Hinderink P. Reduction in the cost of pre-combustion CO2 capture through advancements in sorption-enhanced water-gas-shift. Energy Procedia 2009;1(1):707−14.

[198] Van Selow E, Cobden P, Van den Brink R, Hufton J, Wright A. Performance of sorption-enhanced water-gas shift as a pre-combustion CO2 capture technology. Energy Procedia 2009;1(1):689−96.

[199] Van Selow E, Cobden P, Van den Brink R, Wright A, White V, Hinderink P, et al. Pilot-scale development of the sorption enhanced water gas shift process. In: Eide LI, editor. Carbon dioxide capture for storage in deep geologic formations. Berks; 2009. p. 157−80.

[200] Najmi B, Bolland O, Westman SF. Simulation of the cyclic operation of a PSA-based SEWGS process for hydrogen production with CO2 capture. Energy Procedia 2013;37:2293−302.

[201] Liu Z, Green WH. Analysis of adsorbent-based warm CO2 capture technology for integrated gasification combined cycle (IGCC) power plants. Industrial & Engineering Chemistry Research 2014;53(27):11145−58.

[202] Boon J, Cobden P, Van Dijk H, van Sint Annaland M. High-temperature pressure swing adsorption cycle design for sorption-enhanced water−gas shift. Chemical Engineering Science 2015;122:219−31.

[203] Jansen D, van Selow E, Cobden P, Manzolini G, Macchi E, Gazzani M, et al. SEWGS technology is now ready for scale-up! Energy Procedia 2013;37:2265−73.

[204] Gazzani M, Macchi E, Manzolini G. CO2 capture in integrated gasification combined cycle with SEWGS−Part A: thermodynamic performances. Fuel 2013;105:206−19.

[205] Zhu X, Shi Y, Li S, Cai N. Two-train elevated-temperature pressure swing adsorption for high-purity hydrogen production. Applied Energy 2018;229:1061−71.

[206] Yang J, Han S, Cho C, Lee C-H, Lee H. Bulk separation of hydrogen mixtures by a one-column PSA process. Separations Technology 1995;5(4):239−49.

[207] Ribeiro AM, Grande CA, Lopes FV, Loureiro JM, Rodrigues AE. Four beds pressure swing adsorption for hydrogen purification: case of humid feed and activated carbon beds. AIChE Journal 2009;55(9):2292−302.

[208] You Y-W, Lee D-G, Yoon K-Y, Moon D-K, Kim SM, Lee C-H. H_2 PSA purifier for CO removal from hydrogen mixtures. International Journal of Hydrogen Energy 2012;37(23):18175−86.

[209] Lively RP, Bessho N, Bhandari DA, Kawajiri Y, Koros WJ. Thermally moderated hollow fiber sorbent modules in rapidly cycled pressure swing adsorption mode for hydrogen purification. International Journal of Hydrogen Energy 2012;37(20):15227−40.

[210] Rahimpour M, Ghaemi M, Jokar SM, Dehghani O, Jafari M, Amiri S, et al. The enhancement of hydrogen recovery in PSA unit of domestic petrochemical plant. Chemical Engineering Journal 2013;226:444−59.

[211] Moon D-K, Lee D-G, Lee C-H. H_2 pressure swing adsorption for high pressure syngas from an integrated gasification combined cycle with a carbon capture process. Applied Energy 2016;183:760−74.

[212] Grande CA, Ribeiro RP, Rodrigues AE. Challenges of electric swing adsorption for CO2 capture. ChemSusChem 2010;3(8):892−8.

[213] Grande CA, Ribeiro RP, Rodrigues AE. CO2 capture from NGCC power stations using electric swing adsorption (ESA). Energy & fuels 2009;23(5):2797−803.

[214] Grande CA, Rodrigues AE. Electric swing adsorption for CO2 removal from flue gases. International Journal of Greenhouse Gas Control 2008;2(2):194–202.
[215] Lillia S, Bonalumi D, Grande C, Manzolini G. A comprehensive modeling of the hybrid temperature electric swing adsorption process for CO2 capture. International Journal of Greenhouse Gas Control 2018;74:155–73.
[216] Zhao Q, Wu F, Men Y, Fang X, Zhao J, Xiao P, et al. CO2 capture using a novel hybrid monolith (H-ZSM5/activated carbon) as adsorbent by combined vacuum and electric swing adsorption (VESA). Chemical Engineering Journal 2019;358:707–17.
[217] Liang Z, Marshall M, Chaffee AL. CO2 adsorption-based separation by metal organic framework (Cu-BTC) versus zeolite (13X). Energy & Fuels 2009;23(5):2785–9.
[218] Cavenati S, Grande CA, Rodrigues AE. Adsorption equilibrium of methane, carbon dioxide, and nitrogen on zeolite 13X at high pressures. Journal of Chemical & Engineering Data 2004;49(4):1095–101.
[219] Joss L, Gazzani M, Mazzotti M. Rational design of temperature swing adsorption cycles for post-combustion CO2 capture. Chemical Engineering Science 2017;158:381–94.
[220] Wang L, Liu Z, Li P, Yu J, Rodrigues AE. Experimental and modeling investigation on post-combustion carbon dioxide capture using zeolite 13X-APG by hybrid VTSA process. Chemical Engineering Journal 2012;197:151–61.
[221] Wurzbacher JA, Gebald C, Steinfeld A. Separation of CO2 from air by temperature-vacuum swing adsorption using diamine-functionalized silica gel. Energy & Environmental Science 2011;4(9):3584–92.
[222] Shen C, Yu J, Li P, Grande CA, Rodrigues AE. Capture of CO2 from flue gas by vacuum pressure swing adsorption using activated carbon beads. Adsorption 2011;17(1):179–88.
[223] Ko D, Siriwardane R, Biegler LT. Optimization of pressure swing adsorption and fractionated vacuum pressure swing adsorption processes for CO2 capture. Industrial & Engineering Chemistry Research 2005;44(21):8084–94.
[224] Wang L, Yang Y, Shen W, Kong X, Li P, Yu J, et al. CO2 capture from flue gas in an existing coal-fired power plant by two successive pilot-scale VPSA units. Industrial & Engineering Chemistry Research 2013;52(23):7947–55.
[225] Qu D, Yang Y, Qian Z, Li P, Yu J, Ribeiro AM, et al. Enrichment of low-grade methane gas from nitrogen mixture by VPSA with CO2 displacement process: modeling and experiment. Chemical Engineering Journal 2020;380:122509.
[226] Liu H, Ding W, Zhou F, Yang G, Du Y. An overview and outlook on gas adsorption: for the enrichment of low concentration coalbed methane. Separation Science and Technology 2020;55(6):1102–14.
[227] Hu S, Guo X, Li C, Feng G, Yu X, Zhang A, et al. An approach to address the low concentration methane emission of distributed surface wells. Industrial & Engineering Chemistry Research 2018;57(39):13217–25.
[228] Maté VIÁ, Dobladez JAD, Álvarez-Torrellas S, Larriba M, Rodríguez ÁM. Modeling and simulation of the efficient separation of methane/nitrogen mixtures with [Ni3 (HCOO) 6] MOF by PSA. Chemical Engineering Journal 2019;361:1007–18.
[229] Qian Z, Yang Y, Li P, Wang J, Rodrigues AE. An improved vacuum pressure swing adsorption process with the simulated moving bed operation mode for CH4/N2 separation to produce high-purity methane. Chemical Engineering Journal 2021;419:129657.

[230] Delgado JA, Águeda VI, García J, Álvarez-Torrellas S. Simulation of the recovery of methane from low-concentration methane/nitrogen mixtures by concentration temperature swing adsorption. Separation and Purification Technology 2019;209:550−9.

[231] Saleman TL, Li GK, Rufford TE, Stanwix PL, Chan KI, Huang SH, et al. Capture of low grade methane from nitrogen gas using dual-reflux pressure swing adsorption. Chemical Engineering Journal 2015;281: 739−48.

[232] Kennedy D, Khanafer M, Tezel F. The effect of Ag+ cations on the micropore properties of clinoptilolite and related adsorption separation of CH4 and N2 gases. Microporous and Mesoporous Materials 2019;281: 123−33.

[233] Bae J-S, Su S, Yu XX. Enrichment of ventilation air methane (VAM) with carbon fiber composites. Environmental Science & Technology 2014; 48(10):6043−9.

[234] Majumdar B, Bhadra S, Marathe R, Farooq S. Adsorption and diffusion of methane and nitrogen in barium exchanged ETS-4. Industrial & Engineering Chemistry Research 2011;50(5):3021−34.

[235] Niu Z, Cui X, Pham T, Lan PC, Xing H, Forrest KA, et al. A metal−organic framework based methane nano-trap for the capture of coal-mine methane. Angewandte Chemie International Edition 2019;58(30): 10138−41.

[236] Chen Y, Wu H, Yuan Y, Ly D, Qiao Z, An D, et al. Highly rapid mechanochemical synthesis of a pillar-layer metal-organic framework for efficient CH4/N2 separation. Chemical Engineering Journal 2020;385: 123836.

[237] Hu G, Zhao Q, Manning M, Chen L, Yu L, May EF, et al. Pilot scale assessment of methane capture from low concentration sources to town gas specification by pressure vacuum swing adsorption (PVSA). Chemical Engineering Journal 2022;427:130810.

[238] Grande CA, Rodrigues AE. Layered vacuum pressure-swing adsorption for biogas upgrading. Industrial & Engineering Chemistry Research 2007; 46(23):7844−8.

[239] Liu Z, Wang L, Kong X, Li P, Yu J, Rodrigues AE. Onsite CO2 capture from flue gas by an adsorption process in a coal-fired power plant. Industrial & Engineering Chemistry Research 2012;51(21):7355−63.

[240] Lu W, Ying Y, Shen W, Li P, Yu J. Experimental evaluation on CO2 capture from flue gas by two successive VPSA units in an existing coal power plant. In: 11th international conference on fundamentals of adsorption; 2013. Baltimore, MD.

12

Zeolite sorbents and nanosorbents for natural gas sweetening

Maryam Koohi-Saadi and Mohammad Reza Rahimpour
Department of Chemical Engineering, Shiraz University, Shiraz, Iran

1. Introduction

Fossil fuels, such as coal, natural gas (NG), and oil are still dominant source of energy [1,2] though they suffer from severe environmental issues [3,4]. They can all be used interchangeably but to varied degrees of efficiency, depending on the need. In the past, coal has been the most cost-effective material for producing power, but since it is the most polluting, new coal-fired plants are likely to have to pay additional carbon costs through cap-and-trade systems or carbon capture and sequestration (CCS). However, coal can be converted into gas at an additional expense to provide NG for consumption by diverse energy end users with fewer emissions. NG may power combined-cycle turbines to generate electricity efficiently, heat spaces on a variety of scales, and, with additional infrastructure costs, fuel road vehicles. Through a number of gas-to-liquid (GTL) conversion processes, it can be transformed from a gas to unleash its hydrogen and produce extended hydrocarbon liquid fuel molecules [5,6].

Natural gas specifically, a fossil fuel, has a huge range of applications. In addition to helping to offset their prices as a fuel, residues from fossil fuels also make them crucial to a variety of nonfuel petrochemical industries [5].

The location, kind, and depth of the subsurface sediment, as well as the local geostrategy, all affect the constitution of the naive NG that is taken from generated wells. NG and oil are frequently found in the similar tank. NG extracted from petroleum wells is typically categorized as "associated-dissolved," which simply means it is dissolved in or related to crude oil [7].

Production of NG that is not connected to the generation of crude petroleum is referred to as "nonassociated" [8].

1.1 NG constitution

The constitution of NG differs depending on the source from where it is collected. NG's constitution is never stable because it might include various hydrocarbons and nonhydrocarbons components. The usual composition of NG is tabulated in Table 12.1 [9].

Ethane, propane, butanes, and a minor amount of C_5+ hydrocarbons are examples of paraffinic hydrocarbons that methane may associate with them. Additionally, it might include certain aromatics including benzene, toluene, and xylenes [10]. Diluents, pollutants, and solid matter can all be categorized as nonhydrocarbon components of NG [11–13]. The dilutors are nonflammable gases that lower the gas's heating worthiness, such as CO_2, N_2, and He [14]. When the gas's heat content needs to be reduced, it could be employed as fillers [11].

Table 12.1 Natural gas' (NG's) general chemical composition [9].

Constituents	Composition (vol%)
Hydrocarbons	
Methane	84.6
Ethane	6.4
Propane	5.3
Isobutane	1.2
n-Butane	1.4
Isopentane	0.4
n-Pentane	0.2
Hexanes	0.4
Heptanes	0.1
Nonhydrocarbons	
Nitrogen	≤10
Carbon dioxide	≤5
Helium	≤0.5
Argon	≤0.05
Radon, krypton, xenon	Traces
Hydrogen sulfide	≤5

In addition to being unpleasant pollutants, the contaminants are gases that harm industrial and transportation machinery [11]. These pollutants include:

  I. **Sulfur species:** A hydrogen sulfide (H_2S) can be produced when sulfur-reducing bacteria in last fouling use a biochemical pathway to reduce sulfate ions at sulfates dissolved in water. It can also be created through the thermal decomposition of deep-seated, sulfur-rich kerogen. The gas that is produced may contain a variety of sulfur species, including elemental sulfur, carbonyl sulfide (COS), carbon disulfide (CS_2), and mercaptans (RSH, where R demonstrates an alkane category, such as methyl mercaptan CH_3SH, ethyl mercaptan C_2H_5SH) [15].

 II. **Mercury:** The devaluation or thermic disintegration of mercury sulfide (cinnabar) in the presence of hydrocarbons is assumed to be the process that produces mercury. It was claimed that mercury has a strong proximity for sulfides and carbon. Mercury can be found in a different type of form, containing mineral mercury compounds like HgCl and $HgCl_2$, as well as fundamental organometallic compounds like dimethylmercury, methylethylmercury, and dimethylmercury [16,17].

III. **Arsenic:** It is thought that the creation of geological gas is the cause of the arsenic that can be detected in petroleum and gas processes. NG and gas condensate contain it in a variety of forms. Trimethylarsine (Me3As), dimethylethylarsine (Me2EtAs), methyldiethylarsine (MeEt2As), triethylarsine (Et3As), and triphenylarsine are just a few of the trialkylarsines that can be found in addition to arsine (H3As) (Ph3As). Minerals containing arsenic sulfide may be found in sour NG systems [18,19].

 IV. **Naturally occurring radioactive material (NORM):** By removing radon decay products with filter assemblies in gas lines, NORM emissions are reduced. A different approach involves injecting scale inhibitors into the system as soon as formation water starts to be created.

  V. **Solid matter:** Due to scaling in the pipe, there might be solid material in the form of tiny silica (sand) and black flour [20]. Iron oxides and iron sulfides make up the majority of the black powder. But it could also have deposits of polonium (Po-210), lead oxides, metallic lead, barium sulfate, zinc sulfide, lead sulfide, and calcium carbonate [21].

In the following, different procedures for NG purification are described in detail.

2. Natural gas purification techniques

Different procedures are required for cleaning raw NG to ensure it meets pipeline (pipeline-quality NG) requirements and clean gas burns to the environment [8,22], schematic of which is demonstrated in Fig. 12.1.

Natural gas is processed after it is extracted in order to accomplish the following:

I. Removing impediments to the usage of the gas as combustion from the raw gas (residential or industrial).

II. The process of separating useful ingredients from crude gas that can be used as petrochemical feedstocks, combustion (like propane), or commercial gases (helium, ethane).

III. NG that will be stored or transferred is liquefied [11].

The origin and composition of the wellhead generation current determine how many steps and what kinds of procedures

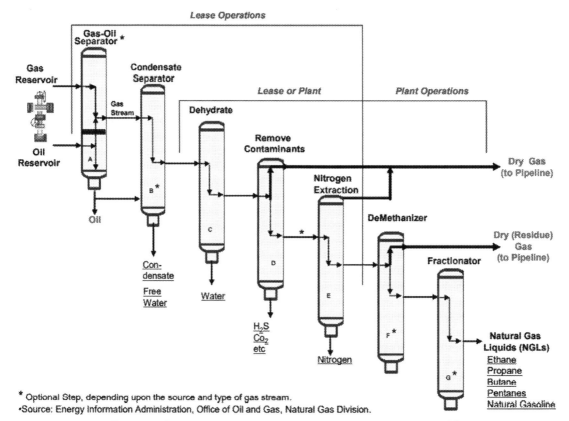

Figure 12.1 The general gas processing units are shown schematically [7].

are utilized to produce NG that is suitable for pipeline usage. Some of the stages in Fig. 12.1 may be combined into a single operation, carried out in a different sequence or at a different location (lease/plant), or they may not even be necessary.

Gas—oil separator, dehydration, extraction of trace components, and elimination of acid gases are a few of the various phases (as denoted by the letters in Fig. 12.1) of gas processing/treatment [7].

To get rid of the water that could result in the production of hydrates, a dehydration procedure is required. When a liquid or gas including gratis water is exposed to particular temperature/pressure conditions, hydrates are formed. There are numerous ways to dehydrate, which is the process of removing the water from the NG that has been created [8]. Extraction of trace elements such as nitrogen, helium, oxygen, and arsenic should be done for the following reasons:

I. Because a great nitrogen concentration lowers the NG's ability to heat buildings, nitrogen should be eliminated from the created gas. It can be eliminated by membrane separation, pressure swing adsorption, or cryogenic distillation [11,23—25].

II. A useful byproduct of the processing of NG is helium. Helium should be present in significant concentrations, unlike the other trace component. Nitrogen injections are able to get rid of it [26].

III. When oxygen is present in low concentrations, it can be taken out by nonregenerative scavengers. A catalytic reaction at greater concentrations can remove it from the gas and create water that is then expelled during the dehydration procedure [26].

IV. NG must be purified of any arsenic compounds, and they should not exist for the following reasons:
Using fire to pollute the environment.
Risk to one's health and safety.

A future processing plant's catalysts becoming contaminated palladium and platinum catalysts [19,27].

A nonregenerative adsorption technique can be used to remove the gas's arsenic components. The sediments that are often made of iron sulfide can also be eliminated by cleaning the pipelines [19,28,29].

2.1 Elimination of acid gases

The phrase "acid gases" refers to the existence of CO_2 and H_2S, both of which combine with water to generate weak acids. These acidic solutions are extremely corrosive, therefore getting rid of them means great care. However, carbon dioxide has negligible heating value, whereas hydrogen sulfide has a significant level of toxicity and a strong disagreeable odor [30]. On the other hand, when CO_2 or H_2S and water are present, solid hydrate can develop under specific pressure and temperature circumstances.

With increasing H_2S content, the temperature needed for hydrate establishment rises, whereas with CO_2, the converse is true [31].

The amount of gas provided, the content of the crude gas being refined, and the characteristics needed for the remaining gas all have a considerable effect on the acid gas removal procedure [15]. The elimination of acid gases often involves either adsorption or absorption [32]. In the following (Fig. 12.2), different methods of acid gas removal are listed.

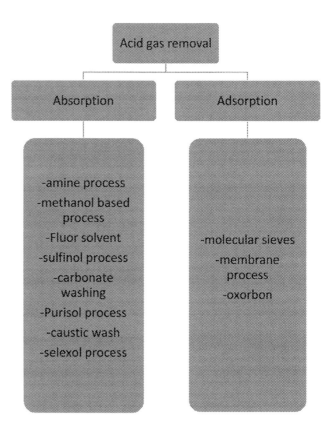

Figure 12.2 Different methods of acid gas removal.

2.1.1 Absorption processes

Chemical procedures for solvents: Through a chemical interaction with a substance in the solvent solution, CO_2 and H_2S may be eliminated from the gas current in these procedures. This process may or may not be reversible [26,33].

I. Amine process: NG is subjected to a procedure well-known as "amine washing" in which amine compounds and acid gases (CO_2 and H_2S) undergo a chemical reaction. Monoethanolamine (MEA), diethanolamine (DEA), triethanolamine (TEA), methyldiethanolamine (MDEA), di-isopropanolamine (DIPA), and diglycolamine (DGA) are examples of commercial amine-derivative chemicals that have been employed [10,32,34]. To face the requirements for the created gas, several amines can be chosen based on the make-up and usage of the feed gas. Using sterically hindered amines can increase the selectivity of H_2S/CO_2. The spatial obstacle in the amine structure prevents the creation of carbamate, a byproduct of the interaction among CO_2 and the amine blend, when these amines respond with CO_2. As a result, the interaction between CO_2 and H_2S is delayed but not stopped [35].

II. Water washing and carbonate washing: In this procedure, potassium carbonate, a weak alkali, is used to wash away the acid gases. H_2S is also absorbed using this technique, which was initially utilized for CO_2 [12,32].

III. Caustic wash: NG can be cleaned with caustic (NaOH) scrubbing systems to remove CS_2, CO_2, H_2S, and mercaptans (RSH, where R is an alkyl group). After a caustic wash procedure, NG is often washed with water to eliminate any abrasive materials entered in the gas before anhydride [33].

Physical solvent processes: These procedures are based on physical sorption. Here is a list of some of these procedures:

I. Methanol-based procedure: Methanol was the prime commercially available organic physical solvent. It has been utilized for liquid recovery, gas softening, hydrate inhibition, and dehydration [10,36]. This procedure, also known as Rectisol, uses pure, chilled methanol as the solvent. This technique was created and granted a license by Lurgi oel gas chemie gmbH and Linde AG [33].

II. Selexol procedure: Polyethylene glycol dimethyl ether is the solvent utilized in this procedure. Allied chemical corporation invented this method [37].

III. Purisol procedure: N-methyl-2-pyrrolidone (NMP or N-Pyrol), that has a great seethe dot and a strong selectivity for H_2S, is the solvent utilized for the Purisol method. This procedure was created and granted a license by Lurgi oel gas chemie [33].

IV. Fluor solvent: Anhydrous propylene carbonate has been applied in this procedure. It is mostly utilized in situations with high carbon dioxide concentrations. Fluor corporation filed a patent for this method [12].

V. Hybrid solvent processes (sulfinol process): H_2S, CO_2, COS, CS_2, mercaptans, and polysulfides can all be eliminated from natural and synthetic gases using this method. Diisopropanolamine (DIPA) or MDEA are two chemical solvents that are combined with water and create a physical solvent (sulfolane) and sulfinol. Shell global solutions holds a license for the sulfinol process [11,33,36].

2.1.2 Adsorption processes

The H_2S, CO_2, and other sulfur pollutants are subjected to an adsorption procedure using a solid substance known as an adsorbent [32]. Adsorbents are either nonregenerative, like impregnated activated carbon, or regenerable, like molecular sieves.

I. Molecular sieves: H_2S (like other sulfur mixtures) can be removed from gas currents with excellent selectivity using molecular sieves [38].

II. Oxorbon: This rigid adsorbent is made of potassium iodide-impregnated activated carbon [39,40]. Such carbon is sold by Donau carbon to remove mercaptans and H_2S [33].

III. Membrane process: The removal of carbon dioxide has been the exclusive application of membranes for gas separation [10]. The principal drawbacks of adopting the membrane procedure for NG softening include weak mechanical strength and plasticization, which are brought on by disposal to great temperature or pressure during operation. The creation of a high-performance membrane system has been attempted [41]. These experiments revealed that crosslinking the membrane material could prevent membrane plasticization, while adding inorganic fillers to the polymer matrix could increase the membrane's mechanical strength [42–44].

2.2 The significance of eliminating H_2S gas

One of the main issues for the majority of the energy professions, including NG, tail gas, liquefied petroleum gas, combust cells, and haul gases including jet fuels, diesel and gasoline, is H_2S. H_2S elimination is necessary because, even at low concentrations, it causes erosion in haul paths and poisons plenty of catalysts [45]. When it interacts with a Ni catalyst in fuel cell applications, the electrolyte may degrade, reducing the life cycle of the cell. H_2S should be eliminated from the environment because it is another dangerous substance that can oxidized to SO_2 and results in acid rain. Regarding the environment, it is crucial that environmental rules keep the amounts of hydrogen sulfide in transportation fuels to extremely low levels. A significant operational and financial problem for the petroleum refining sector is ensuring that H_2S concentrations in the combustions match the standards [46,47]. Additionally, the integrated gasification combined cycle (IGCC) procedure that is utilized for gas turbines and energy origins uses syngas produced by the gasification of coal. H_2S should therefore be eliminated from syngas because it negatively impacts pipes and machinery [48].

In several industrial processes, including the Claus process and NG sweetening, the usage of H_2S in down temperatures is rising [49]. These technologies come with some benefits and drawbacks. It is appropriate to remove hydrogen sulfide using biological treatment, which also lowers energy, executive, and chemical expenses. In other words, compared to the dry-based techniques, it requires a higher capital investment. Although regeneration procedures are appropriate, liquid-based and membrane technologies demand a disproportionately high capital and energy cost [50]. H_2S can also be removed from biogas using water and polyethylene glycol (PEG) washing by solving it into the water or the dissolvent [51]. However, because H_2S corrodes the equipment of this operation, it is not typically [52]. The arid adsorption procedure is a more accessible, environmentally responsible, and cost-effective way among these various approaches utilized to eliminate H_2S from gases at bottom [53]. The removal of H_2S has been studied using a variety of techniques and adsorbents, including zeolites, metal oxides, and activated carbons. These spongy substances are typically employed as catalyst supports, segregation media, or adsorbents. Activated carbons have a great inclusion to adsorb H_2S, but because of their limited mechanical stability and abundance of micropores, they suffer from significant curvature and the generation of forfeit within processes [54,55].

2.3 NG sweetening via zeolite-based adsorbents

The elimination of H_2S involved the use of a number of metal oxides, including Cu, Zn, Fe, Co, Mo, Ce, Sn, Mn, Ni, and W. However, prior study demonstrates that due to the thermodynamic equilibrium, metal oxides were not effective adsorbents for H_2S

elimination. As a result, blended metal oxide adsorbents made by altering the metal oxide ratios were more effective at removing sulfides. But blended metal oxides' adsorption capability was less than that of adsorbents based on zeolite [56]. Due to their cage-like shape, zeolites can trap gases including H_2S well. This issue is clearly visible in Table 12.2.

Zeolites are very porous materials that are used as molecular sieves to collect molecules. Zeolites are efficient at removing some substances, including water and H_2S. There are around 40 naturally occurring zeolites and 194 distinct zeolite frameworks. There is a rising need for natural zeolites including mordenite, clinoptilolite, erionite, phillipsite, and ferrierite in the gas separation industry today. First, it is necessary to activate natural zeolites before employing them. So in terms of commercial separation, natural zeolites have superior qualities than synthetic ones. Zeolites have a great capacity to adsorb H_2S, and they are also very simple and inexpensive to regenerate. Zeolites were also employed in IGCC power plants, commonly known as "clean coal technology" (CCT) facilities. Zeolite Na-X, for instance, has been investigated for the elimination of H_2S from IGCC. On zeolites, a sulfur output of %86 has been reported [57,58]. Accordingly, several experiential and theoretic methods for the elimination of H_2S have been applied. Various kinds of zeolite-based adsorbents (both modified and nude structures) that have large surface space, adsorption valence, renewability, selectivity, and nice resilience to wide temperatures can be a good choice as H_2S adsorbents [56]. Table 12.3 summarizes all experimental methods for H_2S elimination by zeolites, including zeolite kinds, selectivity/yield worthinesses, and correction kinds.

2.3.1 Natural zeolites for acid gas removal

Due to their accessibility and affordability, clinoptilolite, the most prevalent natural zeolite in the world, has been employed as an adsorbent for the adsorption of H_2S. A constant floor technique was applied to examine the adsorption of H_2S on this natural zeolite at various temperatures between 100 and 600°C. Natural zeolite was discovered to have a capacity of 0.03 g S/g natural zeolite. In conclusion, it was discovered that natural zeolite was more efficient at removing H_2S at temperatures less than 600°C [60].

The good separation of fresh zeolite was demonstrated by Yasyerli et al. [60]. Due to the interval between studies, the outcomes of this investigation may be split in two groups: up and down yield. The medium was only subjected to pristine air via

Table 12.2 Different shapes of zeolites.

Zeolite types	Shapes
Fresh zeolite	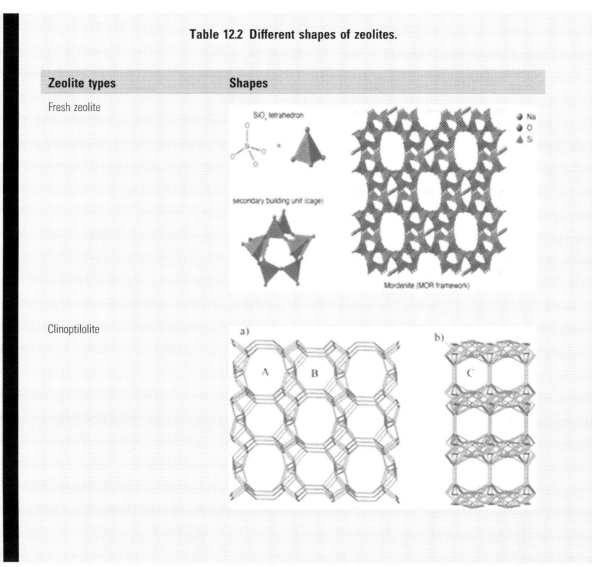
Clinoptilolite	

Continued

Table 12.2 Different shapes of zeolites.—*continued*

Zeolite types	Shapes
13X	
5A	

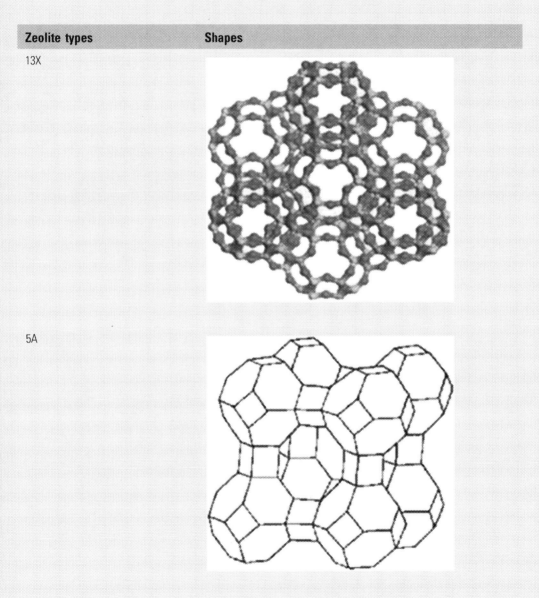

Table 12.2 Different shapes of zeolites.—*continued*

Zeolite types	Shapes
4A	
Mordenite	

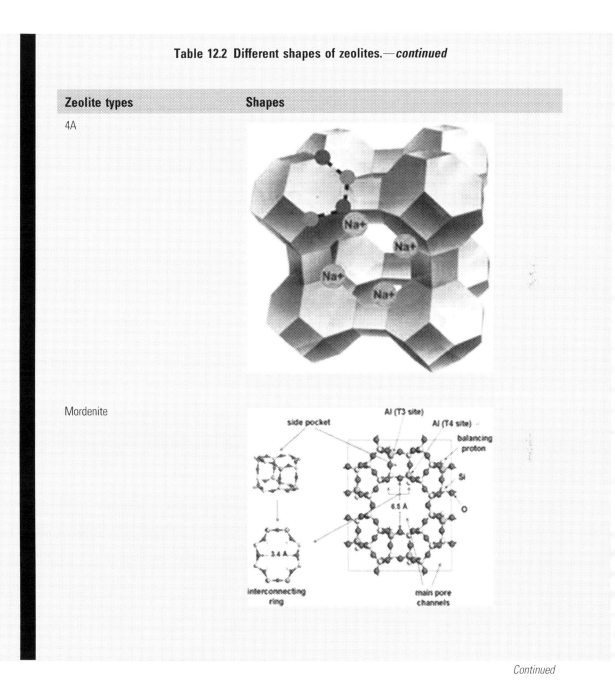

Continued

Table 12.2 Different shapes of zeolites.—*continued*

Zeolite types	Shapes
ZSM-5	
Na-X	

Table 12.2 Different shapes of zeolites.—*continued*

Zeolite types	Shapes
SP-115	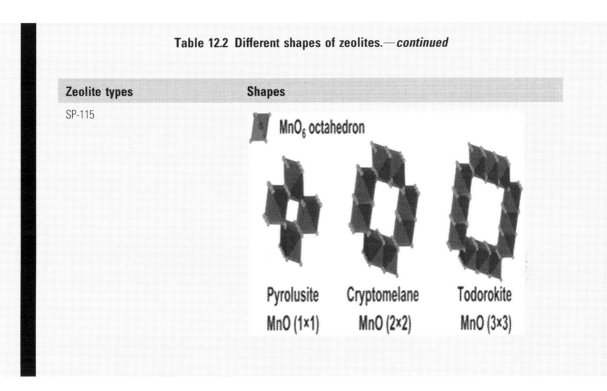

the upward and downward direction of the pillar for the 17 days in between trials. High efficiency was reportedly attained for four or more days, and this zeolite was renewed without the use of an additional regeneration method. Yasyerli et al. also reported the same outcome. The experiment's findings indicated that natural zeolite has a 94% valence to remove H_2S.

2.3.2 13X, 4A, and 5A zeolites for acid gas removal

Natural zeolites like zeolites and synthetic zeolites alike 13X and 5A were researched for purification and upgrading by Alonso-Vicario et al. [61]. According to the methods described in the literature, activation was done on these zeolites before use to get rid of soluble impurities and volatile chemicals. Following washing to get rid of these contaminants, calcination was used to get rid of vaporizable chemicals for the clinoptilolites. Zeolites 5A and 13X could only be operated through calcination.

Zeolite 13X was employed by Melo et al. [68] as an adsorbent to eliminate H_2S from NG. Matching to X-ray diffraction peaks of zeolite 13X that are identical to the Faujasite structure, zeolites have a composition of $Na_2O:Al_2O_3:2.8 \pm 0.2SiO_2:XH_2O$, where X depends on how dry or activated the substance is. Zeolite 13X

Table 12.3 Application and efficiency of various zeolites for H_2S elimination.

Zeolite types	Modification type	Selectivity/efficiency	Reference
Fresh zeolite	—	94%	[59]
Clinoptilolite	—	0.03 g S/g zeolite	[60]
Clinoptilolite	—	1.39 mg H_2S/g zeolite	[61]
Mo-SP-115	Wet impregnation	0.34 wt.%	[62]
Mn-SP-115	Wet impregnation	0.47 wt.%	[62]
Cu-13X	Ion exchange	1.17 mmol/g adsorbed H_2S	[63]
Cu-13X	Wet impregnation	0.20 mmol/g adsorbed H_2S	[63]
Na-Mordenite	Ion exchange	97%	[64]
Zn-13X	Ion exchange	0.22 mmol/g adsorbed H_2S	[63]
Zn-13X	Wet impregnation	0.15 mmol/g adsorbed H_2S	[63]
Cu-SP-115	Wet impregnation	1.11 wt.%	[62]
Cu-ETS-2	Ion exchange	6.9 ppm H_2S at bed exit	[65,66]
Cu-ETS-2	Ion exchange	47 mg H_2S/g of adsorbent	[67]
13X	—	53 mg H_2S/g of adsorbent	[68]
13X	—	0.09 mmol/g adsorbed H_2S	[63]
13X	—	84%–85%	[69]
13X	—	1.0 mg H_2S/g 13X	[61]
5A	—	84%–85%	[69]
5A	—	0.52 mg H_2S/g 5A	[61]
4A	—	61%	[69]
5A	—	66.6%	[63]
Cu:Ce—Y	Ion exchange	1.55 mg S/g sorbent —97.8% desorbed/adsorbed H_2S	[70]
Ce—Na—Y-2	Ion exchange	39% sulfur reduction	[71]
Ce—Na—Y	Ion exchange	42% sulfur reduction	[71]
Ca-X and Na-X	Ion exchange	10 wt.% S	[72]
Na—Y	—	31% sulfur reduction	[71]
La—Na—Y	Ion exchange	37% sulfur reduction	[71]
H—Y	—	59% sulfur reduction	[71]
Ag-MFI	Ion exchange	0% after 8 h	[73]
LTA (Zeolite-A)	—	20% after 24 h	[73]
MFI (ZSM-5)	—	82% after 24 h	[73]
Ag-LTA	Ion exchange	0% after 4 h	[73]
Fe—Na-A	Ion exchange	0.2%–1.5%	[74]

had a surface space of 16.15 m^2/g, making it suitable for adsorbing H_2S from NG. At 25°C, the greatest adsorption valence was 53 mg H_2S/g of adsorbent. At this work, the adsorption balance information was correlated with a few adsorption patterns, including Langmuir, Freundlich, Langmiur-Friendlich, and Toth. These models indicated that Toth's model fit the data the best.

Zeolite 5A made from kaolin clay that was studied by Muhammend and Nasrullah [75] to lower hydrogen sulfide levels in liquefied petroleum gases (LPG). Ion interchanged zeolite 4A was created at 90°C for 5 h while being exposed to 2 N NaOH and 1.5 N $CaCl_2$. In this investigation, a fixed bed adsorption method was used, and the breakthrough curves were used to determine the zeolite's properties. On 66.6% of the ion exchanged zeolite, H_2S was not substantially adsorbed, but greater adsorption capacities were visible with higher flow rates. Water was forcefully adsorbed on the zeolite due to the increased adsorption tendency of water because of its tiny diameter structure and the fact that LPG contains H_2O molecules.

After that, H_2S adsorption was inhibited by water molecules that stuck to the zeolite area, and the valence dropped. 500 ppm H_2O and 209 ppm H_2S were introduced to the system at current rate of 3, 4, and 5 L/min, respectively, based on the adsorption studies, and 2.69 and 1.76 g/g zeolite were determined to be the system's capacity values. With rising flow rates, H_2S adsorption rose while water adsorption dropped. Therefore, 5 L/min was determined to be the ideal flow rate for both compounds. Surface space and pile shrinking intensity of the zeolite both enhanced with the application of the kaolin clay connector.

2.3.3 Zeolite X interchanged by Na and Ca for acid gas removal

Due to its unique crystal structure and consistent pore channels, zeolite X is a type of microstructural features with skeleton structure that has a high specific volume and surface space. Zeolite X has significant adhesion and election susceptibilities for vaporizable organic molecules as a result (VOCs). In comparison to the Y zeolite, the X zeolite has a shorter silica-alumina ratio. Moreover, the X zeolite is less stable thermally and hydrothermally than the Y zeolite.

Zeolites with Ca- and Na-metal exchanges have been explored by Ratnasamy et al. [72] to refine NG for polymer electrode membrane (PEM) combustion cells. Compound metal oxides maintained on aluminum (Cu–Mn, Fe–Mn) were also studied in addition to all these zeolites. The perusal's objective was to eliminate all sulfur mixtures, including tertiary butyl mercaptan (TBM), carbonyl sulfide (COS), dimethyl sulfide (DMS), and ethyl mercaptan (EM). Unsurprisingly, the sulfur composition would not have been entirely removed by zeolites or metal oxides. This study showed that the alkali exchange zeolites' adsorption efficiency was Ca-X > Na-X. Al supported Cu—Mn and Fe—Mn adsorbents were found to be the most effective at removing H_2S.

2.3.4 Zeolite Y interchanged by Ce and Cu for acid gas removal

Zeolite Y is typically produced by hydrothermal crystal growth of reflexive alkali metal aluminosilicate gels or lucid dilutions at down temperature (70–300°C, typically 100°C) and pressure (autogenous) under alkaline circumstances. With pores measuring 7.4 in diameter and a three-dimensional pore structure, zeolite Y is a faujasite molecular sieve. The sodalite coops that are collocated to create supercages big enough to hold spheres with 1.2 nm diameter are the fundamental structural components of zeolites Y.

The removal of H_2S in the range of some 1000 ppm in logistic combustions like JP-8 has been studied through regeneracy and sulfidation trials using Cu—Y and Ce—Y zeolites prepared by ion-interchange procedure [70]. By interchanging Cu ions with Ce ions and sodium ions with copper ions in zeolite, ion exchange was achieved. The sulfidation capacity was lowest when bare zeolite was compared to swapped zeolites. Regeneracy trials were conducted at 50% hydrogen and 10% H_2O with moderate helium prices, whereas sulfidation trials were done under the circumstances of 0.1% H_2S, 50% H_2 and 10% H_2O with moderate helium prices. During sulfidation testing, the reduction in zeolite Y's surface area was lessened with the introduction of Cu and Ce ions, and these incorporations boosted the material's ability to adsorb H_2S. Cu and Ce were also utilized in the trials at a proportion of 0.5. The sequence of the metal insertions was Ce < Cu: Ce < Cu after that. When regeneration tests were conducted, the zeolite surface crumbled and Cu ion was converted to Cu metal at wide temperatures. As a result, at 800°C, Ce—Y and Cu: Ce—Y displayed greater surface area stability than Cu—Y [56].

2.3.5 Zeolite Na-Y interchanged by various metals for acid gas removal

Despite having a wide energy valence and being simple and secure to carry, diesel and gasoline nevertheless contain sulfur compounds that can damage catalysts, even in little amounts. Hoguet et al. [71] examined Na—Y and H—Y zeolites with SiO_2/Al_2O_3 mole proportions of 5.1 and 80 in order to avoid this issue. Used nitrate dilutions of any metal, Na—Y, Pr-Na-Y, Ga-Na-Y, Y—Na—Y, La—Na—Y, Ce—Na—Y, and Ce—Na—Y-2 zeolites were produced when the liquid interchange procedure was used. One of the crucial components of this interchange was the calcination of sorbents among 400 and 600°C in order to eliminate the nitrate contaminations. The connection among the replacement metal ions of the zeolite and the sulfide could be used to explain the adsorption mechanism of H_2S. Ion swapped zeolites outperformed Na–Y in terms of adsorption effectiveness, and all other exchanged zeolites performed best. Due to their consistency in efficacy on sulfur feed, La–Na–Y and Ce–Na–Y-2 zeolites were also considered suitable adsorbents in light of this outcome. The removal of sulfur was shown to be highly challenging if the sulfur concentration in the fuel was very low when these adsorbents were used for three rates of sulfur included in diesel (vast, middle, and little).

Lee et al. [74] have looked at sodium aluminate, which is the main ingredient in sodium silicate solution synthesized from slag melting and containing iron embedded Na–Y zeolite. In this experiment, H_2S was removed using a packed floor pillar. A gas detector was applied to analyze the egress gases, and the breakthrough concentration was observed after the column. When the H_2S gas density attained 2000 ppm, the test was declared complete. Adsorption capability was shown to be dependent on the solution's iron concentration and calcination temperature, according to XRD analyses. With incrementing Fe^{3+} density from 0 to 78 mM, the surface space rose from 20 m^2/g to 85 m^2/g. At 78 mM Fe^{3+} solution calcined at 200°C, the maximum adsorption valence was discovered. According to the FT-IR and XRD measurements, the zeolite network structure might be destroyed when the solution's Fe^+ content exceeds 90 mM. Because crystal stability declines with increasing iron concentrations in solution, zeolite's crystallinity is reduced. The surface area of the zeolite has a significant impact on its ability to absorb gases. Because of the acidic area characteristics of Fe–Na-A, H_2S has a higher capacity for adsorption. It has been determined that iron-incorporated Na-A zeolite has a premier adsorption valence toward 4A zeolite and a less valence toward activated carbons when compared to other adsorbents.

2.3.6 ZSM-5 and Zeolite-A for acid gas removal

The elimination of vaporizable H_2S of aqueous and gaseous district was accomplished using MFI and LTA zeolites, also known as ZSM-5 and zeolite-A, respectively. These two zeolite materials were created using a wet process and had differing Si/Al proportions (1 and 60, respectively). As is well known, the tetrahedrons of SiO_4 and AlO_4 were joined in three dimensions in zeolites that contained alkali or alkali-earth metal atoms. Both of them were effective at purifying the surrounding air, and the swap of an Ag metal increased their H_2S adsorption resistance. The initial H_2S concentration in the experiments was 30 ppm, and specimens were analyzed using the powder X-ray diffraction procedure. Despite using equal amounts of zeolites, LTA had more hydrogen sulfide adsorbed onto it than MFI did. The diversity on plane potentials among LTA and MFI arose because Na ions in the LTA are abroad of the zeolite crystals and MFI had incredibly high Si/AL proportions. The speed of H_2S adsorption was lower than that of the LTA zeolite frame because of the small plane potential worthiness of MFI. While LTA reduced H_2S concentration by roughly 20% after 24 h, Ag-doped LTA had a 0% reduction after 4 h. The H_2S quantity was 82% for naked MFI after 24 h, but Ag-doped MFI only achieved success after 8 h. In an intriguing finding, the authors found that 24% H_2S on MFI was desorbed, while 46% H_2S on LTA was dropped when both were warmed to 400°C for desorption [73].

2.3.7 Zeolite SP-115 changed by Mo, Cu, and Mn for acid gas removal

In order to remove H_2S from charcoal gas at 871°C and 205 kPa, Gasper-galvin et al. [62] examined vast silica-having zeolite SP-115 produced by wet impregnation with the Mo, Cu, and Mn oxides. This temperature was chosen in order to prohibit gas chilling of a liquefied floor gasifier, to reduce sulfate production, and to match the temperature levels of sulfidation and regeneration. Ammonium molybdate, manganese acetate, and cupric acetate were the solutions employed to wet impregnate zeolite. For each of the sorbents, $5^{1/2}$ course sulfidation/regeneration trials were conducted. To examine the adsorption capabilities of Cu, Mn, and Mo oxides on zeolite, each compound was utilized alone and in combination. The copper acetate was impregnated three times into the metal oxide supported zeolites in the shape of SP3C7, where SP stands for the zeolite support and C7 for the copper molar concentration. The sorbents created by Cu, Mo, and Mn were SP3C7, SP3MO7, and SP3M7, respectively. The development curves showed that CuO was the maximum effective material for removing H_2S because it brought the level of the gas below 100 ppmv while maintaining its reactivity. However, MoO_3 failed in this process because its thermodynamics and those of H_2S were incompatible. As a result, 86% of the Mo vapourized at 871°C and escaped from the sorbent. Mn had a significant capacity for adsorption for the first 5 min, but the breakthrough curve revealed minimal sulfur absorption for the next 15 to 35 min. Because both Mo and Mn improved the ability of Cu oxide to remove hydrogen and sulfur, they were combined with Cu. At higher temperatures, Cu and Mn oxides hindered the vaporization of Mo oxide. Mn oxide increased shrink intensity by 30%, while Mo oxide enhanced the first desulfurization stage. With the help of this investigation, it was discovered that employing Cu, Mn, and Mo in the right proportions improved sulfur elimination yield.

2.3.8 Engelhard Titanosilicate zeolite for acid gas removal

A molecular sieve zeolite with pores that are nearly 8 in size and a titania/silica mole proportion of 2.5–25 is called Engelhard Titanosilicate (ETS). The titanium in the ETS structure comes from solid Ti_2O_3, and the silica is derived from sodium silicate. As it can be characterized as the impressive diameter of the biggest gas molecule within the structure, pore size is a crucial metric for the features of ETS. ETS zeolites are thus adsorptive to molecules with pore diameters up to 8. The ETS framework can be reversibly dehydrated by changing the pore size. Surface area is important for zeolites' adsorption properties as well as for other zeolites since it increases the amount of active ions that are widely spread and accessible to gas molecules. Rezai et al. [56] used this substance, which has been described as both an excellent adsorbent for H_2S and an effective support for metals, to remove hydrogen sulfide down to 0.5 ppm at room temperature. In the perusal, ETS-2, ETS-4, and ETS-10 kinds were created and utilized in adsorption trials using a mixture containing 10 ppm H_2S. The plane space of the adsorbents was computed utilizing the BET procedure, and their crystallinity was examined using powder XRD equipment. Atomic absorption spectroscopy was applied to examine the amount of Cu. The impact of copper on the morphology of copper exchanged ETS-2 was examined using TEM pictures, and it was discovered that ion interchange had no impact on the structure even though Cu ions displayed a vast atomic partition in an XRD examination. When contrasting the development curves of several Titanosilicate adsorbents, it was found that ETS-4 had small Cu sorption because of blocked holes, while ETS-10 had higher copper uptake due to bigger pore surface areas. However, Cu-ETS-2 was the one with the greatest H_2S development capability when copper absorption and distribution on the structure properties were to work together. In the wake of these tests, the team also investigated Cu-ETS-2 for the adsorption of H_2S among 250 and 950°C to look at the effects of temperature fluctuations at the functionality and crystal structure of copper sulfide crystals. With evaluations made at intervals of 100°C between 250 and 950°C, the capacity of H$_2$S adsorption dropped by half (of 0.7 mol H$_2$S per Cu to 0.35 mol H$_2$S per Cu) because of the change in the oxidation condition of Cu^{+2}. Additionally, due to temperature variations, titania in the adsorbent changed forms, such as tetragonal anatase among 350 and 550°C, and rutile upper 750°C. The active adsorption sites of the copper sulfide crystals were thought to be responsible for the appearance of various crystal shapes [65].

In a different investigation by Rezai et al. [66], ETS-2 was substituted with several metals, including Ag, Cu, Ca, and Zn, due to examine the efficacy of these metals at H$_2$S sorption and contrast with a productive trading H$_2$S adsorbent R3-11G. For ion interchange, the ETS-2 was mixed with the nitrate salts of Ag and Cu, as well as the chloride salts of Zn and Ca. The H$_2$S adsorption valence of various materials was investigated using a GC equipped with an MXT-1 pillar and a flare photometric tracer have the ability to discover H$_2$S even at 200 ppb. The H$_2$S development curves were calculated using a number of variables, including the H$_2$S concentration in the incoming gas, gas current speeds, floor volume with vacant deduction, mass of adsorbent, and H$_2$S molar mass. Pursuant to the development curves, Cu-ETS-2 demonstrated the greatest H$_2$S uptake, even greater than that of the commercially adsorbent (see Fig. 12.3). The materials were arranged as follows: Cu-ETS-2 > Ag-ETS-2 > Zn-ETS-2 ~ R3- 11G > Ca-ETS-2 ~ Na-ETS-2. The metallic points and more hard structures of these metals were thought to be the cause of the reduced sulfidation efficiency, which resulted in the lower sulfur adsorption capabilities. The plane space worthiness and metal compositions of the specimens were contained in this perusal together with the ground-breaking performances of these metals. Consequently, Ag-smaller ETS-2's surface area may account for its poor adsorption properties.

2.3.9 Other zeolites for acid gas removal

For the Claus reaction, Lee and Whan Chi [64] developed the process for producing zeolite catalysts such as the hydrogen shapes of zeolites, sodium ion interchanged hydrogen zeolites, and particularly, the sodium interchanged hydrogen shapes of mordenite. In that reaction, a portion of the H$_2$S is burned to produce sulfur dioxide, and the remaining H$_2$S and SO$_2$ are allowed to pass over an aluminum catalyst. By mixing zeolite with a solution containing hydrogen ions, hydrogen shapes of zeolites were produced in an alkali shape. This form was then exchanged for

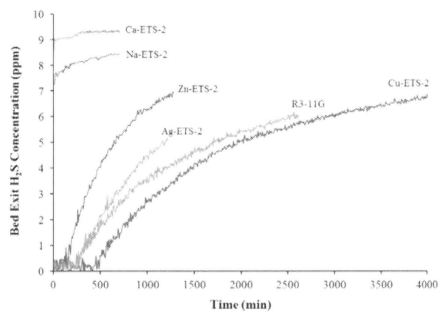

Figure 12.3 Breakthrough curves for the exchange of various metals with ETS-2 for the sake of adsorption of H_2S [76].

sodium by contacting the zeolite with a solution containing sodium salt. Sodium swapped hydrogen zeolites (mordenite) that are extremely acidic, and thermally permanent and hydrogen versions of zeolites were employed to adsorb H_2S. In order to improve its resistance to environments with sulfuric acid, mordenite, which was in the form of sodium, was treated with hydrochloric acid. It was claimed that the Claus process turned 97% of H_2S and SO_2 into basic sulfur and water, and this flat could be raised by joining Claus reactors in a sequence.

In a study on supported substances used to eliminate H_2S of combustion gas, the wet impregnation approach was employed to modify the structure of the support substance with cupric acetate and manganese acetate. To test the zeolites' ability to regenerate, trials were conducted at temperatures of 538, 677, and 871°C for sulfidation and regeneration with a value of 2000 h^{-1} space velocity. In order to comprehend how adsorption works, the loss of zeolite surface area brought on by the interaction of the material's pores and the creation of acids at wide temperatures was assessed. The crush strength of the zeolite rose from 11.1 to 29.6 kg/cm when it was full of metals (Mn and Cu). Fuel gas H_2S concentration was shown to decrease from 2000 to

200 ppm for 60–70 min on supported zeolite substances by examining breakthrough curves [76].

3. Conclusion and future outlooks

A hydrocarbon-rich gas largely made up of methane is NG. NG is helpful as a combustion because of the tremendous energy it produces when burned. In contrast to other fossil fuels, it emits less gases that could harm the environment, and it is also simpler to transport, use, and store. For these reasons, it is regarded as a clean fuel. As a result, it is regarded as a secure origin of vigor. In the industrial sector, NG is also employed for heating, the manufacture of power, fertilizers, and petrochemicals. The current literature on the use of zeolites in experimental methods to remove H_2S from various sources has been the main emphasis of this chapter. Zeolites must have a significant sulfur lading valence, well regenerability, and a firm construction to be an efficient adsorbent for the elimination of H_2S. The faujasite kind zeolite familiar as 13X, designated as FAU, was found to have the best selectivity values. The 4 and 5A zeolites LTA and MFI were assessed to have the lowest selectivity values. Zeolites have a wide sulfur valence that increases with metal or metal oxide alteration. They are therefore promising materials because they have wide surface-to-volume ratios that are one of the key elements of adsorption.

Abbreviations and symbols

C_2H_5SH	Ethyl mercaptan
C7	Copper molar concentration
CCS	Carbon capture and sequestration
CCT	Clean coal technology
CH_3SH	Methyl mercaptan
CO_2	Carbon dioxide
COS	Carbonyl sulfide
CS_2	Carbon disulfide
DEA	Diethanolamine
DGA	Diglycolamine
DIPA	Di_isopropanolamine
DMS	Dimethyl sulfide
EM	Ethyl mercaptan
ET_3AS	Triethylarsine
ETS	Engelhard titanosilicate
GTL	Gas-to-liquid
H_2S	Hydrogen sulfide
He	Helium
IGCC	Integrated gasification combined cycle

LPG	Liquefied petroleum gases
MDEA	Methyldiethanolamine
Me$_2$ETAS	Dimethylethylarsine
Me$_3$AS	Trimethylarsine
MEA	Monoethanolamine
MeET$_2$AS	Methyldiethylarsine
N$_2$	Nitrogen
NaOH	Sodium hydroxide
NG	Natural gas
Ni	Nickel
NMP or N-PYROL	N_methyl_2_pyrrolidone
NORM	Naturally occurring radioactive material
PEG	Polyethylene glycol
PEM	Polymer electrode membrane
Ph$_3$AS	Triphenylarsine
Po_210	Polonium
RSH	Mercaptans
SO$_2$	Sulfur dioxide
SP	Zeolite support
TBM	Tertiary butyl mercaptan
TEA	Triethanolamine
VOCs	Volatile organic compounds

References

[1] Rahimpour MR, Makarem MA, Meshksar M. In: Rahimpour MR, Makarem MA, Meshksar M, editors. Advances in synthesis gas : methods, technologies and applications. Syngas production and preparation, vol. 1. Elsevier; 2023. https://doi.org/10.1016/C2021-0-00292-3.

[2] Rahimpour MR, Makarem MA, Meshksar M. In: Rahimpour MR, Makarem MA, Meshksar M, editors. Advances in synthesis gas : methods, technologies and applications. Syngas products and usages, vol. 3. Elsevier; 2023. https://doi.org/10.1016/C2021-0-00380-1.

[3] Makarem MA, Farsi M, Rahimpour MR. CFD simulation of CO$_2$ removal from hydrogen rich stream in a microchannel. International Journal of Hydrogen Energy 2021;46(37):19749–57. https://doi.org/10.1016/j.ijhydene.2020.07.221.

[4] Makarem MA, Kiani MR, Abbaspour M, Farsi M, Rahimpour MR. Nanoparticle-enhanced hydrogen separation from CO$_2$ in cylindrical and cubic microchannels: a 3D computational fluid dynamics simulation. International Journal of Hydrogen Energy 2023;48(32):12045–55. https://doi.org/10.1016/j.ijhydene.2022.07.157.

[5] Economides MJ, Wood DA. The state of natural gas. Journal of Natural Gas Science and Engineering 2009;1(1–2):1–13.

[6] Kiani MR, Meshksar M, Makarem MA, Rahimpour MR. Preparation, stability, and characterization of nanofluids. In: Nanofluids and mass transfer. Elsevier; 2022. p. 21–38.

[7] Tobin J, Shambaugh P, Mastrangelo E. Natural gas processing: the crucial link between natural gas production and its transportation to market. Energy Information Administration, Office of Oil and Gas; 2006.

[8] Gas P-QN. Natural gas processing: the crucial link between natural gas production and its transportation to market. Los Alamos; 2006. p. 20–2.

[9] Speight J. Liquid fuels from natural gas. In: Handbook of alternative fuel technologies, vol. 153; 2007.
[10] Mokhatab S, Poe WA, Mak JY. Handbook of natural gas transmission and processing: principles and practices. Gulf Professional Publishing; 2018.
[11] Faramawy S, Zaki T, Sakr A-E. Natural gas origin, composition, and processing: a review. Journal of Natural Gas Science and Engineering 2016;34:34—54.
[12] Rojey A, Jaffret C. Natural gas: production, processing, transport. Editions Technip; 1997.
[13] Speight JG. Thermal cracking, chapter 14. In: The chemistry and technology of petroleum. 3rd ed. New York: Marcel Dekker Inc.; 1999. p. 565—84.
[14] Speight JGX. The chemistry and technology of petroleum. CRC Press; 2006.
[15] Younger AH, Eng P. Natural gas processing principles and technology-part I. Tulsa Oklahoma: Gas Processors Association; 2004.
[16] Fitzgerald WF, Lamborg CH. Geochemistry of mercury in the environment. Treatise on Geochemistry 2003;9:612.
[17] Rios JA, Coyle DA, Durr CA, Frankie BM. Removal of trace mercury contaminants from gas and liquid streams in the LNG and gas processing industry. Tulsa, OK (United States): Gas Processors Association; 1998.
[18] Caumette G, Lienemann C-P, Merdrignac I, Bouyssiere B, Lobinski R. Element speciation analysis of petroleum and related materials. Journal of Analytical Atomic Spectrometry 2009;24(3):263—76.
[19] Trahan DO. Arsenic compounds in natural gas pipeline operations. Pipeline and Gas Journal 2008;235(3):95—7.
[20] Speight JG. Handbook of petroleum product analysis. John Wiley and Sons; 2015.
[21] Godoy JM, Carvalho F, Cordilha A, Matta LE, Godoy ML. 210Pb content in natural gas pipeline residues ("black-powder") and its correlation with the chemical composition. Journal of Environmental Radioactivity 2005;83(1):101—11.
[22] Speight JG. Shale gas production processes. Gulf Professional Publishing; 2013.
[23] Cheng HC, Hill FB. Separation of helium-methane mixtures by pressure swing adsorption. AIChE Journal 1985;31(1):95—102.
[24] Mitariten M. New technology improves nitrogen-removal economics. Oil and Gas Journal 2001;99(17).
[25] Hoffman EJ. Membrane separations technology: single-stage, multistage, and differential permeation. Elsevier; 2003.
[26] Kidnay AJ, Parrish WR. Fundamentals of natural gas processing. CRC Press; 2006.
[27] Krupp EM, Johnson C, Rechsteiner C, Moir M, Leong D, Feldmann J. Investigation into the determination of trimethylarsine in natural gas and its partitioning into gas and condensate phases using (cryotrapping)/gas chromatography coupled to inductively coupled plasma mass spectrometry and liquid/solid sorption techniques. Spectrochimica Acta Part B: Atomic Spectroscopy 2007;62(9):970—7.
[28] Hennico A, Barthel Y, Cosyns J, Courty P. Mercury and arsenic removal in the natural gas, refining and petrochemical industries. Oil, Gas (Hamburg) 1991;17(2):36—8.
[29] Sarrazin P, Cameron CJ, Barthel Y, Morrison ME. Processes prevent detrimental effects from as and Hg in feedstocks. Oil and Gas Journal 1993;91(4).

[30] Carroll JJ. Acid gas injection and carbon dioxide sequestration. John Wiley and Sons; 2010.
[31] Song KY, Kobayashi R. Water content values of a CO2-5.31 mol percent methane mixture. Gas Processors Association; 1989.
[32] Speight JG. Gas processing: environmental aspects and methods. Butterworth-Heinemann; 1993.
[33] Gudmundsson JS, Nazir A, Ismailpour A, Saleem F, Idrees MU, Zaidy SAH. Natural gas sweetening and effect of declining pressure. TPG4140 Project Report. Trondheim, Norway: NTNU; 2013.
[34] Rho S-W, Yoo K-P, Lee JS, Nam SC, Son JE, Min B-M. Solubility of CO_2 in aqueous methyldiethanolamine solutions. Journal of Chemical and Engineering Data 1997;42(6):1161–4.
[35] Weinberg HN, Heinzelmann FJ, Savage DW. New gas treating alternatives for saving energy in refining and naturel gas production. OnePetro; 1984.
[36] Rooney PC, DuPart M. Corrosion in alkanolamine plants: causes and minimization. OnePetro; 2000.
[37] Johnson JE, Homme Jr AC. SELEXOL solvent process reduces lean, high-CO/sub 2/natural gas treating costs. Energy Program;(United States) 1984; 4(4).
[38] Mokhatab S, Towler BF, Manning FS, Thompson RE. Oilfield processing of petroleumvol. 1. Tulsa, OK: PennWell Publishing Company; 2006. Natural Gas, Published 1991, 408pp., ISBN: 0-87814-342-2; vol. 2—Crude Oil, Published 1995, 434pp., ISBN: 0-87814-354-8.
[39] Abatzoglou N, Boivin S. A review of biogas purification processes. Biofuels, Bioproducts and Biorefining 2009;3(1):42–71.
[40] Marsh H, Rodriguez-Reinonso F. Activated carbon. Amsterdam: Elsevier Science and Technology Books; 2006. p. 89–100.
[41] Kiani MR, Meshksar M, Makarem MA, Rahimpour E. Catalytic membrane micro-reactors for fuel and biofuel processing: a mini review. Topics in Catalysis 2021:1–20.
[42] Adams RT, Lee JS, Bae T-H, Ward JK, Johnson JR, Jones CW, et al. CO_2–CH_4 permeation in high zeolite 4A loading mixed matrix membranes. Journal of Membrane Science 2011;367(1–2):197–203.
[43] He X, Hägg M-B. Membranes for environmentally friendly energy processes. Membranes 2012;2(4):706–26.
[44] Wind JD, Staudt-Bickel C, Paul DR, Koros WJ. The effects of crosslinking chemistry on CO_2 plasticization of polyimide gas separation membranes. Industrial and Engineering Chemistry Research 2002;41(24):6139–48.
[45] Hamed P, Mohd H, Shamsul I. Review of H_2S sorbents at low-temperature desulfurization of biogas. International Journal of Chemical and Environmental Engineering 2014;5(1):22–8.
[46] Cooper BH, Knudsen KG. Ultra deep desulfurization of diesel: how an understanding of the underlying kinetics can reduce investment costs. Practical Advances in Petroleum Processing. Springer; 2006. p. 297–316.
[47] Petrov SG, Donov AM, Stratiev DS. Challenges facing European refiners today. Journal of International Research Publication 2002;3:1311.
[48] Koryabkina NA, Phatak AA, Ruettinger WF, Farrauto RJ, Ribeiro FH. Determination of kinetic parameters for the water–gas shift reaction on copper catalysts under realistic conditions for fuel cell applications. Journal of Catalysis 2003;217(1):233–9.
[49] Gabriel D, Deshusses MA. Retrofitting existing chemical scrubbers to biotrickling filters for H_2S emission control. Proceedings of the National Academy of Sciences 2003;100(11):6308–12.

[50] Bandosz TJ. On the adsorption/oxidation of hydrogen sulfide on activated carbons at ambient temperatures. Journal of Colloid and Interface Science 2002;246(1):1−20.
[51] Meshksar M, Kiani MR, Mozafari A, Makarem MA, Rahimpour MR. Promoted nickel−cobalt bimetallic catalysts for biogas reforming. Topics in Catalysis 2021:1−13.
[52] Zhao Q, Leonhardt E, MacConnell C, Frear C, Chen S. In: Purification technologies for biogas generated by anaerobic digestion, vol. 24. Compressed Biomethane, CSANR; 2010.
[53] Bagreev A, Rahman H, Bandosz TJ. Thermal regeneration of a spent activated carbon previously used as hydrogen sulfide adsorbent. Carbon 2001;39(9):1319−26.
[54] Nhut J-M, Vieira R, Pesant L, Tessonnier J-P, Keller N, Ehret G, et al. Synthesis and catalytic uses of carbon and silicon carbide nanostructures. Catalysis Today 2002;76(1):11−32.
[55] Taghaddom K, Saadi MK, Kiani P, Rahimpour HR, Rahimpour MR. Activated carbon for syngas purification. In: Advances in synthesis gas: methods, technologies and applications: syngas purification and separation. Elsevier; 2022. p. 229.
[56] Ozekmekci M, Salkic G, Fellah MF. Use of zeolites for the removal of H_2S: a mini-review. Fuel Processing Technology 2015;139:49−60.
[57] Ackley MW, Rege SU, Saxena H. Application of natural zeolites in the purification and separation of gases. Microporous and Mesoporous Materials 2003;61(1−3):25−42.
[58] Siriwardane RV, Shen M-S, Fisher EP, Poston JA. Adsorption of CO_2 on molecular sieves and activated carbon. Energy and Fuels 2001;15(2):279−84.
[59] Pourzolfaghar H, Ismail MHS. Study of H_2S removal efficiency of virgin zeolite in POME biogas desulfurization at ambient temperature and pressure. In: Developments in sustainable chemical and bioprocess technology; 2013. p. 295−301.
[60] Yasyerli S, Ar I, Doğu G, Doğu T. Removal of hydrogen sulfide by clinoptilolite in a fixed bed adsorber. Chemical Engineering and Processing: Process Intensification 2002;41(9):785−92.
[61] Alonso-Vicario A, Ochoa-Gómez JR, Gil-Río S, Gómez-Jiménez-Aberasturi O, Ramírez-López CA, Torrecilla-Soria J, et al. Purification and upgrading of biogas by pressure swing adsorption on synthetic and natural zeolites. Microporous and Mesoporous Materials 2010;134(1−3):100−7.
[62] Gasper-Galvin LD, Atimtay AT, Gupta RP. Zeolite-supported metal oxide sorbents for hot-gas desulfurization. Industrial and Engineering Chemistry Research 1998;37(10):4157−66.
[63] Micoli L, Bagnasco G, Turco M. H_2S removal from biogas for fuelling MCFCs: new adsorbing materials. International Journal of Hydrogen Energy 2014;39(4):1783−7.
[64] Lee H, Chi CW. Zeolite catalyst for dilute acid gas treatment via claus reaction. Google Patents; 1976.
[65] Yazdanbakhsh F, Blaesing M, Sawada JA, Rezaei S, Mueller M, Baumann S, et al. Copper exchanged nanotitanate for high temperature H_2S adsorption. Industrial and Engineering Chemistry Research 2014;53(29):11734−9.
[66] Rezaei S, Jarligo MOD, Wu L, Kuznicki SM. Breakthrough performances of metal-exchanged nanotitanate ETS-2 adsorbents for room temperature desulfurization. Chemical Engineering Science 2015;123:444−9.
[67] Rezaei S, Tavana A, Sawada JA, Wu L, Junaid ASM, Kuznicki SM. Novel copper-exchanged titanosilicate adsorbent for low temperature H_2S

removal. Industrial and Engineering Chemistry Research 2012;51(38): 12430−4.
[68] Melo DMdA, De Souza JR, Melo MAdF, Martinelli AE, Cachima GHB, Cunha JDd. Evaluation of the zinox and zeolite materials as adsorbents to remove H_2S from natural gas. Colloids and Surfaces A: Physicochemical and Engineering Aspects 2006;272(1−2):32−6.
[69] Tomadakis MM, Heck HH, Jubran ME, Al-Harthi K. Pressure-swing adsorption separation of H_2S from CO_2 with molecular sieves 4A, 5A, and 13X. Separation Science and Technology 2011;46(3):428−33.
[70] Rong C, Chu D, Hopkins J. Test and characterization of some zeolite supported gas phase desulfurization sorbents. Army Research Lab Adelphi MD Sensors and Electron Devices Directorate; 2009.
[71] Hoguet J-C, Karagiannakis GP, Valla JA, Agrafiotis CC, Konstandopoulos AG. Gas and liquid phase fuels desulphurization for hydrogen production via reforming processes. International Journal of Hydrogen Energy 2009;34(11): 4953−62.
[72] Ratnasamy C, Wagner JP, Spivey S, Weston E, et al. Removal of sulfur compounds from natural gas for fuel cell applications using a sequential bed system. Catalysis Today 2012;198(1):233−8.
[73] Yokogawa Y, Sakanishi M, Morikawa N, Nakamura A, Kishida I, Varma HK. VSC adsorptive properties in ion exchanged zeolite materials in gaseous and aqueous medium. Procedia Engineering 2012;36:168−72.
[74] Lee S-K, Jang Y-N, Bae I-K, Chae S-C, Ryu K-W, Kim J-K. Adsorption of toxic gases on iron-incorporated Na-A zeolites synthesized from melting slag. Materials Transactions 2009;50(10):2476−83.
[75] Mohammed A-HA, Nassrullah ZK. Preparation and formation of zeolite 5A from local kaolin clay for drying and desuphurization of liquefied petroleum gas. Iraqi Journal of Chemical and Petroleum Engineering 2013; 14(1):1−13.
[76] Atimtay AT. Development of supported sorbents for hydrogen sulfide removal from fuel gas. Springer; 1998.

13

Porous metal structures, metal oxides, and silica-based sorbents for natural gas sweetening

Mohammad Rahmani and Fatemeh Boshagh
Department of Chemical Engineering, Amirkabir University of Technology (Tehran Polytechnic), Tehran, Iran

1. Introduction

Natural gas, as the most efficient fossil fuel, is mainly composed of hydrocarbons, including methane (CH_4), ethane (C_2H_6), propane (C_3H_8), butane (C_4H_{10}), pentane (C_5H_{12}), and hexane (C_6H_{14}), and trace amounts of impurities, such as carbon dioxide (CO_2), hydrogen sulfide (H_2S), sulfur (mercaptans, carbon disulfide, and carbonyl sulfide), nitrogen (N_2), and helium (He) [1]. The main component of natural gas is methane. Based on the location of a field, the composition of natural gas can vary greatly [2]. Since the amount of acidic gases (H_2S and CO_2) in natural gas is substantial, it is crucial to remove these components from gas streams. Besides the environmental concerns, these gases also cause many problems during transportation, such as gas hydrate formation that may block pipelines [3]. In addition, the presence of acid gases leads to the corrosion of equipment, the reduction of thermal efficiency, and the formation of harmful combustion products [4]. Natural gas reserves can be classified as sweet or sour according to their sulfur contents. Natural gas with a hydrogen sulfide content of more than 4 ppm is classified as sour gas [5]. H_2S is a water-soluble, corrosive, highly toxic, flammable, odorous, and colorless pollutant that is found in biogas, natural gas, coal/biomass gasification gas, syngas, and crude oil [6]. The main purpose of natural gas sweetening is to remove H_2S; however, it is also possible to remove part of CO_2. There are four methods—absorption, adsorption, membranes, and cryogenic distillation—for removing H$_2$S from the gas mixture (Fig. 13.1) [7]. Each method has its own advantages and drawbacks (Table 13.1). Commercially, liquid amine solutions are used to desulfurize natural gas. Monoethanolamine (MEA), methyldiethanolamine (MDEtA), and diethanolamine (DEA) are the most commonly used amines [9].

Adsorption is a surface-based exothermic process that involves the transfer of molecules from a fluid bulk to the adsorbent's solid surface. Depending on the interactions between molecules and surfaces, adsorption can be divided into physical and chemical adsorption [6]. Adsorption with sorbents is an

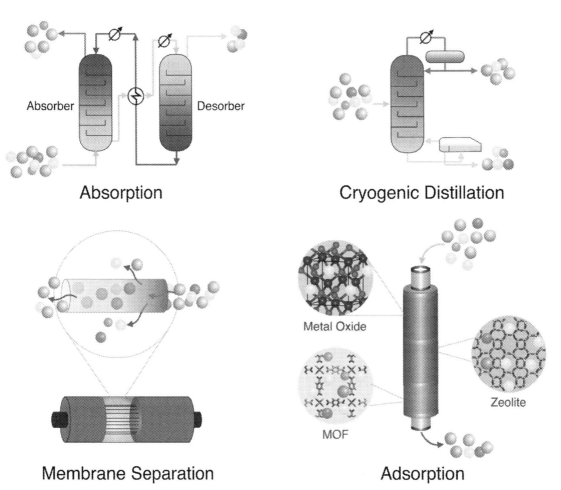

Figure 13.1 Methods of sour gas sweetening [7].

Table 13.1 Techniques of natural gas purification, their advantages, and drawbacks [1,8].

Method	Advantages	Limitations
Absorption	Selective H_2S elimination using regenerative solvent	The requirement of high weight and space
		Inflexible operation
		Costly step for chemical solvent regeneration
Membrane	Without moving parts	Low selectivity of H_2S/CO_2
	Linear scale-up	Uncertainty
	Small footprint	
Cryogenic distillation	High purity of the upgraded gas	Limited to separate CO_2 (H_2S removal has been rarely reported)
		Very energy intensive (needs very low temperatures)
		High cost
Adsorption	Scalable and compact design	Complicated control
	Deep elimination	Treatment of spent sorbent
		Uncertainty over regeneration

effective removal method for H_2S due to its simplicity, ease of handling, low energy cost, high product recovery, simultaneous dehydration and removal of acid gases, the possibility of reusing the adsorbent material, and environmentally friendly sorbents [10,11]. With solid adsorbents, it is possible to selectively separate H_2S or/and CO_2 from mixtures containing CH_4. The adsorption technologies commercially applied are pressure swing adsorption (PSA) [12], vacuum swing adsorption (VSA) [13], and temperature swing adsorption (TSA) [14]. PSA is based on the gas adsorption on an adsorbent at high pressure and gas recovery at low pressure [15]. In VSA, the adsorption at atmospheric pressure and desorption at vacuum conditions are performed, whereas in TSA, gas is released via an increase in temperature at a constant pressure [1]. These methods have high operating costs, which make them economically unattractive. Therefore, various studies attempted to achieve H_2S selective adsorption using more cost-effective techniques. Adsorbents such as zeolites [16], metal oxides [17], metal−organic frameworks (MOFs) [18], carbon materials [19], and mesoporous silica [20] have been studied for H_2S removal from natural gas.

The present chapter focuses on applying metal oxides, MOFs, and silica-based sorbents (mesoporous silica structures and zeolites) for natural gas sweetening.

2. Metal-based sorbents
2.1 Metal—organic frameworks

MOFs, which are composed of metal clusters and organic linkers (pyridine, pyrazole, imidazole), have been extensively studied for various applications, including oxidative desulfurization as catalysts and adsorptive desulfurization (ADS) as adsorbents [21]. MOFs are ideal candidates for gas adsorption, owing to their excellent properties, including large surface area, high porosity, adjustable pore size, the diversity of metal centers and ligands, and easy modification and functionalization [22,23]. ADS of gases has been conducted using MOFs, including Mg-based [24,25], Sc-based [26], materials from the Institute Lavoisier (MIL)—based [18,27—29], rare-earth—based [30], Zr-based [31—34], Cu-based [35,36], Zn-based [37,38], zeolite-like (zeolitic imidazolate frameworks [ZIFs]) [39,40], and fluorinated MOF [41].

MILs are derived using trivalent cations like chromium (III), vanadium (III), and iron (III), developed by using p-block elements such as indium (III), gallium (III), and aluminum (III) [42]. Isoreticular metal—organic frameworks (IRMOFs) as zinc-based MOF, Hong Kong University of Science and Technology (HKUST-1), or MOF-199; Cu-BTC as a copper-based MOF; and ZIF as a kind of MOFs composed of Co^{2+}/Zn^{2+} and imidazole ligands have been used in natural gas sweetening. The functionalization of the surface of MOFs can also enhance their capacity to adsorb H_2S.

The desulfurization of gases using MOF has been extensively investigated (Table 13.2). The various MOFs such as MIL-53(Fe) [27], MIL-53(Cr) [27,29], MIL-53(Al) [27,49], MIL-127(Fe) [50], MIL-125(Ti) [18], MIL-47(V) [27,29,51], MIL-101(Cr) [18,27], MIL-100(Cr) [27], Zr-MOF [31], MOF-199 [44], UiO-66(Zr) [18,33], UiO-66-NH_2 [33], ZIf-8 [40,50,52], ZIF-67 [39,50], CO/Zn-ZIF [50], HKUST-1 [52], Mg-CUK-1 [25], MFM-300(Sc) [26], IRMOF-3 [37], ionic liquids/IRMOF-1 [38], ionic liquids/Cu-TDPAT [53], HKUST-1 [33], HKUST-1/graphene oxide (GO) [54,55], Zn-MOF/ZnO [47], Zn-COP-27 [43], Ni-CPO-27 [43,56], UiO-66/GO [32], PMO_{12}@UiO-66@ H_2S-MIPs [34], and MOF-199—activated carbon [57] have been studied for the H_2S removal.

For the first time, Hamon et al. [27] used six MOF-based sorbents, including MIL-100(Cr), MIL-101(Cr), MIL-47(V), and MIL-53(Cr, Al, Fe) (Fig. 13.2) to study the H_2S removal from natural gas at room temperature. The H_2S adsorption maximum for

Table 13.2 Metal−organic framework sorbents used for H_2S removal from gas streams.

Adsorbent	Feed gas composition	Adsorption temperature (°C)	H_2S breakthrough capacity (mmol/g)	Reference
Ni-CPO-27	H_2S/N_2	30	12	[43]
MIL-53(Fe)	H_2S/CH_4	30	1.18	[27]
MIL-53(Cr)	H_2S/CH_4	30	3.02	[27]
MIL-53(Al)	H_2S/CH_4	30	3.24	[27]
MIL-47(V)	H_2S/CH_4	30	14.6	[27]
MIL-100(Cr)	H_2S/CH_4	30	16.7	[27]
MIL-101(Cr)	H_2S/CH_4	30	38.4	[27]
MIL-53(Cr)-NP	H_2S/CH_4	30	3.7	[29]
MIL-53(Cr)-LP	H_2S/CH_4	30	13.1	[29]
MIL-47(V)	H_2S/CH_4	30	14.6	[29]
MOF-199	H_2S/N_2	30	1.67	[44]
MEA/MOF-199	H_2S/N_2	30	0.83	[44]
DEA/MOF-199	H_2S/N_2	30	0.58	[44]
TEA/MOF-199	H_2S/N_2	30	2.74	[44]
MIL-125(Ti)	H_2S/CH_4	20	~0.21	[18]
MIL-125-NH_2(Ti)	H_2S/CH_4	20	0.56	[18]
MIL-101(Cr)	H_2S/CH_4	20	~0.48	[18]
MIL-101-NH_2(Cr)	H_2S/CH_4	20	~0.5	[18]
UiO-66(Zr)	H_2S/CH_4	20	~0.24	[18]
UiO-66-NH_2(Zr)	H_2S/CH_4	20	~0.38	[18]
MFM-300(Sc)	H_2S/N_2	25	16.55	[26]
MOF-5/GO	H_2S/N_2	20	130.1[a]	[45]
HKUST-1/GO	H_2S/air	Room	200[a]	[46]
Zn-MOF/ZnO	H_2S/N_2	Room	14.2[a]	[47]
IRMOF-3	H_2S/N_2	30	10.61[a]	[37]
ZiF-8	H_2S/N_2	30	3720[a]	[48]

[a]mg/g

MIL-101, MIL-100, and MIL-47(V), at 2 MPa, was 38.4, 16.7, and 14.6 mmol/g, respectively. The adsorption isotherms for MIL-53(Al, Cr, Fe) included two steps. For MIL-53(Fe, Cr, and Al), the H_2S adsorption maximum was 1.18, 3.02, and 3.24 mmol/g in the first step and 8.53, 13.12, and 11.77 mmol/g in the second

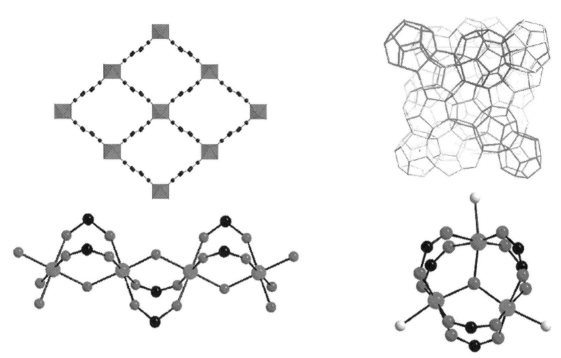

Figure 13.2 (*top left*) MIL-47(V) and MIL-53(Al, Cr, Fe); (*top right*) MIL-100(Cr) and MIL-101(Cr). Atoms of metal, carbon, and oxygen are colored *light gray*, *black*, and *dark gray*, respectively and terminal water molecules and fluorine are colored *white* [27].

phase, respectively. There was a difference in the adsorption behavior among the materials due to the rigid structures of MIL-47(V), MIL-101(Cr), and MIL-100(Cr) compared to the flexible structures of MIL-53(Cr, Al, and Fe). To demonstrate the reversibility and recycling of these MOFs, CH_4 adsorption at room temperature was carried out before and after H_2S adsorption. They reported that with the exception of MIL-53(Fe), all of these MOF materials were chemically stable to H_2S adsorption. When MIL-53(Fe) was treated with H_2S, its crystal structure was destroyed, and a black powder of iron sulfide was formed. According to the achieved results, the metal type in the MOFs plays a critical role in structural stability, and adsorption and regeneration performance. The presence of hydroxyl groups at the MIL-53(Fe) pore opening would cause strong interactions with polar molecules of H_2S and pore blockage.

Joshi et al. [18] studied the desulfurization of natural gas simulant mixture using three various MOFs, including MIL-125(Ti), MIL-101(Cr), and UiO-66(Zr), and their amine-functionalized analogs, including MIL-125-NH$_2$, MIL-101-NH$_2$, and UiO-66-NH$_2$. Two natural gas mixtures of H$_2$S/CH$_4$ (1/99%) and H$_2$S/CO$_2$/CH$_4$ (1/10/89%) were simulated. According to adsorption experiments (Fig. 13.3), the mesoporous MIL-101(Cr) was more effective in adsorbing H$_2$S than microporous UiO-66(Zr) and MIL-125(Ti). Based on this finding, H$_2$S selectivity could be improved by installing H$_2$S binding sites in MOFs with high pores. Linker-based amines also increased the capacity of all investigated materials to absorb H$_2$S. In natural gas mixtures containing and without CO$_2$, MIL-101-NH$_2$(Cr) and MIL-125-NH$_2$(Ti) indicated the greatest H$_2$S saturation capacities with 0.45 and 0.56 mmol H$_2$S/g, respectively. After being exposed to acid gas, pelletized MIL-101-NH$_2$(Cr) shows degrading evidence, indicating it may not be an effective adsorbent for large-scale adsorption operations.

MOF-199 was modified with amines of tertiary amine triethanolamine (TEA), secondary amine DEA, and primary amine MEA to be used as adsorbents for removing H$_2$S at room temperature. When MOF-199 was modified with TEA, it was able to maintain its structural integrity due to the moderate interaction with the TEA. However, when it was modified with MEA or DEA, its structure may be destroyed owing to the strong interaction with MOF-199.

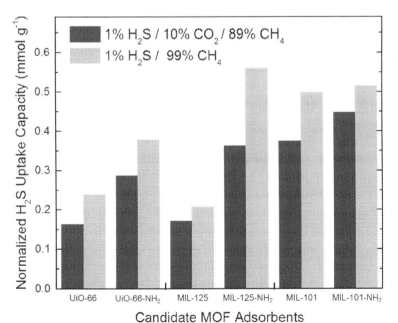

Figure 13.3 H$_2$S saturation capacity for different metal–organic frameworks at 20°C [18].

Therefore, TEA was selected as an appropriate agent for the modification of MOF-199. Based on the achieved results, at the temperature of 30°C, the H₂S adsorption capacities of TEA-MOF-199, MOF-199, MEA-MOF-199, and DEA-MOF-199 were 2.74, 1.67, 0.83, and 0.58 mmol/g, respectively. As a result, MOF-199 functionalized with DEA and MEA displayed lower H₂S adsorption capacities than MOF-199. However, TEA-MOF-199 demonstrated higher H₂S adsorption capacities than MOF-199 [44]. The adsorption models of H₂S on TEA-MOF-199 are demonstrated in Fig. 13.4. The H atom of H₂S and the O atom of loaded TEA had a smaller interaction distance than that between H₂S and bare TEA, resulting in a strong O...H-HS bond between them.

Allan et al. [43] employed MOFs of Zn-COP-27 and Ni-CPO-27 as sorbents for H₂S adsorption, where Ni/Zn cations are linked by organic linkers of 2,5-dihydroxyterephthalic acid. The Brunauer–Emmett–Teller (BET) surface area for Zn-COP-27 and Ni-CPO-27 was 379 and 1193 m²/g, respectively. The Ni-CPO-27 structure was not degraded by exposure to H₂S, whereas the Zn-CPO-27

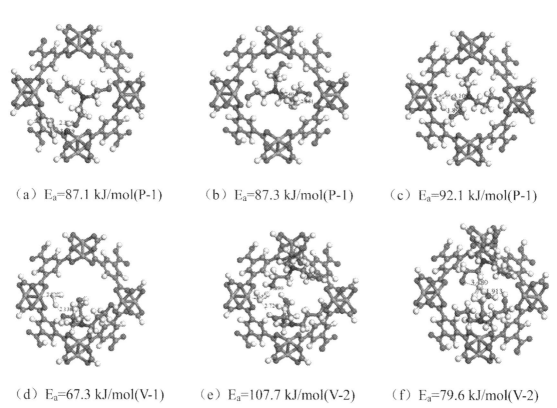

(a) E_a=87.1 kJ/mol(P-1) (b) E_a=87.3 kJ/mol(P-1) (c) E_a=92.1 kJ/mol(P-1)

(d) E_a=67.3 kJ/mol(V-1) (e) E_a=107.7 kJ/mol(V-2) (f) E_a=79.6 kJ/mol(V-2)

Figure 13.4 The adsorption models of H₂S on TEA-MOF-199 [44].

structure was amorphized. Ni-CPO-27 exhibited high uptake and adsorption enthalpy and structural stability. The H_2S adsorption for Ni-MOF-74 at room temperature and low pressure was 6.4 mmol H_2S/g. The highest H_2S uptake for Ni-MOF-74 at 100 kPa and 25°C was 12 mmol/g (Fig. 13.5). After regeneration, in the second run, H_2S adsorption capacity decreased, confirming that H_2S binding on Ni sites is irreversible. Over 6 months, the Ni-CPO material indicates no crystalline structure degradation and just a small loss of deliverable gas capacity.

Wang et al. [37] used IRMOF-3 to remove ethyl mercaptan, hydrogen sulfide, and dimethyl sulfide. With regard to their results, IRMOF-3 was the most effective in removing H_2S, followed by ethyl mercaptan and dimethyl sulfide. The sulfur breakthrough capacity for dimethyl sulfide, ethyl mercaptan, and H_2S was 0.58%, 0.59%, and 1.06%, respectively. For H_2S, the main mechanism was sulfur atoms interacting with amino groups and zinc sites in MOFs. The latter, as opposed to the former, generated new ZnS and H_2O products and severely damaged MOFs.

Liu et al. [58] investigated different MOFs with varying ligands, metal sites, porous structures, and surface areas for H_2S adsorption. They used 11 MOFs, including Mg-MOF-74, Zn-MOF-74, MOF-5, Cu-BDC(ted)0.5, UiO-66, UiO-66-NH_2, Ce-BTC, Fe-BTC, Cu-BTC, ZIF-8, and MIL-101(Cr), with BET surface areas of

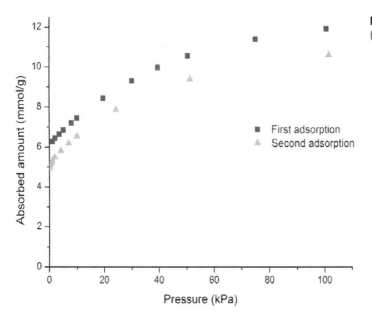

Figure 13.5 Adsorption isotherms for H_2S adsorption on CPO-27 [43].

1244, 920, 2250, 1045, 1322, 1097, 930, 816, 1590, 1602, and 3203 m^2/g, respectively. Among these MOFs, UiO-66, MIL-101(Cr), and Mg-MOF-74 were selected as effective adsorbents for H$_2$S capture (Fig. 13.6). The H$_2$S uptake was decreased in the order: Cu-BDC (1.78) > Zn-MOF-74 (1.64) > MOF-5 (1.11) > Cu-BTC (1.1) > UiO-66-NH$_2$ (0.909) > Fe-BTC (0.9) > MIL-101(Cr) (0.4) > UiO-66 (0.234) > Mg-MOF (0.24) > Ce-BTC (0.126) > ZiF-8 (0.05 mmol/g). The reversible physical sorption of H$_2$S on ZIF-8, UiO-66, Mg-MOF-74, Ce-BTC, and MIL-101(Cr) was observed, where MIL-101(Cr), with its high surface area and open metal sites, exhibited high H$_2$S sorption. Due to irreversible H$_2$S chemisorption in MOF-5, Cu-BTC, and Zn-MOF-74, high H$_2$S adsorption and CO$_2$/H$_2$S selectivity were achieved, although metal sulfide formed during reaction with hydrogen sulfide can damage their structure. For MIL-100(Fe), as a result of the oxidation–reduction reaction of Fe(III) to Fe(II) in the presence of H$_2$S, S8 was formed. The situations were similar for Cu-BDC(ted)0.5 and UiO-66-NH$_2$. However, they were able to maintain their original structures. In Fig. 13.7, H$_2$S adsorption sites on UiO-66, MIL-101(Cr), Mg-MOF-74, ZIF-8, Cu-BDC(ted)0.5, and Ce-BTC are indicated.

H$_2$S adsorptive removal using two MOFs, HKUST-1 and ZIF-8, was looked into by Ethiraj et al. [52]. They found that ZIF-8 had

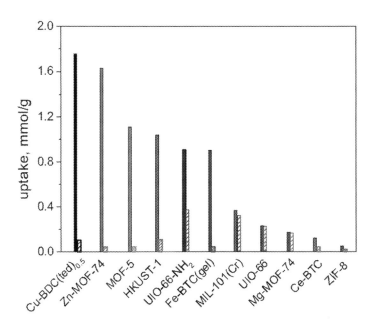

Figure 13.6 H$_2$S uptake on different metal–organic frameworks [58].

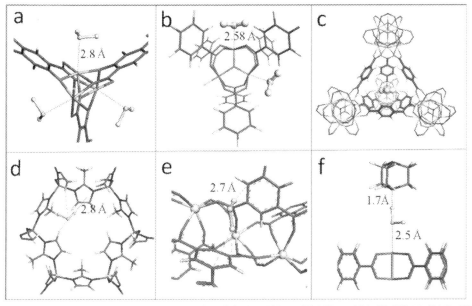

Figure 13.7 Adsorption sites of H_2S on different MOFs: (A) Mg-MOF-74; (B) MIL-101(Cr); (C) UiO-66; (D) Cu-BDC(ted) 0.5; (E) ZIF-8; and (F) Ce-BTC [58].

higher H_2S stability when compared to HKUST-1. Owing to the formation of a covellite CuS phase and the subsequent disintegration of the framework, all electronic, vibrational, and structural fingerprints were severely disturbed in HKUST-1.

In another study by Qiao et al. [4], a computational analysis for the screening of 6013 CoRE-MOFs for natural gas upgrading was conducted. A total of 606 hydrophobic MOFs were chosen based on the H_2O Henry constants and tested for the separation of H_2S and CO_2 from the gas stream of $CO_2/H_2S/H_2O/C_3H_8/C_2H_6/CH_4$. Between MOF characteristics such as the largest cavity diameter (LCD), void fraction (Φ), volumetric surface (VS), and isosteric heat (Q^o_{st}) and performance metrics like adsorption capacity ($N_{H2S + CO2}$) and selectivity ($S_{H2S + CO2/C1-C3}$), structure–performance relationships were developed. TSN (trade-off between $N_{H2S + CO2}$ and $S_{H2S + CO2/C1-C3}$) was also offered as a new performance metric. The TSN had substantial connections with the MOF descriptors, as indicated by the Pearson correlation technique. Metals such as Zn, Cu, Cd, Ag, and Co were utilized to make these MOFs. Based on the authors, the type of metal used in this work cannot be regarded as a significant parameter in measuring CO_2 and H_2S adsorption, especially for hydrophobic MOFs. Organic linkers with N atoms were found in 39 of the 45 best MOFs. Pyridine was found in 23 of the 45 samples (51.1%), whereas azoles were found in 12 of the 45 samples (26.7%). Acid gases can interact aggressively with organic linkers containing nitrogen. Fig. 13.8 indicates the MOF number with various metals and linkers. Table 13.3 lists the 14 best MOFs identified based on the TSN >2 benchmark. According to the authors' suggestion, MOFs designed for CO_2 and H_2S separation should include N-rich organic linkers such as pyridine and azoles.

2.2 Metal oxides

The desulfurization of gases using single and mixed metal oxides has been extensively studied (Table 13.4). Metal oxides exhibit a high capacity for sulfur removal. However, pure metal oxides have limitations such as low surface area, insufficient porosity, large particle size, and metal evaporation, which result in low sulfurization kinetics and grain agglomeration [75]. Therefore, mixed metal oxides and multicomponent sorbents were used for ADS. Support materials like carbon nanofiber, activated carbon, graphite oxide, and mesoporous silica were also employed. The general reaction between the metal oxide and H_2S is as given in Eq. (13.1):

$$M_xO_y\,(S) + yH_2S(g) \rightarrow M_xS_y\,(S) + yH_2O(g) \quad (13.1)$$

The various metal oxides such as Ag_2O [62], CuO [62], ZnO [62], Co_3O_4 [62], NiO [62], CaO [62], Mn_3O_4 [62], SnO [62], Cr_2O_3 [61], Fe_2O_3 [61], CuO [61], Mn_2O_3 [61], Co_3O_4 [60], Fe−Mn−Zn−Ti−O [65], NiO/TiO_2 [82], CoO/TiO_2 [82], CuO/TiO_2 [82], CuO/

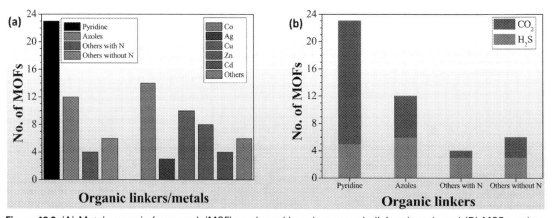

Figure 13.8 (A) Metal−organic framework (MOF) number with various organic linkers/metals and (B) MOF number with various organic linkers for the adsorption of CO_2 and H_2S [4].

Table 13.3 The best 14 metal–organic framework characteristics [4].

CSD No.	CSD code	$N_{H2S + CO2}$ TSN (mol/kg)	$S_{H2S + CO2}$ C1–C3	$Q°_{st, H2S}$ (KJ/mol)	Φ	VS (m²/cm³)	LCD (A°)	Metal	Organic linker
1	OFUSAL	4.26 2.48	52.34	36.13	0.2	0	3.61	Ag	1,3,5-Triaza-7-phosphaadamantane-7-sulfide (PTA=S)
2	ECIWUJ	3.48 0.73	55,709.8	24.31	0.15	0	3.25	Zn	Pyridine
3	NABCIE	3.33 1.81	69.30	39.56	0.15	0	3.56	Co	Pyridine
4	ZONBAH	3.03 2.03	31.04	33.48	0.2	0	3.78	Zr	Other groups without N
5	FIKQIB	2.47 1.72	27.35	35.09	0.17	31.77	4.07	Pb	Other groups without N
6	PEJRON	2.46 0.89	597.61	39.66	0.07	0	3.58	Co	Imidazole
7	ILUFUR	2.33 2.15	12.14	27.35	0.31	335.15	7.14	Co	Pyridine
8	TAKYEL	2.32 2.09	13.02	27.26	0.3	319.89	7.08	Co	Pyridine
9	POXLEV	2.32 0.77	1088.35	31.51	0.16	0	3.45	Th	Other groups without N
10	ILUFIF	2.19 2.01	12.36	27.23	0.31	326.74	7.10	Co	Pyridine
11	ILUFOL	2.19 2.05	11.73	27.62	0.31	333.14	7.11	Co	Pyridine
12	TAKYOV	2.09 1.95	11.83	27.90	0.31	324.14	7.07	Co	Pyridine
13	ZONCAI	2.05 0.69	966.8	28.99	0.09	0	3.55	Zn	Other groups without N
14	HIQPEE	2.04 1.81	13.24	40.97	0.15	7.68	3.84	Co	Pyrazole

SiO_2 [72,83], ZnO/SiO_2 [68,72,84], γ-Fe_2O_3 [64,85], α-Fe_2O_3 [64], γ-Fe_2O_3/SiO_2 [85], α-FeOOH [63], α-FeOOH–activated carbon [63], γ-FeOOH [64], CoOOH [59], Cu–Zn–Al mixed metal oxide [86], γ-Fe_2O_3/SiO_2 [85], ZnO/CuO [87], ZnO/Co_3O_4 [87], Fe_2O_3/Al_2O_3 [69], Mn_2O_3/Fe_2O_3 [70], CoOOH/graphite oxide [59], TiO_2/GO [78], ZnO/reduced graphite oxide [88], Fe_3O_4/graphite oxide [89], $Fe_5O_7.4H_2O$/graphite oxide [89], $Cd(OH)_2$/graphite oxide [90], ZnO/carbon nanofiber [91], $ZnFe_2O_4$/activated carbon [81], Zn/mesoporous silica molecular seive (MSU-1) [75], Ni-doped ZnO/Al_2O_3 [77], ZnO–MgO/activated carbon [79], and Cu–Zn mixed oxide/SBA-15 [92], ZnO/SBA-15 [73,93,94], Fe_2O_3/SBA-15 [73], ZnO-SBA-16 [95], ZnO-MCM-48 [95], ZnO-KIT-6 [95], CeO_2-MnO_x/ZSM-5 [96], ZnO/MCM-48 [94], ZnO/MCM-41 [94], and $ZnO/Co_3O_4/SiO_2$ [97] were investigated in H_2S adsorptive removal.

A series of metal oxides such as Ag_2O, CuO, ZnO, Co_3O_4, NiO, CaO, Mn_3O_4, and SnO and mixed oxides of Zn containing Al, Mn, Co, Fe, Ni, Ti, Zr, and Cu were studied as adsorbents for H_2S

Table 13.4 Metal oxide sorbents used for H$_2$S removal from gas streams.

Adsorbent	Feed gas composition	Adsorption temperature (°C)	H$_2$S breakthrough capacity (mg/g)	Reference
CoOOH	H$_2$S/moist air	Room	121.8	[59]
Co$_3$O$_4$	H$_2$S/moist N$_2$	30	189	[60]
Cr$_2$O$_3$	H$_2$S/He	200	196	[61]
Fe$_2$O$_3$	H$_2$S/He	200	248	[61]
CuO	H$_2$S/He	200	246	[61]
CuO	H$_2$S/N$_2$	Room	200	[62]
Mn$_2$O$_3$	H$_2$S/He	200	431	[61]
CoOOH	H$_2$S/dry air	Room	69.1	[59]
α-FeOOH	H$_2$S/N$_2$	Room	65	[63]
γ-Fe$_2$O$_3$	H$_2$S/N$_2$	25	63	[64]
γ-FeOOH	H$_2$S/N$_2$	25	46.9	[64]
Fe–Mn–Zn–Ti–O	H$_2$S/H$_2$/H$_2$O/CO$_2$	25	85	[65]
Fe–Mn–ZnO/SiO$_2$	H$_2$S/H$_2$	25	37	[66]
Fe–Cu–Al–O	H$_2$S/CO$_2$	40	113.9	[67]
Fe–Cu–Al–O	H$_2$S/N$_2$	40	125.3	[67]
ZnO/SiO$_2$	H$_2$S/N$_2$	20	96.4	[68]
Fe$_2$O$_3$/Al$_2$O$_3$	H$_2$S/N$_2$/H$_2$/CO/CO$_2$	700	44.7	[69]
Mn$_2$O$_3$/Fe$_2$O$_3$	H$_2$S/N$_2$	25	11.97	[70]
Cu–ZnO/SiO$_2$	H$_2$S/air	Room	77	[71]
ZnO/SiO$_2$	H$_2$S/moist N$_2$	30	108.9	[72]
CuO/SiO$_2$	H$_2$S/moist N$_2$	30	145.6	[72]
Co$_3$O$_4$/SiO$_2$	H$_2$S/moist N$_2$	30	114.3	[72]
ZnO/SBA-15	H$_2$S/He	300	53	[73]
Fe$_2$O$_3$/SBA-15	H$_2$S/He	300	401	[73]
Fe$_2$O$_3$/MCM-41	H$_2$S/He	300	38	[74]
Zn/MSU-1	H$_2$S/CH$_4$	25	42.3	[75]
Cu/MSU-1	H$_2$S/CH$_4$	25	19.2	[75]
Fe–Ti bimetal oxide (Fe$_8$Ti$_1$)	H$_2$S/N$_2$	20	222.8	[76]
Ni-doped ZnO/Al$_2$O$_3$	H$_2$S/moist N$_2$	30	92	[77]
TiO$_2$/graphene oxide	H$_2$S/N$_2$	220	250	[78]
CoOOH/graphite oxide	H$_2$S/dry air	Room	66.3	[59]
CoOOH/graphite oxide	H$_2$S/moist air	Room	108.2	[59]
α-FeOOH—activated carbon	H$_2$S/N$_2$	Room	171	[63]

Table 13.4 Metal oxide sorbents used for H$_2$S removal from gas streams.—*continued*

Adsorbent	Feed gas composition	Adsorption temperature (°C)	H$_2$S breakthrough capacity (mg/g)	Reference
ZnO—MgO/activated carbon	H$_2$S/moist N$_2$	30	113.4	[79]
ZnO—CuO/activated carbon	H$_2$S/N$_2$	30	50	[80]
ZnFe$_2$O$_4$/activated carbon	H$_2$S/moist air	Room	122.5	[81]

elimination at room temperature by Xue et al. [62]. According to their results, adsorbents such as CuO, Zn/Al type, Zn/Mn type, Zn/Ti/Zr type, and Zn/Co type (Fig. 13.9) indicated high capacities for H$_2$S removal (between 107 and 283 mg/g). Among investigated metal oxides, CuO achieved the highest H$_2$S uptake amount of 283 mg/g. Also, mixed metal oxides that contained small hexagonal crystallites of ZnO showed high H$_2$S uptake.

Nanocomposites of CoO/TiO$_2$, NiO/TiO$_2$, and CuO/TiO$_2$ were used by Orojlou et al. [82] to study H$_2$S removal at relatively high

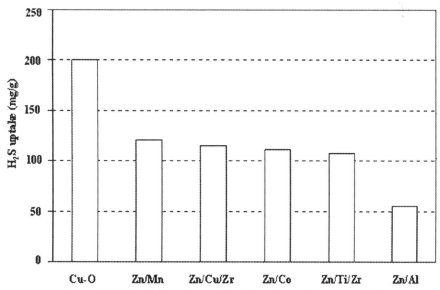

Figure 13.9 H$_2$S uptake using metal oxides at room temperature [62].

temperatures. The performance of sorbents was investigated using different factors such as sorbent mass, temperature, and promoter to TiO$_2$ ratio. The best results were achieved when a promoter to TiO$_2$ ratio of 2.5/5 was applied. The CoO/TiO$_2$ nanocomposite displayed the best performance at 480°C. A temperature reduction to 400°C resulted in significant changes in sample performance. At temperature of 400°C, the NiO/TiO$_2$ nanocomposite showed the best results, followed by the CoO/TiO$_2$ and CuO/TiO$_2$ nanocomposites.

To study H$_2$S adsorption, TiO$_2$ and N-doped TiO$_2$ nanoparticles were loaded over and within GO nanosheets to prepare a nanoadsorbent. At temperatures ranging from 200 to 250°C, the adsorption capacity was investigated. At 220°C, optimum H$_2$S uptake capacities of 250, 200, and 170 mg S/g sorbent were achieved for 5% TiO$_2$/GO, 10 %TiO/GO, and 5%N-TiO$_2$/GO, respectively. The best nanoadsorbent was 5%N-TiO$_2$/GO, as it had a lower band gap, larger surface area, synergic effects, and N-TiO$_2$ active sites [78]. Fig. 13.10 shows the adsorption of H$_2$S on the N–TiO$_2$/GO nanoadsorbent.

Another work was performed by Cao et al. [64], who studied H$_2$S adsorption using γ-Fe$_2$O$_3$, α-Fe$_2$O$_3$, and γ-FeOOH sorbents. Based on their results, γ-Fe$_2$O$_3$ was the most suitable iron oxide for the low-temperature desulfurization process, whereas α-Fe$_2$O$_3$ was the worst. They suggested that the hydroxyl groups,

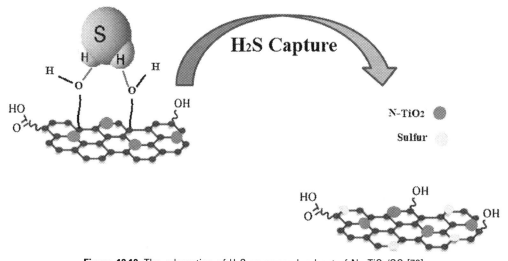

Figure 13.10 The adsorption of H$_2$S on nanoadsorbent of N–TiO$_2$/GO [78].

specific surface area, and oxygen vacancies can play an important role in desulfurization performance at room temperature. At low temperatures, oxygen vacancies were not as important due to poor ion diffusion in the lattice. Therefore, desulfurization was affected less by the number of oxygen vacancies than by the specific surface area and number of hydroxyl groups.

Mureddu et al. [73] used Fe_2O_3/SBA-15, ZnO/SBA-15, and commercial sorbent to investigate the removal of H_2S at 300°C. According to their outcome, both of these sorbents were significantly more efficient than commercial sorbent. The H_2S uptake was decreased in the order: Fe_2O_3/SBA-15 (401) > ZnO/SBA-15 (53) > commercial sorbent (6 mg/g). The zinc oxide formed a thin homogeneous layer inside the SBA-15 channels, whereas the Fe_2O_3 was dispersed as small maghemite crystals.

The moisture is a key factor in keeping the activity of metal oxides for removing H_2S at room temperature. According to Florent and Bandosz [59], water can be helpful in removing H_2S using metal oxides. Based on their findings, when water was present in the system, the adsorption capacity of H_2S utilizing cobalt oxyhydroxide/graphite oxide adsorbent increased. The H_2S breakthrough capacities of 108.2 and 66.3 mg/g were obtained with and without moisture. However, H_2S uptake capacity can be reduced by high moisture due to the thick film of H_2O deposited on the solid surface, which can clog the pores and prevent access to the activated sites.

3. Silica-based sorbents

3.1 Mesoporous silica structures

Mesoporous silica materials (MSMs) have several interesting properties, including big pore size, large surface area, large porosity, adjustable structures, favorable thermal and mechanical durability, etc. [98]. Due to their unique properties, MCMs were applied in various fields like catalyst, polymer filler, drug delivery, adsorption, and biochemical applications.

MSMs have a limited affinity for hydrogen sulfide due to their neutral structures. Therefore, functionalizing or immobilizing MSM with/on active components could provide superior materials with improved desulfurization capabilities [22]. Among the mesoporous silica structures, Santa Barbara Amorphous (SBA-15) and Mobil Composition of Matter No. 41 (MCM-41) and No. 48 (MCM-48) are the most commonly employed for H_2S adsorptive

Table 13.5 Mesoporous silica sorbents used for H_2S elimination from gas streams.

Adsorbent	Feed gas composition	Adsorption temperature (°C)	H_2S breakthrough capacity (mmol/g)	Reference
MAPS/SBA-15	H_2S/N_2	30	0.16	[99]
APS/SBA-15	H_2S/N_2	30	~0.135	[99]
APS/MCM-41	H_2S/N_2	25	134.4[a]	[100]
DMAPS/SBA-15	H_2S/N_2	30	0.055	[99]
MDEA-SBA-15	H_2S/CH_4	25	0.109	[101]
MCM-41@ZiF-8	H_2S/N_2	30	4920[a]	[48]
SBA-15@ZiF-8	H_2S/N_2	30	5210[a]	[48]
UVM-7@ZiF-8	H_2S/N_2	30	6120[a]	[48]
PEI/SBA-15	$H_2S/N_2/H_2$	22	0.79	[102]
PEI/MCM-48	$H_2S/N_2/H_2$	22	0.81	[102]
PEI/MCM-41	$H_2S/N_2/H_2$	22	0.46	[102]
Si-PEI-800-50	H_2S/N_2	30	0.45	[103]

[a]mg/g

removal (Table 13.5). They were amine functionalized and employed as adsorbents for H_2S removal. Amine types of methyldiethylamine (MDEA) [101], polyallylamine (PA) [104], (N, N-dimethylaminopropyl)trimethoxysilane (DMAPS) [99], aminopropyltrimethoxysilane (APS) [99], N-methylaminopropyltrimethoxysilane (MAPS) [99], 3-(triethoxysilyl)propylamine [105], N-(2-aminoethyl)-3-aminopropyltrimethoxysilane [105], n-(3-trimethoxysilylpropyl)diethylenetriamine (TRI) [103,105], polyethylenimine (PEI) [103,104], tetramethyl hexanediamine (TMHDA) [104], TEA [106], and hexamine [107] have been used for the amine functionalization of mesoporous silica structures. The sorbents of SBA-15 [107], hexamine/SBA-15 [107], amine/SBA-15 [105], amine/MCM-41 [105], MDEA-SBA-15 [101], triamine-grafted pore-expanded MCM-41 (TRI-PE-MCM-41) [108], Zn-MCM-41 [109], TEA-SBA-15 [106], PEI/MCM-41 [102], PEI/MCM-48 [102], PEI/SBA-15 [102], PEI/SBA-15 [104], TMHDA/SBA-15 [104], PA/SBA-15 [104], MAPS/SBA-15 [99], APS/SBA-15 [99], DMAPS/SBA-15 [99], NH_2-SBA-15-IL [110], CuCl/MCM-41 [111], and CuCl/SBA-15 [111] have been used for H_2S elimination from gas streams.

Quan et al. [104] reported an interesting study in which they applied various types of amines to functionalize SBA-15. TMHDA/SBA-15, PA/SBA-15, and PEI/SBA-15 were employed as adsorbents for H_2S removal. Among these sorbents, TMHDA/SBA-15 was selected as the best sorbent for H_2S selective adsorption in highly concentrated CO_2. The interaction between the amine and silanol groups on the SBA-15 surface caused TMHDA to be dispersed within pore channels. Under mild regeneration conditions, this sorbent showed good regenerability, and other components of the gas stream like H_2O and CH_4 did not affect its capacity. The H_2S sorption on TMHDA/SBA-15 is indicated in Fig. 13.11. Standing and lying loaded TMDDA molecules are thought to exist on SBA-15's surface. The former indicates that the surface silanol groups interact with only one TMHDA's tertiary amine, whereas the other remains available for H_2S adsorption. For the standing type, two TMHDA molecules are involved in the adsorption of one molecule of H_2S. The interaction with silanol groups of the lying type occupies both tertiary amine groups, causing the TMHDA molecule to rest on top of SBA15. It seems that the lying form is inactive to remove H_2S. Hence, at low loading levels, the huge unoccupied surface area would promote TMHDA molecule dispersion at the expense of consuming silanol groups that may otherwise aid in the sorption of H_2S.

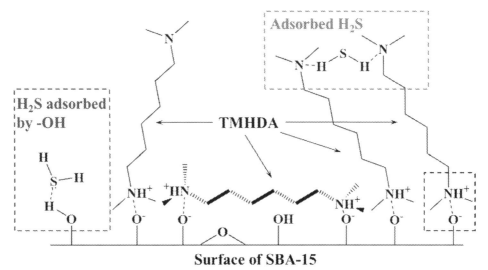

Figure 13.11 H_2S adsorption model on TMHDA/SBA-15 [104].

The sorbents of SBA-15, MCM-48, and MCM-41 were functionalized by PEI and employed to remove H_2S from $H_2S/H_2/N_2$ gas mixture. The PEI content on these materials significantly affected the sorbents' sorption performance. The highest breakthrough capacity was achieved with the amount of 50 wt% PEI on SBA-15, whereas the highest saturation capacity was obtained with the 65 wt% PEI content on SBA-15. According to the researchers, the PEI/SBA-15 sorbent had a saturation capacity of 3.02 mmol H_2S g-sorb^{-1} and a breakthrough capacity of 0.79 mmol H_2S g-sorb^{-1} for a model gas containing 4000 ppmv H_2S at 22°C and a gas hourly space velocity (GHSV) of 674 h^{-1} [102].

3.2 Zeolites

The zeolites have well-defined porous structures and pores ranging from 0.3 to 2 nm. They are made up of a three-dimensional framework of SiO_4^{4-} and AlO_4^{5-} tetrahedra connected by oxygen atoms [112]. Zeolites, unlike many other adsorbents, can generally withstand high temperatures and pressures, as well as harsh chemical environments. The adsorption applicability of zeolites is determined by their Si/Al ratio. Low-silica zeolites are hydrophilic and adsorb polar substances better, whereas zeolites with higher Si/Al ratios are hydrophobic and hydrothermally stable, have fewer structural defects, and are preferred in the separation of nonpolar gases [113]. All-silica zeolites such as 386 zeolitic frameworks [114], aluminosilicate zeolites such as Zeolite Socony Mobil-5 (ZSM-5) [115], and titanosilicate zeolites such as Engelhard titanosilicate (ETS) [116] have been studied for H_2S adsorptive removal. Zeolites, both natural such as clinoptilolite and synthetic such as zeolite-X (FAU), zeolite-A (LTA), and zeolite-Y (FAU), were used for the ADS of gases (Table 13.6). The adsorption capacity of zeolites was increased by adding metal or metal oxides. Several zeolites, including 3A [126], 4A [126], 5A [15], NaY [11], NaA [113], ZnO/NaA [120], CaNaA [127], 13X [15,16,117,118] Cu-ETS-2 [128], Ag-ETS-2 [128], 4A molecular sieve [124], industrial molecular sieve (IMS) [125], Cu-X [129], Cu-Y [111,129], Ag-X [129], Ag-Y [111,129], ZSM-5 [130], Cu-ZSM-12 [131], SAPO-43 [132], clinoptilolite [15,123], and Cu-Ce-zeolite [119] have been used as adsorbents to eliminate H_2S from gas streams.

Rezaei et al. [128] looked into the adsorption performance of metal-exchanged ETS-2 adsorbents with Ca, Zn, Ag, and Cu for H_2S adsorption against a commercial adsorbent. Based on their outcome, the Cu-exchanged ETS-2 exhibited the highest capacity

Table 13.6 Zeolite sorbents used for H$_2$S removal from gas streams.

Adsorbent	Feed gas composition	Adsorption temperature (°C)	H$_2$S breakthrough capacity (mg/g)	Reference
13X	H$_2$S/CH$_4$	25	53	[16]
13X	H$_2$S/N$_2$/H$_2$O/SO$_2$	150	179.7	[117]
13X Ex-Cu	H$_2$S/N$_2$	120	40.12	[118]
13X Ex-Cu	H$_2$S/CH$_4$/CO$_2$	30	18.17	[118]
13X Ex-Cu	H$_2$S/CH$_4$	30	23.36	[118]
13X Ex-Cu	H$_2$S/CO$_2$	30	14.24	[118]
Zeolite-Y	H$_2$S/H$_2$/H$_2$O/He	800	0.26	[119]
ZnO/NaA	H$_2$S/N$_2$	28	15.75	[120]
Cu-Y	H$_2$S/H$_2$/H$_2$O/He	800	1.80	[119]
Ce-Y	H$_2$S/H$_2$/H$_2$O/He	800	1.04	[119]
TiO$_2$-zeolite	H$_2$S/CH$_4$/N$_2$	25	0.13[a]	[121]
Cu-Ce-Y	H$_2$S/H$_2$/H$_2$O/He	800	1.78	[119]
CuO/CeO$_2$-Y	H$_2$S/H$_2$/H$_2$O/He	800	2.45	[119]
Zinox 380	H$_2$S/CH$_4$	25	9.5	[16]
Cu-ETS-2	H$_2$S/N$_2$	100	47	[122]
Cu-ETS-10	H$_2$S/N$_2$	100	45	[122]
Clinoptilolite	H$_2$S/He	600	30	[123]
Clinoptilolite	H$_2$S/He	100	87	[123]
Clinoptilolite	H$_2$S/CH$_4$/CO$_2$	25	1.39	[15]
4A	H$_2$S/N$_2$	50	8.36	[124]
IMS	H$_2$S/CH$_4$/CO$_2$/Ar	25	193.3	[125]

[a] mmol/g

of 29.7 mg H$_2$S/g adsorbent for H$_2$S removal. The order of H$_2$S uptake capacities at room temperature was Cu-ETS-2 > Ag-ETS-2 > Zn-ETS-2 ~ R3-ETS-2 > Ca-ETS-2 ~ Na-ETS-2. CuO is thermodynamically more favorable for sulfidation. A rigid physical structure and the presence of inactive sites like metallic silver dots resulted in less thermodynamically favorable sulfidation processes, and it led to the reduction of H$_2$S removal capabilities for zinc, calcium, and silver.

Liu et al. [124] examined H$_2$S removal from coal gasification gas, natural gas, and digester gas using 4A zeolite made from attapulgite in various conditions. The experimental results indicated that, at 50°C, the maximum saturation and breakthrough sulfur sorption capacities were 12.4 and 8.36 mg/g, respectively. After sorbent regeneration, the adsorption capacity was reduced because the particle structure was significantly damaged during the sorbent regeneration process.

SAPO-43 as a microporous sorbent was used to separate H_2S, H_2O, and CO_2 from natural gas. The main limitation of this sorbent was its low thermal stability over 300°C. Element analysis and CO_2 adsorption heats showed that amine-containing compounds entrapped in the SAPO-43 surface kept the framework from collapsing. The adsorption capacities of H_2S, H_2O, and CO_2 at ambient temperature and pressure were 2.52, 4.93, and 1.1 mmol/g, respectively. The sorbent regeneration at 180°C using pure helium restored 40% of captured H_2S and removed nearly all the adsorbed water. The authors suggested that sorbent regeneration utilizing chemicals like H_2O_2 could be an alternative to thermal regeneration.

Another study carried out by Barelli et al. [118] was devoted to 13X zeolite modified with Cu to remove H_2S from the gas matrix. Several parameters, including reactor temperature (30–120°C), GHSV (850–16,941 h^{-1}), particle size, and gas stream composition (CO_2, CH_4, N_2, and CH_4/CO_2), were examined for the identification of optimal conditions. According to the achieved results, the greatest H_2S adsorption capacity was 40.12 mg/g for GHSV equal to 850/h at the temperature of 120°C in a dry N_2 matrix. At the temperature of 30°C, the H_2S adsorption capacities of 14, 23, 20, and 18 mg/g were obtained in the presence of matrixes of pure CO_2, pure CH_4, pure N_2, and CH_4 (60%)/CO_2 (40%), respectively. Owing to the competition for adsorption between H_2S and CO_2 in the pure CO_2 matrix, the adsorption capacity was low.

Eighteen different porous materials, such as zeolites, MOFs, and carbons, were used to investigate the decarburization and desulfurization of natural gas, biogas, and flue gas (Fig. 13.12). The most promising candidates, including zeolite-like MOF (zMOF) and Na-5A in the $H_2S/CO_2/CH_4$ systems (biogas and natural gas) and MIL-47 and Na-13X in the $SO_2/CO_2/N_2$ systems (flue gas), were screened due to their excellent selectivity and capacity in removing sulfide. For the simultaneous removal of sulfide and carbon dioxide from flue gases, MOF-74-Zn was the best candidate, whereas zMOF was the best choice for biogas and natural gas. In addition, the effect of temperature on adsorption was investigated to determine the ability of adsorbents to regenerate. According to their findings, the Na-13X and Na-5A zeolites cannot be readily regenerated compared to zMOFs and MIL-47, owing to the problem of desorbing sulfide at high temperatures, which may be caused by zeolites' strong adsorbent–adsorbate interactions. It was also investigated how sulfide concentration influences the

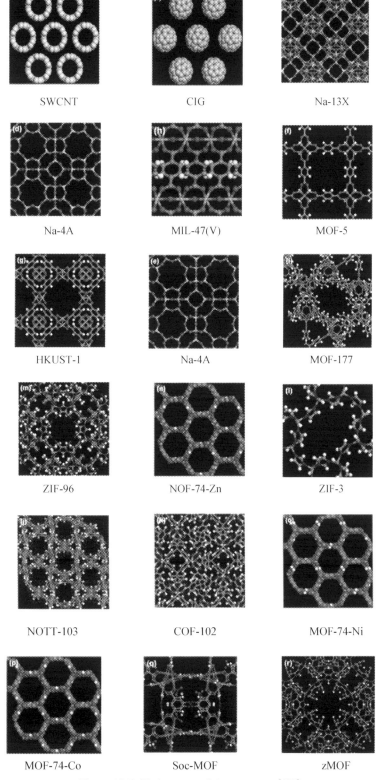

Figure 13.12 Various materials structures [133].

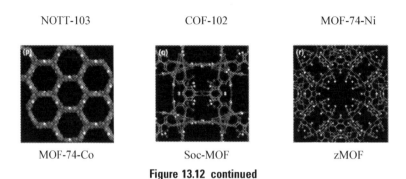

Figure 13.12 continued

adsorption properties of adsorbents. For desulfurizing gas mixtures containing high sulfide concentrations, the MIL-47 and zMOF performed better than Na-13X and Na-5A due to their larger pore volume [133].

4. Conclusion and future outlooks

The present study reviews metal oxides and silica-based sorbents for natural gas sweetening. Introducing sorbents with high H_2S adsorption and selectivity, full regeneration capability, long-term stability, and low operational cost is needed for industrial applications. The sorbents of MOFs, metal oxides, zeolites, and mesoporous silica structures have been extensively applied to eliminate H_2S from gas streams under different working conditions.

Since MOFs have high porosity and surface area and can be changed in nature and size without alteration of their underlying topology, they were broadly employed for the omission of H_2S from gas mixtures, although they have limitations. The structure of MOF with open sites loses its crystal structure when it treats with H_2S and forms metal sulfide. Using MOF composite may solve this problem, although these materials have poor H_2S adsorption. Open metal sites and H_2S interact mildly, allowing MOFs to maintain their structural integrity and improve the H_2S uptake, whereas strong interaction can destroy the MOFs structure. Thus, moderate interactions present the opportunity for reversible adsorption processes. MOFs and H_2S form strong and often irreversible bonds, which can limit their application. The formation of noncovalent bonding between H_2S and MOFs improves reversibility after H_2S sulfidation. The functionalization of the surface of MOFs can also enhance their capacity to adsorb H_2S. Developing stable porous MOFs with high H_2S adsorption capacities and selectivity, low cost, and regenerable is a challenge for industrial applications.

Metal oxides and mixed metal oxides sorbents were extensively applied for H$_2$S removal. Among different metal oxides, ZnO-based adsorbents were widely used for H$_2$S removal from gas streams and they reported the best performance due to favorable sulfidation thermodynamics. The development of regeneration methods for metal oxide sorbents that retain both structural reversibility and adsorption capacity reversibility is necessary. Also, the elucidation of sulfidation mechanisms in the presence of metal oxides as adsorbents should be performed.

Mesoporous silicas with big pore volumes and large surface areas were used as adsorbents or supports for H$_2$S adsorption. The amount and type of amine functionalities strongly affect H$_2$S adsorption by amine-modified mesoporous silicas. Various types of amines, including primary, secondary, and tertiary, were used to functionalize the mesoporous silica structures. Modifying mesoporous silica with secondary and tertiary amines showed higher H$_2$S adsorption and good regenerability under mild regeneration.

Zeolites are promising materials for H$_2$S adsorption, owing to their high surface-volume ratios. Modifying zeolites with metals or metal oxides can increase their adsorption capacity.

Zeolites may not be simply regenerated because of the problem of desorbing sulfide at high temperatures, caused by strong interactions between adsorbent and adsorbate in zeolites. They typically require energy-demanding regeneration processes (generally above 450°C). Therefore, more studies on adsorption at low temperatures and adsorbent regeneration are suggested.

Moisture is a key factor that influences H$_2$S adsorption by adsorbents. The presence of H$_2$O can have a negative or positive effect on H$_2$S removal. The ability to maintain structural integrity in the presence of moisture is an important issue for developing these sorbents.

Abbreviations and symbols

ADS	Adsorptive desulfurization
APS	Aminopropyltrimethoxysilane
BET	Brunauer–Emmett–Teller
DEA	Diethanolamine
DMAPS	(N,N-dimethylaminopropyl)trimethoxysilane
ETS	Engelhard titanosilicate
GHSV	Gas hourly space velocity
GO	Graphene oxide
H$_2$S	Hydrogen sulfide

HKUST-1	Hong Kong University of Science and Technology
IRMOF	Isoreticular metal–organic frameworks
LCD	Largest cavity diameter
MAPS	N-methylaminopropyltrimethoxysilane
MCM-41	Mobil Composition of Matter No. 41
MCM-48	Mobil Composition of Matter No. 48
MDEA	Methyldiethylamine
MDEtA	Methyldiethanolamine
MEA	Monoethanolamine
MILs	Materials from the Institute Lavoisier
MOFs	Metal–organic frameworks
MSM	Mesoporous silica materials
ODS	Oxidative desulfurization
PA	Polyallylamine
PEI	Polyethylenimine
PSA	Pressure swing adsorption
$Q°_{st}$	Isosteric heat
SBA-15	Santa Barbara Amorphous
TEA	Triethanolamine
TMHDA	Tetramethyl hexanediamine
TRI	N-(3-trimethoxysilylpropyl)diethylenetriamine
TSA	Temperature swing adsorption
TSN	Trade-off between $N_{H2S + CO2}$ and $S_{H2S + CO2/C1-C3}$
VS	Volumetric surface
VSA	Vacuum swing adsorption
Φ	Void fraction
ZIF	Zeolitic imidazolate framework
ZSM-5	Zeolite Socony Mobil-5

References

[1] Tengku Hassan TNA, Shariff AM, Mohd Pauzi MM, Khidzir MS, Surmi A. Insights on cryogenic distillation technology for simultaneous CO_2 and H_2S removal for sour gas. Molecules 2022;27:1. https://doi.org/10.3390/molecules27041424.

[2] Mokhatab S, Poe WA, Mak JY. Handbook of natural gas transmission and processing: principles and practices. Netherlands: Elsevier; 2018. https://doi.org/10.1016/C2017-0-03889-2.

[3] Karadas F, Atilhan M, Aparicio S. Review on the use of ionic liquids (ILs) as alternative fluids for CO_2 capture and natural gas sweetening. Energy and Fuels 2010;24:5817–28. https://doi.org/10.1021/ef1011337.

[4] Qiao Z, Xu Q, Jiang J. Computational screening of hydrophobic metal-organic frameworks for the separation of H_2S and CO_2 from natural gas. Journal of Materials Chemistry A 2018;6:18898–905. https://doi.org/10.1039/c8ta04939d.

[5] Goodwin MJ, Musa OM, Steed JW. Problems associated with sour gas in the oilfield industry and their solutions. Energy & Fuels 2015;29:4667–82. https://doi.org/10.1021/acs.energyfuels.5b00952.

[6] Georgiadis A, Charisiou N, Goula M. Removal of hydrogen sulfide from various industrial gases: a review of the most promising adsorbing materials. Catalysts 2020;10:521. https://doi.org/10.3390/catal10050521.

[7] Shah MS, Tsapatsis M, Siepmann JI. Hydrogen sulfide capture: from absorption in polar liquids to oxide, zeolite, and metal−organic framework adsorbents and membranes. Chemistry Review 2017;117:9755−803. https://doi.org/10.1021/acs.chemrev.7b00095.

[8] Pudi A, Rezaei M, Signorini V, Peter Andersson M, Giacinti Baschetti M, Soheil Mansouri S. Hydrogen sulfide capture and removal technologies: a comprehensive review of recent developments and emerging trends. Separation and Purification Technology 2022;298:121448. https://doi.org/10.1016/j.seppur.2022.121448.

[9] Wang L, Yang RT. New nanostructured sorbents for desulfurization of natural gas. Frontiers of Chemical Science and Engineering 2014;8:8−19. https://doi.org/10.1007/s11705-014-1411-4.

[10] Ahmad W, Sethupathi S, Kanadasan G, Lau LC, Kanthasamy R. A review on the removal of hydrogen sulfide from biogas by adsorption using sorbents derived from waste. Reviews in Chemical Engineering 2021;37:407−31. https://doi.org/10.1515/revce-2018-0048.

[11] de Oliveira LH, Meneguin JG, Pereira MV, da Silva EA, Grava WM, do Nascimento JF, et al. H_2S adsorption on NaY zeolite. Microporous and Mesoporous Materials 2019;284:247−57. https://doi.org/10.1016/j.micromeso.2019.04.014.

[12] Liu T, First EL, Faruque Hasan MM, Floudas CA. A multi-scale approach for the discovery of zeolites for hydrogen sulfide removal. Computers & Chemical Engineering 2016;91:206−18. https://doi.org/10.1016/j.compchemeng.2016.03.015.

[13] Demir H, Keskin S. Computational insights into efficient CO_2 and H_2S capture through zirconium MOFs. Journal of CO2 Utilization 2022;55:101811. https://doi.org/10.1016/j.jcou.2021.101811.

[14] Berg F, Pasel C, Eckardt T, Bathen D. Temperature swing adsorption in natural gas processing: a concise overview. ChemBioEng Reviews 2019;6. https://doi.org/10.1002/cben.201900005.

[15] Alonso-Vicario A, Ochoa-Gómez JR, Gil-Río S, Gómez-Jiménez-Aberasturi O, Ramírez-López CA, Torrecilla-Soria J, et al. Purification and upgrading of biogas by pressure swing adsorption on synthetic and natural zeolites. Microporous and Mesoporous Materials 2010;134:100−7. https://doi.org/10.1016/j.micromeso.2010.05.014.

[16] Melo DMA, De Souza JR, Melo MAF, Martinelli AE, Cachima GHB, Cunha JD. Evaluation of the zinox and zeolite materials as adsorbents to remove H_2S from natural gas. Colloids and Surfaces A: Physicochemical and Engineering Aspects 2006;272:32−6. https://doi.org/10.1016/j.colsurfa.2005.07.005.

[17] Nabipoor Hassankiadeh M, Hallajisani A. Application of Molybdenum oxide nanoparticles in H_2S removal from natural gas under different operational and geometrical conditions. Journal of Petroleum Science & Engineering 2020;190. https://doi.org/10.1016/j.petrol.2020.107131.

[18] Joshi JN, Zhu G, Lee JJ, Carter EA, Jones CW, Lively RP, et al. Probing metal-organic framework design for adsorptive natural gas purification. Langmuir 2018;34:8443−50. https://doi.org/10.1021/acs.langmuir.8b00889.

[19] Daneshyar A, Ghaedi M, Sabzehmeidani MM, Daneshyar A. H_2S adsorption onto Cu-Zn−Ni nanoparticles loaded activated carbon and Ni-Co nanoparticles loaded γ-Al_2O_3: optimization and adsorption isotherms. Journal of Colloid and Interface Science 2017;490:553−61. https://doi.org/10.1016/j.jcis.2016.11.068.

[20] Belmabkhout Y, De Weireld G, Sayari A. Amine-bearing mesoporous silica for CO_2 and H_2S removal from natural gas and biogas. Langmuir 2009;25: 13275–8. https://doi.org/10.1021/la903238y.

[21] Hao L, Hurlock MJ, Ding G, Zhang Q. Metal-organic frameworks towards desulfurization of fuels. Topics in Current Chemistry 2020;378:17. https://doi.org/10.1007/s41061-020-0280-1.

[22] Khabazipour M, Anbia M. Removal of hydrogen sulfide from gas streams using porous materials: a review. Industrial & Engineering Chemistry Research 2019;58:22133–64. https://doi.org/10.1021/acs.iecr.9b03800.

[23] Bhadra BN, Jhung SH. Oxidative desulfurization and denitrogenation of fuels using metal-organic framework-based/-derived catalysts. Applied Catalysis B: Environmental 2019;259:118021. https://doi.org/10.1016/j.apcatb.2019.118021.

[24] Reynolds JE, Bohnsack AM, Kristek DJ, Gutiérrez-Alejandre A, Dunning SG, Waggoner NW, et al. Phosphonium zwitterions for lighter and chemically-robust MOFs: highly reversible H_2S capture and solvent-triggered release. Journal of Materials Chemistry A 2019;7:16842–9. https://doi.org/10.1039/c9ta05444h.

[25] Sánchez-González E, Mileo PGM, Sagastuy-Breña M, Álvarez JR, Reynolds JE, Villarreal A, et al. Highly reversible sorption of H_2S and CO_2 by an environmentally friendly Mg-based MOF. Journal of Materials Chemistry A 2018;6:16900–9. https://doi.org/10.1039/c8ta05400b.

[26] Flores JG, Zárate-Colín JA, Sánchez-González E, Valenzuela JR, Gutiérrez-Alejandre A, Ramírez J, et al. Partially reversible H_2S adsorption by MFM-300(Sc): formation of polysulfides. ACS Applied Materials & Interfaces 2020;12:18885–92. https://doi.org/10.1021/acsami.0c02340.

[27] Hamon L, Serre C, Devic T, Loiseau T, Millange F, Férey G, et al. Comparative study of hydrogen sulfide adsorption in the MIL-53(Al, Cr, Fe), MIL-47(V), MIL-100(Cr), and MIL-101(Cr) metal-organic frameworks at room temperature. Journal of the American Chemical Society 2009;131: 8775–7. https://doi.org/10.1021/ja901587t.

[28] Heymans N, Vaesen S, De Weireld G. A complete procedure for acidic gas separation by adsorption on MIL-53 (Al). Microporous and Mesoporous Materials 2012;154:93–9. https://doi.org/10.1016/j.micromeso.2011.10.020.

[29] Hamon L, Leclerc H, Ghoufi A, Oliviero L, Travert A, Lavalley J-C, et al. Molecular insight into the adsorption of H_2S in the flexible MIL-53(Cr) and rigid MIL-47(V) MOFs: infrared spectroscopy combined to molecular simulations. Journal of Physical Chemistry C 2011;115:2047–56. https://doi.org/10.1021/jp1092724.

[30] Bhatt PM, Belmabkhout Y, Assen AH, Weseliński ŁJ, Jiang H, Cadiau A, et al. Isoreticular rare earth fcu-MOFs for the selective removal of H_2S from CO_2 containing gases. Chemical Engineering Journal 2017;324: 392–6. https://doi.org/10.1016/j.cej.2017.05.008.

[31] Zhou F, Zheng B, Liu D, Wang Z, Yang Q. Large-scale structural refinement and screening of zirconium metal-organic frameworks for H_2S/CH_4 separation. ACS Applied Materials & Interfaces 2019;11: 46984–92. https://doi.org/10.1021/acsami.9b17885.

[32] Daraee M, Ghasemy E, Rashidi A. Synthesis of novel and engineered UiO-66/graphene oxide nanocomposite with enhanced H_2S adsorption capacity. Journal of Environmental Chemical Engineering 2020;8. https://doi.org/10.1016/j.jece.2020.104351.

[33] Pokhrel J, Bhoria N, Wu C, Reddy KSK, Margetis H, Anastasiou S, et al. Cu- and Zr-based metal organic frameworks and their composites with

graphene oxide for capture of acid gases at ambient temperature. Journal of Solid State Chemistry 2018;266:233–43. https://doi.org/10.1016/j.jssc.2018.07.022.

[34] Huang Y, Wang R. Highly selective separation of H_2S and CO_2 using a H_2S-imprinted polymers loaded on a polyoxometalate@Zr-based metal-organic framework with a core-shell structure at ambient temperature. Journal of Materials Chemistry A 2019;7:12105–14. https://doi.org/10.1039/c9ta01749f.

[35] Li Y, Wang LJ, Fan HL, Shangguan J, Wang H, Mi J. Removal of sulfur compounds by a copper-based metal organic framework under ambient conditions. Energy and Fuels 2015;29:298–304. https://doi.org/10.1021/ef501918f.

[36] Zhang HY, Zhang ZR, Yang C, Ling LX, Wang BJ, Fan HL. A Computational study of the adsorptive removal of H_2S by MOF-199. Journal of Inorganic and Organometallic Polymers and Materials 2018;28:694–701. https://doi.org/10.1007/s10904-017-0740-4.

[37] Wang XL, Fan HL, Tian Z, He EY, Li Y, Shangguan J. Adsorptive removal of sulfur compounds using IRMOF-3 at ambient temperature. Applied Surface Science 2014;289:107–13. https://doi.org/10.1016/j.apsusc.2013.10.115.

[38] Ishak MAI, Jumbri K, Daud S, Abdul Rahman MB, Abdul Wahab R, Yamagishi H, et al. Molecular simulation on the stability and adsorption properties of choline-based ionic liquids/IRMOF-1 hybrid composite for selective H_2S/CO_2 capture. Journal of Hazardous Materials 2020;399. https://doi.org/10.1016/j.jhazmat.2020.123008.

[39] Liu X, Wang B, Cheng J, Meng Q, Song Y, Li M. Investigation on the capture performance and influencing factors of ZIF-67 for hydrogen sulfide. Separation and Purification Technology 2020;250:117300. https://doi.org/10.1016/j.seppur.2020.117300.

[40] Jameh AA, Mohammadi T, Bakhtiari O, Mahdyarfar M. Synthesis and modification of Zeolitic Imidazolate Framework (ZIF-8) nanoparticles as highly efficient adsorbent for H_2S and CO_2 removal from natural gas. Journal of Environmental Chemical Engineering 2019;7. https://doi.org/10.1016/j.jece.2019.103058.

[41] Belmabkhout Y, Bhatt PM, Adil K, Pillai RS, Cadiau A, Shkurenko A, et al. Natural gas upgrading using a fluorinated MOF with tuned H_2S and CO_2 adsorption selectivity. Nature Energy 2018;3:1059–66. https://doi.org/10.1038/s41560-018-0267-0.

[42] Janiak C, Vieth JK. MOFs, MILs and more: concepts, properties and applications for porous coordination networks (PCNs). New Journal of Chemistry 2010;34:2366. https://doi.org/10.1039/c0nj00275e.

[43] Allan PK, Wheatley PS, Aldous D, Mohideen MI, Tang C, Hriljac JA, et al. Metal-organic frameworks for the storage and delivery of biologically active hydrogen sulfide. Dalton Transactions 2012;41:4060–6. https://doi.org/10.1039/c2dt12069k.

[44] Zhang HY, Yang C, Geng Q, Fan HL, Wang BJ, Wu MM, et al. Adsorption of hydrogen sulfide by amine-functionalized metal organic framework (MOF-199): an experimental and simulation study. Applied Surface Science 2019;497:143815. https://doi.org/10.1016/j.apsusc.2019.143815.

[45] Huang ZH, Liu G, Kang F. Glucose-promoted Zn-based metal-organic framework/graphene oxide composites for hydrogen sulfide removal. ACS Applied Materials & Interfaces 2012;4:4942–7. https://doi.org/10.1021/am3013104.

[46] Petit C, Levasseur B, Mendoza B, Bandosz TJ. Reactive adsorption of acidic gases on MOF/graphite oxide composites. Microporous and Mesoporous Materials 2012;154:107–12. https://doi.org/10.1016/j.micromeso.2011.09.012.

[47] Gupta NK, Bae J, Kim S, Kim KS. Fabrication of Zn-MOF/ZnO nanocomposites for room temperature H_2S removal: adsorption, regeneration, and mechanism. Chemosphere 2021;274. https://doi.org/10.1016/j.chemosphere.2021.129789.

[48] Saeedirad R, Taghvaei Ganjali S, Bazmi M, Rashidi A. Effective mesoporous silica-ZIF-8 nano-adsorbents for adsorptive desulfurization of gas stream. Journal of the Taiwan Institute of Chemical Engineers 2018;82:10–22. https://doi.org/10.1016/j.jtice.2017.10.037.

[49] Antonio Zárate J, Sánchez-González E, Jurado-Vázquez T, Gutiérrez-Alejandre A, González-Zamora E, Castillo I, et al. Outstanding reversible H_2S capture by an Al(iii)-based MOF. Chemical Communications 2019;55:3049–52. https://doi.org/10.1039/c8cc09379b.

[50] Ploymeerusmee T, Janke W, Remsungnen T, Hannongbua S, Chokbunpiam T. Porous material adsorbents ZIF-8, ZIF-67, Co/Zn-ZIF and MIL-127(Fe) for separation of H_2S from an H_2S/CH_4 mixture. Molecular Simulation 2022;48:417–26. https://doi.org/10.1080/08927022.2021.2025232.

[51] Chanajaree R, Sailuam W, Seehamart K. Molecular self-diffusivity and separation of CH_4/H_2S in metal organic framework MIL-47(V). Microporous and Mesoporous Materials 2022;335:111783. https://doi.org/10.1016/j.micromeso.2022.111783.

[52] Ethiraj J, Bonino F, Lamberti C, Bordiga S. H_2S interaction with HKUST-1 and ZIF-8 MOFs: a multitechnique study. Microporous and Mesoporous Materials 2015;207:90–4. https://doi.org/10.1016/j.micromeso.2014.12.034.

[53] Li Z, Xiao Y, Xue W, Yang Q, Zhong C. Ionic liquid/metal–organic framework composites for H_2S removal from natural gas: a computational exploration. Journal of Physical Chemistry C 2015;119:3674–83. https://doi.org/10.1021/acs.jpcc.5b00019.

[54] Bhoria N, Basina G, Pokhrel J, Kumar Reddy KS, Anastasiou S, Balasubramanian VV, et al. Functionalization effects on HKUST-1 and HKUST-1/graphene oxide hybrid adsorbents for hydrogen sulfide removal. Journal of Hazardous Materials 2020;394. https://doi.org/10.1016/j.jhazmat.2020.122565.

[55] Ebrahim AM, Jagiello J, Bandosz TJ. Enhanced reactive adsorption of H_2S on Cu-BTC/S- and N-doped GO composites. Journal of Materials Chemistry A 2015;3:8194–204. https://doi.org/10.1039/c5ta01359c.

[56] Chavan S, Bonino F, Valenzano L, Civalleri B, Lamberti C, Acerbi N, et al. Fundamental aspects of H_2S adsorption on CPO-27-Ni. Journal of Physical Chemistry C 2013;117:15615–22. https://doi.org/10.1021/jp402440u.

[57] Fan HL, Shi RH, Zhang ZR, Zhen T, Shangguan J, Mi J. Cu-based metal–organic framework/activated carbon composites for sulfur compounds removal. Applied Surface Science 2017;394:394–402. https://doi.org/10.1016/j.apsusc.2016.10.071.

[58] Liu J, Wei Y, Li P, Zhao Y, Zou R. Selective H_2S/CO_2 separation by metal-organic frameworks based on chemical-physical adsorption. Journal of Physical Chemistry C 2017;121:13249–55. https://doi.org/10.1021/acs.jpcc.7b04465.

[59] Florent M, Bandosz TJ. Effects of surface heterogeneity of cobalt oxyhydroxide/graphite oxide composites on reactive adsorption of hydrogen sulfide. Microporous and Mesoporous Materials 2015;204:8−14. https://doi.org/10.1016/j.micromeso.2014.11.001.

[60] Wang J, Yang C, Zhao YR, Fan HL, Wang Z De, Shangguan J, et al. Synthesis of porous cobalt oxide and its performance for H_2S removal at room temperature. Industrial & Engineering Chemistry 2017;56:12621−9. https://doi.org/10.1021/acs.iecr.7b02934.

[61] Pahalagedara LR, Poyraz AS, Song W, Kuo CH, Pahalagedara MN, Meng YT, et al. Low temperature desulfurization of H_2S: high sorption capacities by mesoporous cobalt oxide via increased H_2S diffusion. Chemistry of Materials 2014;26:6613−21. https://doi.org/10.1021/cm503405a.

[62] Xue M, Chitrakar R, Sakane K, Ooi K. Screening of adsorbents for removal of H_2S at room temperature. Green Chemistry 2003;5:529−34. https://doi.org/10.1039/b303167p.

[63] Lee S, Lee T, Kim D. Adsorption of hydrogen sulfide from gas streams using the amorphous composite of α-FeOOH and activated carbon powder. Industrial & Engineering Chemistry 2017;56:3116−22. https://doi.org/10.1021/acs.iecr.6b04747.

[64] Cao Y, Zheng X, Du Z, Shen L, Zheng Y, Au C, et al. Low-Temperature H_2S removal from gas streams over γ-FeOOH, γ-Fe_2O_3, and α-Fe_2O_3: effects of the hydroxyl group, defect, and specific surface area. Industrial & Engineering Chemistry 2019;58:19353−60. https://doi.org/10.1021/acs.iecr.9b03430.

[65] Polychronopoulou K, Efstathiou AM. Effects of sol–gel synthesis on 5Fe–15Mn–40Zn–40Ti–O mixed oxide structure and its H_2S removal efficiency from industrial gas streams. Environmental Science & Technology 2009;43:4367−72. https://doi.org/10.1021/es803631h.

[66] Dhage P, Samokhvalov A, Repala D, Duin EC, Tatarchuk BJ. Regenerable Fe-Mn-ZnO/SiO_2 sorbents for room temperature removal of H_2S from fuel reformates: performance, active sites, Operando studies. Physical Chemistry Chemical Physics 2011;13:2179−87. https://doi.org/10.1039/c0cp01355b.

[67] Liu D, Chen S, Fei X, Huang C, Zhang Y. Regenerable CuO-based adsorbents for low temperature desulfurization application. Industrial & Engineering Chemistry 2015;54:3556−62. https://doi.org/10.1021/acs.iecr.5b00180.

[68] Liu G, Huang ZH, Kang F. Preparation of ZnO/SiO_2 gel composites and their performance of H_2S removal at room temperature. Journal of Hazardous Materials 2012;215−216:166−72. https://doi.org/10.1016/j.jhazmat.2012.02.050.

[69] Su Y-M, Huang C-Y, Chyou Y-P, Svoboda K. Sulfidation/regeneration multi-cyclic testing of Fe_2O_3/Al_2O_3 sorbents for the high-temperature removal of hydrogen sulfide. Journal of the Taiwan Institute of Chemical Engineers 2017;74:89−95. https://doi.org/10.1016/j.jtice.2016.12.011.

[70] Kim S, Gupta NK, Bae J, Kim KS. Fabrication of coral-like Mn_2O_3/Fe_2O_3 nanocomposite for room temperature removal of hydrogen sulfide. Journal of Environmental Chemical Engineering 2021;9:105216. https://doi.org/10.1016/j.jece.2021.105216.

[71] Dhage P, Samokhvalov A, Repala D, Duin EC, Bowman M, Tatarchuk BJ. Copper-promoted ZnO/SiO_2 regenerable sorbents for the room temperature removal of H_2S from reformate gas streams. Industrial &

Engineering Chemistry 2010;49:8388—96. https://doi.org/10.1021/ie100209a.

[72] Yang C, Kou J, Fan H, Tian Z, Kong W, Shangguan J. Facile and versatile sol—gel strategy for the preparation of a high-loaded ZnO/SiO$_2$ adsorbent for room-temperature H$_2$S removal. Langmuir 2019;35:7759—68. https://doi.org/10.1021/acs.langmuir.9b00853.

[73] Mureddu M, Ferino I, Musinu A, Ardu A, Rombi E, Cutrufello MG, et al. MeOx/SBA-15 (Me= Zn, Fe): highly efficient nanosorbents for mid-temperature H$_2$S removal. Journal of Materials Chemistry A 2014;2:19396—406. https://doi.org/10.1039/c4ta03540b.

[74] Cara C, Rombi E, Musinu A, Mameli V, Ardu A, Sanna Angotzi M, et al. MCM-41 support for ultrasmall γ-Fe$_2$O$_3$ nanoparticles for H$_2$S removal. Journal of Materials Chemistry A 2017;5:21688—98. https://doi.org/10.1039/C7TA03652C.

[75] Montes D, Tocuyo E, González E, Rodríguez D, Solano R, Atencio R, et al. Reactive H$_2$S chemisorption on mesoporous silica molecular sieve-supported CuO or ZnO. Microporous and Mesoporous Materials 2013;168:111—20. https://doi.org/10.1016/j.micromeso.2012.09.018.

[76] Guo Z, Zhang Z, Cao X, Feng D. Fe—Ti bimetal oxide adsorbent for removing low concentration H$_2$S at room temperature. Environmental Technology 2021;43(24):1—13. https://doi.org/10.1080/09593330.2021.1931472.

[77] Yang C, Wang J, Fan HL, Shangguan J, Mi J, Huo C. Contributions of tailored oxygen vacancies in ZnO/Al$_2$O$_3$ composites to the enhanced ability for H$_2$S removal at room temperature. Fuel 2018;215:695—703. https://doi.org/10.1016/j.fuel.2017.11.037.

[78] Daraee M, Ghasemy E, Rashidi A. Effective adsorption of hydrogen sulfide by intercalation of TiO$_2$ and N-doped TiO$_2$ in graphene oxide. Journal of Environmental Chemical Engineering 2020;8:103836. https://doi.org/10.1016/j.jece.2020.103836.

[79] Yang C, Wang Y, Fan H, de Falco G, Yang S, Shangguan J, et al. Bifunctional ZnO-MgO/activated carbon adsorbents boost H$_2$S room temperature adsorption and catalytic oxidation. Applied Catalysis B: Environmental 2020;266:118674. https://doi.org/10.1016/j.apcatb.2020.118674.

[80] Balsamo M, Cimino S, de Falco G, Erto A, Lisi L. ZnO-CuO supported on activated carbon for H$_2$S removal at room temperature. Chemical Engineering Journal 2016;304:399—407. https://doi.org/10.1016/j.cej.2016.06.085.

[81] Yang C, Florent M, de Falco G, Fan H, Bandosz TJ. ZnFe$_2$O$_4$/activated carbon as a regenerable adsorbent for catalytic removal of H$_2$S from air at room temperature. Chemical Engineering Journal 2020;394:124906. https://doi.org/10.1016/j.cej.2020.124906.

[82] Orojlou SH, Zargar B, Rastegarzadeh S. Metal oxide/TiO$_2$ nanocomposites as efficient adsorbents for relatively high temperature H$_2$S removal. Journal of Natural Gas Science and Engineering 2018;59:363—73. https://doi.org/10.1016/j.jngse.2018.09.016.

[83] Karvan O, Sirkecioğlu A, Atakül H. Investigation of nano-CuO/mesoporous SiO$_2$ materials as hot gas desulphurization sorbents. Fuel Processing Technology 2009;90:1452—8. https://doi.org/10.1016/j.fuproc.2009.06.027.

[84] Wang LJ, Fan HL, Shangguan J, Croiset E, Chen Z, Wang H, et al. Design of a sorbent to enhance reactive adsorption of hydrogen sulfide. ACS

Applied Materials & Interfaces 2014;6:21167−77. https://doi.org/10.1021/am506077j.
[85] Huang G, He E, Wang Z, Fan H, Shangguan J, Croiset E, et al. Synthesis and characterization of γ-Fe$_2$O$_3$ for H$_2$S removal at low temperature. Industrial & Engineering Chemistry 2015;54:8469−78. https://doi.org/10.1021/acs.iecr.5b01398.
[86] Jiang D, Su L, Ma L, Yao N, Xu X, Tang H, et al. Cu−Zn−Al mixed metal oxides derived from hydroxycarbonate precursors for H$_2$S removal at low temperature. Applied Surface Science 2010;256:3216−23. https://doi.org/10.1016/j.apsusc.2009.12.008.
[87] Yu T, Chen Z, Wang Y, Xu J. Synthesis of ZnO-CuO and ZnO-Co$_3$O$_4$ materials with three-dimensionally ordered macroporous structure and its H$_2$S removal performance at low-temperature. Processes 2021;9. https://doi.org/10.3390/pr9111925.
[88] Song HS, Park MG, Kwon SJ, Yi KB, Croiset E, Chen Z, et al. Hydrogen sulfide adsorption on nano-sized zinc oxide/reduced graphite oxide composite at ambient condition. Applied Surface Science 2013;276:646−52. https://doi.org/10.1016/j.apsusc.2013.03.147.
[89] Arcibar-Orozco JA, Wallace R, Mitchell JK, Bandosz TJ. Role of surface chemistry and morphology in the reactive adsorption of H$_2$S on Iron (Hydr)oxide/graphite oxide composites. Langmuir 2015;31:2730−42. https://doi.org/10.1021/la504563z.
[90] Florent M, Wallace R, Bandosz TJ. Removal of hydrogen sulfide at ambient conditions on cadmium/GO-based composite adsorbents. Journal of Colloid and Interface Science 2015;448:573−81. https://doi.org/10.1016/j.jcis.2015.02.021.
[91] Kim S, Bajaj B, Byun CK, Kwon S-J, Joh H-I, Yi KB, et al. Preparation of flexible zinc oxide/carbon nanofiber webs for mid-temperature desulfurization. Applied Surface Science 2014;320:218−24. https://doi.org/10.1016/j.apsusc.2014.09.093.
[92] Zhang H, Wang J, Liu T, Zhang M, Hao L, Phoutthavong T, et al. Cu-Zn oxides nanoparticles supported on SBA-15 zeolite as a novel adsorbent for simultaneous removal of H$_2$S and Hg0 in natural gas. Chemical Engineering Journal 2021;426:131286. https://doi.org/10.1016/j.cej.2021.131286.
[93] Mureddu M, Ferino I, Rombi E, Cutrufello MG, Deiana P, Ardu A, et al. ZnO/SBA-15 composites for mid-temperature removal of H$_2$S: synthesis, performance and regeneration studies. Fuel 2012;102:691−700. https://doi.org/10.1016/j.fuel.2012.05.013.
[94] Geng Q, Wang LJ, Yang C, Zhang HY, Zhao YR, Fan HL, et al. Room-temperature hydrogen sulfide removal with zinc oxide nanoparticle/molecular sieve prepared by melt infiltration. Fuel Processing Technology 2019;185:26−37. https://doi.org/10.1016/j.fuproc.2018.11.013.
[95] Li L, Sun TH, Shu CH, Zhang HB. Low temperature H$_2$S removal with 3-D structural mesoporous molecular sieves supported ZnO from gas stream. Journal of Hazardous Materials 2016;311:142−50. https://doi.org/10.1016/j.jhazmat.2016.01.033.
[96] Liu D, Zhou W, Wu J. CeO$_2$-MnOx/ZSM-5 sorbents for H$_2$S removal at high temperature. Chemical Engineering Journal 2016;284:862−71. https://doi.org/10.1016/j.cej.2015.09.028.
[97] Yang C, Yang S, Fan HL, Wang J, Wang H, Shangguan J, et al. A sustainable design of ZnO-based adsorbent for robust H$_2$S uptake and

secondary utilization as hydrogenation catalyst. Chemical Engineering Journal 2020;382. https://doi.org/10.1016/j.cej.2019.122892.

[98] Chircov C, Spoială A, Păun C, Crăciun L, Ficai D, Ficai A, et al. Mesoporous silica platforms with potential applications in release and adsorption of active agents. Molecules 2020;25:3814. https://doi.org/10.3390/molecules25173814.

[99] Okonkwo CN, Okolie C, Sujan A, Zhu G, Jones CW. Role of amine structure on hydrogen sulfide capture from dilute gas streams using solid adsorbents. Energy & Fuels 2018;32:6926–33. https://doi.org/10.1021/acs.energyfuels.8b00936.

[100] Zhang J, Song H, Chen Y, Hao T, Li F, Yuan D, et al. Study on the preparation of amine-modified silicate MCM-41 adsorbent and its H_2S removal performance. Progress in Reaction Kinetics and Mechanism 2019;45. https://doi.org/10.1177/1468678319825900.

[101] Xue Q, Liu Y. Removal of minor concentration of H_2S on mdea-modified SBA-15 for gas purification. Journal of Industrial and Engineering Chemistry 2012;18:169–73. https://doi.org/10.1016/j.jiec.2011.11.005.

[102] Wang X, Ma X, Xu X, Sun L, Song C. Mesoporous-molecular-sieve-supported polymer sorbents for removing H_2S from hydrogen gas streams. Topics in Catalysis 2008;49:108–17. https://doi.org/10.1007/s11244-008-9072-5.

[103] Jaiboon V, Yoosuk B, Prasassarakich P. Amine modified silica xerogel for H_2S removal at low temperature. Fuel Processing Technology 2014;128:276–82. https://doi.org/10.1016/j.fuproc.2014.07.032.

[104] Quan W, Wang X, Song C. Selective removal of H_2S from biogas using solid amine-based "molecular Basket" sorbent. Energy and Fuels 2017;31:9517–28. https://doi.org/10.1021/acs.energyfuels.7b01473.

[105] Abdouss M, Hazrati N, Miran Beigi AA, Vahid A, Mohammadalizadeh A. Effect of the structure of the support and the aminosilane type on the adsorption of H_2S from model gas. RSC Advances 2014;4:6337–45. https://doi.org/10.1039/c3ra43181a.

[106] Chu X, Cheng Z, Zhao Y, Xu J, Zhong H, Zhang W, et al. Study on sorption behaviors of H_2S by triethanolamine-modified mesoporous molecular sieve SBA-15. Industrial & Engineering Chemistry 2012;51:4407–13. https://doi.org/10.1021/ie202360h.

[107] Anbia M, Babaei M. Novel amine modified nanoporous SBA-15 sorbent for the removal of H_2S from gas streams in the presence of CH_4. International Journal of Engineering 2014;27(11):1697–704. https://doi.org/10.5829/idosi.ije.2014.27.11b.07.

[108] Belmabkhout Y, Heymans N, De Weireld G, Sayari A. Simultaneous adsorption of H_2S and CO_2 on triamine-grafted pore-expanded mesoporous MCM-41 silica. Energy & Fuels 2011;25:1310–5. https://doi.org/10.1021/ef1015704.

[109] Hazrati N, Beigi AAM, Abdouss M, Vahid A. One-step synthesis of zinc-encapsulated MCM-41 as H_2S adsorbent and optimization of adsorption parameters. Analytical Methods in Environmental Chemistry Journal 2020;3:74–81. https://doi.org/10.24200/amecj.v3.i02.104.

[110] Wang Y, Yang RT. Template removal from SBA-15 by ionic liquid for amine grafting: applications to CO_2 capture and natural gas desulfurization. ACS Sustainable Chemistry & Engineering 2020;8:8295–304. https://doi.org/10.1021/acssuschemeng.0c01941.

[111] Crespo D, Qi G, Wang Y, Yang FH, Yang RT. Superior sorbent for natural gas desulfurization. Industrial & Engineering Chemistry 2008;47:1238−44. https://doi.org/10.1021/ie071145i.

[112] Koohsaryan E, Anbia M. Nanosized and hierarchical zeolites: a short review. Chinese Journal of Catalysis 2016;37:447−67. https://doi.org/10.1016/S1872-2067(15)61038-5.

[113] Kristóf T. Selective removal of hydrogen sulphide from industrial gas mixtures using zeolite NaA. Hungarian Journal of Industry and Chemistry 2018;45:9−15. https://doi.org/10.1515/hjic-2017-0003.

[114] Shah MS, Tsapatsis M, Siepmann JI. Identifying optimal zeolitic sorbents for sweetening of highly sour natural gas. Angewandte Chemie International Edition 2016;55:5938−42. https://doi.org/10.1002/anie.201600612.

[115] Iravani H, Jafari MJ, Zendehdel R, Khodakarim S, Rafieepour A. Removing H_2S gas from the air stream using zeolite ZSM-5 substrate impregnated with magnetite and ferric nanoparticles. Journal of Health & Safety at Work 2020;10:9−13.

[116] Yazdanbakhsh F, Bläsing M, Sawada JA, Rezaei S, Müller M, Baumann S, et al. Copper exchanged nanotitanate for high temperature H_2S adsorption. Industrial & Engineering Chemistry 2014;53:11734−9. https://doi.org/10.1021/ie501029u.

[117] Yang K, Su B, Shi L, Wang H, Cui Q. Adsorption mechanism and regeneration performance of 13X for H_2S and SO_2. Energy and Fuels 2018;32:12742−9. https://doi.org/10.1021/acs.energyfuels.8b02978.

[118] Barelli L, Bidini G, Micoli L, Sisani E, Turco M. 13X Ex-Cu zeolite performance characterization towards H_2S removal for biogas use in molten carbonate fuel cells. Energy 2018;160:44−53. https://doi.org/10.1016/j.energy.2018.05.057.

[119] Rong C, Chu D, Hopkins J. Test and characterization of some zeolite supported gas phase desulfurization sorbents. Adelphi, MD: Army Research Laboratory; 2009.

[120] Abdullah AH, Mat R, Somderam S, Abd Aziz AS, Mohamed A. Hydrogen sulfide adsorption by zinc oxide-impregnated zeolite (synthesized from Malaysian kaolin) for biogas desulfurization. Journal of Industrial and Engineering Chemistry 2018;65:334−42. https://doi.org/10.1016/j.jiec.2018.05.003.

[121] Liu C, Zhang R, Wei S, Wang J, Liu Y, Li M, et al. Selective removal of H_2S from biogas using a regenerable hybrid TiO_2/zeolite composite. Fuel 2015;157:183−90. https://doi.org/10.1016/j.fuel.2015.05.003.

[122] Rezaei S, Tavana A, Sawada JA, Wu L, Junaid ASM, Kuznicki SM. Novel copper-exchanged titanosilicate adsorbent for low temperature H_2S removal. Industrial & Engineering Chemistry 2012;51:12430−4. https://doi.org/10.1021/ie300244y.

[123] Yasyerli S, Ar İ, Doğu G, Doğu T. Removal of hydrogen sulfide by clinoptilolite in a fixed bed adsorber. Chemical Engineering and Processing: Process Intensification 2002;41:785−92. https://doi.org/10.1016/S0255-2701(02)00009-0.

[124] Liu X, Wang R. Effective removal of hydrogen sulfide using 4A molecular sieve zeolite synthesized from attapulgite. Journal of Hazardous Materials 2017;326:157−64. https://doi.org/10.1016/j.jhazmat.2016.12.030.

[125] Georgiadis AG, Charisiou ND, Gaber S, Polychronopoulou K, Yentekakis IV, Goula MA. Adsorption of hydrogen sulfide at low temperatures using an industrial molecular sieve: an experimental and

theoretical study. ACS Omega 2021;6:14774−87. https://doi.org/10.1021/acsomega.0c06157.

[126] Wynnyk KG, Hojjati B, Marriott RA. Sour gas and water adsorption on common high-pressure desiccant materials: zeolite 3A, zeolite 4A, and silica gel. Journal of Chemical & Engineering Data 2019;64:3156−63. https://doi.org/10.1021/acs.jced.9b00233.

[127] Starke A, Pasel C, Bläker C, Eckardt T, Zimmermann J, Bathen D. Impact of Na+ and Ca^{2+} cations on the adsorption of H_2S on binder-free LTA zeolites. Adsorption Science and Technology 2021;2021:1−12. https://doi.org/10.1155/2021/5531974.

[128] Rezaei S, Jarligo MOD, Wu L, Kuznicki SM. Breakthrough performances of metal-exchanged nanotitanate ETS-2 adsorbents for room temperature desulfurization. Chemical Engineering and Science 2015;123:444−9. https://doi.org/10.1016/j.ces.2014.11.041.

[129] Kumar P, Sung CY, Muraza O, Cococcioni M, Al Hashimi S, McCormick A, et al. H_2S adsorption by Ag and Cu ion exchanged faujasites. Microporous and Mesoporous Materials 2011;146:127−33. https://doi.org/10.1016/j.micromeso.2011.05.014.

[130] Rahmani M, Mokhtarani B, Mafi M, Rahmanian N. Acid gas removal by superhigh silica ZSM-5: adsorption isotherms of hydrogen sulfide, carbon dioxide, methane, and nitrogen. Industrial & Engineering Chemistry 2022;61(19). https://doi.org/10.1021/acs.iecr.2c00196.

[131] Fellah MF. Adsorption of hydrogen sulfide as initial step of H_2S removal: a DFT study on metal exchanged ZSM-12 clusters. Fuel Processing Technology 2016;144:191−6. https://doi.org/10.1016/j.fuproc.2016.01.003.

[132] Hernández-Maldonado AJ, Yang RT, Chinn D, Munson CL. Partially calcined gismondine type silicoaluminophosphate SAPO-43: isopropylamine elimination and separation of carbon dioxide, hydrogen sulfide, and water. Langmuir 2003;19:2193−200. https://doi.org/10.1021/la026424j.

[133] Peng X, Cao D. Computational screening of porous carbons, zeolites, and metal organic frameworks for desulfurization and decarburization of biogas, natural gas, and flue gas. AIChE Journal 2013;59:2928−42. https://doi.org/10.1002/aic.14046.

14

Natural gas CO$_2$-rich sweetening via adsorption processes

Syed Ali Ammar Taqvi[1], Durreshehwar Zaeem[1] and Haslinda Zabiri[2,3]

[1]*Department of Chemical Engineering, NED University of Engineering and Technology, Karachi, Pakistan;* [2]*Department of Chemical Engineering, Universiti Teknologi PETRONAS, Bandar Seri Iskandar, Perak, Malaysia;* [3]*CO$_2$RES, Institute of Contaminant Management (ICM), Universiti Teknologi PETRONAS, Bandar Seri Iskandar, Perak, Malaysia*

1. Introduction

Due to the increase in CO$_2$ levels in the atmosphere, CO$_2$ capture systems have recently piqued researcher interest. According to the Global Greenhouse Gas Emissions (GHG) statistics, the combustion of fossil fuels is responsible for 78% of GHG emissions [1]. Ocean levels are rising, hurricanes are becoming more violent, and droughts are becoming more frequent due to global warming. Despite methane and chlorofluorocarbons in GHGs, the primary focus is on reducing CO$_2$ emissions. According to the Intragovernmental Panel on Climate Change (IPCC), the earth's atmosphere might contain 570 ppm CO$_2$ by 2100, generating a 1.9°C rise in global temperature and a 3.8 m rise in mean sea level [1]. International goals have been made to reduce global warming. After the Paris Agreement on setting the maximum limit on the rise of global average temperature as 2°C, research on CO$_2$ capture techniques has paced. To achieve this goal, numerous countries, cities, regions, and huge companies are trying to get the least carbon emissions [2].

CO$_2$ is removed from natural gas during the precombustion process to lower GHG emissions and to increase its calorific value. Because the calorific value of natural gas is determined by its methane content, CO$_2$ is removed from natural gas to get a high methane concentration per unit mass of natural gas [3]. Apart from its calorific value, CO_2-rich natural gas is costly to transport; also, higher carbon content can cause pipe damage such as line choking, corrosion, and other issues. Although most natural gas reserves do not have a large composition of CO_2 and other impurities such as H_2S, a sub-quality standard for natural gas is defined as one having $CO_2 \geq 2\%$, $N_2 \geq 4\%$, or $H_2S \geq 4$ ppmv [4].

2. Adsorption processes in natural gas sweetening

2.1 Process description

A process in which a substance moves out of the gas phase and gets attached to a solid phase (adsorbent) as shown in Fig. 14.1, due to intermolecular forces between a solid surface and gas is known as adsorption [5]. Adsorption is accomplished through adsorbate-surface interactions, exclusion of molecules due to size differences, and diffusion rates of molecules [6].

The regeneration of the adsorbent is done by desorption. Adsorbent regeneration is done either through temperature elevation, pressure reduction, or by the discharge of electric current via adsorbent [7]. The absorption process for CO_2 capture from natural gas suffers from solvent degradation, toxicity, and a small CO_2–solvent contact area; as a result, sequestering acid gas from natural gas through the adsorption method is regarded as an alternative [8]. Natural gas fields with high CO_2 content and high pressures that favor high adsorption capacities can be captured using the adsorption-based method [9].

Figure 14.1 Schematic diagram of an adsorption process.

2.2 Types of adsorption processes

T.E. Rufford Rufford et al. [6] extensively reviewed natural gas purification by providing details about absorption, desorption, membrane, condensation, sublimation, distillation, and hydrates. For adsorption, different adsorption−desorption techniques are discussed. Heating the adsorbent bed to a temperature higher than the adsorption temperature via heating coils or exposure to nonsoluble hot gases is used to regenerate the adsorbent in temperature swing adsorption (TSA) as shown in Fig. 14.2. Cool purge gas is utilized to cool down the adsorbent bed after desorption. However, this heating and cooling of the bed needs more extended periods. Pressure swing adsorption (PSA) is conducted at higher pressures of approximately 400−2000 kPa, followed by desorption at reduced pressures (Fig. 14.3). Since pressure elevation-reduction is a rapid process, PSA cycles require much less time than TSA cycles, resulting in the small adsorbent requirement for PSA. The author identified the economic potential of the adsorption process.

Further study was recommended to be based on the Modification of TSA- and PSA-based process configurations and the cost and performance of adsorbents [6].

Webley [10] examined various modes of adsorption in detail. The TSA process widely employs fluidized and fixed bed arrangements for adsorption. Chemisorbents with adsorption heats of

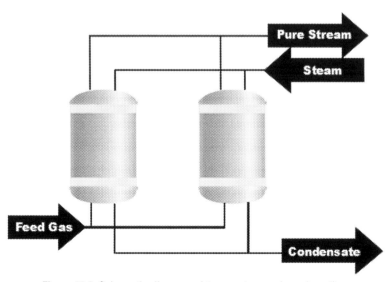

Figure 14.2 Schematic diagram of temperature swing adsorption.

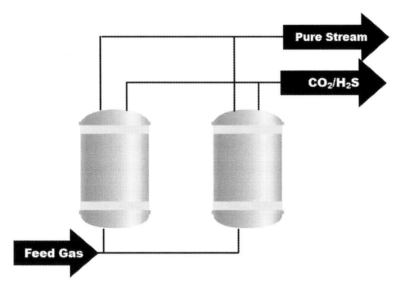

Figure 14.3 Schematic diagram of pressure swing adsorption.

40–70 kJ/mol are suitable for TSA because they are sensitive to temperatures between 30 and 150°C. Due to lengthy cycle times, the TSA process's fixed bed arrangements are less effective.

Due to its tendency to cycle quickly, pressure swing adsorption overcomes TSA's limitation's, which makes use of physical adsorbents like zeolites and activated carbons, and is not as effective as vacuum swing adsorption (VSA) as shown in Fig. 14.4. Due

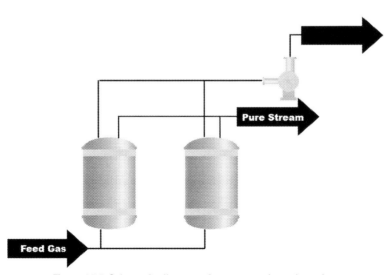

Figure 14.4 Schematic diagram of vacuum swing adsorption.

to the fact that the adsorption of moisture and impurities reduce the adsorption capacity of CO_2, the presence of impurities and water presents a challenge for VSA-based processes. Additionally, VSA processes require high-volume vacuum machinery such as parallel train vacuum blowers [10].

The commonly used beds can be exchanged with microchannels covered with adsorbents to enhance the performance of the TSA-based processes. Four-stage TSA was performed on impure methane containing CO_2 using the bed of Monolith microchannels. Usage of monolith micro walls improved the process's overall and stage performances and competed with the PSA-based cycles. Higher purities and recoveries were achieved with a greater adsorption capacity. In addition, an energy analysis was conducted on the TSA-based process using a 50 μm monolith wall. The calculated specific energy requirement, if the wall is cut in half, is 0.96 kWh/kg CO_2, which is comparable to the monoethanolamine (MEA)-based absorption process but superior to it in terms of cost, durability, and process output [11]. Mondino et al. [12] reviewed the TSA process for natural gas purification. The study went over the adsorbents that are used in TSA and came to the conclusion that more research is needed to improve and optimize TSA for industrial applications. Additionally, the problem must be resolved because the formation of carbamic acid on zeolites causes the destruction of the zeolitic structures in CO_2 adsorption. Temperature and VSA studies were reviewed extensively. Hedin et al. [13] discovered that because a typical coal-fired power plant emits 1000 tons of CO_2 stream per hour, temperature, or a VSA were deemed to be the best separation methods for the postcombustion carbon capture process. Hedin et al. [13] further studied how volume filling and surface coverage overlapped when sorbents had very small pore sizes. Adsorbents' pores can take a variety of shapes, such as tiny slits, cylinders, or cages connected by pore windows. When the size of the window is the same as the size of the molecules, equilibrium, kinetic, or molecular sieving mechanisms are used to separate gas molecules [13].

Vacuum swing adsorption and TSA having rapid cycles make physio sorbents with large pores less effective [13]. Because of the process's technological and economic feasibility, pressure swing adsorption has the edge over temperature and electrical swing adsorption. However, the liability associated with this technology is methane loss. Methane gets adsorbed on the adsorbent surface and is taken along with CO_2. Using pressure swing adsorption, extensive research has been carried out to enrich the biogas. The most important factor in adsorption capacity is the adsorption pressure. PSA also achieved a product purity of 100% by raising the adsorption pressure to 20 bars with just one bed [10]. Manocha et al. [14] found out that physio sorbents are best to use in pressure swing adsorption because of high selectivity, adsorption capacity, surface properties, and ease in regeneration. Large pore sizes and a microporous structure make zeolite-based adsorbents highly selective for CO_2. They also came to the conclusion that zeolite 13X, which has an adsorption capacity of 2.63 mmol/g, is the adsorbent that is used the most frequently. Zeolite NaX's performance in CO_2 capture by vacuum pressure swing adsorption was evaluated and found to have a high adsorption capacity of 7.04 mmol/g at 1 bar and 25°C M.P.G. introduced VPSA, or vacuum pressure swing adsorption; When the captured molecules are released at a pressure lower than atmospheric pressure, this modification speeds up regeneration. Lopes et al. [15] introduced pressure equalization using two columns. The first column deals with the feed process at high pressure, and the second column deals with purge at low pressure. This pressure equalization enhances the recovery of enriched gas. Abd et al. [16] review indicated that for an enhanced adsorption process, adsorption time must be around adsorption breakthrough time, equalization stage is vital in PSA operating with high pressure, rinse stage must be critical when CO_2 is completely adsorbed, and adsorption pressure must be fixed to accomplish high selectivity. Cavenati et al. [17] investigated CO_2 and N_2 removal from methane using the layered pressure swing adsorption process. The process constituted zeolite 13X to separate CO_2 with a layered carbon molecular sieve to remove N_2. The process was carried out in four steps: pressurization, feed, blow down, and purge. A gaseous mixture consisting of 60% CH_4, 20% CO_2, and 20% N_2 was fed, which yielded a purity of 86% and 88.8% at ambient temperatures and 323K, respectively. A ratio of adsorbent layers was reported for maximum purity for both temperatures. Using a multiobjective optimization approach, Perez et al. [18] developed an adsorption process to separate CO_2 from a mixture of 15% CO_2 + 85% N_2 using two different VSA cycles. The isotherms of CO_2 and N_2 on zeolite 13X pellets were estimated using volumetry and dynamic column breakthrough experiments. A mathematical model that utilizes inputs from these experiments was combined with a genetic algorithm (GA) to obtain the Pareto curves for the multiobjective optimization problem that aims to maximize CO_2 purity and recovery. The decision variables corresponding to specific points on the Pareto curve were restated into lab-scale experiments on a two-column VSA rig containing an adsorbent in each column. The experimental performance indicators were CO_2

purity and recovery. The transients of temperatures, outlet flow, and composition perfectly fit with the experiments, affirming multiobjective optimization approaches as a reliable way to develop VSA separations.

Australia's leading CCS and RD company developed a pressure swing adsorption skid to capture CO_2 from natural gas. The 2-month operation showed that the rig was functioning for natural gas sweetening and could treat natural gas containing 30%–50% CO_2 with a 66% recovery. However, the purity of obtained methane was only 80% instead of 95%. It was deduced that the purity was lacking because desorption pressure could not be lowered by more than 5 bars when 1 bar was required. A simulation based on desorption pressure of 1 bar for natural gas sweetening through adsorption showed enhanced purity up to 98% [9]. Chen et al. [19] experimentally examined natural gas purification using volumetric adsorption apparatus. Zeolite 13X adsorbent was employed, which was priorly analyzed through X-ray diffraction, scanning electron microscope, and Micromeritics ASAP 2020 instrument. The zeolite was investigated to have many microchannels with a microporous volume of 0.22 cm^3(STP)/g. The adsorption experiment was performed in static and swing tanks, and the obtained data fitted with the Langmuir model. CO_2/CH_4 selectivity in static and swing adsorption was discovered to be 3.57 and 3.93; it was concluded that swing enhances selectivity.

3. Adsorbent material selection

CO_2 capture through adsorption is typically based on physisorption through physical adsorbents [20]. The adsorbent selection is critical in designing an adsorption-based process [9]. Low-cost materials, low heat capacity, high rates of reactions, high CO_2 adsorption capacity and selectivity, and high thermal, mechanical, and chemical stability to withstand extensive cycling make up an appropriate adsorption material for CO_2 capture [8]. In the following sections, the most frequently utilized adsorbents—zeolites, carbon-based adsorbents, and metal–organic frameworks (MOFs)—are examined.

3.1 Zeolites

Zeolites are a class of physical adsorbents, their adsorption capacity is influenced by their dimensions, chemical compositions, and charge density. They are classified based on Si/Al molar ratio.

Zeolites with a lesser Si/Al ratio (1−5) are much more hydrophilic materials. Up to 191 zeolitic materials with known structures exist; their discovery is a significant breakthrough in acid gas adsorption processes because of the presence of a wide variety of zeolites with different compositions. Zeolites can also be classified based on pore aperture into small pore zeolites(4Å), medium pore zeolites (5−6Å), large pore zeolites (7Å), and extra-large pore zeolites (>7Å). Another important criterion is the shape of pores, as there are different zeolites with the same pore aperture, but their absorption capacity is different because of pore shape. The dimensionality of channels is another essential for the adsorption of molecules. One-dimensional, two-dimensional, and three-dimensional zeolitic materials can exist with different pore diameters. Channel connection is another course of classifying zeolites. The existence of zeolites with interconnected and independent pores and with structured cages is reported to be found [21]. A comparative analysis was performed on three natural zeolites (clinoptilolite, erionite, and mordenite). Clinoptilolite was found best for the adsorption of CO_2 from natural gas [21]. The low cost of natural zeolites provides a significant benefit in using these adsorbents because of the low cost of raw material. A comparative study on synthesized zeolite with commercial one (NaA) was performed. The synthetic zeolite (NaA-S) was bound with montmorillonite and then shaped into granules to check the adsorption capacity of CO_2 from the CO_2/CH_4 gaseous mixture and then compared with commercial. Adsorption tests were performed at 277, 290, and 310 K by going up to 10 bar pressures. The ideal selectivity of synthesized and commercial zeolite was 7.1 and 6.4 at atmospheric pressure and 290 K, respectively. Moreover, the breakthrough time on CO_2 was 649 and 545s, respectively. In light of the results, synthesized zeolite was suggested to be a better adsorbent for CO_2 removal from natural gas [22].

Zhu et al. [23] modified zeolites with metal to remove sulphides from natural gas. Different zeolites studied were 13X, Ni+13X, Zn+13X, Cu+13X, Ce+13X, and Ag+13X. The adsorption performance for 13X was most enhanced by 162% by adding Ag+. The adsorption performance of five studied zeolites was in the order Ag+13X > Ce+13X > Cu+13X > Zn+13X > Ni+13X > 13X.

Studies are focused on enhancing the CO_2 adsorption of zeolites by altering their compositions; however, the selectivity of CO_2 is low. In addition, their hydrophilic nature hinders adsorption in the presence of moisture; therefore, a high regeneration temperature is needed [8]. Mofarahi and Gholipour [24] did an experimental study on CO_2 adsorption in CO_2/CH_4 mixtures using zeolite 5A at a wide range of temperatures and pressures. Isosteric heat and selectivity of CO_2 were calculated. Isotherm data found that CO_2 are absorbed much more than methane. Further, the ideal selectivity for the CO_2/CH_4 system was directly proportional to temperature and inversely proportional to pressure. It was concluded that zeolite 5A was an effective adsorbent for purifying biogas and landfill gas [24]. Li et al. [25] studied six types of industrially used zeolites. The six adsorbents were NaX, CaX, NaA, CaA, ZSM-5, and Y. It was found that NaX and CaX outperformed the other four zeolites on separation performance. NaX indicated the most significant adsorption amount of CO_2 at 0.1 MPa, and CaX showed the highest affinity for CO_2. The IAST model predicted the selectivity of NaX and CaX as 76 and 74 with $y_{CH4} = 0.5$ and $y_{CH4} = 0.9$, respectively. It was further concluded that the existence of moistness deteriorated the working of zeolites [25]. McEwen et al. [26] conducted a comparative study on ZIF-8, zeolite 13X, and BPL activated carbon. Adsorption isotherms were determined from gravimetric analysis up to 1 bar at a constant temperature of 25°C for CO_2, CH_4, and N_2. Zeolite 13X showed a greater adsorption capacity than ZIF-8 and BPL-activated carbon. It also showed a higher selectivity of carbon dioxide over methane; however, it is more susceptible to degradation by impurities. The potential of ZIF-8 for carbon dioxide sequestration is also discussed at higher pressures due to its large pore volume. Acid gas separation from natural gas using Na-SSZ-13 was done by gravimetric, dynamic column breakthrough and volumetric techniques. The gas mixtures contained CO_2, C_2H_6, and H_2S. The ideal adsorption model fitted well for experimental breakthrough curves. Na-SSZ-13 consists of small pores and exhibits adsorption of CO_2 in H_2S and COS containing systems; this zeolite tends to have a more comprehensive application than other high aluminum zeolites such as A, X, and Y [27]. Shang et al. [28] performed an experiment using a molecular trapdoor chabazite adsorbent to separate CO_2 from CH_4 through pressure swing adsorption. CO_2/CH_4 showed selectivity in adsorption equilibrium experiments. The experiment demonstrated that a cyclic adsorption process with the molecule trapdoor chabazite yielded methane products with a purity of 100% and a recovery of more than 90%. High CO_2 selectivity over CH_4 at high pressures and temperatures below 303 K is the reason for the impressive performance; however, the temperature-VSA process is recommended for this kind of material in the future. Zhang Z et al. [13] investigated the kinetics and capacity of zeolite NaX. At low pressures, CO_2 adsorption capacity on zeolite NaX was much higher than on activated carbon.

3.2 Carbonaceous adsorbents

Carbonaceous adsorbents have been widely used for CO_2 capture due to their high thermal stability, low cost, and resistance to moisture. Their porosities are well-developed. Their surface chemistry is based on groups of neutral, acidic, or basic heteroatoms. Activated carbons, carbon molecular sieves, activated carbon fibers, and carbon-based nanomaterials are the four main types of carbon adsorbents. Adsorption sweetens natural gas only with carbon molecular sieves and activated carbon. The current study aims to increase alkalinity or improve the pore structure of these adsorbents to increase their adsorption capacity [21].

3.2.1 Activated carbons

Activated carbons are created by carbonizing at lower temperatures (approximately 773 K) and then activating at higher temperatures (approximately 1273 K). They are primarily utilized for purifying recycled amines in natural gas sweetening. According to the literature, the incorporation of nitrogen-based functional groups into the structures of activated carbons can enhance their capacity for CO_2 adsorption [21].

Polymeric and pitch fibers are used to make activated carbon fibers. They consist of small but uniformly organized pores with regular pore diameters, enhancing interaction with adsorbates. They are mainly used for producing high-purity hydrogen gas [21]. Hydrocarbon or carbon monoxide decomposition is employed to design carbon-based nanomaterials; however, their industrial applications have not been undiscovered. Only activated carbon and carbon molecular sieves are used for natural gas sweetening via adsorption. The present research focuses on enhancing the adsorption capacity of these adsorbents by improving pore structure or increasing alkalinity. The literature suggests CO_2 adsorption capacity of activated carbons can be improved by the addition of nitrogen-based functional groups into their structures [21].

Saxena et al. [29] studied about carbon dioxide sequestration through adsorption using activated carbon packed in a cylindrical reaction bed, which included a numerical analysis to find out the rate of adsorption, and a parametric analysis was done to determine the influence of essential adsorption parameters including bed temperature, bed radius, heat transfer coefficient, and cooling fluid temperature on the adsorption capacity. Dubinin's theory of volume of filling of micropores was found to be the best model for adsorption through activated carbons. Lower bed radii with high heat transfer coefficient and high initial temperature-favored adsorption. In addition, a low temperature of the cooling fluid was preferred for a high adsorption rate. In another study, a crushed microwave-activated carbon with mesh size ranging between 40 and 60 was used as an adsorbent for CO_2, CH_4, and N_2 systems. Adsorption equilibrium data were obtained from the static volume instrument experiment. For activated carbon adsorption capacity in mmol/g was highest for CO_2 (2.13), then CH_4 (0.98), and lowest for N_2 (0.33) at 298 K and 100 kPa; the highest separation factor was obtained for CO_2/N_2 (14.6) and CO_2/CH_4 (4.37) mixtures [30]. A similar experimental and model-based study on two types of activated carbons, AC1(PK-360) and AC2 (BG-1240), was done for adsorption of CO_2, CH_4, and N_2 at elevated temperatures and pressures. Two activated carbons differed on structural parameters, including surface area, total volume, micro, and meso volumes, and average pore radius. The results showed that AC2 had a higher absorption capacity than AC1; its higher specific area explained it. Activated carbon had the highest and lowest adsorption amount of CO_2 and N_2, respectively. However, the results showed that both types of activated carbons used in the study did not have good potential for sequestering CO_2 from CO_2/CH_4 systems [31]. Adsorption of CH_4, C_2H_4, CO_2, H_2, and N_2 on activated carbon and zeolite-5A was analyzed using MLP-ANN, hybrid ANFIS, PSO-ANFIS, and CSA-LSSVM models; it was found that these models concurred with experimental adsorption data with accuracy and reliability [32]. Olajire [7] conducted a study on activated anthracite which showed that the adsorption capacity of anthracite declined with the adsorption temperature. Anthracite with the highest surface area of 1071 m^2/g only had a CO_2 adsorption capacity of 40 mg CO_2/g adsorbent. In addition, primary groups such as ammonia and polyethyleneimine were induced in the structure of anthracite, which showed enhanced CO_2 capture capacity. Jiang et al. [33] assessed CO_2 capture from natural gas combined power plants using activated carbon as the physical adsorbent. Energy, exergy, and economic analyses are performed on the adsorption process, and then the results are compared with the MEA-based absorption process and a chemical adsorbent polyethyleneimine/silica. The net efficiency of the process will increase from 49.67% to 51.09% by using activated carbon in place of MEA. The study can predict the behavior of three adsorbents for natural gas sweetening.

3.2.2 Molecular sieves

Carbon molecular sieves are developed either by pyrolysis of carbon precursor or by additional treatments, mainly carbon vapor deposition [34]. CVD consists of pore size tuning. The commercial CMSs are prepared with pyrolytic Carbon. They separate molecules based on size, shape, or difference in adsorption rates. Natural gas sweetening by molecular sieves through pressure swing adsorption has been widely studied. It has been reported that CMSs are advantageous due to their higher stability toward alkaline and acidic media and hydrophobic nature [34]. Heck et al. [35] performed a study on molecular sieves 4A, 5A, and 13X was performed on gas mixtures containing H_2S, CO_2, and CH_4. Pressure swing adsorption was carried out. Selective adsorption of CO_2 was noted for all three adsorbents. Experimental examination indicated that molecular sieves 13X and 5A yielded up to 98% pure methane. Molecular sieve 4A showed the highest recovery in the desorbed product. The obtained results were in accord with the experimental results in the publications. Sarker et al. [36] characterized the equilibrium and kinetic behavior of six commercial adsorbents. The equilibrium behavior indicated the maximum CO_2 adsorption capacity of the adsorbents, whereas kinetic data predicted the rates of adsorption reactions. Adsorbents included zeolite 4A, zeolite 5A, zeolite 13X, carbon molecular sieve (MSC-3R), and activated carbons (GCA-830 and GCA-1240). Experiments were performed on the gas adsorption apparatus, CO_2 adsorption rate was noted to increase rapidly with increasing equilibrium pressure up to a particular point, after which it begins to stabilize. The highest CO_2 adsorption capacity and mass transfer rate were found for GCA-1240 and the lowest for zeolite 4A. The pore properties and adsorption capacities of carbon molecular sieves designed by polymeric precursors at different temperatures were evaluated. The study performed at three different temperatures (273, 298, and 323 K) at atmospheric pressure showed that a well-developed microporosity with micropores of high volume yields a high adsorption amount of CO_2 [34]. Rocha et al. [37] reported the working of commercial carbon molecular sieves on CO_2 and CH_4 adsorption at a range of 5–70 bar. The equilibrium data of these adsorbents concurred with the Langmuir models at 298 and 343 K. It was found that the equilibrium selectivity of this adsorbent was low, while its kinetic selectivity was high. The pore diffusion rates for CO_2 were found to be higher than CH_4. Further, this material was suggested to be better suited for biogas purification than natural gas due to enhanced working at low-pressure conditions.

Song et al. [38] analyzed three commercial carbon molecular sieves' pore structures and chemical properties were analyzed. Adsorption isotherms of CO_2 and CH_4 were studied at three different temperatures up to 1 MPa, and the Langmuir model was used to fit the experimental data. For all of the CMSs studied, the adsorption capacity of CO_2 was found to be greater than CH_4. Using intelligent gravimetric analysis, some other thermodynamic properties were also determined, such as the changes in Standard Enthalpy, Gibbs Free Energy, and Entropy. Park et al. [39] utilized a gravimetric method to analyze CO_2, CO, and N_2 adsorption on carbon molecular sieves at different temperatures going up to 1 bar. Another experimental volumetric analysis verified the kinetic and equilibrium results and fitted Langmuir and Sips models. Adsorption capacity and heat were found to be most significant for CO_2 and lowest for N_2. CO_2, CO, and N_2 adsorption rates were highly affected by electrical properties rather than kinetic properties. In addition, large surface area and pore size of ACs resulted in a more significant heat of adsorption and adsorption capacity for ACs than CMS.

3.2.3 Activated carbon fibers and carbon-based nanomaterials

Polymeric and pitch fibers are used to make activated carbon fibers. They consist of small but uniformly organized pores with regular pore diameters, enhancing interaction with adsorbates. They are mainly used for producing high-purity hydrogen gas [21]. Hydrocarbon or carbon monoxide decomposition is employed to design carbon-based nanomaterials; however, their industrial applications have not been undiscovered. Only activated carbon and carbon molecular sieves are used for natural gas sweetening via adsorption. The present research focuses on enhancing the adsorption capacity of these adsorbents by improving pore structure or increasing alkalinity [21].

3.3 Metal-organic frameworks

Yu et al. [40] have found metal-organic frameworks as a significant area of research because of the possibility of infinite structures. They have controllable pore properties due to their varying metallic clusters of ligands. The metal-organic frameworks that adsorb CO_2 have lower heats of adsorption and can be advantageous for PSA-based processes; however, they have low CO_2/CH_4 selectivity that results in the loss of methane [21].

Microporous metal-organic frameworks have been researched widely and have been promising adsorbents to be used for natural gas purification. Wang et al. [41] analyzed the adsorption performance of two microporous metal-organic frameworks, JLU-Liu5 and JLU-Liu6. The IAST model was used to predict the adsorption selectivity. At 298K and 1 bar, the selectivity of carbon dioxide over methane for JLU-Liu5 and JLU-Liu6 was found to be 4.6 and 7.3, which was higher than most carbon-based adsorbents and metal-organic frameworks [41]. Joshi et al. [42] H_2S evaluated the adsorption capacity of parent and amine-functionalized metal−organic frameworks, including MIL-125(Ti), MIL-101(Cr), and UiO-66(Zr), from natural gas mixtures with and without the presence of H_2S with compositions of 89% CH_4+10% CO_2+1% H_2S and 99% CH_4+1% H_2S. The presence of H_2S resulted in a partial deterioration of MIL-101(Cr) materials, which was not observed in other materials studied, due to which MIL-101(Cr) and its linker amine were not recommended to be used industrially despite the most remarkable adsorption capacity of MIL-101-NH_2(Cr). Experimental investigations showed that MOFs' adsorption capacities and breakthrough times increased with the addition of amine functional groups. Qiao et al. [43] performed screening on 6013 hydrophobic MOFs for simultaneous CO_2 and H_2O removal in the presence of moisture. Henry's Law was initially used to select 606 MOFs, and further screening was done by adsorbing CO_2 and H_2S from natural gas. Among top 45 adsorbents, 39 had N-containing organic linkers, 23 were pyridine, and 12 were azoles. These MOFs were superior to other adsorbents due to their high CO_2 and H_2S affinity. TSN (selectivity) was considered more appropriate for determining the performance of MOFs. 14 adsorbents with TSN > 2 were characterized as the best MOFs [43]. Pham et al. [44] performed a simulation-based study on CO_2 and H_2 absorption using UTSA-20. UTSA-20 is made of Cu^{+2} ions coordinated with benzene hexa-benzoate. It was reported that the adsorption isotherms and isosteric heat obtained from the simulation agreed with the experimental data on this MOF. Experimental observations indicated that by changing pressure, CO_2 and H_2 both show distinct behaviors for occupying sites. This study provides information about microscopic details of UTSA-20, which can help researchers develop new MOFs [44]. Huang et al. [45] conducted a study to uncover the impact of water on the adsorption selectivity of frameworks. CO_2/CH_4 selectivity under humid conditions depended on the framework's interaction with water; for weak framework−water interactions, water had negligible effects. The selectivity is affected when the forces of attraction between water and the adsorption sites are strong enough to adsorb water molecules on the surface [45]. Vicent-Luna et al. [46] investigated the impact of using ionic liquids and their mixtures with MOFs on the adsorption capacity of CO_2, CH_4, and N_2 using molecular dynamics simulations. Since CO_2 is highly soluble in ionic liquids, the CO_2 selectivity of ionic liquid-soaked MOFs is enhanced. The use of ionic liquids can control the pore size of frameworks, which helps in the exclusion of less adsorbed substances, leading to the removal of only desired components [46]. Roztocki et al. [47] discussed the structure and performance of a flexible water-stable MOF JUK-8 using X-ray diffraction (XRD), infrared spectroscopy (IR), and CO_2- nuclear magnetic resonance (NMR) techniques. The single crystal X-ray diffraction technique unveils the presence of two energy minima in MOF profile. In contrast, IR and NMR techniques give insight into the positions of adsorbed molecules in the framework. High-pressure co-adsorption experiments confirmed the results at various temperatures. It had been concluded that JUK-8 was a highly favorable CO_2 adsorbent for CO_2/CH_4 mixtures.

Copolymerization of zinc with benzene tricarboxylic acid results in synthesis of metal-organic frameworks with rigid porosity; however, these MOFs have not been commercialized due to costly solvents used; MOFs were synthesized using unconventional solvent systems such as dimethyl formamide (DMF), DMF/MeOH(methanol)DMF/isopropanol (IPA)/deionized water (DI), and EtOH(ethanol)/DI. The CO_2 adsorption capacities of prepared MOFs were then analyzed using X-ray diffraction (XRD), scanning electron microscopy (SEM), and thermogravimetric analysis (TGA). Brunauer-Emmett-Teller (BET) technique was employed for surface area analysis. DMF systems had more suitable adsorption performance; economic analysis showed that DMF/MeOH and DMF/IPA/DI were more economically viable and could be substituted by DMF. Aftab et al. [48] further considered including amine-based functional groups in Zn-BTC (1,3,5-benzene tricarboxylic acid), a promising area for enhancing CO_2 adsorption [48]. Table 14.1 shows the CO_2 and H_2S adsorption capacity and selectivity of different adsorbents in literature from 2012 to 2021.

4. Conclusion and future outlooks

The presence of acid in sour natural gas has been a growing problem, and the weaknesses in conventional MEA-based absorption processes demand advancement in CO_2 capture technologies, including adsorption. In this view, this chapter provides in-depth details on the technological advancement,

Table 14.1 CO_2 and H_2S adsorption capacity and selectivity of different adsorbents in literature from 2012 to 2021.

Year	Technique	Procedure adopted	Adsorbent material	Removal CO_2	Removal H_2S	Feed conditions (°C, KPa)	Adsorption capacity	Selectivity CO_2/CH_4	References
2016	Zeolite	XRD, SEM, FTIR and BET analyses	NaS	Y	N	(17, 100)	5.2 mol/kg	7.1	[22]
			NaA	Y	N	(17, 100)	2.6 mol/kg	6.4	
2019		Dynamic fixed bed adsorption method	13X	Y	N	(450, 100)	24.7 mg S/g	—	[23]
			Ni/13X	Y	N	(450, 100)	60.8 mg S/g	—	
			Zn/13X	Y	N	(450, 100)	54.9 mg S/g	—	
			Cu/13X	Y	N	(450, 100)	51.3 mg S/g	—	
			Ce/13X	Y	N	(450, 100)	60.8 mg S/g	—	
			Ag/13X	Y	N	(450, 100)	64.8 mg S/g	—	
2014		Volumetric method, Gas chromatograph (GC) Ideal adsorbed solution theory (IAST), and vacancy solution model (VSM)	5A	Y	N	(30, 74)	1.66 mol/kg	81	[24]
2013		IAST model	NaX	Y	N	(30, 86)	4.00 mol/kg	76	[25]
			CaX	Y	N	(30, 86)	3.20 mol/kg	23	
			NaA	Y	N	(30, 86)	2.30 mol/kg	38	
			CaA	Y	N	(30, 86)	2.40 mol/kg	74	
			ZSM-5	Y	N	(30, 86)	1.50 mol/kg	4	
			Y	Y	N	(30, 86)	1.00 mol/kg	4	
2012		Gravimetric analysis	ZIF-8	Y	N	(25, 100)	0.80 mol/kg	3.5	[26]
			13X	Y	N	(25, 100)	1.70 mol/kg	32	
		Activated carbon	BPL AC	Y	N	(25, 100)	2.10 mol/kg	4	
2020	Zeolite	Gravimetric, dynamic column breakthrough and volumetric techniques	A	Y	N	—	—	—	[27]
			X	Y	N	—	—	—	
			Y	Y	N	—	—	—	
			Na-SSZ13	Y	Y	(30, 3000)	—	—	

Year	Material	Method	Adsorbent			Conditions	Capacity	Value	Ref
2012	Activated carbon	Static volume instrument method	MAC	Y	N	(25, 50)	1.40 mol/kg	4.8	[30]
2019		Experimental and model-based study	PK-360	Y	N	(27, 4000)	1.83 mol/kg	2.01	[31]
			BG-1240	Y	N	(27, 4000)	2.80 mol/kg	1.65	
			PEI/Silica	Y	N	(293, 30)	0.04 mol/kg	—	[33]
2017	Carbon molecular sieves	Pressure swing adsorption experiments	4A	Y	Y	(20, 580)	95.0, 0.2(%)	2.0, 1.8	[35]
			5A	Y	Y	(30, 580)	79.6, 0.2(%)	5.0, 3.9	
			13X	Y	Y	(30, 580)	64.0, 0.7(%)	19.8, 20.8	
	Zeolite	Equilibrium and kinetic study	4A	Y	N	(60, 3000)	2.90 mol/kg	—	[36]
			5A	Y	N	(60, 3000)	3.50 mol/kg	—	
	Carbon molecular sieves		MSC-3R	Y	N	(60, 30)	3.00 mol/kg	—	
	Activated carbon		GCA-830	Y	N	(60, 3000)	6.70 mol/kg	—	
2014			GCA-1240	Y	N	(60, 3000)	6.80 mol/kg	—	
	Metal–organic frameworks	Adsorption experiments	JLU-Liu5	Y	N	(25, 100)	2.25 mol/kg	4.6	[41]
			JLU-Liu6	Y	N	(25, 100)	2.00 mol/kg	7.3	
2018		Adsorption experiments	UiO-66(Zr)	N	Y	(20, 550)	0.24 mol/kg	—	[42]
			MIL-125(Ti)	N	Y	(20, 550)	0.21 mol/kg	—	
			MIL-101(Cr)	N	Y	(20, 550)	0.50 mol/kg	—	
			MIL-101 NH$_2$(Cr)	N	Y	(20, 550)	0.52 mol/kg	—	
			UiO-66-NH$_2$(Zr)	N	Y	(20, 550)	0.38 mol/kg	—	
			MIL-125-NH$_2$(Ti)	N	Y	(20, 550)	0.56 mol/kg	—	
2021		XRD, IR, and 13C NMR	JUK-8	Y	N	(25, 2000)	100 cm^3/g	1.99	[47]
		Adsorption experiments	Zn-BTC	Y	N	(120, 1500)	5.70 mol/kg	—	[48]

material prepositions of adsorbents, and their selection criteria which can provide effective CO_2 separation from natural gas. It was observed that pressure swing adsorption is the commercially most used technology which provides an avenue for enhanced adsorption with less economic intensity. Furthermore, it is critically important to have fewer hydrocarbons lost during the process. The literature findings show temperature and vacuum-based adsorption of CO_2 is gaining attention in recent times and more focused has been done on the advancement of the adsorbent material such as chabazite, which is thought to be highly selective and to provide a larger surface area for CO_2 selective separation at a higher rate. Yet this technology still requires further research in order to scale it up for industry. The chapter also presented a detailed comparative analysis of various adsorbents being used for the natural gas sweetening process and assessed their performance against the conventional processes such as the amine-based process.

Yet still, it has been observed that very little work has been done in the field of process system engineering, which can provide a detailed in-depth assessment of the avenues an adsorption-based process can provide based on energy utilization, eco-friendly prospects, and most importantly, economic viability for its industrialization at the commercial level. As a result, it is critical to continue working on the characterization aspects of adsorbent material and scalability for the use of highly relevant technology of MOF, which has the potential to revolutionize adsorption technology in the future. Additionally, for the future, it is important to gain some knowledge about adsorption as a natural gas sweetening technology via life cycle assessment (LCA) that quantitatively assesses adsorbent capability for elevating it to the industrial level. Hence, LCA can be a prosperous tool that evaluates the sustainability of an adsorption-based natural gas sweetening process by quantifying its benefits, risks, ecological impact, and performance and also lays the groundwork for identifying and testing viable alternatives. As for future prospects, it is also essential to have the characterization studies on the material properties of adsorbent with more development, especially on the kinetics aspects. Hence, it can be concluded that adsorption is still in its early phase of development and requires extensive work experimentally, process viability, and implementation.

Abbreviations and symbols

BET	Brunauer-Emmett-Teller
CCS	carbon capture and storage
CO_2	carbon dioxide
DI	deionized water
DMF	dimethyl formamide
GA	genetic algorithm
GHG	greenhouse gas
IPA	isopropanol
IPCC	intragovernmental panel on climate change
IR	infrared spectroscopy
LCA	life cycle assessment
MEA	monoethanol amine
MOF	metal organic framework
NMR	nuclear magnetic resonance
PSA	pressure swing adsorption
SEM	scanning electron microscopy
STP	standard temperature and pressure
TGA	thermogravimetric analysis
TSA	temperature swing adsorption
VPSA	vacuum pressure swing adsorption
VSA	vacuum swing adsorption
XRD	X-ray diffraction

References

[1] Tiseo I. Historic average carbon dioxide (CO2) levels in the atmosphere worldwide from 1959 to 2021 (in parts per million). https://www.statista.com/statistics/1091926/atmospheric-concentrationofco2historic/#:~:text=Global%20atmospheric%20concentration%20of%20carbon%20dioxide%201959%2D2021&text=The%20atmospheric%20level%20of%20carbon,about%20316.91%20parts%20per%20million.

[2] Paris agreement, UNFCCC.

[3] Kazmi B, Raza F, Taqvi SAA, Ali SI, Suleman H. Energy, exergy and economic (3E) evaluation of CO2 capture from natural gas using pyridinium functionalized ionic liquids: a simulation study. Journal of Natural Gas Science and Engineering 2021;90:103951. https://doi.org/10.1016/j.jngse.2021.103951.

[4] Arthur WRP, Kidnay J, McCartney DG. Fundamentals of natural gas processing. 2019.

[5] Sifat NS, Haseli Y. A critical review of CO2 capture technologies and prospects for clean power generation. Energies 2019;12(21):4143. https://doi.org/10.3390/en12214143.

[6] Rufford TE, Smart S, Watson GC, Graham BF, Boxall J, Da Costa JD, et al. The removal of CO2 and N2 from natural gas: a review of conventional and emerging process technologies. Journal of Petroleum Science and Engineering 2012;94:123–54. https://doi.org/10.1016/j.petrol.2012.06.016.

[7] Olajire AA. CO2 capture and separation technologies for end-of-pipe applications—a review. Energy 2010;35(6):2610–28. https://doi.org/10.1016/j.energy.2010.02.030.

[8] Yu C, Huang C, Tan C. A review of CO2 capture by absorption and adsorption. Aerosol and Air Quality Research 2012;12(5):745−69.

[9] Tao L, Xiao P, Qader A, Webley PA. CO2 capture from high concentration CO2 natural gas by pressure swing adsorption at the CO2CRC Otway site, Australia. International Journal of Greenhouse Gas Control 2019;83:1−10. https://doi.org/10.1016/j.ijggc.2018.12.025.

[10] Webley PA. Adsorption technology for CO2 separation and capture: a perspective. Adsorption 2014;20(2):225−31. https://doi.org/10.1007/s10450-014-9603-2.

[11] Pahinkar DG, Garimella S, Robbins TR. Feasibility of temperature swing adsorption in adsorbent-coated microchannels for natural gas purification. Industrial & Engineering Chemistry Research 2017;56(18):5403−16. https://doi.org/10.1021/acs.iecr.7b00389.

[12] Mondino G, Grande CA, Blom R, Nord LO. Moving bed temperature swing adsorption for CO2 capture from a natural gas combined cycle power plant. International Journal of Greenhouse Gas Control 2019;85:58−70. https://doi.org/10.1016/j.ijggc.2019.03.021.

[13] Hedin N, Andersson L, Bergström L, Yan J. Adsorbents for the post-combustion capture of CO2 using rapid temperature swing or vacuum swing adsorption. Applied Energy 2013;104:418−33. https://doi.org/10.1016/j.apenergy.2012.11.034.

[14] Manocha SM. Porous carbons. Sadhana 2003;28(1):335−48. https://doi.org/10.1007/BF02717142.

[15] Lopes FV, Grande CA, Rodrigues AE. Activated carbon for hydrogen purification by pressure swing adsorption: multicomponent breakthrough curves and PSA performance. Chemical Engineering Science 2011;66(3): 303−17. https://doi.org/10.1016/j.ces.2010.10.034.

[16] Abd A, Othman M, Naji S, Hashim A. Methane enrichment in biogas mixture using pressure swing adsorption: process fundamental and design parameters. Materials Today Sustainability 2021;11:100063. https://doi.org/10.1016/j.mtsust.2021.100063.

[17] Cavenati S, Grande CA, Rodrigues AE. Separation of CH4/CO2/N2 mixtures by layered pressure swing adsorption for upgrade of natural gas. Chemical Engineering and Science 2006;61(12):3893−906. https://doi.org/10.1016/j.ces.2006.01.023.

[18] Perez LE, Sarkar P, Rajendran A. Experimental validation of multi-objective optimization techniques for design of vacuum swing adsorption processes. Separation and Purification Technology 2019;224:553−63. https://doi.org/10.1016/j.seppur.2019.05.039.

[19] Chen S, Fu Y, Huang Y, Tao Z, Zhu M. Experimental investigation of CO2 separation by adsorption methods in natural gas purification. Applied Energy 2016;179:329−37. https://doi.org/10.1016/j.apenergy.2016.06.146.

[20] Creamer AE, Gao B. Carbon-based adsorbents for postcombustion CO2 capture: a critical review. Environmental Science & Technology 2016;50(14): 7276−89. https://doi.org/10.1021/acs.est.6b00627.

[21] Tagliabue M, Farrusseng D, Valencia S, Aguado S, Ravon U, Rizzo C, et al. Natural gas treating by selective adsorption: material science and chemical engineering interplay. Chemical Engineering Journal 2009;155(3):553−66. https://doi.org/10.1016/j.cej.2009.09.010.

[22] Pour AA, Sharifnia S, Salehi RN, Ghodrati M. Adsorption separation of CO2/CH4 on the synthesized NaA zeolite shaped with montmorillonite clay in natural gas purification process. Journal of Natural Gas Science and Engineering 2016;36:630−43. https://doi.org/10.1016/j.jngse.2016.11.006.

[23] Zhu L, Lv X, Tong S, Zhang T, Song Y, Wang Y, et al. Modification of zeolite by metal and adsorption desulfurization of organic sulfide in natural gas. Journal of Natural Gas Science and Engineering 2019;69:102941. https://doi.org/10.1016/j.jngse.2019.102941.

[24] Mofarahi M, Gholipour F. Gas adsorption separation of CO2/CH4 system using zeolite 5A. Microporous and Mesoporous Materials 2014;200:1—10. https://doi.org/10.1016/j.micromeso.2014.08.022.

[25] Li Y, Yi H, Tang X, Li F, Yuan Q. Adsorption separation of CO2/CH4 gas mixture on the commercial zeolites at atmospheric pressure. Chemical Engineering Journal 2013;229:50—6. https://doi.org/10.1016/j.cej.2013.05.101.

[26] McEwen J, Hayman J-D, Yazaydin AO. A comparative study of CO2, CH4 and N2 adsorption in ZIF-8, Zeolite-13X and BPL activated carbon. Chemical Physics 2013;412:72—6. https://doi.org/10.1016/j.chemphys.2012.12.012.

[27] Thompson JA. Acid gas adsorption on zeolite SSZ-13: equilibrium and dynamic behavior for natural gas applications. AIChE Journal 2020;66(10): e16549. https://doi.org/10.1002/aic.16549.

[28] Shang J, Hanif A, Li G, Xiao G, Liu JZ, Xiao P, et al. Separation of CO2 and CH4 by pressure swing adsorption using a molecular trapdoor chabazite adsorbent for natural gas purification. Industrial & Engineering Chemistry Research 2020;59(16):7857—65.

[29] Saxena R, Singh VK, Kumar EA. Carbon dioxide capture and sequestration by adsorption on activated carbon. Energy Procedia 2014;54:320—9. https://doi.org/10.1016/j.egypro.2014.07.275.

[30] Yi H, Li F, Ning P, Tang X, Peng J, Li Y, et al. Adsorption separation of CO2, CH4, and N2 on microwave activated carbon. Chemical Engineering Journal 2013;215:635—42. https://doi.org/10.1016/j.cej.2012.11.050.

[31] Ghalandari V, Hashemipour H, Bagheri H. Experimental and modeling investigation of adsorption equilibrium of CH4, CO2, and N2 on activated carbon and prediction of multi-component adsorption equilibrium. Fluid Phase Equilibria 2020;508:112433. https://doi.org/10.1016/j.fluid.2019.112433.

[32] Dashti A, Raji M, Azarafza A, Baghban A, Mohammadi AH, Asghari M. Rigorous prognostication and modeling of gas adsorption on activated carbon and Zeolite-5A. Journal of Environmental Management 2018;224: 58—68. https://doi.org/10.1016/j.jenvman.2018.06.091.

[33] Jiang L, Gonzalez-Diaz A, Ling-Chin J, Roskilly A, Smallbone A. Post-combustion CO2 capture from a natural gas combined cycle power plant using activated carbon adsorption. Applied Energy 2019;245:1—15. https://doi.org/10.1016/j.apenergy.2019.04.006.

[34] Wahby A, Silvestre-Albero J, Sepúlveda-Escribano A, Rodríguez-Reinoso F. CO2 adsorption on carbon molecular sieves. Microporous and Mesoporous Materials 2012;164:280—7. https://doi.org/10.1016/j.micromeso.2012.06.034.

[35] Heck HH, Hall ML, dos Santos R, Tomadakis MM. Pressure swing adsorption separation of H2S/CO2/CH4 gas mixtures with molecular sieves 4A, 5A, and 13X. Separation Science and Technology 2018;53(10):1490—7. https://doi.org/10.1080/01496395.2017.1417315.

[36] Sarker AI, Aroonwilas A, Veawab A. Equilibrium and kinetic behaviour of CO2 adsorption onto zeolites, carbon molecular sieve and activated carbons. Energy Procedia 2017;114:2450—9. https://doi.org/10.1016/j.egypro.2017.03.1394.

[37] Rocha LA, Andreassen KA, Grande CA. Separation of CO2/CH4 using carbon molecular sieve (CMS) at low and high pressure. Chemical Engineering Science 2017;164:148−57. https://doi.org/10.1016/j.ces.2017.01.071.
[38] Song X, Ma X, Zeng Y. Adsorption equilibrium and thermodynamics of CO2 and CH4 on carbon molecular sieves. Applied Surface Science 2017;396:870−8. https://doi.org/10.1016/j.apsusc.2016.11.050.
[39] Park Y, Moon D-K, Park D, Mofarahi M, Lee C-H. Adsorption equilibria and kinetics of CO2, CO, and N2 on carbon molecular sieve. Separation and Purification Technology 2019;212:952−64. https://doi.org/10.1016/j.seppur.2018.11.069.
[40] Yu C-H, Huang C-H, Tan C-S. A review of CO2 capture by absorption and adsorption. Aerosol and Air Quality Research 2012;12(5):745−69. https://doi.org/10.4209/aaqr.2012.05.0132.
[41] Wang D, Zhao T, Cao Y, Yao S, Li G, Huo Q, et al. High performance gas adsorption and separation of natural gas in two microporous metal−organic frameworks with ternary building units. Chemical Communications 2014;50(63):8648−50. https://doi.org/10.1039/c4cc03729d.
[42] Joshi JN, Zhu G, Lee JJ, Carter EA, Jones CW, Lively RP, et al. Probing metal−organic framework design for adsorptive natural gas purification. Langmuir 2018;34(29):8443−50. https://doi.org/10.1021/acs.langmuir.8b00889.
[43] Qiao Z, Xu Q, Jiang J. Computational screening of hydrophobic metal−organic frameworks for the separation of H 2 S and CO 2 from natural gas. Journal of Materials Chemistry A 2018;6(39):18898−905. https://doi.org/10.1039/C8TA04939D.
[44] Pham T, Forrest KA, Franz DM, Guo Z, Chen B, Space B. Predictive models of gas sorption in a metal−organic framework with open-metal sites and small pore sizes. Physical Chemistry Chemical Physics 2017;19(28):18587−602. https://doi.org/10.1039/c7cp02767b.
[45] Huang H, Zhang W, Liu D, Zhong C. Understanding the effect of trace amount of water on CO2 capture in natural gas upgrading in metal−organic frameworks: a molecular simulation study. Industrial & Engineering Chemistry Research 2012;51(30):10031−8. https://doi.org/10.1021/ie202699r.
[46] Vicent-Luna JM, Gutierrez-Sevillano JJ, Hamad S, Anta J, Calero S. Role of ionic liquid [EMIM]+[SCN]− in the adsorption and diffusion of gases in metal−organic frameworks. ACS Applied Materials & Interfaces 2018;10(35):29694−704. https://doi.org/10.1021/acsami.8b11842.
[47] Roztocki K, Rauche M, Bon V, Kaskel S, Brunner E, Matoga D. Combining in situ techniques (XRD, IR, and 13C NMR) and gas adsorption measurements reveals CO2-induced structural transitions and high CO2/CH4 selectivity for a flexible metal−organic framework JUK-8. ACS Applied Materials & Interfaces 2021;13(24):28503−13.
[48] Aftab L, Iqbal N, Asghar A, Noor T. Synthesis, characterization and gas adsorption analysis of solvent dependent Zn-BTC metal organic frameworks. Separation Science and Technology 2021;56(13):2159−69.

SECTION IV

Membrane technology for natural gas sweetening

15

Polymeric membranes for natural gas sweetening

Abdul Latif Ahmad[1], Muhd Izzudin Fikry Zainuddin[1] and Meor Muhammad Hafiz Shah Buddin[1,2]

[1]School of Chemical Engineering, Universiti Sains Malaysia Engineering Campus, Nibong Tebal, Pulau Pinang, Malaysia; [2]School of Chemical Engineering, College of Engineering, Universiti Teknologi MARA, Shah Alam, Selangor, Malaysia

1. Introduction

Natural gas sweetening is a process where the content of H_2S and CO_2 is reduced to meet the typical specifications for pipeline quality. The common requirement for H_2S is less than 4 ppm, while the CO_2 content must be less than 2% [1]. Otherwise, the gas mixture is referred to as the sour gas. Although other components are present in natural gas, these two gases (CO_2 and H_2S) must be removed due to environmental restrictions and technical problems during gas transportation and commercialization [2]. The pipeline gets corroded over time by both gases, and the presence of extremely toxic and hazardous properties of H_2S even at a low concentration toward living beings is a significant concern [3]. To meet such requirements, plenty of technologies have been used in the industry—for instance, the absorption and cryogenic processes [4]. Despite the efficiency offered by these technologies, they are relatively expensive and involve high energy requirements and complex equipment with large carbon footprints compared to membrane technology. On the other hand, membrane technology is seen as an interesting alternative for gas separation applications due to its easy processability and simple operation procedure in which the separation is governed by a simple pressure gradient between the feed and permeate side [5]. In fact, there are multiple reports on the utilization of membrane technology for natural gas sweetening applications [6,7]. Furthermore, the simulation by He et al. [8] estimated that the minimum specific cost to sweeten the natural gas is less than

2.71×10^{-3} \$/m^3 to achieve the separation requirement of <2.5 mol% CO_2 in purified natural gas together with captured high purity CO_2 (>95 mol%). More importantly, the technology was proven effective for natural gas sweetening. For instance, the glassy polymer of intrinsic microporosity (PIM) membrane fabricated by Yi et al. [9] recorded H_2S/CH_4 selectivity up to 75 with H_2S permeability >4000 Barrers in a ternary feed mixture (20% H_2S:20% CO_2:60% CH_4).

The development of membranes for gas separation has a long history of nearly 200 years. The first observation of the penetration of gas through a film was first recorded by Mitchell (1831) [10]. However, in 1866, the first systematic study on gas separation through a membrane was studied by Graham [11]. The fundamental gas separation mechanism known as the solution-diffusion model was first proposed by the author [12,13]. At that time, no polymeric membrane as we know it today existed. Afterward, various reports on the gas separation properties of polymeric materials were available. In the 1960s, the first ever anisotropic membrane made up of cellulose acetate for reverse osmosis application was discovered by Loeb and Sourirajan. Their discovery has made a huge breakthrough in membrane fabrication technique, which allowed membranes to be operated for high flux separation compared to the typical isotropic membrane, which posed high resistance for permeation [14]. Regardless, polymeric membranes exhibit an intrinsic performance that is characterized by an "upper boundary" introduced by Robeson [15]. Breaking through the limit of the upper bound has become a crucial interest among researchers in overcoming the intrinsic limit of polymeric membrane to enhance the gas separation performance for commercialization. In the next subtopic, the fundamentals of polymeric membrane for gas separation will be covered in depth. It also discusses the membrane fabrication process, its modification, and its prospect for natural gas sweetening.

2. Fundamentals of polymeric membrane gas separation

2.1 Mechanism

The polymeric membrane is a semipermeable material that is selective toward targeted gas molecules. The diffused gas molecules will be on the permeate side; hence, high purity of the permeable gas species can be obtained. Gas separation across a polymeric membrane can be governed by several mechanisms. They are known as convective flow, Knudsen diffusion, molecular

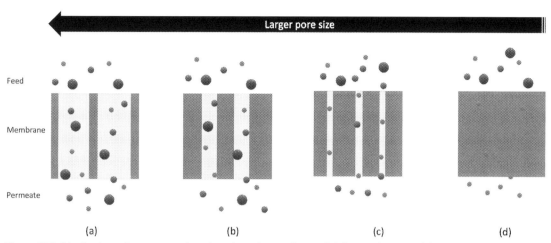

Figure 15.1 Mechanism of gas separation via polymeric membrane. (a) Convective flow, (b) Knudsen diffusion, (c) molecular sieving, and (d) solution-diffusion.

sieving, and solution diffusion. The membrane's structure and pore size dictate the separation mechanism, as illustrated in Fig. 15.1. However, only separation by molecular sieving and solution diffusion results in a polymeric membrane's selective separation.

For a highly porous membrane, all gas molecules can easily pass through the polymer due to the pore size reaching 10μm. Such pore size is bigger than the gas molecules by several magnitudes. Hence, it failed to separate the gas mixture effectively. In this case, the gas molecules' transportation mechanism is known as convective flow, and no selectivity could be achieved. This condition is also accompanied by the high permeability of all gases in the system as they can pass through the membrane with very minimal resistance. On the other hand, Knudsen diffusion involves the collision of the gas molecules with the pore wall due to the smaller distance than the nearest gas molecules [16]. The path of the molecules which are known as the mean free path (λ) is an essential feature in modeling Knudsen diffusion. This is because the molecular mean free path is much larger than the diameter of the pore. According to Rahimpour et al. [17], the Knudsen number (K_n) is calculated by taking the ratio between the mean free path of molecules to the membrane pore diameter (d). λ is a function of the gas viscosity (η), temperature (T), pressure (P), and molecular weight of the gas (M) [18]. The mean free path can be calculated using the equation below:

Table 15.1 Ideal selectivity of gas pairs in the case of Knudsen diffusion.

Gas pair	Ideal selectivity
CO_2/N_2	0.80
CO_2/CH_4	0.60
H_2/CO_2	4.69
He/CO_2	4.69
H_2S/CO_2	0.88
H_2S/CH_4	1.46

$$\lambda = \frac{3\eta(\pi RT)^{0.5}}{2P(2M)} \tag{15.1}$$

As a result, the separation took place due to the difference in the mean free path of the gas molecules as each gas species has its own λ value. This separation mechanism could also occur due to the membrane's defects. It is generally accepted that convective flow results in selectivity lower than unity, while the selectivity of separation by Knudsen diffusion can be easily characterized by the inverse of the square root of the molecular weight of the gas species. Table 15.1 lists the ideal selectivity of the gas pair in the case of Knudsen diffusion.

As the pore size approaches the size of gas molecules or becomes extremely small, better separation performance can be expected due to the molecular sieving ability of the membrane. Size exclusion allows the separation process to take place according to the kinetic diameter of the gas molecules. As the targeted gas molecules are equally small to the pore size, they can pass through the pores. In return, the permeability and selectivity are simultaneously enhanced. For a pristine polymeric membrane, the pore size can be manipulated by several ways, including altering the dope formulation and membrane fabrication method. Increasing the polymer concentration is the most straightforward approach to improve the polymer chain entanglement in producing membranes with smaller pores. The membrane fabrication method will be discussed in detail in the subsequent subsection.

Finally, the gas molecules can only permeate through a dense membrane via the solution diffusion mechanism. The dense structure possesses large resistance for the gas molecules to permeate. Solution diffusion is widely accepted as the primary mechanism of gas permeation through polymeric membranes. In fact, most commercialized polymeric membranes for gas separation obey this mechanism. It is a three-step process where the gas on the feed side is first adsorbed onto the membrane's outer skin, diffused through the polymer matrix, and finally desorbs on the permeate side [19]. These steps took place due to the concentration difference between the feed and permeate side of the membrane and the diffusion property of the gas molecules [20]. Moreover, the movement of the gas molecules across the dense membrane relies on the motion of the polymer chain.

Based on this mechanism, an expression of $P = D \bullet S$ was developed to characterize the separation properties of a polymeric membrane, where P, D, and S are the expressions of gas permeability ($cm^3(STP) \bullet cm/(cm^2 \bullet s \bullet cmHg)$), diffusivity ($cm^2/s$), and solubility ($cm^3(STP)/cm^3 \bullet cmHg$), respectively. The solubility term is primarily dependent on the chemical nature of the permeating molecules and polymer [21]. Koros and Fleming [22] suggested that the diffusivity is closely related to the size of the molecules, while the condensability of the gas molecules dictates the solubility factor. The permeability is calculated in terms of gas flux, according to the following equation:

$$P_i = \frac{Q\,l}{A(\Delta p)} \quad (15.2)$$

Where Q is the flux of the gas in cm^3/s, P_i is the permeability of the gas species i in Barrer ($1 \times 10^{-10}\,cm^3(STP) \cdot cm/cm^2 \cdot s \cdot cmHg$), A is the area of the permeation site for the membrane, l is the thickness of the membrane, and Δp is the differential pressure between the feed and permeate side. Eq. (15.2) is also used to calculate the permeance of a gas species ($\frac{P_i}{l}$) by considering the thickness of the membrane. Meanwhile, the following equation is derived to calculate the diffusivity and solubility selectively:

$$\alpha_{i/j} = \frac{P_i}{P_j} = \left[\frac{D_i}{D_j}\right] \bullet \left[\frac{S_i}{S_j}\right] \quad (15.3)$$

Where $\alpha_{i/j}$ is the selectivity between the gas i and j. Although separation via solution-diffusion offers excellent selectivity, the process is slow, hence impractical for many industrial applications. Based on Eq. (15.2), higher flux can be achieved by (i) increasing the pressure of the feed, (ii) creating a larger area for separation, and (iii) fabricating a thinner membrane. However, method (i) and (ii) are not feasible especially on an industrial scale, while the thin membrane is not easily fabricated and has low mechanical strength. It is widely accepted that diffusion is the rate-limiting step in the overall process. In this regard, altering a polymer's separation via the formulation and fabrication of membrane is among the strategies to enhance the diffusivity coefficient of CO_2 in the polymer. Meanwhile, method (iii) is plausible as an asymmetric membrane can be synthesized by controlling several dope formulations and fabrication methods.

Note that the separation process could also take place via the synergy between two mechanisms, for instance, solution diffusion/molecular sieving and Knudsen/solution diffusion. In this case, the solubility and diffusivity of the membrane were evaluated by $P = D \cdot S$ to confirm the effective mechanism that led to the separation. In the case of mixed gas as the feed, the performance of the membrane deviates from ideality due to the competitive dual-sorption mode. The Henry's Law mode sorption and Langmuir mode sorption were considered in developing the model for gas sorption across a glassy polymer [23].

Furthermore, the D and S values, hence the permeability of gas will change with respect to temperature, as described by the Arrhenius equation. The activation energy of the permeation in a nonporous material can be determined using Eqs. (15.4)–(15.6) [24]. The relation between E_P, E_d and ΔH_s is given by Eq. (15.7). The activation energy also scales linearly with the square of penetrant diameter as described in Eq. (15.5) [25].

$$P = P_o \exp \exp\left(-\frac{E_P}{RT}\right) \tag{15.4}$$

$$D = D_o \exp \exp\left(-\frac{E_D}{RT}\right) \tag{15.5}$$

$$S = S_o \exp \exp\left(-\frac{E_S}{RT}\right) \tag{15.6}$$

$$E_P = E_D + \Delta H_S \tag{15.7}$$

Where P_o, D_o and S_o are the pre-exponential factors obtained from Arrhenius plot, T is temperature, R is gas constant, E_P is activation energy of permeation, E_S is activation energy of sorption, and E_D is activation energy of diffusion.

$$E_D = md_i^2 + y \tag{15.8}$$

Where m and y are empirical parameters and d_i is the diameter of the gas penetrant.

In general, the pressure does not affect the permeability of the gas penetrant when the membrane is subjected to pressure lower than its plasticization pressure. However, it is observed that the temperature plays a major role in affecting the gas permeance and selectivity as observed by Acharya et al. [26]. With higher temperature, the gas permeance typically increases in a nonporous membrane. The phenomenon is ascribed to the enhanced polymer chain mobility at higher operating temperature, thus facilitating the gas diffusivity across the membrane at higher temperature yielding higher gas flux [24]. However, the segmental motion of the polymer chain with increasing temperature will reduce the capability of the gas to discriminate the penetrant based on the dimension thus resulting in minimized gas selectivity. Apart from diffusivity, the solubility of the penetrant which is governed by the nature of the chemical penetrant and the polymer-penetrant interaction is also affected by the temperature which in most cases increases with temperature [27]. For flue gas separation, the choice of CO_2-philic material with high thermal and mechanical stability and understanding the behavior of the gas at the targeted operating temperature conditions are crucial for the commercialization step.

2.2 Subclass and fabrication method of polymeric membrane

The fabrication method of polymeric membrane has evolved since its first introduction. Manipulating the fabrication process of a polymeric membrane resulted in the distinctive polymer structure, characteristics, and separation properties. In general, polymeric membranes are grouped according to their structure, which is symmetric and asymmetric. The illustration of the membrane structure is shown in Fig. 15.2.

Unlike nonporous dense membranes, microporous structure is not preferred for gas separation as the pore size does not allow selective separation. Meanwhile, the asymmetric membrane consists of thin dense skin and porous substructure where the first is responsible for selective separation, while the latter is vital to keep the integrity of the membrane [29]. An asymmetric membrane is attractive as it could reduce the thickness of the membrane while maintaining its mechanical strength to achieve higher flux.

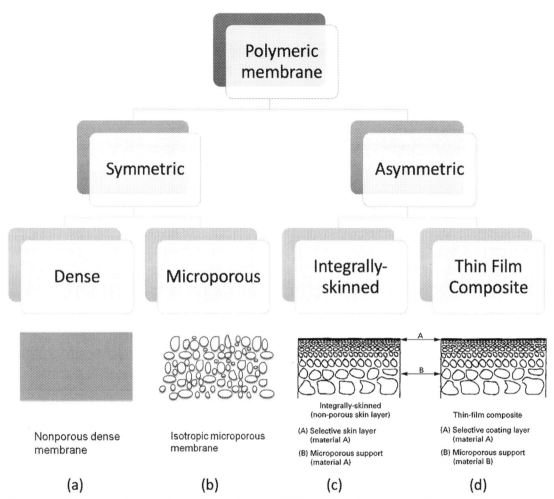

Figure 15.2 Subclasses of polymeric membrane structure. (a) Dense, (b) microporous, (C) integrally-skinned, and (d) thin-film composite. Adopted from Pinnau [28]. Reproduced with permission.

Asymmetric and symmetric structures can be obtained via different routes of fabrication. Among them are nonsolvent induced phase separation (NIPS) and evaporation induced phase separation (EIPS). Membranes prepared via the NIPS are exposed to a strong nonsolvent for phase inversion. In such a situation, the demixing process is faster, resulting in a less packed and porous membrane to be produced. NIPS is a method adopted from the Loeb-Sourirajan, which involves the immersion of the casted dope solution into a coagulation bath to allow the demixing of the polymer into polymer lean and rich phases. The thermodynamics and kinetics aspects are important factors to be considered during membrane fabrication, which was translated into a ternary diagram by researchers. The binodal and spinodal curves from the ternary diagram are useful indicators for forming a membrane with desired structure or morphology [30,31].

The EIPS method on the other hand is rather simple where the casted dope is subjected to heat and/or air convection to allow the solvent to escape from the casted polymer film. EIPS allows slow evaporation of solvent from the dope and typically used for dense membrane fabrication. Controlling the solvent evaporation time results in the formation of membranes with asymmetric structure.

According to Wang et al. [32], there are two subclasses of asymmetric membranes. The first consists of chemical compositions that are alike and fabricated by phase inversion (Loeb-Sourirajan membrane). The other type is formed using two distinct types of polymers or better known as the thin-film composite (TFC). Their differences are illustrated in Fig. 15.2C and D. TFC is also referred to as the ultrathin or thin polymeric layer assembled on polymer support. In many cases, TFC was introduced due to the surface defects that exist on the Loeb-Sourirajan membrane. Surface defects such as pinholes significantly deteriorate the membrane's performance in gas separation application as the size is large enough for convective flow. A typical example of TFC is the PES/PDMS membrane. There are many new formulations of TFC known today. For example, Langmuir–Blodgett (LB) films and layer-by-layer (LbL), and deposited polyelectrolyte multilayers (PEMs) [32].

NIPS and EIPS are well-established methods for flat sheet and hollow fiber polymeric membrane fabrications. However, the hollow fiber membrane is preferred for industrial applications as it offers a larger surface area with high self-mechanical support, easy handling during module fabrication, maintenance, and system operation [33]. Hollow fiber module is particularly attractive because it can bundle up thousands of fibers within a shell. However, the fabrication of hollow fiber membrane is a complex process, and many factors must be taken into consideration such as air gap, bore fluid composition, and flowrate and extrusion rate, among others. These parameters are important to be controlled to achieve the desired morphology of a hollow fiber membrane.

Many researchers have undermined the tunable trait of polymeric membranes by introducing fillers in the polymer matrix to form mixed matrix membrane (MMM) [34]. Materials such as zeolites, graphene, and metal-organic frameworks are highly selective toward CO_2 where its incorporation in the polymer matrix was aimed to enhance the permeability and selectivity of the

428 Chapter 15 Polymeric membranes for natural gas sweetening

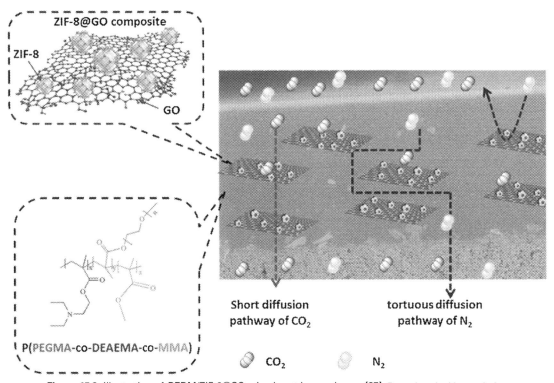

Figure 15.3 Illustration of PEDM/ZIF-8@GO mixed matrix membrane [37]. Reproduced with permission.

membrane [35]. The material could be introduced in the dope by direct mixing or priming method (in the case of highly viscous dope solution) [19]. The MMM can be formed either as symmetric (dense) or asymmetric. Besides altering the free volume and mechanical as well as thermal properties of the polymer matrix [36], the presence of the fillers impacted the diffusion path of the gas (illustrated in Fig. 15.3). In this regard, the resistance to the permeation is enhanced. Although excellent performance can be expected from this combination, many MMM formulations suffer from the trade-off effect. This is because MMMs possess several inherent limitations mainly caused by the filler and polymer matrix incompatibility.

3. Ideal membrane

An ideal polymeric membrane must exhibit high selectivity and permeability. However, the intrinsic property of the polymer restricts such cases. The trade-off between the permeability and selectivity is common in polymeric membranes. Traditionally, polymeric membrane is fabricated using glassy and/or rubbery polymer. Both types of polymers have distinctive separation properties. Glassy is known to have excellent selectivity at the expense of permeability, while the case is the opposite for rubbery polymer. This is because glassy polymer has high free volume due to its bulky constituents, rigid main chains, and low cohesive energies [38]. Typical glassy polymer used for natural gas processing includes cellulose acetate (CA), polyphenylene oxide (PPO), polyether sulfone (PES), polyimide (PI), and 6FDA-DAM, which has been marked as CO_2-philic polymers [39]. However, cellulose triacetate (CTA) membrane dominates the global market by 80% for natural gas removal due to its cost, processability, durability, reliability, high pressure resistance, and competitive gas-separation performance [40]. On the other hand, rubbery polymer has a more flexible polymer chain structure. As a result, the material has higher permeability [41]. However, rubbery polymer has low mechanical strength. Unlike glassy polymer, the rubbery able to work at temperature higher than the glass transition temperature (T_g). The typical example of rubbery polymer is polyethylene oxide (PEO), polyvinylidene fluoride (PVDF), and polydimethylsiloxane (PDMS).

In 1991, Robeson established a correlation of separation factor [15]. The study gathered the data of separation performance of numerous polymers to generate the correlation that we know now as Robeson's upper limit. The empirical correlation was updated in 2008 due to the myriad of data on many new polymers. The upper bound correlation can be used to qualitatively determine where the permeability process changes from solution-diffusion to Knudsen diffusion [42], whereas for H_2S, the fitted upper bound can be found in Ref. [39]. With the introduction of new materials for gas separation, the required high-performance membranes and the values of k and n to construct the upper bound are progressing.

Among newly introduced polymers for gas separation is Pebax. It combines rigid polyamide blocks (PA) and flexible polyether (PE) segments, which are glassy and rubbery at room temperature, respectively [43]. Many published works indicate that Pebax can easily approach Robeson's upper limit if not surpass it. On the other hand, polymers of intrinsic microporosity or PIMs have been taken into account to update the upper limit by Robeson. PIMs have exceptionally high permeability with moderate selectivity. The incorporation of bulky, rigid contortion centers disrupts the polymer chain packing and results in high excess free volume hence, high gas permeability. With an appropriate narrow bimodal distribution of micropores (<20 Å) and ultramicropores (<7 Å) in the polymer backbone, PIMs can transcend the permeability and selectivity trade-off [44].

Another relatively new polymer for efficient CO_2 separation is thermally rearranged (TR) polymer. Heat is an essential element in fabricating membrane using TR polymer to rearrange the molecules of polyimide and/or polyamides with ortho-functional groups (-OH, NH_2, and many more) [45]. The thermal treatment allows the formation of high free volume where it can be tuned to separate large gas molecules from small ones [46]. Similar to PIMs, TR allows high permeability of CO_2 due to the high free volume of the polymer following the reaction as subjected to heat. These materials performed exceptionally well due to the incorporation of intrinsically microporous units and/or increasing chain rigidity that can enhance microporosity in conventional polymer membrane materials [47]. The data of permeability and selectivity of pristine polymeric membrane can be found in Table 15.2.

4. Current application

Polymeric membrane is a promising technology for gas separation. The demand for the technology is reflected by the introduction of commercialized polymeric membrane by several organizations. This includes PRISM Membrane Separators and SEPURAN, which were fabricated by Air Products and Evonik, respectively. The industrial application of polymeric membrane includes reducing CO_2 content in flue gas and natural gas processing as well as upgrading biogas content. The demand for the technology for each application is unique due to specific requirements of the permeate. As an example, removal of CO_2 in natural gas processing is important to meet the pipeline specifications and regulatory standards on calorific value [66]. Farnam et al. [41] claimed that the membrane technology is the most suitable method for CO_2 separation from natural gas. The technology is particularly attractive to be used offshore due to its compactness, low labor intensity, ease of expansion, low maintenance, and low cost. Chu and He [67] on the other hand have proved the feasibility of polymeric membrane for natural gas processing. However, most lab-scale study of the membrane material is subjected to pure gas permeation at typically low pressure (<10 bar)

Table 15.2 Permeability and selectivity of pristine polymeric membrane.

Polymer class	Polymer	Permeation conditions	CO_2 permeability (Barrer)	H_2S permeability (Barrer)	He permeability (Barrer)	CO_2/CH_4	H_2S/CH_4	H_2/CO_2	He/CH_4	References
Glassy	Polyethersulfone (PES)	30°C, 10 atm	3.22			32.0				[48]
	6FDA-DAM:DABA	35°C, 2 bar	~20				8.3 ± 0.5			[49]
	6FDA-DAF	35°C, 1 −20 at,			98.5				406	[50]
	Poly(1-trimethylsilyl-1-propyne) (PTMSP)	23°C, 20 psig	18,200[a]	21,400[a]				~1.17		[51]
	Cellulose triacetate (CTA) hollow fiber	25°C, 1 atm 31.3 bar		~144 GPU	4100			~1.3	0.98	[52] [53]
	Cellulose acetate (CA)	22°C, 14 atm			14				97	[54]
	Polycarbonate	30°C, 14 atm			67				14	[54]
	Poly(methyl methacrylate) (PMMA)—atactic	35°C, 1 atm			9.43				1715	[55]
	Poly(methyl methacrylate) (PMMA)—syndiotactic				9.57				1495	
Rubbery	Polydimethylsiloxane (PDMS)	23°C, 20 psig	3200[a]	5100[a]				~1.59		[51]
		30°C, 14 atm			230				0.4	[54]
	Poly(ether urethanes) (PU)	35°C, 10 atm	44.7	199			74	4.45		[56]
	poly(ester urethane urea) (PEUU)	35°C, 10 bar		~140GPU[a]			43			[57]
Polymer blend	PSf/PEG	25°C, 1.5 atm	6.4			43.0				[58]
	Polysulfone/poyetherimide (PSF/PEI)	25°C, 10 bar	3.5			13.06				[59]
Composite	PES/PDMS hollow fiber	25°C, 30 bar		32.8GPU[a]			11.04			[60]
Pebax	Pebax	4.2 bar	100	487			49	10		[61]
	SA01 MV 300									

Continued

Table 15.2 Permeability and selectivity of pristine polymeric membrane.—continued

Polymer class	Polymer	Permeation conditions	CO_2 permeability (Barrer)	H_2S permeability (Barrer)	He permeability (Barrer)	CO_2/ CH_4	H_2 S/CH_4	H_2 S/CO_2	He/ CH_4	References
Polymers of intrinsic microporosity (PIM)	PIM-1	30°C, 2 bar	4216[a]			70.5				[62]
		25°C, 65 psi	6500			15.1				[63]
Thermally rearranged (TR)	Polypyrolone (PPL-450)		193.7			29.4				[64]
Ion-exchange membrane	Nafion 117	35°C, 1–20 atm			40.9				401	[65]

[a] mixed gas data.

4.1 CH$_4$ enrichment

Over the past decade, various literature on CH$_4$ separation using membranes were studied for both natural and biogas upgrading for its purification to replace the conventional cryogenic distillation, pressure swing adsorption, or amine scrubbing as these technologies require large carbon footprint and energy intensive. The composition of biogas content depends on the characteristics of the process such as the substrate or pH of the reactor. The main constituent consists of 60% of CH$_4$ with 40% of CO$_2$ with other traces of contaminants. In a typical process, CH$_4$ is composed of about 50%–70%, CO$_2$ at 30%–50%, while the rest consist of a trace amount of other gases such as volatile organic compounds (VOCs), carbon monoxide (CO), and ammonia (NH$_3$) [69]. Meanwhile, the typical concentration of natural gas is shown in Table 15.3, while the composition of the natural gas specification utilized for commercial application is shown in Table 15.4.

The main working principle of membrane for CH$_4$ enrichment application is that the feed stream with high impurities (which majorly consist of CO$_2$, N$_2$, and other traces) is fed to the retentate side of the membrane where CO$_2$ will permeate through the membrane and captured while CH$_4$ will flow to the retentate stream. As such, for CH$_4$ enrichment application, the separation quality is always described by the ratio of CO$_2$

Table 15.3 Typical composition of natural gas [70].

Gas components	Composition
Helium	0.0–1.8
Nitrogen	0.21–26.10
Carbon dioxide	0.06–42.66
Hydrogen sulfide	0.0 – 3.3
Methane	29.98–90.12
Ethane	0.55–14.22
Propane	0.23–12.54
Butane	0.14–8.12
Pentane and heavier hydrocarbon chain	0.037–3.0

Table 15.4 Natural gas specification for commercial application [70].

Gas component	Minimum mol%	Maximum mol%
Methane	75	None
Ethane	None	10
Propane	None	5
Butane	None	2
Pentane and heavier hydrocarbon chain	None	0.5
Nitrogen and other inerts	None	3
Carbon dioxide	None	2—3
Total diluent gases	None	4—5

permeate to CH_4 permeate (selectivity). CO_2 will permeate faster through the membrane due to its smaller gas kinetic diameter than CH_4. Although numerous polymeric materials were reported exhibiting high CO_2/CH_4 ideal selectivity that could reach beyond 50 along with high reported CO_2 permeability, CA or polyimide-based membrane is the material of choice in real industrial plant with CO_2/CH_4 selectivity ranging from 10 to 15 [68]. In a mixed-gas system, the permeation behaves differently from pure gas permeation as the gas is no longer governed by simple diffusion mechanism across the nonporous wall but also experiences competitive dual-sorption mode among the gas mixture, which will reduce the solubility coefficient of CO_2 of the membrane materials, thus contributing to lower CO_2/CH_4 selectivity as compared to pure gas performance [71]. Additionally, the pressure of natural gas typically ranges from 200 to 1500 psi. In a mixed-gas separation concerning the separation of condensable components, in high-pressure conditions such condensable components may liquefy and dissolve in the membrane matrix thus leading to plasticization phenomena. Plasticization will cause swelling of the membrane matrix due to sorption of the penetrant, thus causing an irreversible increase in the membrane matrix free volume. Such phenomenon is typically observed when there is a sudden increase of gas permeability with reduced gas selectivity when the membrane is operated exceeding a certain pressure. Such pressure is called the plasticizing pressure, where its influence on the permeability is illustrated in Fig. 15.4 [72—74]. Thermal treatment, cross-linking, polymer sulfonation, and thermal rearrangement are

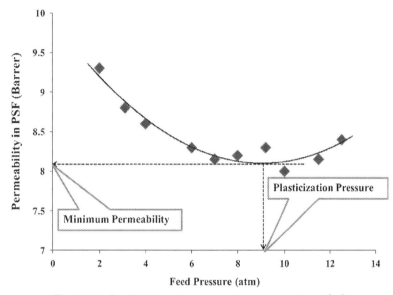

Figure 15.4 Plasticization phenomena in a gas membrane [72].

among the common techniques that are employed to overcome the swelling phenomenon [75].

4.2 Acid gas removal unit

In natural gas sweetening, the removal of acidic gases from natural gas liquids is also important in order to meet the pipeline requirement and prevent pipeline corrosion. H_2S and CO_2 are the acidic gases that can be found in natural gas liquids. The contaminant concentration should be lower than 4 ppm for H_2S, <2 wt.% of CO_2 and <0.1 g/m^3 of water vapor [76]. The presence of these gases alone is not capable of causing corrosion; however, with the presence of water vapor, these gases may form by products and dissociate into hydrogen ions (or hydronium), which may cause corrosion to the pipeline. For instance, H_2S can dissociate into H^+ and S^{2-} ions when it is dissolved in water vapor which can cause pipeline corrosion even when it present in low concentration level [77,78] while CO_2 in aqueous condition forms carbonic acid which further dissociates into two H^+ and CO_3^{2-}. The equation for the dissociation of the gases is shown below:

For H_2S:

$$H_2S(aq) \rightleftharpoons HS^- + H^+ \qquad (15.9)$$

$$HS^-(aq) \rightleftharpoons S^{2-} + H^+ \qquad (15.10)$$

For CO_2:

$$CO_2(aq) + H_2O(aq) \rightleftharpoons H_2CO_3(aq) \quad (15.11)$$

$$H_2CO_3(aq) \rightleftharpoons H^+ + HCO_3^- \quad (15.12)$$

$$HCO_3^- \rightleftharpoons H^+ + CO_3^{2-} \quad (15.13)$$

H_2S can act as a catalyst in the absorption of atomic hydrogen in steel, promoting sulfide stress cracking (SSC) in high-strength steels. The corrosion products are iron sulfides and hydrogen. Iron sulfide forms a scale at low temperatures and can act as a barrier to slow corrosion. The absence of chloride salts strongly promotes this condition, and the absence of oxygen is absolutely essential. At higher temperatures, the scale is cathodic in relation to the casing and galvanic corrosion starts. The chloride forms a layer of iron chloride, which is acidic and prevents the formation of an FeS layer directly on the corroding steel, enabling the anodic reaction to continue. Hydrogen produced in the reaction may lead to hydrogen embrittlement [79].

Typically, Girbotol process is used to remove H_2S from the natural gas liquid by reacting H_2S with ethanolamine at low temperature, and the H_2S will be released and captured when it is heated to high temperature. This process is chemical absorption which is similarly used for carbon capture. The process is energy demanding as it requires heat to heat the ethanolamine containing H_2S to release the H_2S and capture it. Yahaya et al. [80] synthesized copolyimide membrane based on with various combinations of homopolymers for H_2S separation from CH_4. They discovered that polyimide made from 6FDA-DAM/CARDO (6FDA = 4,4'-(Hexafluoroisopropylidene)diphthalic anhydride, DAM = 2,4,6-trimethylbenzene-1,3-diamine, CARDO = 4,4'-(9H-fluorene-9,9-diyl)dianiline) with a ratio of 1:3 shows the highest H_2S/CH_4 selectivity when the membrane is subjected to 24.1 bar at 22°C. The selectivity remained relatively constant when the pressure is further increased to 34.5 bar. Harrigan et al. [81] showed that the properties of the PEG membrane can be tuned by crosslinking different polyethylene glycol (PEG) with different molecular weight. Crosslinked PEG membrane with lower molecular weight favored the separation of both H_2S/CH_4 and CO_2/CH_4 with separation factor greater than 60, while crosslinked PEG with higher molecular weight shows an increased separation factor of H_2S/CH_4 with reduced CO_2/CH_4 performance. The resulting membranes span both rubbery and glassy regimes, rubbery favoring H_2S/CH_4 and glassy favoring CO_2/CH_4. Liu et al. [53] showed in their study that the plasticization effect on asymmetric cellulose triacetate membrane is beneficial for the simultaneous removal of H_2S and CO_2 from CH_4 with the presence of toluene and other hydrocarbon compounds. For H_2S/CH_4 where it is sorption-dominated process, the effect of plasticization leads to increased free volume in the glassy polymer, hence providing more free volume for more sorption capacity, regardless of the type of glassy polymer.

Membrane separation of bulk H_2S removal from natural gas pipeline stream is established in commercial scales [82]. However, it is important to highlight that H_2S is a condensable gas where the vapor pressure is approximately 17.5 barg in room temperature condition. Hence, for a gas separation membrane where a glassy polymer is concerned, the high-pressure operation to separate H_2S may lead to plasticization, which will deformed the membrane. As such, rubbery polymer is more favorable where the separation is favored by solubility (S) selectivity of the condensable gas rather than by diffusivity (D) through the polymer matrix.

4.3 Hydrocarbon separations

The use of membrane technology to separate hydrocarbons is rather challenging. Currently, the separation of hydrocarbons is done by fractional distillation [1]. Hydrocarbons with longer chains have low dew points, thus easily condensable at high pressure. Therefore, concerning hydrocarbon separation, rubbery polymer is favorable due to its ability to separate based on condensability of the gas. Typical glass polymers used for CO_2 capture are very unlikely to be utilized for hydrocarbon separation as they are selective based on diffusivity of the gas.

Polydimethylsiloxane (PDMS) is a common rubbery polymer that is often utilized for hydrocarbon separation as it is readily available, cheap and can be easily fabricated as a membrane. Moreover, PDMS shows good selectivity toward methane relative to other heavier hydrocarbons as summarized in Table 15.5. However, the use of PDMS alone is not sufficient as the presence of these heavier hydrocarbons will cause the membrane to swell

Table 15.5 Pure gas permeability of hydrocarbons in PDMS (at 35°C) [83].

Gas	Permeability (Barrer)	Ideal selectivity (against CH_4)
Methane	800	—
Ethane	2100	2.625
Propane	3500	4.375
n-Butane	7500	9.375
n-Pentane	16,700	20.875

over time. Subsequently, the performance of the membrane easily degrades. To minimize methane loss while maximizing the hydrocarbon recovery, multiple stages of membrane separation with recycle stream design is required.

Gou et al. [84] prepared PDMS embedded with MFI nanosheet MMM for isomer butane separation. In their study, the incorporation of 1 wt.% open pore structure MFI nanosheet into PDMS achieved an n-butane/i-butane separation factor of 15.6 with 15,615 Barrer of n-butane at room temperature. It is hypothesized that the incorporation of MFI nanosheet zeolite increased the trapped free volume of PDMS and tortuous path of nondiffusive gas, hence providing simultaneous improvement of permeability and selectivity.

5. Challenges

Despite wide literature published on the polymeric membrane for gas separation application, the commercialization aspect is still far from realization. It is desirable for the membrane to exhibit high gas flux with high degree of separation to be utilized for commercial use. However, it is recognized that the polymeric material is often bounded by the Robeson upper boundary. As such, various modification strategies to the polymeric material have been utilized to overcome the upper boundary and improve the separation performance. Hollow fiber is more suitable for industrial commercialization due to their high effective permeation area to volume ratio that could exceed 1000 m^2/m^3 [48,85]. However, the fabrication of asymmetric hollow fiber membrane is more complex than asymmetric flat sheet membrane, which involves complex interplay between various factors such as the polymer dope concentration and viscosity, the temperature of dope spinning solution, the air gap distance, the type and temperature of coagulant bath, the composition of the bore fluid, the ratio of bore fluid flow rate and dope extrusion speed, the drawing speed ratio, and the effect of nonsolvent in the dope solution [86]. Often time, the fabrication of asymmetric hollow fiber membrane leads to incomplete polymer coalescence in the skin layer and formation of pinholes on the surface of the membrane as the polymer dope is subjected to fast liquid—liquid demixing [29]. The presence of pinholes is not favorable for gas separation application as they can detrimentally affect the selectivity ratio of the desired gas, which usually drops close to 1 as the membrane cannot discriminate against the gas penetrant as the gas will simply pass through the pinhole. In such a case, the gas separation will be dominated by a viscous flow mechanism. Therefore, highly permeable rubbery polymer such as PDMS is

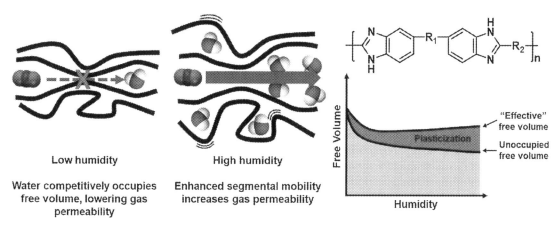

Figure 15.5 Effect of humidity on polymer chain arrangement [88].

used to seal the defective surface of the membrane to improve the selectivity performance [87].

Other than defective surfaces, humidity may also affect the separation performance. Presence of water vapor will alter the polymeric chain material as they are also a condensable penetrant as illustrated in Fig. 15.5. Additionally, vapor permeation is much higher than other gas penetrants; hence, polymeric membrane is commonly studied with dry pure or mixed gas separation in a lab scale study. The inclusion of humid air in the literature study as one of the factors to gauge the performance of the polymeric membrane in the long run is crucial.

Polymeric membrane is susceptible to aging over the long run. As such, at some point in the operation, the polymeric membrane needs to be replaced with a new module. Physical aging of membranes made out of polymeric material is a common occurrence and one of the hurdles that need to be overcome to prolong the operational time and reduce capital cost. Typically, physical aging can be detected by the performance of the polymeric membrane where the gas permeability reduces with respect to time after the long run. This was ascribed to the relaxation of the glassy polymeric chain as they achieved the equilibrium state, thus reducing the free fractional volume of the glassy polymer, imposing more resistance to the penetrant. However, in most cases, the gas selectivity remains relatively unaffected. Plasticization is usually observed in the highly permeable glassy polymeric membranes [89,90]. In the long run, the membrane may also lose the rigidity of the polymeric chain as they are constantly subjected to harsh conditions and penetrants, thus accelerating the physical aging phenomenon.

Overcoming the intrinsic limitation of polymeric material has become the main focus of membrane research. There are several ways of improving the separation capability of polymeric membranes such as by surface modification [91], polymer cross-linking [92], polymer blending [93] or combining inorganic–organic material to fabricate MMM [94].

The blending of glassy and rubbery polymer draws attention as this approach is able to produce a material with new and excellent characteristics. Blend of glassy and rubbery polymers should combine the advantages of both materials and improved selectivity may be expected compared to the neat polymer [95]. However, the main issue to be addressed is the miscibility between the components. There are three classes of polymer blends which are miscible, partially miscible and immiscible blend. The first is named due to the complete miscibility of both polymers with each other and form a homogeneous system, while partially miscible blend is a result of a component with a smaller, but sufficient quantity dissolved in another to alter its properties [96]. An immiscible blend is separated by an interface between the two phases as the polymers do not dissolve in each other [97]. The miscibility of polymer blends is determined by the solubility parameter calculation and T_g, where a miscible blend recorded a single T_g value [41]. Among excellent polymer blends for CO_2 separation are PES/PEG [98] and PSF/PEI [59].

MMM is widely studied due to its easy and facile preparation procedure in the laboratory thus potentially having the widest option for modification. Moreover, MMM not only improves the separation performance but also improves the characteristics of the membrane such as their free fractional volume and tensile strength. Various polymer fillers have been studied for MMM fabrication such as metal oxides [99], graphene oxide (GO) [100], carbon nanotube (CNT) [101], zeolites [102], MOF [103] and clays [104]. Nevertheless, even though the MMM effectively improves gas separation, there are also several drawbacks such as agglomeration, poor particle distribution, incompatible polymer–filler interface, and many more. The complication arising from these drawbacks could cause polymer chain rigidification or voidage formation inside the polymer matrix, thus deterring the gas separation performance instead of improving it. The use of porous fillers such as zeolites, MOF, carbon nanotubes, or carbon molecular sieves is interesting as these materials possessed pore sizes that could discriminate the gas molecule through molecular sieving. For example, SAPO-34 has a pore window of 3.8 Å, which falls between the kinetic diameter of CO_2 (3.3 Å) and CH_4 (3.8 Å). Zeolitic imidazolate framework-8 (ZIF-8) has a pore window of 3.4 Å that falls between the kinetic diameter of CO_2 and N_2 (3.64 Å). The capability of these porous materials to discriminate the gas molecules by molecular sieving is appealing as they are highly selective while offering the highest gas flux [105]. However, during the physical blending in MMM preparation, there is a tendency for the polymer to block the pores of the fillers thus reducing the gas flux. Nonporous fillers such as metal oxides or mineral clays will alter the polymeric chain of the membrane, thus, increasing the free fractional volume of the polymer chain. Therefore, the permeability of the gas increases. Regardless, this phenomenon does not always happen as there is a tendency for voidage to form around the polymer–filler interface forming a nonselective layer, thus affecting the selectivity of the gas [106]. There is also a possibility that the polymeric chain around the filler is rigidified thus reducing the permeability of the gas in the MMM. The common defects that arise from MMM are shown in Fig. 15.6. Other than that, aggregation of filler particles also forms

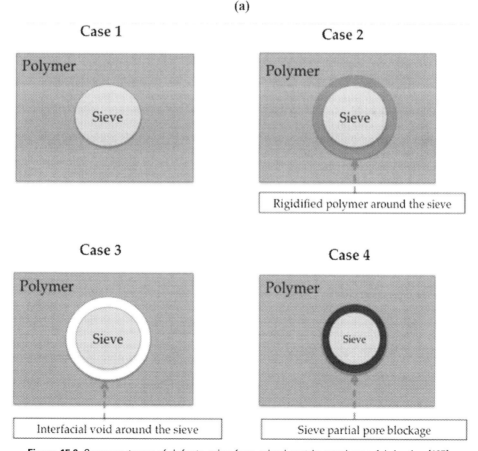

Figure 15.6 Common types of defects arise from mixed matrix membrane fabrication [107].

nonselective voids which jeopardize the selectivity of the membrane [108]. To overcome these limitations, the inorganic filler was modified by various means to simultaneously enhance the compatibility between polymer and inorganic filler interface, hence the separation performance. Common modification methods include amine-functionalization, surface grafting, pre or postimpregnation, dual-filler hybrid, or chemical etching [19,109,110]. Moreover, recently it has been reviewed that the potential of using 2D-based inorganic filler in polymeric membranes can potentially increase the separation performance due to the barrier properties of the 2D structure [111–113].

6. Conclusion and future outlooks

Although various literature reports on membranes for gas separation are currently blooming, the realization of these findings for industrial use is yet to come. This is mainly because in a real plant, the composition of the gas component could vary and fluctuate during the separation process thus causing the performance of the membrane to vary as well. Additionally, the condition subjected to the membrane on a lab scale also varies from the one in a real plant. The presence of condensable gases such as light hydrocarbon, CO_2, and water vapor may cause the membrane to perform lower than its intrinsic performance.

In MMM fabrication, it has long been stated that the inorganic filler with a lamellar structure that is oriented perpendicular to the gas penetration pathway exhibits barrier properties that can improve the selectivity of the gas separation performance. However, the challenge to fabricate the lamellar nanosheet fillers at a large scale still remains elusive. As such, despite the good prospect offered by 2D lamellar material such as exhibiting barrier properties, which can improve the separation performance, fabrication of MMM with 2D lamellar filler remains a challenge. Additionally, most of the 2D nanosheet materials utilize the hydrothermal method which requires a long fabrication time to obtain the materials. As such, scaling up the fabrication process for large-scale MMM production is also another challenge that needs to be overcome. Additionally, new polymeric materials that exhibit highly porous and selective properties such as polymer of intrinsic microporosity (PIM) have paved a new perspective on polymeric material development to further improve the gas flux and selectivity to meet industrial demand.

Abbreviations and symbols

6FDA	4,4′ -(Hexafluoroisopropylidene)diphthalic anhydride
AGRU	Acid gas removal unit (AGRU)
CA	cellulose acetate
CARDO	4,4′ - (9H-fluorene-9,9-diyl)dianiline)
CTA	cellulose triacetate
DAM	2,4,6-trimethylbenzene-1,3-diamine
EIPS	evaporation induced phase separation
GO	graphene oxide
LB	Langmuir–Blodgett
LbL	layer-by-layer
MMM	mixed matrix membrane
NG	natural gas
NIPS	nonsolvent induced phase separation
PA	polyamide blocks
PDMS	polydimethylsiloxane
PE	flexible polyether
PEG	polyethylene glycol
PEM	polyelectrolyte multilayers
PES	polyether sulfone
PESO	polyethylene oxide
PEUU	poly(ester urethane urea)
PI	polyimide
PIM	polymer of intrinsic microporosity
PMMA	poly(methyl methacrylate)
PPO	polyphenylene oxide
PTMSP	poly(1-trimethylsilyl-1-propyne)
PU	poly(ether urethanes)
PVDF	polyvinylidene fluoride
SSC	sulfide stress cracking
TFC	thin-film composite
TR	thermally rearranged
ZIF-8	zeolitic imidazolate framework-8

Symbols

A	area of the permeation site
d	membrane pore diameter
D	diffusivity (cm^2/s)
d_i	diameter of the gas penetran
D_o	pre-exponential factors
E_D	activation energy of diffusion
E_P	activation energy of permeation
E_S	activation energy of sorption
K_n	Knudsen number
L	thickness of the membrane
M	molecular weight of the gas
m,y	empirical parameters
P	pressure
P	permeability (cm^3(STP) • cm/(cm^2 • s • cmHg))
P_o	pre-exponential factors
Q	flux of the gas in cm^3/s
R	gas constant

S	solubility (cm³(STP)/cm³ • cmHg)
S_o	pre-exponential factors
T	temperature
η	gas viscosity
$\alpha_{i/j}$	selectivity between the gas i and j
λ	mean free path
Δp	differential pressure between the feed and permeate side

References

[1] Baker RW, Lokhandwala K. Natural gas processing with membranes: an overview. Industrial & Engineering Chemistry Research 2008;47:2109−21. https://doi.org/10.1021/ie071083w.

[2] Karadas F, Atilhan M, Aparicio S. Review on the use of ionic liquids (ILs) as alternative fluids for CO 2 capture and natural gas sweetening. Energy & Fuels 2010;24:5817−28. https://doi.org/10.1021/ef1011337.

[3] Tikadar D, Gujarathi AM, Guria C. Safety, economics, environment and energy based criteria towards multi-objective optimization of natural gas sweetening process: an industrial case study. Journal of Natural Gas Science and Engineering 2021;95:104207. https://doi.org/10.1016/j.jngse.2021.104207.

[4] Ababneh H, AlNouss A, Karimi IA, Al-Muhtaseb SA. Natural gas sweetening using an energy-efficient, state-of-the-art, solid−vapor separation process. Energies 2022;15:1−12. https://doi.org/10.3390/en15145286.

[5] Mustafa J, Farhan M, Hussain M. CO2 separation from flue gases using different types of membranes. Journal of Membrane Science & Technology 2016;6. https://doi.org/10.4172/2155-9589.1000153.

[6] Kadirkhan F, Sean GP, Ismail AF, Wan Mustapa WNF, Halim MHM, Kian SW, et al. CO2 plasticization resistance membrane for natural gas sweetening process: defining optimum operating conditions for stable operation. Polymers 2022;14. https://doi.org/10.3390/polym14214537.

[7] Liu Y, Liu Z, Kraftschik BE, Babu VP, Bhuwania N, Chinn D, et al. Natural gas sweetening using TEGMC polyimide hollow fiber membranes. Journal of Membrane Science 2021;632:119361. https://doi.org/10.1016/j.memsci.2021.119361.

[8] He X, Kumakiri I, Hillestad M. Conceptual process design and simulation of membrane systems for integrated natural gas dehydration and sweetening. Separation and Purification Technology 2020;247:116993. https://doi.org/10.1016/j.seppur.2020.116993.

[9] Yi S, Ghanem B, Liu Y, Pinnau I, Koros WJ. Ultraselective glassy polymer membranes with unprecedented performance for energy-efficient sour gas separation. Science Advances 2019;5:1−12. https://doi.org/10.1126/sciadv.aaw5459.

[10] Mitchell JK. On the penetrativeness of fluids. Journal of Membrane Science 1995;100:11−6. https://doi.org/10.1016/0376-7388(94)00227-P.

[11] Graham T, Lv. On the absorption and dialytic separation of gases by colloid septa. London, Edinburgh, Dublin Philosophical Magazine and Journal of Science 1866;32:401−20. https://doi.org/10.1080/14786446608644207.

[12] Böddeker KW. Commentary: tracing membrane science. Journal of Membrane Science 1995;100:65—8. https://doi.org/10.1016/0376-7388(94)00223-L.
[13] Fauzi A, Djoko T, Mustafa A, Hasbullah H. Understanding the solution-diffusion mechanism in gas separation membrane for engineering students. Physics 2005:155—9.
[14] Loeb S, Sourirajan S. Sea water demineralization by means of an osmotic membrane. In: Sea water deminer. By means an osmotic membr; 1963. p. 117—32. https://doi.org/10.1021/ba-1963-0038.ch009.
[15] Robeson LM. Correlation of separation factor versus permeability for polymeric membranes. Journal of Membrane Science 1991;62:165—85. https://doi.org/10.1016/0376-7388(91)80060-J.
[16] Gitis V, Rothenberg G. Fundamentals of membrane separation. In: Ceram. Membr. New oppor. Pract. Appl. Wiley; 2016. p. 91—148. https://doi.org/10.1002/9783527696550.ch2.
[17] Rahimpour MR, Kazerooni NM, Parhoudeh M. Water treatment by renewable energy-driven membrane distillation. In: Current trends and future developments on (bio-) membranes. Elsevier; 2019. p. 179—211. https://doi.org/10.1016/B978-0-12-813545-7.00008-8.
[18] Basu A, Akhtar J, Rahman MH, Islam MR. A review of separation of gases using membrane systems. Petroleum Science and Technology 2004;22:1343—68. https://doi.org/10.1081/LFT-200034156.
[19] Shah Buddin MMH, Ahmad AL. A review on metal-organic frameworks as filler in mixed matrix membrane: recent strategies to surpass upper bound for CO2 separation. Journal of CO2 Utilization 2021;51:101616. https://doi.org/10.1016/j.jcou.2021.101616.
[20] Gugliuzza A, Basile A. Membrane processes for biofuel separation: an introduction. In: Membranes for clean and renewable power applications. Elsevier; 2014. p. 65—103. https://doi.org/10.1533/9780857098658.2.65.
[21] Dudek G, Borys P. A simple methodology to estimate the diffusion coefficient in pervaporation-based purification experiments. Polymers 2019;11. https://doi.org/10.3390/polym11020343.
[22] Koros WJ, Fleming GK. Membrane-based gas separation. Journal of Membrane Science 1993;83:1—80. https://doi.org/10.1016/0376-7388(93)80013-N.
[23] Biondo LD, Duarte J, Zeni M, Godinho M. A dual-mode model interpretation of CO2/CH4 permeability in polysulfone membranes at low pressures. Anais da Academia Brasileira de Ciencias 2018;90:1855—64. https://doi.org/10.1590/0001-3765201820170221.
[24] Car A, Stropnik C, Yave W, Peinemann KV. Pebax®/polyethylene glycol blend thin film composite membranes for CO2 separation: performance with mixed gases. Separation and Purification Technology 2008;62:110—7. https://doi.org/10.1016/j.seppur.2008.01.001.
[25] Lasseuguette E, Malpass-Evans R, Carta M, McKeown NB, Ferrari MC. Temperature and pressure dependence of gas permeation in a microporous Tröger's base polymer. Membranes 2018;8:1—11. https://doi.org/10.3390/membranes8040132.
[26] Acharya NK, Yadav PK, Vijay YK. Study of temperature dependent gas permeability for polycarbonate membrane. Indian Journal of Pure & Applied Physics 2004;42:179—81.
[27] Kulshrestha V, Awasthi K, Acharya NK, Singh M, Vijay YK. Effect of temperature and α-irradiation on gas permeability for polymeric

membrane. Bulletin of Materials Science 2005;28:643—6. https://doi.org/10.1007/BF02708532.
[28] Pinnau I. Membrane separations | membrane preparation. In: Encycl. Sep. Sci. Elsevier; 2000. p. 1755—64. https://doi.org/10.1016/B0-12-226770-2/05241-8.
[29] Ismail AF, Yean LP. Review on the development of defect-free and ultrathin-skinned asymmetric membranes for gas separation through manipulation of phase inversion and rheological factors. Journal of Applied Polymer Science 2003;88:442—51. https://doi.org/10.1002/app.11744.
[30] Wang D, Li K, Teo WK. Polyethersulfone hollow fiber gas separation membranes prepared from NMP/alcohol solvent systems. Journal of Membrane Science 1996;115:85—108. https://doi.org/10.1016/0376-7388(95)00312-6.
[31] Barzin J, Sadatnia B. Theoretical phase diagram calculation and membrane morphology evaluation for water/solvent/polyethersulfone systems. Polymer (Guildf) 2007;48:1620—31. https://doi.org/10.1016/j.polymer.2007.01.049.
[32] Wang M, Zhao J, Wang X, Liu A, Gleason KK. Recent progress on submicron gas-selective polymeric membranes. Journal of Materials Chemistry A 2017;5:8860—86. https://doi.org/10.1039/C7TA01862B.
[33] Bazhenov SD, Bildyukevich AV, Volkov AV. Gas-liquid hollow fiber membrane contactors for different applications. Fibers 2018;6. https://doi.org/10.3390/fib6040076.
[34] Chung TS, Jiang LY, Li Y, Kulprathipanja S. Mixed matrix membranes (MMMs) comprising organic polymers with dispersed inorganic fillers for gas separation. Progress in Polymer Science 2007;32:483—507. https://doi.org/10.1016/j.progpolymsci.2007.01.008.
[35] Li S, Liu Y, Wong DA, Yang J. Recent advances in polymer-inorganic mixed matrix membranes for CO 2 separation. Polymers (Basel) 2021;13(15).
[36] Muthukumaraswamy Rangaraj V, Wahab MA, Reddy KSK, Kakosimos G, Abdalla O, Favvas EP, et al. Metal organic framework — based mixed matrix membranes for carbon dioxide separation: recent advances and future directions. Frontiers in Chemistry 2020;8:1—25. https://doi.org/10.3389/fchem.2020.00534.
[37] Chen B, Wan C, Kang X, Chen M, Zhang C, Bai Y, et al. Enhanced CO2 separation of mixed matrix membranes with ZIF-8@GO composites as fillers: effect of reaction time of ZIF-8@GO. Separation and Purification Technology 2019;223:113—22. https://doi.org/10.1016/j.seppur.2019.04.063.
[38] Arya RK, Thapliyal D, Sharma J, Verros GD. Glassy polymers—diffusion, sorption, ageing and applications. Coatings 2021;11:1049. https://doi.org/10.3390/coatings11091049.
[39] Ma Y, Guo H, Selyanchyn R, Wang B, Deng L, Dai Z, et al. Hydrogen sulfide removal from natural gas using membrane technology: a review. Journal of Materials Chemistry A 2021;9:20211—40. https://doi.org/10.1039/D1TA04693D.
[40] Lu HT, Liu L, Kanehashi S, Scholes CA, Kentish SE. The impact of toluene and xylene on the performance of cellulose triacetate membranes for natural gas sweetening. Journal of Membrane Science 2018;555:362—8. https://doi.org/10.1016/j.memsci.2018.03.045.

[41] Farnam M, bin Mukhtar H, bin Mohd Shariff A. A review on glassy and rubbery polymeric membranes for natural gas purification. ChemBioEng Reviews 2021;8:90–109. https://doi.org/10.1002/cben.202100002.

[42] Robeson LM. The upper bound revisited. Journal of Membrane Science 2008;320:390–400. https://doi.org/10.1016/j.memsci.2008.04.030.

[43] Embaye AS, Martínez-Izquierdo L, Malankowska M, Téllez C, Coronas J. Poly(ether- block -amide) copolymer membranes in CO 2 separation applications. Energy & Fuels 2021;35:17085–102. https://doi.org/10.1021/acs.energyfuels.1c01638.

[44] Wang Y, Ghanem BS, Han Y, Pinnau I. State-of-the-art polymers of intrinsic microporosity for high-performance gas separation membranes. Current Opinion in Chemical Engineering 2022;35:100755. https://doi.org/10.1016/j.coche.2021.100755.

[45] Tong Z, Sekizkardes A. Recent developments in high-performance membranes for CO2 separation. Membranes 2021;11:156. https://doi.org/10.3390/membranes11020156.

[46] Sanders DF, Smith ZP, Guo R, Robeson LM, McGrath JE, Paul DR, et al. Energy-efficient polymeric gas separation membranes for a sustainable future: a review. Polymer (Guildf). 2013;54:4729–61. https://doi.org/10.1016/j.polymer.2013.05.075.

[47] Lee WH, Seong JG, Hu X, Lee YM. Recent progress in microporous polymers from thermally rearranged polymers and polymers of intrinsic microporosity for membrane gas separation: pushing performance limits and revisiting trade-off lines. Journal of Polymer Science 2020;58:2450–66. https://doi.org/10.1002/pol.20200110.

[48] Chen XY, Kaliaguine S, Rodrigue D. Polymer hollow fiber membranes for gas separation: a comparison between three commercial resins. In: AIP Conference Proceedings; 2019. p. 070003. https://doi.org/10.1063/1.5121669.

[49] Kraftschik B, Koros WJ, Johnson JR, Karvan O. Dense film polyimide membranes for aggressive sour gas feed separations. Journal of Membrane Science 2013;428:608–19. https://doi.org/10.1016/j.memsci.2012.10.025.

[50] Kim T-H, Koros WJ, Husk GR. Advanced gas separation membrane materials: rigid aromatic polyimides. Separation Science and Technology 1988;23:1611–26. https://doi.org/10.1080/01496398808075652.

[51] Merkel T, Gupta RP, Turk B, Freeman BD. Mixed-gas permeation of syngas components in poly(dimethylsiloxane) and poly(1-trimethylsilyl-1-propyne) at elevated temperatures. Journal of Membrane Science 2001;191:85–94. https://doi.org/10.1016/S0376-7388(01)00452-5.

[52] Takada K, Matsuya H, Masuda T, Higashimura T. Gas permeability of polyacetylenes carrying substituents. Journal of Applied Polymer Science 1985;30:1605–16. https://doi.org/10.1002/app.1985.070300426.

[53] Liu Y, Liu Z, Morisato A, Bhuwania N, Chinn D, Koros WJ. Natural gas sweetening using a cellulose triacetate hollow fiber membrane illustrating controlled plasticization benefits. Journal of Membrane Science 2020;601:117910. https://doi.org/10.1016/j.memsci.2020.117910.

[54] Gantzel PK, Merten U. Gas separations with high-flux cellulose acetate membranes. Industrial and Engineering Chemistry Process Design and Development 1970;9:331–2. https://doi.org/10.1021/i260034a028.

[55] Min KE, Paul DR. Effect of tacticity on permeation properties of poly(methyl methacrylate). Journal of Polymer Science, Part B: Polymer Physics 1988;26:1021–33. https://doi.org/10.1002/polb.1988.090260507.

[56] Chatterjee G, Houde AA, Stern SA. Poly(ether urethane) and poly(ether urethane urea) membranes with high H2S/CH4 selectivity. Journal of Membrane Science 1997;135:99–106. https://doi.org/10.1016/S0376-7388(97)00134-8.

[57] Mohammadi T, Moghadam MT, Saeidi M, Mahdyarfar M. Acid gas permeation behavior through poly(ester urethane urea) membrane. Industrial & Engineering Chemistry Research 2008;47:7361–7. https://doi.org/10.1021/ie071493k.

[58] Jujie L, He X, Si Z. Polysulfone membranes containing ethylene glycol monomers: synthesis, characterization, and CO2/CH4 separation. Journal of Polymer Research 2017;24:1. https://doi.org/10.1007/s10965-016-1163-6.

[59] Mukhtar H, Mannan HA, Minh D, Nasir R, Moshshim DF, Murugesan T. Polymer blend membranes for CO2 separation from natural gas. IOP Conference Series: Earth and Environmental Science 2016;36. https://doi.org/10.1088/1755-1315/36/1/012016.

[60] Saedi S, Madaeni SS, Shamsabadi AA. PDMS coated asymmetric PES membrane for natural gas sweetening: effect of preparation and operating parameters on performance. Canadian Journal of Chemical Engineering 2014;92:892–904. https://doi.org/10.1002/cjce.21947.

[61] Vaughn JT, Koros WJ. Analysis of feed stream acid gas concentration effects on the transport properties and separation performance of polymeric membranes for natural gas sweetening: a comparison between a glassy and rubbery polymer. Journal of Membrane Science 2014;465:107–16. https://doi.org/10.1016/j.memsci.2014.03.029.

[62] Gao Z, Wang Y, Wu H, Ren Y, Guo Z, Liang X, et al. Surface functionalization of Polymers of Intrinsic Microporosity (PIMs) membrane by polyphenol for efficient CO2 separation. Green Chemical Engineering 2021;2:70–6. https://doi.org/10.1016/j.gce.2020.12.003.

[63] Thomas S, Pinnau I, Du N, Guiver MD. Pure- and mixed-gas permeation properties of a microporous spirobisindane-based ladder polymer (PIM-1). Journal of Membrane Science 2009;333:125–31. https://doi.org/10.1016/j.memsci.2009.02.003.

[64] AlQahtani MS, Mezghani K. Thermally rearranged polypyrrolone membranes for high-pressure natural gas separation applications. Journal of Natural Gas Science and Engineering 2018;51:262–70. https://doi.org/10.1016/j.jngse.2018.01.011.

[65] Chiou JS, Maeda Y, Paul DR. Gas permeation in polyethersulfone. Journal of Applied Polymer Science 1987;33:1823–8. https://doi.org/10.1002/app.1987.070330533.

[66] Scholes CA, Stevens GW, Kentish SE. Membrane gas separation applications in natural gas processing. Fuel 2012;96:15–28. https://doi.org/10.1016/j.fuel.2011.12.074.

[67] Chu Y, He X. Process simulation and cost evaluation of carbon membranes for CO 2 removal from high-pressure natural gas. Membranes 2018;8. https://doi.org/10.3390/membranes8040118.

[68] Baker RW, Low BT. Gas separation membrane materials: a perspective. Macromolecules 2014;47:6999–7013. https://doi.org/10.1021/ma501488s.

[69] Adnan AI, Ong MY, Nomanbhay S, Chew KW, Show PL. Technologies for biogas upgrading to biomethane: a review. Bioengineering 2019;6:1–23. https://doi.org/10.3390/bioengineering6040092.

[70] Adewole JK, Ahmad AL, Ismail S, Leo CP. Current challenges in membrane separation of CO2 from natural gas: a review. International

[70] Journal of Greenhouse Gas Control 2013;17:46–65. https://doi.org/10.1016/j.ijggc.2013.04.012.
[71] Mubashir M, Dumée LF, Fong YY, Jusoh N, Lukose J, Chai WS, et al. Cellulose acetate-based membranes by interfacial engineering and integration of ZIF-62 glass nanoparticles for CO2 separation. Journal of Hazardous Materials 2021;415. https://doi.org/10.1016/j.jhazmat.2021.125639.
[72] Ahmad AL, Adewole JK, Leo CP, Ismail S, Sultan AS, Olatunji SO. Prediction of plasticization pressure of polymeric membranes for CO2 removal from natural gas. Journal of Membrane Science 2015;480:39–46. https://doi.org/10.1016/j.memsci.2015.01.039.
[73] Wessling M, Schoeman S, van der Boomgaard T, Smolders CA. Plasticization of gas separation membranes. Gas Separation & Purification 1991;5:222–8. https://doi.org/10.1016/0950-4214(91)80028-4.
[74] Bos A, Pünt IGM, Wessling M, Strathmann H. CO2-induced plasticization phenomena in glassy polymers. Journal of Membrane Science 1999;155:67–78. https://doi.org/10.1016/S0376-7388(98)00299-3.
[75] Suleman MS, Lau KK, Yeong YF. Plasticization and swelling in polymeric membranes in CO2 removal from natural gas. Chemical Engineering & Technology 2016;39:1604–16. https://doi.org/10.1002/ceat.201500495.
[76] Tabe-Mohammadi A. A review of the applications of membrane separation technology in natural gas treatment. Separation Science and Technology 1999;34:2095–111. https://doi.org/10.1081/SS-100100758.
[77] Khabazipour M, Anbia M. Removal of hydrogen sulfide from gas streams using porous materials: a review. Industrial & Engineering Chemistry Research 2019;58:22133–64. https://doi.org/10.1021/acs.iecr.9b03800.
[78] Latosov E, Loorits M, Maaten B, Volkova A, Soosaar S. Corrosive effects of H2S and NH3 on natural gas piping systems manufactured of carbon steel. Energy Procedia 2017;128:316–23. https://doi.org/10.1016/j.egypro.2017.08.319.
[79] Bai Y, Bai Q. Subsea corrosion and scale. In: Subsea engineering handbook. Elsevier; 2019. p. 455–87. https://doi.org/10.1016/B978-0-12-812622-6.00017-8.
[80] Yahaya GO, Hayek A, Alsamah A, Shalabi YA, Ben Sultan MM, Alhajry RH. Copolyimide membranes with improved H2S/CH4 selectivity for high-pressure sour mixed-gas separation. Separation and Purification Technology 2021;272:118897. https://doi.org/10.1016/j.seppur.2021.118897.
[81] Harrigan DJ, Lawrence JA, Reid HW, Rivers JB, O'Brien JT, Sharber SA, et al. Tunable sour gas separations: simultaneous H2S and CO2 removal from natural gas via crosslinked telechelic poly(ethylene glycol) membranes. Journal of Membrane Science 2020;602:117947. https://doi.org/10.1016/j.memsci.2020.117947.
[82] Chan YH, Lock SSM, Wong MK, Yiin CL, Loy ACM, Cheah KW, et al. A state-of-the-art review on capture and separation of hazardous hydrogen sulfide (H2S): recent advances, challenges and outlook. Environmental Pollution 2022;314:120219. https://doi.org/10.1016/j.envpol.2022.120219.
[83] Watler KG. Process for separating higher hydrocarbons from natural or produced gas streams. 1989. No. US 4857078.
[84] Gou Y, Xiao L, Yang Y, Guo X, Zhang F, Zhu W, et al. Incorporation of open-pore MFI zeolite nanosheets in polydimethylsiloxane (PDMS) to isomer-selective mixed matrix membranes. Microporous and Mesoporous

Materials 2021;315:110930. https://doi.org/10.1016/j.micromeso.2021.110930.

[85] Li G, Kujawski W, Válek R, Koter S. A review - the development of hollow fibre membranes for gas separation processes. International Journal of Greenhouse Gas Control 2021;104:103195. https://doi.org/10.1016/j.ijggc.2020.103195.

[86] Ahmad AL, Otitoju TA, Ooi BS. Hollow fiber (HF) membrane fabrication: a review on the effects of solution spinning conditions on morphology and performance. Journal of Industrial and Engineering Chemistry 2019;70: 35–50. https://doi.org/10.1016/j.jiec.2018.10.005.

[87] Choi S-H, Tasselli F, Jansen JC, Barbieri G, Drioli E. Effect of the preparation conditions on the formation of asymmetric poly(vinylidene fluoride) hollow fibre membranes with a dense skin. European Polymer Journal 2010;46:1713–25. https://doi.org/10.1016/j.eurpolymj.2010.06.001.

[88] Moon JD, Borjigin H, Liu R, Joseph RM, Riffle JS, Freeman BD, et al. Impact of humidity on gas transport in polybenzimidazole membranes. Journal of Membrane Science 2021;639:119758. https://doi.org/10.1016/j.memsci.2021.119758.

[89] Budd PM, McKeown NB, Ghanem BS, Msayib KJ, Fritsch D, Starannikova L, et al. Gas permeation parameters and other physicochemical properties of a polymer of intrinsic microporosity: polybenzodioxane PIM-1. Journal of Membrane Science 2008;325:851–60. https://doi.org/10.1016/j.memsci.2008.09.010.

[90] Lasseuguette E, Ferrari MC, Brandani S. Humidity impact on the gas permeability of PIM-1 membrane for post-combustion application. Energy Procedia 2014;63:194–201. https://doi.org/10.1016/j.egypro.2014.11.020.

[91] Konruang S, Sirijarukul S, Wanichapichart P, Yu L, Chittrakarn T. Ultraviolet-ray treatment of polysulfone membranes on the O $_2$/N $_2$ and CO $_2$/CH $_4$ separation performance. Journal of Applied Polymer Science 2015;132. https://doi.org/10.1002/app.42074. n/a-n/a.

[92] Dilshad MR, Islam A, Haider B, Sajid M, Ijaz A, Khan RU, et al. Effect of silica nanoparticles on carbon dioxide separation performances of PVA/PEG cross-linked membranes. Chemical Papers 2021;75:3131–53. https://doi.org/10.1007/s11696-020-01486-7.

[93] Yong WF, Zhang H. Recent advances in polymer blend membranes for gas separation and pervaporation. Progress in Materials Science 2021;116: 100713. https://doi.org/10.1016/j.pmatsci.2020.100713.

[94] Winarta J, Meshram A, Zhu F, Li R, Jafar H, Parmar K, et al. Metal-organic framework-based mixed-matrix membranes for gas separation: an overview. Journal of Polymer Science 2020;58:2518–46. https://doi.org/10.1002/pol.20200122.

[95] Mosleh S, Mozdianfard MR, Hemmati M, Khanbabaei G. Synthesis and characterization of rubbery/glassy blend membranes for CO2/CH4 gas separation. Journal of Polymer Research 2016;23:120. https://doi.org/10.1007/s10965-016-1005-6.

[96] Mazinani S, Ramezani R, Darvishmanesh S, Molelekwa GF, Di Felice R, Van Der Bruggen B. A ground breaking polymer blend for CO2/N2 separation. Journal of CO2 Utilization 2018;27:536–46. https://doi.org/10.1016/j.jcou.2018.08.024.

[97] Mannan HA, Mukhtar H, Murugesan T, Nasir R, Mohshim DF, Mushtaq A. Recent applications of polymer blends in gas separation

[97] membranes. Chemical Engineering & Technology 2013;36:1838–46. https://doi.org/10.1002/ceat.201300342.
[98] Juber FAH, Jawad ZA, Chin BLF, Yeap SP, Chew TL. The prospect of synthesis of PES/PEG blend membranes using blend NMP/DMF for CO2/N2 separation. Journal of Polymer Research 2021;28. https://doi.org/10.1007/s10965-021-02500-6.
[99] Molki B, Aframehr WM, Bagheri R, Salimi J. Mixed matrix membranes of polyurethane with nickel oxide nanoparticles for CO2 gas separation. Journal of Membrane Science 2018;549:588–601. https://doi.org/10.1016/j.memsci.2017.12.056.
[100] Dai Y, Ruan X, Yan Z, Yang K, Yu M, Li H, et al. Imidazole functionalized graphene oxide/PEBAX mixed matrix membranes for efficient CO2 capture. Separation and Purification Technology 2016;166:171–80. https://doi.org/10.1016/j.seppur.2016.04.038.
[101] Borgohain R, Jain N, Prasad B, Mandal B, Su B. Carboxymethyl chitosan/carbon nanotubes mixed matrix membranes for CO2 separation. Reactive and Functional Polymers 2019;143:104331. https://doi.org/10.1016/j.reactfunctpolym.2019.104331.
[102] Zarshenas K, Raisi A, Aroujalian A. Mixed matrix membrane of nano-zeolite NaX/poly (ether-block-amide) for gas separation applications. Journal of Membrane Science 2016;510:270–83. https://doi.org/10.1016/j.memsci.2016.02.059.
[103] Sabetghadam A, Seoane B, Keskin D, Duim N, Rodenas T, Shahid S, et al. Metal organic framework crystals in mixed-matrix membranes: impact of the filler morphology on the gas separation performance. Advances in Functional Materials 2016;26:3154–63. https://doi.org/10.1002/adfm.201505352.
[104] Jamil A, Oh PC, Shariff AM. Polyetherimide-montmorillonite mixed matrix hollow fibre membranes: effect of inorganic/organic montmorillonite on CO2/CH4 separation. Separation and Purification Technology 2018;206:256–67. https://doi.org/10.1016/j.seppur.2018.05.054.
[105] Haider B, Dilshad MR, Atiq Ur Rehman M, Vargas Schmitz J, Kaspereit M. Highly permeable novel PDMS coated asymmetric polyethersulfone membranes loaded with SAPO-34 zeoilte for carbon dioxide separation. Separation and Purification Technology 2020;248:116899. https://doi.org/10.1016/j.seppur.2020.116899.
[106] Matteucci S, Raharjo RD, Kusuma VA, Swinnea S, Freeman BD. Gas permeability, solubility, and diffusion coefficients in 1,2-polybutadiene containing magnesium oxide. Macromolecules 2008;41:2144–56. https://doi.org/10.1021/ma702459k.
[107] Castro-Muñoz R, Fíla V. Progress on incorporating zeolites in matrimid® 5218 mixed matrix membranes towards gas separation. Membranes 2018;8:1–23. https://doi.org/10.3390/membranes8020030.
[108] Natarajan P, Sasikumar B, Elakkiya S, Arthanareeswaran G, Ismail AF, Youravong W, et al. Pillared cloisite 15A as an enhancement filler in polysulfone mixed matrix membranes for CO2/N2 and O2/N2 gas separation. Journal of Natural Gas Science and Engineering 2021;86:103720. https://doi.org/10.1016/j.jngse.2020.103720.
[109] Zhu W, Li X, Sun Y, Guo R, Ding S. Introducing hydrophilic ultra-thin ZIF-L into mixed matrix membranes for CO2/CH4 separation. RSC Advances 2019;9:23390–9. https://doi.org/10.1039/c9ra04147h.
[110] Cheng Y, Ying Y, Zhai L, Liu G, Dong J, Wang Y, et al. Mixed matrix membranes containing MOF@COF hybrid fillers for efficient CO2/CH4

separation. Journal of Membrane Science 2019;573:97—106. https://doi.org/10.1016/j.memsci.2018.11.060.

[111] Kamble AR, Patel CM, Murthy ZVP. A review on the recent advances in mixed matrix membranes for gas separation processes. Renewable and Sustainable Energy Reviews 2021;145:111062. https://doi.org/10.1016/j.rser.2021.111062.

[112] Zainuddin MIF, Ahmad AL. Mixed matrix membrane development progress and prospect of using 2D nanosheet filler for CO2 separation and capture. Journal of CO2 Utilization 2022;62:102094. https://doi.org/10.1016/j.jcou.2022.102094.

[113] Al-Rowaili FN, Khaled M, Jamal A, Zahid U. Mixed matrix membranes for H2/CO2 gas separation- a critical review. Fuel 2023;333:126285. https://doi.org/10.1016/j.fuel.2022.126285.

16

Natural gas sweetening by ionic liquid membranes

Girma Gonfa[1,2] and Sami Ullah[3]

[1]Department of Chemical Engineering, Addis Ababa Science and Technology University, Addis Ababa, Ethiopia; [2]Biotechnology and Bioprocess Centre of Excellence, Addis Ababa Science and Technology University, Addis Ababa, Ethiopia; [3]Department of Chemistry, College of Science, King Khalid University, Abha, Saudi Arabia

1. Introduction

Natural gas comes from wells containing mainly methane and to some extent other hydrocarbons such as ethane, propane, and butane [1]. It may also contain water vapor and acid gases such as carbon dioxide (CO_2), hydrogen sulfide (H_2S), and other sulfur-containing components [2], as well as trace quantities of other components such mercury [3]. Composition of natural gas widely depends on the location of the field. Depending on the location of its reservoir location, natural gas may contain significant amount of acid gases (CO_2 and H_2S). The CO_2 concentration of raw natural gas may vary from 5% to 90% [4]. For instance, the natural gas reservoirs of about 13 trillion standard cubic feet of CO_2 content of 87 mol% [5]. High CO_2 content in natural gas may form gas hydrate and restrict gas flow in the gas transfer lines. Moreover, the presence of high concentration of CO_2 in natural gas may cause pipe corrosion resulting in its damage. Further, the presence of CO_2 in natural gas reduces the heat value of the gas. Therefore, the acceptable CO_2 concentration in natural gas is usually 2% [4].

Sulfur may exist in natural gas in the form of H_2S, mercaptans, carbonyl sulfide, carbon disulfide, and elemental sulfur [6]. The occurrence of sulfur in natural gas may cause deposition of elemental sulfur in the pipeline causing problems in gas transportations. Sulfur is highly corrosive to pipelines and equipment in the presence of water. Moreover, sulfur in natural gas delivered to customers should fulfill the maximum allowable sulfur concentration as it possesses health and safety risks. Although the maximum sulfur content is usually determined by specific country's national legislation, the maximum total sulfur content (including carbonic sulfides, disulfides) ranges from 10 to 20 ppmv [6].

Natural gas can be classified as sweet or sour based on their sulfur and CO_2 contents. Sweet natural gas contains sulfur and CO_2 below maximum allowable concentration. Whereas sour natural gasses contain high sulfur and CO_2 contents and require gas treatment for removal of these components. Hence, gas sweetening refers to the removal of acidic impurities (sulfur and CO_2) to fulfill the requirements for gas transportations and gas supply standards. Natural gas sweetening can be carried out using absorption, adsorption, cryogenic, and membrane technologies [4]. Currently, natural gas sweetening is performed mainly through amine solution absorption process. However, amine-based absorption processes have several drawbacks, such as cause of corrosion, degradation and loss of amine, and requirement of high energy for amine regeneration. Moreover, amine-based natural gas sweeting technology is not economically viable for gas with high acid gas concentrations. Therefore, various separation technologies have been under instigation to improve natural gas sweetening processes. In the last 2 decades, ionic liquids (ILs) have attracted attention as promising alternative for sweetening of natural due to its desirable properties [7,8].

Ionic liquids are organic salts that comprise organic/inorganic anions and bulky organic cations and with melting point of less than 100°C [9]. Most of the ILs have melting points below room temperature. The properties of the ILs can be tuned for specific applications by tailoring combinations of cations and anions. Moreover, introduction of structural functionalities in the cation and/or anion enables to synthesize task specific ILs. Hence, there are various classes of ILs that can be used for different applications. Fig. 16.1 shows some of the common cations and anions.

ILs have drawn considerable attention for removal of acid gasses from natural gas due to their desirable properties such as negligible vapor pressure, high thermal stability, nonflammable characteristics, wide liquid phase ranges, and tuneable properties for high affinity of acid gases (sulfur and CO_2). The solubilities of CO_2 in ILs were first observed by Blanchard et al. [10], and since then, various works have been reported for removal of acid gases from natural gas. ILs have been investigated for sweetening of natural gas through absorption of CO_2 and sulfur compounds [8,11]. Significant CO_2/H_2S solubility data are available in literature and open data sources [7,8,12]. Further some ILs showed better CO_2/H_2S performance in comparison to amine-based absorption

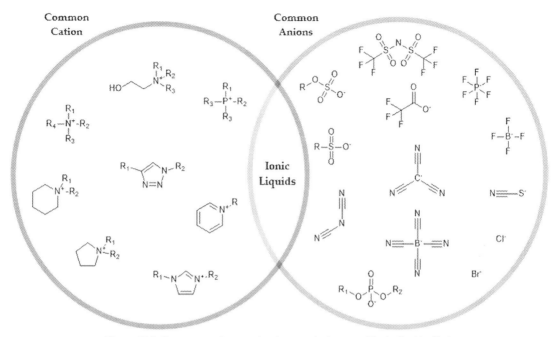

Figure 16.1 Common cations and anions and classes of ionic liquids (ILs).

processes [13–15]. However, drawbacks such as requirement of large amount of ILs for the absorption column and high energy requirement for recycling of ILs limited application of ILs for natural gas sweetening. These drawbacks could be overcome by applying ionic liquid membrane (ILM) techniques. ILM comprises the feed and permeate phases that are separated by membrane containing ILs. Hence, this chapter presents principle and procedures of ILM, process and applications involving ILM-based gas sweetening, and conclusion and future outlooks of the separation technology for sweetening of natural gas.

2. Ionic liquid membranes

Ionic liquid membrane–based acid gas separation process depends on the solubility of the acid gases in the ILs and the permeability of the natural gas in the membrane system. Depending on the nature of the ILs and membrane configurations, ILs can be used in different forms of membrane-based gas sweetening processes. Recently, various IL immobilizing techniques and methods for modifications solid supports were reported. Based on ILs immobilization techniques, ILM configuration can be mainly classified into supported ionic liquid membrane (SILM) and quasi-solidified ionic liquid membrane (QSILM). Fig. 16.2 shows the most common configurations of ILMs. The first ILM configuration is made by stabilizing the ILs by impregnating it inside the pores of the support membrane. The second configuration is made by quasi-solidification of ILs to produce a material with good mechanical strength [11,16]. Both SILM and QSILM configurations can be prepared with or without additional silicone coating. Moreover, polymerizable ILs can be used as self-standing membranes or they can be used in mixed matric membranes. Further ILs are used in liquid membrane contactors. The following section presents preparation of ILMs, long-term stability, and transport mechanisms in the membranes.

2.1 Supported ionic liquid membranes

Application of various liquids for supported liquid membranes existed before ILs is widely known for various applications [17–19]. Application of supported liquid membrane provides higher gas permeability than polymer membranes since

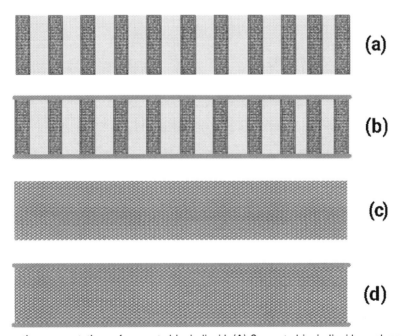

Figure 16.2 Schematic representations of supported ionic liquid. (A) Supported ionic liquid membrane (SILM), (B) SILM with silicon coating, (C) Quasi-solidified ionic liquid membrane (QSILM), and (D) QSILM with silicon coating.

diffusion through a dense liquid film is often much more rapid than that of polymers. However, application of conventional solvent–based supported liquid suffers from loss of solvent to gas stream through evaporation, which causes degradation of membrane integrity and loss membrane performance [18]. Following booming research in ILs and due to its desirable physical and chemical properties, ILs are applied in supported liquid membranes. ILs overcome the limitations of conventional solvent–based supported liquid membranes. Moreover, SILM required less quantity of ILs compared to its bulk application, such as absorption, which reduces the cost of ILs. Moreover, SILM avoids additional steps for the recovery of ILs, which is an energy-intensive process. Application of SILM for gas separation was reported in 2004 [20], and since then, it has been the object of interest of many research groups. Supported ILM is a simple straightforward approach to use ILs in a membrane-based gas sweetening. Fig. 16.3 shows the schematic representation of SILMs.

2.1.1 Preparation methods of SILMs

In supported IL-based membranes, the ILs are immobilized within polymer/inorganic supports and held by capillary forces in the pores of support material [21]. SILM supports are may vary depending on their geometrical configurations (hollow fiber or flat sheet) [22,23], pore size, homogeneity (symmetric/asymmetric) [22,24], and surface free energy (hydrophobicity/

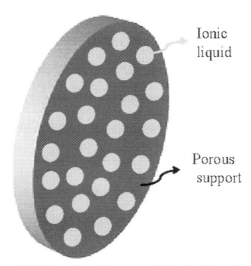

Figure 16.3 Illustration of supported ionic liquid membrane.

hydrophilicity) of the surfaces [25]. These factors affect the deposition of ILs on the pores of the support, which, in turn, affects the performance and structural stability of the SILMs. The performance and structural stability of the SILMs also largely depend the nature of the ILs and its thermophysical properties such as viscosity, surface tension, and their stability in a given environment [26]. Inorganic supports such as graphene oxide [27], ceramic [28], molybdenum disulfide, and activated carbon [29] were investigated as a support for SILMs. Organic supports, mostly polymers such as polyvinylidene fluoride [22,25], polyacrylonitrile [30], tetrafluoroethylene-vinylidene fluoride [31], and polyimide [23] are studied for SILMs. Inorganic supports offer better thermal stability; hence, they can be used for removal of acid gasses from natural gas at elevated temperatures. Polymeric supports only lower temperature process; however, they can be obtained at lower costs. Moreover, composite supports have been reported to improve the performance and stability of the SILMs [32,33].

Supported IL-based membranes are generally prepared though direct immersion method, vacuum method, and pressurization method. Direct immersion involves immobilization of ILs by immersing the support material in the ILs at ambient pressure for some time, usually 24 h [34,35]. Excess ILs are removed by softly weeping with tissues [35] or leaving the ILs dipped support to drip excess ILs overnight [36]. Direct immersion is the simplest and straightforward technique for immobilization of ILs on support materials. In the vacuum SILM preparation technique, the support material may be placed inside a vacuum tight chamber for a certain time to remove air from its pores, and then, ILs can be spread out at the membrane surface while keeping vacuum in the chamber [21]. Both direct immersion and vacuum methods are suitable for the low-viscosity ILs since they are flowable can be immobilized without external forces. The amount of immobilized ILs increases with increasing viscosity of the ILs since the capillary forces in the pores increases with reducing the viscosity [37]. In the pressurization technique, the support materials and ILs are placed in ultrafiltration unit nitrogen pressure is applied to force the ILs to flow into the pores of the materials [34,38]. Once a thin layer of ILs covered the surface of the membrane, the pressure is related, and excess ILs are removed using tissue or using dripping technique. The amount of ILs immobilized by pressure method may be independent of the nature of ILs this approach can be applicable for all ILs. Pressure method can be used for immobilization of viscous ILs on support materials [11]. In all IL immobilization techniques, ILs should be trapped inside the pores of the support by capillary forces, and the support materials should be well wetted by the ILs.

2.1.2 Long-term stability SILMs

One of the challenges related to SILMs is loss of its long-term stability due to leakage of ILs from porous support. The stability of SILMs is characterized by its ability to operate under pressure without weight loss. This means, the weight of IL immobilized in the pores of the support material remains constant. Stabilities of SILMs depend on characteristics of the support martials. Techniques such as membrane surface treatment [39], application of supports with uniform pore sizes [33], and specific membrane preparation methods [40] can further improve the structural and performance stabilities of the SILMs. The stability of the SILMs also depends on the ILs immobilization method used during preparation of the supported membranes. Some studies show SILMs prepared using pressure methods are more stable than prepared by vacuum, and the difference is pronounced when for most viscous ILs [38]. For viscous ILs, it is difficult to flow into the deeper pores of the support materials, and most ILs mainly be immobilized on the most external layer of the membrane, which could be easily removed during applications [41]. Moreover, the affinity between the ILs and the support materials affects the stability of the SILMs. Where there is string interaction between the surface of the support materials and the ILs, the stability of the membrane may enhance. Hence, to enhance the stability of SILMs selection of suitable support materials and ILs is important. Other important factors that affect the stability of SILMs are the operating conditions (pressure and temperature), nature of the gas components, and composition of gas treated by the membrane. Laboratory-scale tests for under binary gas mixtures (CO_2/CH_4 and CO_2/N_2) show longer performance and structural SILMs stabilities. For instance, for 1-methyl-3-propylimidazolium bis(trifluoromethylsulfonyl)imide supported by polytetrafluorethylene polymer used for separation of CO_2 from CH_4 at atmospheric pressure, and no observable changes in its performance was observed after 260 days operation [42]. Similarly, some imidazolium-based ILs supported by polyvinylidene fluoride and polyethersulfone polymers show stabilities from 14 to 106 days under CO_2/CH_4 environment. Some pilot scale under real gas condition was also tested. Some pilot SILMs used for separation of CO_2 from power plant flue gas shows the SILMs can maintain their process and structural stabilities for more than 14 days in the absence of SO_3-containing particulates [30].

2.1.3 Gas transport properties in SILMs

Supported IL membrane is the simplest and straightforward configuration for testing the transport properties of IL–membrane support combinations for gases. The solution-diffusion model of gas transport is usually applied to characterize permeation of gases in the SILMs [43]. According to the solution-diffusion model, the permeability coefficient of a gas is directly proportional to the solubility coefficient of the gas in the IL and its diffusion coefficient. The solubilities of gases in ILs increase with pressure and obeys Henry's law. The gas diffusion through the ILs can be described by Fick's law. An increase in temperature lead to reduction gas solubilities and results in gases diffusion. The gases dissolve in the ILs at the feed side, diffuses through the ILs layers due to the pressure difference developed across the membrane, desorbs from the permeate side, and is swept away by the permeate stream at low pressure. The diffusivity of gases through the ILs is usually correlated with the viscosities of the ILs [44–47]. Some studies indicate that the selectivity may be correlated with the molar volume of the ILs [48,49].

2.2 Quasi-solidified ionic liquid membranes

Quasi-solidified ionic liquid membranes (QSILMs) are obtained by casting solutions containing ILs and special gel to form a thin and stable quasi-solidified film [50]. The mixture of the ILs and gelator agent cast form free-standing membrane films. QSILMs avoid IL leakage, which is the main drawback for SILM technique. Due to the limited solubility of the gelator agents and polymers in ILs, the amount of ILs used is much lower than that used for SILMs [51]. Since the ILs are entrapped into the gelator agent matrixes, QSILMs have long-term stability than SILMs. Moreover, silicone coating can further improve the stability QSILMs. The main gels used for preparation of QSILMs are polymers and low-molecular-weight organic gelator agents. Polymers such as polydimethylsiloxane (PDMS) [16], poly(ethylene glycol) diacrylate [52], and poly(vinylidene fluoride-co-hexafluoropropylene) [53] are used for preparation of QSILMs due to its chemical and thermal stabilities. The polymers should have high thermal and chemical stabilities. Moreover, they should have good miscibility and compatibility with the ILs. Low-molecular-weight organic gelators such as cyclo(L-β-3,7-dimethyloctylasparaginyl-L-phenylalanyl), 12-hydroxystearic acid, cyclo(L-β-2-ethylhexylasparaginyl-L-phenylalanyl) are commonly used for preparation of QSILMs [54].

2.2.1 Preparation methods of QSILMs

QSILMs can be prepared by through direct blending of the ILs and polymers. Usually, the polymers are dissolved in appropriate solvents at elevated temperature, mixed, and then filtered to get the required solutions. Then, the predetermined amount ILs are added to the solution and mixed to get homogenous solutions. Finally, the QSILMs are casted using membrane casting methods. Similarity for low-molecular-weight organic gelator-based QSILMs, a given amount of gelator, is added to the ILs at the required temperature, and the solutions are casted to form the membranes. The membranes are dried to obtained stable QSILMs. The QSILMs may be further improved using different techniques. For instance, Skorikova et al. [55] to improve the characteristics and performance of a quasi-solidified ILM.

2.2.2 Long-term stability QSILMs

Membrane modules for quasi-solidified ILMs are easier to manufacture and control leaking of ILs from the membrane. Hence, QSILMs have higher operational stabilities compared to SILMs. The thermal and chemical stabilities of QSILMs depend on the nature of the gels as well as the ILs used, compatibilities of the gels with the ILs, and concentration of the gels and the ILs in the QSILMs [16]. Concentration of gel and ILs plays an important role on the stability of the QSILMs. Hanabusa et al. [54] observed that the mechanical strength of the QSILMs made from ILs combining with cyclo(L-β-3,7-dimethyloctylasparaginyl-L-phenylalanyl) and cyclo(L-β-2-ethylhexylasparaginyl-Lphenylalanyl) gelators increases nearly proportionally with the concentration of the gelators. QSILMs prepared from ILs, and these gelators are also thermally stable up to 140°C [54]. On the other hand, increases in concentrations of ILs in the QSILMs reduce the mechanical strength of the membranes [56,57]. As it was mentioned earlier, the changes in stabilities of resulting QSILMs depend on the nature of the gel and the ILs. For example, the mechanical stability of QSILM prepared from 1-ethyl-3-methylimidazolium bis(trifluoromethylsulfonyl) imide ILs and poly(vinylidene fluorideco-hexafluoropropylene) gel much lower than the mechanical stability of pure gel at 20% IL concentration [57]. On the other hand, QSILM made from 1-ethylimidazolium bis(trifluoromethanesulfonyl)amide IL and tetra-N-hydroxysuccinimide terminated poly(ethylene glycol)-based gels have high thermal stability at high (94%) IL concentration. Increasing ILs in the QSILMs may increase the performance of QSILMs; however, it may affect the long-term stability of the membranes. Therefore, it is important to optimize the performance and the long-term stability of the QSILMs by selecting suitable ILs and gels and optimizing the proportion of ILs to gels.

2.3 Poly(ionic liquid) (PIL) membranes

Poly(ionic liquid) (PIL) membranes (PILMs) are manufactured from the polymerizable IL monomers. PILs are polyelectrolyte polymers containing cations or anions on the repeating units. PILs have important properties such as high thermal and chemical stability and ion conductivity. They can be classified into two as cationic PILs or anionic PILs based on the moieties in the polymer backbone. Cationic PILs are much more explored than anionic PILs because of their less complicated synthesis procedure and larger number of polymerizable cations. The most studied PILs are based on vinyl imidazolium cationic monomers [58]. PILs mostly show an amorphous solid character or gel-like properties with low transition temperatures.

Application of PILs for preparation of membrane enhances the stability of the membrane and improve their performance compared to corresponding ILs. PILs provide self-standing membranes without additional supporting materials or without blending them with other materials. PILs show higher absorption capacity and higher desorption for CO_2 than their corresponding IL monomers. Hence, applications of PILMs have attracted attention in sweetening of natural gas. The structural variations of the ILs highly affect the properties of the PILs and their performance on gas separations. Some reports indicate ammonium-based PILs show CO_2 sorption capacity of 6–7.6 times higher than their corresponding monomer ILs [59]. Moreover, the CO_2 permeability of PILs membranes is much higher than their corresponding monomer ILs. Therefore, research on application PIL membranes is one of the fast growing areas of application of ILs for gas natural gas sweetening. There are high opportunities for development PILs membranes by varying the structural variations of the ILs monomers, application of different polymerization methods, and post-modification and membrane treatment techniques.

2.4 Ionic liquid mixed-matrix membranes (IL-MMMs)

The mixed-matrix membranes (MMMs) are heterogeneous membranes prepared by dispersing organic/inorganic materials as fillers into polymeric materials as matrixes [60]. MMMs have been widely investigated for separation of gases, including natural gas sweetening [61,62]. In the preparation of mixed-matrix membranes, different polymers and solid fillers such as zeolites, graphite, silica, carbon nanotubes, and metal−organic frameworks were studied [61,62]. ILs were introduced as a third component into the mixed membrane to improve its polymer/filler compatibility and its performances as well as characteristics of solid fillers [63]. Moreover, PILs can be used in place of the conventional polymers in the preparation of MMMs. Introduction of ILs in MMMs used for separation of CO_2 causes the improvement in the performance of the membrane since it enhances the solubility and permeability of the gas in the membrane. ILs also improve the interfacial wetting between the fillers and polymer matrix [64].

Application of ILs for fabrication of mixed membranes fabrication was reported by Hudiono et al. [64]. Since then, several studies have been carried out in fabrication of IL-MMMs using different fillers, polymers and ILs [63]. Fabrication of IL-MMMs can be generally classified into three methods. The first method involves blending of ILs with mixed-matrix membrane dope solutions (fillers and polymer solutions). The membrane films are then formed using the mixture using membrane fabrication techniques and may be further treated. The second technique involves premodification of organic fillers with ILs and then mixing with polymer solution. Treating inorganic fillers with ILs improves the compatibility between the fillers and the polymers [65]. The third methods cover various techniques such as premodification of polymers before embedding them into MMMs (before MMMs synthesis) and postmodifications of fabricated MMMs after they are prepared. Treatments of polymers with ILs before embedding them in MMMs enhance the compatibility between the polymers and the fillers and improve the stability and the performance of the membranes [63]. In the case of MMMs postmodifications, ILs are diffused into the MMMs with aim of solvents and creating interfacial defects and filling the ILs in the void gaps [66]. In all ILM MMM preparations techniques used, introduction of ILs in the conventional MMMs improved the polymers and fillers compatibilities and performance of the membranes in gas separations [66,67].

2.5 Membrane contactors using ILs

Membrane contactor−based gas separation is a hybrid process that integrates membrane separation and solvent absorption. It integrates the advantages of selective liquid absorption and porosity of the membrane that acts as liquid−gas contacting interfaces. Membrane contactors have some benefits. It provides a predetermined interfacial area, which depends on the area of the membrane [68]. Moreover, its performance does not depend on the gas and liquid flowrates [69]. Further, membrane contactor–based gas separation process can be easily scaled-up applying linear up-scaling techniques. For separation gases, it can be applied at lower temperature and pressure, which could lower the operating costs. Solvents such as aqueous amine solutions, salt solutions, and alkaline solutions are studied for gas CO_2 separation using membrane contactors [70].

IL-based membrane contactor for separation of gases is relatively a new concept. It was first introduced for abortion of SO_2 using 1-ethyl-3-methylimidazolium ethylsulfate IL in a membrane contactor [71] and shortly applied for separation of CO_2 [72]. Since then, some works were reported on the use of IL-based membrane contactor for separation of CO_2 and other acid gases [68,73]. ILs offer better advantages than other solvents used in the membrane contactor for acid–gas separation because ILs remove acid gases through physical absorption, which reduce the energy for recovery of ILs. The main drawback of most of the ILs used in membrane contactor is their high viscosities, which reduce the overall gas mass transfer coefficient.

Fig. 16.4 shows the schematic representation of membrane contactor–based gas separation for hollow microporous membrane. The IL flow in the membrane and the gas mixture passes

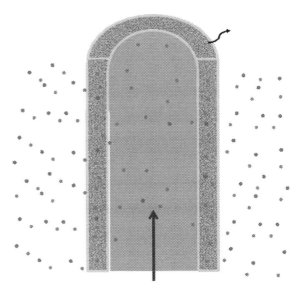

Figure 16.4 Schematic diagram of gas separation with ionic liquid–based membrane contactor.

over the hollow membrane. The microporous membranes serve as a nonselective barrier to provide interfaces between the liquid phase and the gas phase. The gas phase and the liquid phase flow can be concurrent, counter current, or any other arrangement based on the design of the other membrane module. The gases diffuse through the porous membrane and are absorbed in the liquid phases. Separation of acid gases from natural gas with the membrane contactor depends on permeability of the gases in the porous membrane and the affinity of the IL for the acid gases and natural gas components.

3. Conclusion and future outlooks

Membrane-based natural gas sweetening is one of the widely studied and the fast-growing techniques. Application of IL in membrane-based gas separation in various forms further extends attention from academics and industries. IL-based membranes have several advantages compared to conventional membranes. ILs in their various forms can offer an excellent opportunity for the development of membranes for sweetening of natural gas. ILs can be applied as supported liquid membrane, quasi-solidified membrane, in the form of polymerized ILs, or in the form of mixed membrane for fabrication of various forms of membranes. They can be applied in liquid membrane—based gas separation techniques. IL-based membranes have been investigated for separation of CO_2 and other acid gases from natural gas. These studies show that ILs can be applied in different forms of membrane-based gas natural sweetening techniques and can be economically competitive with other techniques. At present, IL-based membrane gas sweetening has only been demonstrated on the laboratory scale, and further works are required to implement in the industry. As the number of possible ILs is unlimited, further work is required to select potential ILs for membrane-based separation processes.

Abbreviations and symbols

IL-MMMs	ionic liquid mixed-matrix membranes
ILM	ionic liquid membrane
ILs	ionic liquids
MMMs	mixed-matrix membranes
PDMS	polydimethylsiloxane
PILMs	poly(ionic liquid) membranes
PILs	polymeric ionic liquids
QSILM	quasi-solidified ionic liquid membrane
SILM	supported ionic liquid membrane

References

[1] Gonfa G, Bustam MA, Sharif AM, Mohamad N, Ullah S. Tuning ionic liquids for natural gas dehydration using COSMO-RS methodology. Journal of Natural Gas Science and Engineering 2015;27:1141−8.

[2] Gonfa G, Bustam MA, Shariff AM, Muhammad N, Ullah S. Quantitative structure−activity relationships (QSARs) for estimation of activity coefficient at infinite dilution of water in ionic liquids for natural gas dehydration. Journal of the Taiwan Institute of Chemical Engineers 2016;66:222−9.

[3] Abbas T, Gonfa G, Lethesh KC, Mutalib MIA, Abai Mb, Cheun KY, et al. Mercury capture from natural gas by carbon supported ionic liquids: synthesis, evaluation and molecular mechanism. Fuel 2016;177:296−303.

[4] Babar M, Bustam MA, Ali A, Maulud AS, Shafiq U, Mukhtar A, et al. Thermodynamic data for cryogenic carbon dioxide capture from natural gas: a review. Cryogenics 2019;102:85−104.

[5] Darman NH, Harun A. Technical challenges and solutions on natural gas development in Malaysia, the petroleum policy and management project. Beijing, China: 4th Workshop of the China-Sichuan Basin Study; 2006.

[6] dos Santos JPL, de Carvalho Lima Lobato AK, Moraes C, de Lima Cunha A, da Silva GF, dos Santos LCL. Comparison of different processes for preventing deposition of elemental sulfur in natural gas pipelines: a review. Journal of Natural Gas Science and Engineering 2016;32:364−72.

[7] Bates ED, Mayton RD, Ntai I, Davis JH. CO_2 capture by a task-specific ionic liquid. Journal of the American Chemical Society 2002;124:926−7.

[8] Karadas F, Atilhan M, Aparicio S. Review on the use of ionic liquids (ILs) as alternative fluids for CO_2 capture and natural gas sweetening. Energy and Fuels 2010;24:5817−28.

[9] Gonfa G, Bustam MA, Muhammad N, Khan AS. Evaluation of thermophysical properties of functionalized imidazolium thiocyanate based ionic liquids. Industrial and Engineering Chemistry Research 2015;54:12428−37.

[10] Blanchard LA, Hancu D, Beckman EJ, Brennecke JF. Green processing using ionic liquids and CO_2. Nature 1999;399:28−9.

[11] Friess K, Izák P, Kárászová M, Pasichnyk M, Lanč M, Nikolaeva D, et al. A review on ionic liquid gas separation membranes. Membranes 2021;11:97.

[12] Dong Q, Kazakov A, Muzny C, Chirico R, Widegren J, Diky V, et al. Ionic liquids database (ILThermo). 2006.

[13] Aghaie M, Rezaei N, Zendehboudi S. A systematic review on CO_2 capture with ionic liquids: current status and future prospects. Renewable and Sustainable Energy Reviews 2018;96:502−25.

[14] Wappel D, Gronald G, Kalb R, Draxler J. Ionic liquids for post-combustion CO_2 absorption. International Journal of Greenhouse Gas Control 2010;4:486−94.

[15] Zhang N, Huang Z, Zhang H, Ma J, Jiang B, Zhang L. Highly efficient and reversible CO_2 capture by task-specific deep eutectic solvents. Industrial and Engineering Chemistry Research 2019;58:13321−9.

[16] Wang J, Luo J, Feng S, Li H, Wan Y, Zhang X. Recent development of ionic liquid membranes. Green Energy and Environment 2016;1:43−61.

[17] Kemperman A, Rolevink H, Bargeman D, Van den Boomgaard T, Strathmann H. Stabilization of supported liquid membranes by interfacial polymerization top layers. Journal of Membrane Science 1998;138:43−55.

[18] Kemperman AJ, Bargeman D, Van Den Boomgaard T, Strathmann H. Stability of supported liquid membranes: state of the art. Separation Science and Technology 1996;31:2733−62.
[19] Teramoto M, Sakaida Y, Fu SS, Ohnishi N, Matsuyama H, Maki T, et al. An attempt for the stabilization of supported liquid membrane. Separation and Purification Technology 2000;21:137−44.
[20] Scovazzo P, Kieft J, Finan DA, Koval C, DuBois D, Noble R. Gas separations using non-hexafluorophosphate [PF6]− anion supported ionic liquid membranes. Journal of Membrane Science 2004;238:57−63.
[21] Kocherginsky NM, Yang Q, Seelam L. Recent advances in supported liquid membrane technology. Separation and Purification Technology 2007;53:171−7.
[22] Cheng L-H, Rahaman MSA, Yao R, Zhang L, Xu X-H, Chen H-L, et al. Study on microporous supported ionic liquid membranes for carbon dioxide capture. International Journal of Greenhouse Gas Control 2014;21:82−90.
[23] Zeh M, Wickramanayake S, Hopkinson D. Failure mechanisms of hollow fiber supported ionic liquid membranes. Membranes 2016;6:21.
[24] Zhao W, He G, Nie F, Zhang L, Feng H, Liu H. Membrane liquid loss mechanism of supported ionic liquid membrane for gas separation. Journal of Membrane Science 2012;411−412:73−80.
[25] Luis P, Neves L, Afonso C, Coelhoso I, Crespo J, Garea A, et al. Facilitated transport of CO2 and SO2 through supported ionic liquid membranes (SILMs). Desalination 2009b;245:485−93.
[26] Tzialla O, Labropoulos A, Panou A, Sanopoulou M, Kouvelos E, Athanasekou C, et al. Phase behavior and permeability of Alkyl-Methyl-Imidazolium Tricyanomethanide ionic liquids supported in nanoporous membranes. Separation and Purification Technology 2014;135:22−34.
[27] Ying W, Cai J, Zhou K, Chen D, Ying Y, Guo Y, et al. Ionic liquid selectively facilitates CO2 transport through graphene oxide membrane. ACS Nano 2018;12:5385−93.
[28] Karousos DS, Labropoulos AI, Tzialla O, Papadokostaki K, Gjoka M, Stefanopoulos KL, et al. Effect of a cyclic heating process on the CO2/N2 separation performance and structure of a ceramic nanoporous membrane supporting the ionic liquid 1-methyl-3-octylimidazolium tricyanomethanide. Separation and Purification Technology 2018;200:11−22.
[29] Chen D, Ying W, Guo Y, Ying Y, Peng X. Enhanced gas separation through nanoconfined ionic liquid in laminated MoS2 membrane. ACS Applied Materials and Interfaces 2017;9:44251−7.
[30] Klingberg P, Wilkner K, Schlüter M, Grünauer J, Shishatskiy S. Separation of carbon dioxide from real power plant flue gases by gas permeation using a supported ionic liquid membrane: an investigation of membrane stability. Membranes 2019;9:35.
[31] Akhmetshina AI, Gumerova OR, Atlaskin AA, Petukhov AN, Sazanova TS, Yanbikov NR, et al. Permeability and selectivity of acid gases in supported conventional and novel imidazolium-based ionic liquid membranes. Separation and Purification Technology 2017;176:92−106.
[32] Chai S-H, Fulvio PF, Hillesheim PC, Qiao Z-A, Mahurin SM, Dai S. "Brick-and-mortar" synthesis of free-standing mesoporous carbon nanocomposite membranes as supports of room temperature ionic liquids for CO2−N2 separation. Journal of Membrane Science 2014;468:73−80.

[33] Tan M, Lu J, Zhang Y, Jiang H. Ionic liquid confined in mesoporous polymer membrane with improved stability for CO2/N2 separation. Nanomaterials 2017;7:299.
[34] Lozano LJ, Godínez C, de los Ríos AP, Hernández-Fernández FJ, Sánchez-Segado S, Alguacil FJ. Recent advances in supported ionic liquid membrane technology. Journal of Membrane Science 2011;376:1−14.
[35] Nosrati S, Jayakumar NS, Hashim MA. Performance evaluation of supported ionic liquid membrane for removal of phenol. Journal of Hazardous Materials 2011;192:1283−90.
[36] Matsumoto M, Panigrahi A, Murakami Y, Kondo K. Effect of ammonium- and phosphonium-based ionic liquids on the separation of lactic acid by supported ionic liquid membranes (SILMs). Membranes 2011;1:98−108.
[37] Fortunato R, Afonso CAM, Reis MAM, Crespo JG. Supported liquid membranes using ionic liquids: study of stability and transport mechanisms. Journal of Membrane Science 2004;242:197−209.
[38] Hernández-Fernández FJ, de los Ríos AP, Tomás-Alonso F, Palacios JM, Víllora G. Preparation of supported ionic liquid membranes: influence of the ionic liquid immobilization method on their operational stability. Journal of Membrane Science 2009;341:172−7.
[39] Dahi A, Fatyeyeva K, Langevin D, Chappey C, Poncin-Epaillard F, Marais S. Effect of cold plasma surface treatment on the properties of supported ionic liquid membranes. Separation and Purification Technology 2017;187:127−36.
[40] Khakpay A, Scovazzo P, Nourian S. Homogeneous and biphasic cellulose acetate/room temperature ionic liquid membranes for gas separations: solvent and phase-inversion casting vs. supported ionic liquid membranes. Journal of Membrane Science 2019;589:117228.
[41] De los Rios A, Hernández-Fernández F, Tomás-Alonso F, Palacios J, Gómez D, Rubio M, et al. A SEM−EDX study of highly stable supported liquid membranes based on ionic liquids. Journal of Membrane Science 2007;300:88−94.
[42] Hanioka S, Maruyama T, Sotani T, Teramoto M, Matsuyama H, Nakashima K, et al. CO2 separation facilitated by task-specific ionic liquids using a supported membrane. Journal of Membrane Science 2008;314:1−4.
[43] Gan Q, Zou Y, Rooney D, Nancarrow P, Thompson J, Liang L, et al. Theoretical and experimental correlations of gas dissolution, diffusion, and thermodynamic properties in determination of gas permeability and selectivity in supported ionic liquid membranes. Advances in Colloid and Interface Science 2011;164:45−55.
[44] Camper D, Bara J, Koval C, Noble R. Bulk-fluid solubility and membrane feasibility of rmim-based room-temperature ionic liquids. Industrial and Engineering Chemistry Research 2006;45:6279−83.
[45] Condemarin R, Scovazzo P. Gas permeabilities, solubilities, diffusivities, and diffusivity correlations for ammonium-based room temperature ionic liquids with comparison to imidazolium and phosphonium RTIL data. Chemical Engineering Journal 2009;147:51−7.
[46] Kilaru PK, Scovazzo P. Correlations of low-pressure carbon dioxide and hydrocarbon solubilities in imidazolium-, phosphonium-, and ammonium-based room-temperature ionic liquids. Part 2. Using activation energy of viscosity. Industrial and Engineering Chemistry Research 2008;47:910−9.

[47] Morgan D, Ferguson L, Scovazzo P. Diffusivities of gases in room-temperature ionic liquids: data and correlations obtained using a lag-time technique. Industrial and Engineering Chemistry Research 2005;44:4815−23.
[48] Horne WJ, Shannon MS, Bara JE. Correlating fractional free volume to CO2 selectivity in [Rmim][Tf2N] ionic liquids. The Journal of Chemical Thermodynamics 2014;77:190−6.
[49] Shannon MS, Tedstone JM, Danielsen SP, Hindman MS, Irvin AC, Bara JE. Free volume as the basis of gas solubility and selectivity in imidazolium-based ionic liquids. Industrial and Engineering Chemistry Research 2012;51: 5565−76.
[50] Hartanto Y, Luis P. Applications of ionic liquid-based materials in membrane-based gas separation. In: Advances in functional separation membranes; 2021. p. 159−83.
[51] Kohoutová M, Sikora A, Hovorka Š, Randová A, Schauer J, Tišma M, et al. Influence of ionic liquid content on properties of dense polymer membranes. European Polymer Journal 2009;45:813−9.
[52] Kusuma VA, Macala MK, Liu J, Marti AM, Hirsch RJ, Hill LJ, et al. Ionic liquid compatibility in polyethylene oxide/siloxane ion gel membranes. Journal of Membrane Science 2018;545:292−300.
[53] Vopička O, Morávková L, Vejražka J, Sedláková Z, Friess K, Izák P. Ethanol sorption and permeation in fluoropolymer gel membrane containing 1-ethyl-3-methylimidazolium bis(trifluoromethylsulfonyl)imide ionic liquid. Chemical Engineering and Processing - Process Intensification 2015;94: 72−7.
[54] Hanabusa K, Fukui H, Suzuki M, Shirai H. Specialist gelator for ionic liquids. Langmuir 2005;21:10383−90.
[55] Skorikova G, Rauber D, Aili D, Martin S, Li Q, Henkensmeier D, et al. Protic ionic liquids immobilized in phosphoric acid-doped polybenzimidazole matrix enable polymer electrolyte fuel cell operation at 200°C. Journal of Membrane Science 2020;608:118188.
[56] Chen HZ, Li P, Chung T-S. PVDF/ionic liquid polymer blends with superior separation performance for removing CO2 from hydrogen and flue gas. International Journal of Hydrogen Energy 2012;37:11796−804.
[57] Jansen JC, Friess K, Clarizia G, Schauer J, Izak P. High ionic liquid content polymeric gel membranes: preparation and performance. Macromolecules 2011;44:39−45.
[58] Eftekhari A, Saito T. Synthesis and properties of polymerized ionic liquids. European Polymer Journal 2017;90:245−72.
[59] Tang J, Tang H, Sun W, Plancher H, Radosz M, Shen Y. Poly (ionic liquid) s: a new material with enhanced and fast CO2 absorption. Chemical Communications 2005:3325−7.
[60] Cheng Y, Ying Y, Japip S, Jiang SD, Chung TS, Zhang S, et al. Advanced porous materials in mixed matrix membranes. Advanced Materials 2018;30: 1802401.
[61] Kamble AR, Patel CM, Murthy Z. A review on the recent advances in mixed matrix membranes for gas separation processes. Renewable and Sustainable Energy Reviews 2021;145:111062.
[62] Wang M, Wang Z, Zhao S, Wang J, Wang S. Recent advances on mixed matrix membranes for CO2 separation. Chinese Journal of Chemical Engineering 2017;25:1581−97.
[63] Ahmad NNR, Leo CP, Mohammad AW, Shaari N, Ang WL. Recent progress in the development of ionic liquid-based mixed matrix membrane for CO2

separation: a review. International Journal of Energy Research 2021b;45: 9800—30.

[64] Hudiono YC, Carlisle TK, Bara JE, Zhang Y, Gin DL, Noble RD. A three-component mixed-matrix membrane with enhanced CO2 separation properties based on zeolites and ionic liquid materials. Journal of Membrane Science 2010;350:117—23.

[65] Hudiono YC, Carlisle TK, LaFrate AL, Gin DL, Noble RD. Novel mixed matrix membranes based on polymerizable room-temperature ionic liquids and SAPO-34 particles to improve CO2 separation. Journal of Membrane Science 2011;370:141—8.

[66] Ahmad N, Leo C, Mohammad A. Enhancement on the CO2 separation performance of mixed matrix membrane using ionic liquid. Materials Letters 2021a;304:130736.

[67] Solangi NH, Anjum A, Tanjung FA, Mazari SA, Mubarak NM. A review of recent trends and emerging perspectives of ionic liquid membranes for CO2 separation. Journal of Environmental Chemical Engineering 2021;9:105860.

[68] Dai Z, Noble RD, Gin DL, Zhang X, Deng L. Combination of ionic liquids with membrane technology: a new approach for CO2 separation. Journal of Membrane Science 2016;497:1—20.

[69] Yan X, Anguille S, Bendahan M, Moulin P. Ionic liquids combined with membrane separation processes: a review. Separation and Purification Technology 2019;222:230—53.

[70] Hafeez S, Safdar T, Pallari E, Manos G, Aristodemou E, Zhang Z, et al. CO2 capture using membrane contactors: a systematic literature review. Frontiers of Chemical Science and Engineering 2021;15:720—54.

[71] Luis P, Garea A, Irabien A. Zero solvent emission process for sulfur dioxide recovery using a membrane contactor and ionic liquids. Journal of Membrane Science 2009a;330:80—9.

[72] Albo J, Luis P, Irabien A. Carbon dioxide capture from flue gases using a cross-flow membrane contactor and the ionic liquid 1-ethyl-3-methylimidazolium ethylsulfate. Industrial and Engineering Chemistry Research 2010;49:11045—51.

[73] Zhang Y, Wang R. Gas—liquid membrane contactors for acid gas removal: recent advances and future challenges. Current Opinion in Chemical Engineering 2013;2:255—62.

17

Application of electrochemical membranes for natural gas sweetening

Fatemeh Haghighatjoo, Behnaz Rahmatmand and Mohammad Reza Rahimpour

Department of Chemical Engineering, Shiraz University, Shiraz, Iran

1. Introduction

Methane, carbon dioxide, nitrogen, and higher hydrocarbons make up the bulk of sour gas, with trace quantities of hydrogen sulfide, helium, oxygen, argon, and water vapor also present. The composition and quality of natural gas vary greatly depending on the level of contamination present. Before natural gas can be sent into a pipeline, it must first undergo purification to remove acid gases (CO_2 and H_2S) and water vapor. Hydrogen sulfide is the most dangerous of all three pollutants, yet they are all corrosive. The impurities must be below the concentration standards for the pipeline, which include 0.02% CO_2, 0.04% H_2S, and 0.01% H_2O [1].

Natural gas may be processed using a variety of methods, including membrane, absorption, adsorption, and cryogenic distillation. Several industrial applications have shown membrane techniques to be technically and economically superior than competing technologies [2,3]. Membrane technology is preferable because it requires little in the way of initial investment, is easy to set up and run, has little in the way of upkeep, is lightweight, and offers a great deal of design freedom [1].

Fig. 17.1 shows the simplest kind of membrane gas separation, which depicts the high-pressure introduction of a gas mixture to the membrane. The rapidly diffusing gas is enriched on the permeate side when it passes through the membrane. The retentate or residue stream is rich in the slower gas [1,4].

Several studies over the last 3 decades have focused on the potential of electrochemical technology for use in cleaning up

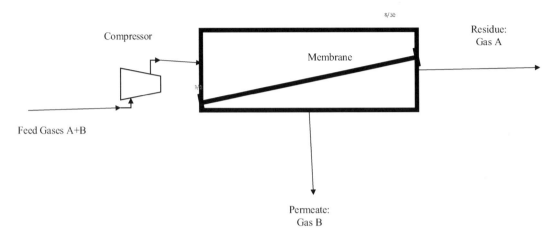

Figure 17.1 An explanation of the gas separation membrane system shown in a simple form [1].

polluted environments [5,6]. Millions of books and articles have been written about the need to create new technologies or enhance old ones. Even after all this time, few innovations are being implemented on a wide scale, and the vast majority of those reviewed are still classified as "promising." Most offer clear advantages, but significant technical and financial disadvantages may be pinpointed due to gaps in the technology's value chain [5].

Some examples of electrolytic processes are the direct oxidation of organic compounds on the anode surface or their mediation by oxidants produced on either the anode or the cathode surface, and the cathodic deposition of metals, as seen in the widely used electrowinning and electrorefining processes. While hydrogen peroxide may be easily created by reducing oxygen using gas diffusion electrode, the second scenario is of major relevance [7] or, more recently, employing cross-flow electrodes despite the application of high pressures [8], leading to more efficient processes. In addition to breaking up emulsions in industrial wastes and removing colloid contaminants in those wastes and during the treatment of surface water, electrocoagulation technologies may be employed since they are also begun by the electrolytic discharge of coagulants from a sacrificial anode [9–11].

The concept of using selective membranes to isolate individual gases from a mixture is gaining momentum. The majority are driven by chemical potentials that are generated by differences in pressure or concentration [12,13]:

$$\Delta \mu_i = \mu_i - \mu_i' = RTln\left[\frac{a_i}{a_i'}\right] \quad (17.1)$$

where a_i and a_i' are component i's activity in the two membrane-separated phases. Selectivity and permeability may be improved by using chemical or surface reactions that facilitate transport.

An electric field may be used as a replacement in certain situations. For charged species, the electrochemical potential difference, z_i, acts as the driving force across a membrane, $\Delta \mu_i$:

$$\Delta \underline{\mu_i} = \underline{\mu_i} - \underline{\mu_i}' = RT\ln\left[\frac{a_i}{a_i'}\right] + z_i F \Delta \varnothing \qquad (17.2)$$

where $\Delta \varnothing$ is the membrane potential difference.

A high-temperature process gas combination is suitable for, and has been used with, this approach. The primary focus of this effort is the removal of H_2S and CO_2 from fuel gases (mostly natural gas). Both sides of the membrane are subjected to the same pressure; hence, there is no theoretical upper limit on how high the pressure may go [14].

2. Removal of H_2S through an electrochemical membrane separator

Several types of industrial and household sewage water include hydrogen sulfide, a poisonous and hazardous gas. Because of its corrosiveness, ecotoxicity, and foul odor, its removal is important for the sake of the environment and human health [15]. Many processes, including chemical oxidation, biological oxidation, and catalytic conversion, convert sulfide to sulfur or sulfate [16–19].

Sulfide is a highly electroactive molecule; therefore, it may be extracted using a fuel cell or electrolysis cell [20,21]. Zhao et al. [22] looked at how fuel cells may remove elemental sulfur while still producing electricity. Sulfur, sulfite, sulfate, and thiosulfate may be produced from sulfide oxidation depending on the parameters of the experiment. While comparing the different electrodes, Pikaar et al. [23] found that the SnO_2 electrode showed superior performance on sulfide oxidation and observed the sulfate production in the anolyte, suggesting that this might be a useful approach for sulfur recovery from household wastewater. The research, however, could not explain where the sulfur buildup on the electrodes came from. Because Ta/Ir and Pt/Ir have a low overpotential for oxygen evolution, they were deemed to be the best electrodes for sulfide oxidation. Dutta et al. [19] observed the deposition of elemental sulfur on the graphite anode while employing Na_2S as the electrolyte; they attributed this to a spontaneous electrochemical sulfur removal mechanism. Around 95% of the oxidation product remained on the electrode's surface after the reaction. For sulfur deposition from $Na_2S.9H_2O$, Dutta et al. [20] also investigated the electrochemical regeneration of sulfur-loaded carbon fiber electrode in a membrane method. Deposition of elemental sulfur on the electrode surface increased the current density, which disrupted the electrochemical process in which the cathode was utilized as the anode in every other batch. Sulfur was broken down into polysulfide and sulfide in this experiment. An effective approach for reactivating the electrode and recovering sulfur from the electrode surface was advocated for by Dutta et al. [19]. At 80°C, Anani et al. [24] electrolyzed hydrogen sulfide to its components in a solution with equal molar concentrations of NaOH and NaHS.

Raw natural gas is heated in a process stream and then passed through a porous electrode to remove H_2S. The reduction of hydrogen sulfide at the cathode results in the formation of a sulfide ion and hydrogen gas. Hydrogen gas from this reaction escapes in the process gas stream, while the sulfide ion travels through a molten salt electrolyte in the direction of the electrical potential gradient. The sulfide ion is oxidized to elemental sulfur at the anode, and this is then removed in an inert sweep gas stream and condensed as a by-product (see Fig. 17.2) [25]. To convert sulfide ions to elemental sulfur, electrons must be transferred from the process gas side to the sweep gas side. As a result, the electrode exposed to the process gas is the cathode and the electrode exposed to the sweep gas is the anode. At the cathode, the reduction of hydrogen sulfide is sought after [12]:

$$H_2S + 2e^- \rightarrow S^{2-} + H_2 \tag{17.3}$$

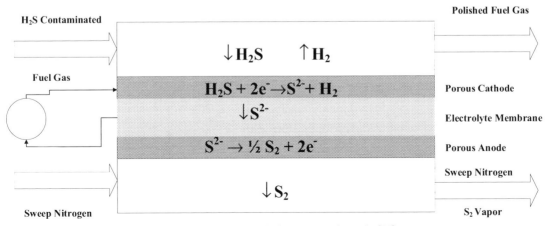

Figure 17.2 Electrochemical process schematic [13].

In the simplest scenario, hydrogen may be delivered to the anode, and the molten sulfide ions in the membrane will react with hydrogen to generate H$_2$S. If the membrane is able to prevent hydrogen from passing through from the cathode side, then it is possible to use an inert sweep gas at the anode to remove oxidized sulfide ions as vaporous sulfur, S$_2$. One example of such a gas would be nitrogen (N$_2$) [13]. The proposed mechanism for this response is as follows:

$$S_2^{2-} + 2e^- \rightarrow 2S^{2-} \qquad (17.4)$$

$$S^{2-} + H_2S + 2e^- \leftrightarrow S_2^{2-} + H_2 \qquad (17.5)$$

for which, S_2^{2-}, polysulfide serves as the catalytic species [26]. This idea has been successfully used to simulate H$_2$S in N$_2$ [25], simulated coal gas [27,28], and simulated natural gas [12]. Very sour natural gas (H$_2$S concentration of 1.5%), moderately sour natural gas (H$_2$S concentration of 2000 ppm), and slightly sour natural gas (H$_2$S concentration of 500 ppm) have all been used to test the efficacy of H$_2$S removal in this chapter (100 ppm H$_2$S).

The sulfide ion is oxidized at the anode in tandem with the reduction of H$_2$S at the cathode, as shown below:

$$S^{2-} \rightarrow 1/2 S_2 + 2e^- \qquad (17.6)$$

3. Removal of CO$_2$ through an electrochemical membrane separator

Carbon dioxide is an acidic gas that must be neutralized before it can be transported [29,30]. Being a heat-trapping gas that is 86 times stronger than CO$_2$, preventing unintentional CH$_4$ escape to the atmosphere is crucial throughout the removal process. Thus, it is technologically and ecologically significant to have a separation process for CO$_2$ removal from raw natural gas that is safe, efficient, and cost-effective. Reversible sorbent/solvent adsorption/absorption procedures are the standard method for separating CO$_2$ from a CO$_2$/CH$_4$ combination. Sorbents like zeolite [31], metal—organic framework (MOF) [32—35], and carbon nanotube [36] are employed in processes like pressure swing adsorption to separate CO$_2$ and CH$_4$. Sorbents used to separate CO$_2$ from CH$_4$ include amine-based liquids like monoethanolamine or methyldiethanolamine [37—40], although these methods often incur a large energy cost and result in CH$_4$ loss [41].

The use of membranes, both organic and inorganic, to separate CO_2 and CH_4 has gained a lot of interest recently [41]. Peters and colleagues [42] removed a mixture of CO_2 and CH_4 using a polyvinyl amine/polyvinyl alcohol membrane with the selectivity of 35–40 achieved, 3.7×10^8 mol s^{-1} m^{-2} Pa^{-1}. In their research on CO_2/CH_4 separation, Xie et al. [43] looked into alumina-supported cobalt–adeninate MOF membranes, finding that they had a high CO_2 permeance (4.55×10^6 mol s^{-1} m^{-2} Pa^{-1}) but a relatively poor selectivity. Nevertheless, Venna and Carreon [44] achieved a high CO_2 permeance of 2.4×10^5 mol s^{-1} m^{-2} Pa^{-1} using a zeolite imidazolate framework membrane at a pressure differential of 40 KPa, although with a selectivity of 5.1. The research shows that these membranes have a significant challenge in overcoming the trade-off between permeability and selectivity, also known as the "Robeson upper limit" [41]. In addition, the organic membranes can only be used for high-pressure and low-temperature CO_2 separation from streams like precombustion products, which generally have pressure of 25–30 bar and temperature of 50°C. In contrast to its solvent and sorbent competitors, membrane technology has less benefits for postcombustion carbon capture in the presence of CO_2-containing streams that are close to ambient pressure and high temperature [41].

Using electrochemical principles, organizations have recently produced a novel form of membranes. The membrane is a hybrid oxide ion and carbon ion conductor made up of a ceramic that conducts oxide ions and a carbonate phase that conducts carbon ions and mixed oxide-ion and carbon-ion conductor (MOCC). Separation of CO_2 from mixtures of CO_2/N_2 (for example, CO_2 flux = 0.13 mL cm^{-2} min^{-1} with a membrane thickness of 1.32 mm at 650°C) and $CO_2/H_2/N_2$ (for example, CO_2 flux = 1.84 mL cm^{-2} min^{-1} with a membrane thickness of 1.2 mm at 700°C) has been demonstrated with high flux and high selectivity using the membranes. Ghezel-Ayagh et al. [45] have shown that at low pressures, MOCC membranes can efficiently separate CO_2 and CH_4 using high flux and selectivity. Samarium-doped ceria was employed as the oxide-ion conducting ceramic matrix in this research, and it included a carbonation conducting eutectic combination of Li_2CO_3 and Na_2CO_3. With CH_4/CO_2 as the feeding gas and helium as the sweeping gas, Fig. 17.3 depicts a schematic of the CO_2 permeation process across an MOCC membrane. The CO_2 separation is propelled by the chemical potential (partial pressure) of CO_2 over the MOCC membrane. CO_2 interacts with O_2 to create CO_3^{2-} at the CH_4-CO_2/MOCC interface. At the MOCC/helium

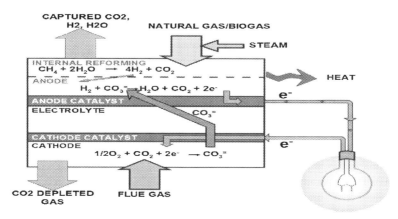

Figure 17.3 The electrochemical membrane cell uses carbon dioxide (CO₂) at the cathode as an oxidant and transports it to the anode in the form of carbonate ions [45].

interface, the newly generated CO_3^{2-} undergoes a reverse reaction to liberate CO_2 and O_2^- [46–48]. In response, O_2^- migrates anticlockwise through the oxide-ion matrix to charge balance the flow of CO_3^{2-}. Obviously, only CO_2 may transit through the membrane, offering exclusive selectivity for CO_2 separation, provided there is no physical leakage. As a result, electrochemical membranes such as MOCC are outside the "Robeson upper boundary" [41].

Processing realistic petrol mixes might further complicate the matter. In the reduction process, carbon dioxide and water vapor compete by:

$$CO_2 + H_2O + 2e^- \rightarrow CO_3^{2-} + H_2 \qquad (17.7)$$

The concentrations and molar mobilities of carbonate and sulfide influence the ionic flow across the membrane. Direct carbonate oxidation happens at a voltage that is 0.70 V higher than what is needed for a sulfide reaction (Eq. 17.6). At a temperature of 900 K, the cell reaction and standard potential are determined by adding up all of the individual half-cell reactions (Eqs. 17.3 and 17.6).

$$CO_3^{2-} \rightarrow CO_2 + 1/2 O_2 + 2e^- \qquad (17.8)$$

$$H_2S \rightarrow H_2 + \frac{1}{2} S_2 \quad E^0 = -0.239V \qquad (17.9)$$

and the cell reaction and standard potential after adding the half-cell reactions (Eqs. 17.7 and 17.8) are:

$$H_2O \rightarrow H_2 + \frac{1}{2}O_2 \quad E^0 = -1.030V \quad (17.10)$$

All of these reactions will take place at the same potential, as anticipated by the Nernst relation, and their relative magnitudes will be set by chemical equilibrium.

$$E = E^0 - \frac{RT}{2F} \ln \left[\frac{(p_{S_2})^{1/2}}{a_{S^{2-}}}\right]_{anode} + \ln \left[\frac{p_{H_2} a_{S^{2-}}}{p_{H_2S}}\right]_{cathode} \quad (17.11)$$

$$E = E^0 - \frac{RT}{2F} \ln \left[\frac{p_{CO_2}(p_{O_2})^{1/2}}{a_{CO_3^{2-}}}\right]_{anode} + \ln \left[\frac{p_{H_2} a_{CO_3^{2-}}}{p_{CO_2} p_{H_2O}}\right]_{cathode} \quad (17.12)$$

Let us say that the cathode is fed a process gas containing 2% H_2S, 1% CO_2, and 12% H_2O. An activity ratio of 665 a $\frac{a_{CO_3^{2-}}}{a_{S^{2-}}}$ will exist in the anolyte before a substantial amount (for example, 1%) of the carbonate is oxidized if 99% of the H_2S is eliminated by the reaction given in Eq. 17.3, and the flow rates of the process gas and the sweep gas are both constant. This presupposes that the cathodic and anodic processes have identical kinetics at the electrodes. Independent studies have revealed that for the current densities at play here, the resistance offered by the electrode kinetics for both processes is negligible [26,49].

As shown by Equation (17.6) and a comparison of the activity ratio of a $\frac{a_{CO_3^{2-}}}{a_{S^{2-}}}$ in the anolyte to the ratio of 26.9 in the catholyte, the oxidation of S_2^- to elemental sulfur is preferred in the absence of a reductant at the anode. This is evidenced by the thermodynamic preference for the oxidation of S_2^- to elemental sulfur. (This method of operation is preferred for commercial use since it does away with the requirement for a Claus reactor and directly produces elemental sulfur vapor.) Under these circumstances, H_2S is continuously stripped out of the process gas, while the gas is simultaneously enriched with hydrogen and elemental sulfur is produced by direct synthesis. Just one reagent is needed, and it is readily available and inexpensive: electricity. The equilibrium potential for a single cell is 0.406 V, which is supplied by Eq. (17.11) for the cathodic and anodic processes (Eqs. 17.3 and 17.6). This potential is applicable to a process gas that contains 2000 ppm H_2S and an anode product of pure sulfur vapor (900 K). Overpotentials for electrode reactions and ohmic loss must be added to this. Graphite electrodes have been used in experiments with free electrolyte to study the electrode reactions

[49]. Experiments with a potential step revealed very fast kinetics, with cathodic and anodic exchange currents approaching 400 A/m². Disulfide was shown to be the electroactive species by cyclic voltammetry, confirming a "catalytic" reaction process. Both the cathode and the anode are considered in Eqs. 17.4 and 17.5:

$$S_2 + 2S^{2-} \leftrightarrow 2S_2^{2-} \qquad (17.13)$$

$$S_2^{2-} \rightarrow S_2 + 2e^- \qquad (17.14)$$

The presence of carbon dioxide and water in the gas surprisingly improved cathodic H₂S removal, perhaps owing to another "catalytic" scheme, Eq. (17.7) followed by:

$$CO_3^{2-} + H_2S \leftrightarrow CO_2 + H_2O + S^{2-} \qquad (17.15)$$

The simultaneous reduction of COS to concentrations below the analytical limit was also an unanticipated outcome (about 2 ppm). At these temperatures, H₂S, CO, CO₂, and COS seem to quickly reach equilibrium with one another:

$$H_2S + CO \leftrightarrow H_2 + COS \qquad (17.16)$$

$$H_2S + CO_2 \leftrightarrow H_2O + COS \qquad (17.17)$$

Some viable alternatives were discovered via research into possible cathode materials [27,28]. The setup used was quite close to what was expected (see Fig. 17.3). Molten carbonate fuel cell (MCFC) "tiles" were employed because, in steady state, the working membrane will consist mostly of carbonate. Acceptable electrode materials include nickel and cobalt powders that are sulfided on-site.

The composition of the electrolyte that would be in equilibrium with a particular process gas at a certain process temperature may be determined by examining the equilibrium of reaction (Eq. 17.15) [12]. Thermodynamic study of the tile equilibrium reaction was used to determine theoretical tile compositions (Eq. 17.18). This study employed membranes comparable to those found in MCFCs, thus the cations present were potassium and lithium in a ratio that matches the low melting carbonate eutectic ($Li_{0.62}K_{0.38}$).

$$(Li_{0.62}K_{0.38})_2CO_3 + H_2S \leftrightarrow (Li_{0.62}K_{0.38})_2S + CO_2 + H_2O \qquad (17.18)$$

In order to carry out this investigation, we first calculated the Gibbs free energy of Eq. (17.18) at the temperature at which it was functioning, and then we linked that value to the constant K_a, using the following formula:

$$-\ln K_a = \frac{\Delta G^0}{RT} \quad (17.19)$$

were K_a is defined as:

$$K_a = \frac{p_{CO_2} p_{H_2O} a_{S^{2-}}}{p_{H_2S} a_{CO_3^{2-}}} \quad (17.20)$$

At a temperature of 883 K, the equilibrium constant for a process gas containing 0.88% CO_2, 1760 parts per million H_2S, 12% water, and the remaining 3% methane is calculated to be 6.9. This corresponds to an electrolyte composition of 19.5% sulfide and 80.5% carbonate, assuming that the activity coefficients of the molten-phase elements (namely the sulfide and carbonate in the electrolyte) are identical [13].

It is unnecessary for a carbonate tile to be "sulfided" or a sulfide tile to be "carbonated" when they are made to be already in equilibrium with the gas to be treated. The idea has been successfully implemented in the coal gasification process cell [28] and the natural gas process cell [12], but the methods for producing such a tile are currently being investigated.

4. Electrode preparation

Many candidates for use as an anode have been discovered [27,28]. CoS_2 anodes have shown promising results in natural gas cells [12]. Ni anodes that are left to reach equilibrium composition in place have also been employed successfully more recently. The chemical breakdown issues that plague carbon electrodes are not present in these alternative anode materials.

4.1 Carbon

Graphite is suitable for bench scale removal cell testing and experiments involving free electrolytes, but it cannot be used in commercial removal cells. Eventually, the cathode will degrade due to the steam or CO_2:

$$C + H_2O \leftrightarrow CO + H_2 \quad (17.21)$$

$$C + CO_2 \leftrightarrow 2CO \quad (17.22)$$

And it may be used as a carbonate reductant at the anode:

$$CO_3^{2-} + C \rightarrow CO + CO_2 + 2e^- \quad (17.23)$$

Stackpole/ultracarbon supplied the carbon electrodes used in the cathode.

4.2 CoS$_2$

Union Carbide Corp.'s hydroxyethyl cellulose (HEC) was combined with the high-purity ingredients purchased from Alfa Chemicals. A combination of 10% HEC and 90% CoS$_2$ powder yielded a void percentage of 60%. A 0.032 m stainless steel die was used to press the mixture with a force of 550,105 N/m^2. The resultant electrode wafer was then subjected to 30 min of heating at 623 K to destroy the HEC. After being prepared for use in the electrochemical cell, this final electrode was cooled, weighed, and stored [50].

4.3 Ni

Energy Research Corporation was the supplier of the porous Ni electrodes used in the MCFC. Nevertheless, because of the proprietary nature of these components, it is not permissible to provide any specifics on the structure. A state of equilibrium between nickel sulfides and nickel oxides was permitted to develop in situ [50].

5. Membrane preparation

One of the two techniques was used in the production of the membranes that were utilized in the study that is being given here. The first approach consisted of producing a sintered ceramic matrix of MgO that was devoid of electrolyte and then "wicking" the molten electrolyte into the matrix by the use of capillary action [50].

The membrane acts to prevent bulk diffusion of gases between the cathode side of the cell and the anode side of the cell, and the inert ceramic matrix that holds the electrolyte in place between the cathode and the anode serves two purposes: it holds the electrolyte by capillary action and it prevents the molten salts from completely flooding the porous electrodes. Both of these functions are performed between the cathode and the anode side of the cell. If the electrolyte were not in chemical equilibrium with the process gas, then localized density changes in the electrolyte that were caused by reaction 18 would cause the membrane to rupture, which would then allow for bulk mixing of the process gas stream and the sweep gas stream [50].

The second way of producing membranes included creating a composite structure by densifying woven zirconia cloths with MgO powder. This approach was used to make the membrane.

The structure was created with three mats of ZYW-30A zirconia fabric (purchased from ZIRCAR Ceramics Inc.), three tapes of MgO ceramic powder suspended in acrylic binder K565-4, and a third mat of ZYW-30A zirconia fabric. All of these components were adhered together with a third mat of ZYW-30A zirconia fabric (purchased from Metoramic Sciences, Inc.). During the assembly process, discs of pressed powder representing the electrolyte were placed into the framework. This gave the impression that the discs contained liquid. Due to the amount of handling that is required outside of the cell run conditions during setup, eutectic Li/K carbonate was initially used in this membrane manufacturing approach [12]. This carbonate is stable in normal room air, despite the fact that it is hygroscopic, and this was done because it is stable in normal room air.

The cell, which had previously been subjected to a nitrogen sweep on both the process side and the sweep side, was then positioned inside of the boiler. Overnight at 648 K, the binder that was on the MgO tapes evaporated. After that, the electrolyte was wicking its way into the MgO particles and the zirconia fabric while the temperature was being brought up to working levels. When the composition of the electrolyte had arrived at the point of equilibrium as determined by reaction 18, process gas was then injected into the cell. Fractures that would have been generated by localized density fluctuations in the electrolyte were avoided due to the fact that the ceramic matrix was no longer a sintered structure [13].

6. Conclusion and future outlooks

The results of this research indicate that it is feasible to remove H_2S from natural gas, but not in an entirely selective manner. Fortunately, the other component that is eliminated for the process gas is CO_2, which is likewise regarded as a pollutant that may be found in natural gas. Consequently, the removal of both species follows similar patterns, provided that the concentration of CO_2 in the process gas does not get so high that it entirely obscures the H_2S removal. It is believed that the lack of selectivity in the removal of species from the cell is due to the bulk diffusion of hydrogen across the membrane of the cell. At this time, the resistance of the gas phase mass transfer is considered to be the limiting element for the total cell performance. It is possible that the issue of H$_2$ transport may be solved by enhancing membrane manufacturing procedures, which would result in larger membrane densities. Moreover, increasing gas velocities should assist in reducing the mass transport resistance and boost removal rates. The process of producing membranes will continue to be improved during the course of this work. In addition to hot pressing, experiments will be conducted in the fields of tape casting and wicking molten electrolyte into a sintered ceramic matrix in situ. Both of these methods are expected to be successful. A membrane whose electrolyte is in equilibrium with the process gas and dense enough to prevent the cross-diffusion of hydrogen will be the end result of this study. Future iterations of the cell will include a process gas recycling pump, which will result in an increase in the cell's mass transfer coefficient. The adjustment will result in an increase in the gas velocity that is achieved per passage through the cell, but it will have no impact on the typical gas residence time [12].

Abbreviations and symbols

ERC	Energy Research Corporation
GDE	Gas diffusion electrode
HEC	Hydroxyethyl cellulose
MCFC	Molten carbonate fuel cell
MOCC	Mixed oxide-ion and carbon-ion conductor
MOF	Metal–organic framework
PSA	Pressure swing adsorption
SDC	Samarium-doped ceria
ZIF	Zeolite imidazolate framework

References

[1] Tabe-Mohammadi A. A review of the applications of membrane separation technology in natural gas treatment. Separation Science and Technology 1999;34(10):2095–111.

[2] Meshksar M, Roostaee T, Rahimpour MR. Membrane technology for brewery wastewater treatment. In: Current trends and future developments on (bio-) membranes. Elsevier; 2020. p. 289–303.

[3] Roostaie T, Meshksar M, Rahimpour MR. Biofuel reforming in membrane reactors. In: Current trends and future developments on (bio-) membranes. Elsevier; 2020. p. 351–66.

[4] Zafarnak S, Meshksar M, Rahimpour HR, Rahimpour MR. Membrane technology for syngas production. In: Advances in synthesis gas: methods, technologies and applications. Elsevier; 2023. p. 291–304.

[5] Lacasa E, Cotillas S, Saez C, Lobato J, Cañizares P, Rodrigo MA. Environmental applications of electrochemical technology. What is needed to enable full-scale applications? Current Opinion in Electrochemistry 2019; 16:149–56.

[6] Al Aukidy M, Verlicchi P, Voulvoulis N. A framework for the assessment of the environmental risk posed by pharmaceuticals originating from hospital effluents. Science of the Total Environment 2014;493:54–64.

[7] Al-Qodah Z, Al-Shannag M, Bani-Melhem K, Assirey E, Yahya MA, Al-Shawabkeh A. Free radical-assisted electrocoagulation processes for wastewater treatment. Environmental Chemistry Letters 2018;16:695–714.

[8] Alshawabkeh AN, Yeung AT, Bricka MR. Practical aspects of in-situ electrokinetic extraction. Journal of Environmental Engineering 1999;125(1): 27–35.

[9] Barba S, Villaseñor J, Cañizares P, Rodrigo MA. Strategies for the electrobioremediation of oxyfluorfen polluted soils. Electrochimica Acta 2019;297:137–44.

[10] Biniaz P, Ardekani NT, Makarem MA, Rahimpour MR. Water and wastewater treatment systems by novel integrated membrane distillation (MD). ChemEngineering 2019;3(1):8.

[11] Rahimpour MR, Kazerooni NM, Parhoudeh M. Water treatment by renewable energy-driven membrane distillation. In: Current trends and future developments on (bio-) membranes. Elsevier; 2019. p. 179–211.

[12] Alexander S, Winnick J. Electrochemical separation of hydrogen sulfide from natural gas. Separation Science and Technology 1990;25(13–15): 2057–72.

[13] Alexander SR, Winnick J. Removal of hydrogen sulfide from natural gas through an electrochemical membrane separator. AIChE Journal 1994;40(4): 613–20.

[14] Meshksar M, Rahimpour MR. Ionic liquid membranes for syngas purification. In: Advances in synthesis gas: methods, technologies and applications. Elsevier; 2023. p. 253–71.

[15] Siavashi F, Saidi M, Rahimpour MR. Purge gas recovery of ammonia synthesis plant by integrated configuration of catalytic hydrogen-permselective membrane reactor and solid oxide fuel cell as a novel technology. Journal of Power Sources 2014;267:104–16.

[16] Selvaraj H, Chandrasekaran K, Gopalkrishnan R. Recovery of solid sulfur from hydrogen sulfide gas by an electrochemical membrane cell. RSC Advances 2016;6(5):3735–41.

[17] Ateya B, AlKharafi F, Al-Azab A. Electrodeposition of sulfur from sulfide contaminated brines. Electrochemical and Solid-State Letters 2003;6(9):C137.

[18] Ateya B, AlKharafi FM, Alazab AS, Abdullah AM. Kinetics of the electrochemical deposition of sulfur from sulfide polluted brines. Journal of Applied Electrochemistry 2007;37(3):395–404.

[19] Dutta PK, Rabaey K, Yuan Z, Keller J. Spontaneous electrochemical removal of aqueous sulfide. Water Research 2008;42(20):4965–75.

[20] Dutta PK, Rozendal RA, Yuan Z, Rabaey K, Keller J. Electrochemical regeneration of sulfur loaded electrodes. Electrochemistry Communications 2009;11(7):1437–40.

[21] Kiani MR, Meshksar M, Makarem MA. Catalytic membrane micro-reactors for fuel and biofuel processing: a mini review. Topics in Catalysis 2021: 1–20.

[22] Zhao Y, Liu Z, Jia Z, Xing X. Elemental sulfur recovery through H_2 regeneration of a SO_2-adsorbed CuO/Al_2O_3. Industrial & Engineering Chemistry Research 2007;46(8):2661–4.

[23] Pikaar I, Rozendal RA, Yuan Z, Keller J, Rabaey K. Electrochemical sulfide oxidation from domestic wastewater using mixed metal-coated titanium electrodes. Water Research 2011;45(17):5381–8.

[24] Anani A, Mao Z, White RE, Srinivasan S, Appleby AJ. Electrochemical production of hydrogen and sulfur by low-temperature decomposition of hydrogen sulfide in an aqueous alkaline solution. Journal of the Electrochemical Society 1990;137(9):2703.

[25] Lim HS, Winnick J. Electrochemical removal and concentration of hydrogen sulfide from coal gas. Journal of the Electrochemical Society 1984;131(3): 562.

[26] White III KA, Winnick J. Electrochemical removal of H_2S from hot coal gas: electrode kinetics. Electrochimica Acta 1985;30(4):511–9.

[27] Weaver D, Winnick J. Electrochemical removal of H_2S from hot gas streams: nickel/nickel-sulfide cathode performance. Journal of the Electrochemical Society 1987;134(10):2451.

[28] Weaver D, Winnick J. Evaluation of cathode materials for the electrochemical membrane H_2S separator. Journal of the Electrochemical Society 1991;138(6):1626.

[29] Meshksar M, Sedghamiz MA, Zafarnak S, Rahimpour MR. CO_2 separation with ionic liquid membranes. In: Advances in carbon capture. Elsevier; 2020. p. 291–309.

[30] Sadatshojaie A, Rahimpour MR. CO_2 emission and air pollution (volatile organic compounds, etc.)–related problems causing climate change. In: Current trends and future developments on (bio-) membranes. Elsevier; 2020. p. 1–30.

[31] Cavenati S, Grande CA, Rodrigues AE. Adsorption equilibrium of methane, carbon dioxide, and nitrogen on zeolite 13X at high pressures. Journal of Chemical and Engineering Data 2004;49(4):1095–101.

[32] Li J, Yang J, Li L, Li J. Separation of CO_2/CH_4 and CH_4/N_2 mixtures using MOF-5 and $Cu_3(BTC)_2$. Journal of Energy Chemistry 2014;23(4):453–60.

[33] Krishna R. Adsorptive separation of $CO_2/CH_4/CO$ gas mixtures at high pressures. Microporous and Mesoporous Materials 2012;156:217–23.

[34] Mishra P, Mekala S, Dreisbach F, Mandal B, Gumma S. Adsorption of CO_2, CO, CH_4 and N_2 on a zinc based metal organic framework. Separation and Purification Technology 2012;94:124–30.

[35] Yousefi S, Meshksar M, Rahimpour HR, Rahimpour MR. MOF mixed matrix membranes for syngas purification. In: Advances in synthesis gas: methods, technologies and applications. Elsevier; 2023. p. 307–23.

[36] Furmaniak S, Terzyk AP, Kowalczyk P, Kaneko K, Gauden PA. Separation of CO_2-CH_4 mixtures on defective single walled carbon nanohorns–tip does matter. Physical Chemistry Chemical Physics 2013;15(39):16468–76.

[37] Rochelle GT. Amine scrubbing for CO_2 capture. Science 2009;325(5948): 1652–4.

[38] Sanz R, Calleja G, Arencibia A, Sanz-Pérez ES. Development of high efficiency adsorbents for CO_2 capture based on a double-functionalization method of grafting and impregnation. Journal of Materials Chemistry A 2013;1(6):1956–62.

[39] Saidi M, Heidarinejad S, Rahimpour HR, Talaghat MR, Rahimpour MR. Mathematical modeling of carbon dioxide removal using amine-promoted hot potassium carbonate in a hollow fiber membrane contactor. Journal of Natural Gas Science and Engineering 2014;18:274–85.

[40] Bakhtyari A, Mohammadi M, Rahimpour MR. Simultaneous production of dimethyl ether (DME), methyl formate (MF) and hydrogen from methanol in an integrated thermally coupled membrane reactor. Journal of Natural Gas Science and Engineering 2015;26:595–607.

[41] Tong J, Zhang L, Fang J, Han M, Huang K. Electrochemical capture of CO_2 from natural gas using a high-temperature ceramic-carbonate membrane. Journal of the Electrochemical Society 2015;162(4):E43.

[42] Peters L, Hussain A, Follmann M, Melin T, Hägg M-B. CO_2 removal from natural gas by employing amine absorption and membrane technology—a technical and economical analysis. Chemical Engineering Journal 2011; 172(2−3):952−60.

[43] Xie Z, Li T, Rosi NL, Carreon MA. Alumina-supported cobalt−adeninate MOF membranes for CO_2/CH_4 separation. Journal of Materials Chemistry A 2014;2(5):1239−41.

[44] Venna SR, Carreon MA. Highly permeable zeolite imidazolate framework-8 membranes for CO_2/CH_4 separation. Journal of the American Chemical Society 2010;132(1):76−8.

[45] Ghezel-Ayagh H, Jolly S, Patel D, Steen W. Electrochemical membrane technology for carbon dioxide capture from flue gas. Energy Procedia 2017; 108:2−9.

[46] Zhang L, Li X, Wang S, Romito KG, Huang K. High conductivity mixed oxide-ion and carbonate-ion conductors supported by a prefabricated porous solid-oxide matrix. Electrochemistry Communications 2011;13(6): 554−7.

[47] Zhang L, Xu N, Li X, Wang S, Huang H, Harris WH, et al. High CO_2 permeation flux enabled by highly interconnected three-dimensional ionic channels in selective CO_2 separation membranes. Energy and Environmental Science 2012;5(8):8310−7.

[48] Zhang L, Mao Z, Thomason JD, Wang S, Huang K. Synthesis of a homogeneously porous solid oxide matrix with tunable porosity and pore size. Journal of the American Ceramic Society 2012;95(6):1832−7.

[49] Banks EK, Winnick J. Electrochemical reactions of H_2S in molten sulphide. Journal of Applied Electrochemistry 1986;16(4):583−90.

[50] Alexander S. Removal of H_2S through an electrochemical membrane separator. In: Studies in environmental science. Elsevier; 1994. p. 535−64.

18

Membrane technology for CO_2 removal from CO_2-rich natural gas

Shaik Muntasir Shovon[1], Faysal Ahamed Akash[1], Minhaj Uddin Monir[1,2], Mohammad Tofayal Ahmed[1,2] and Azrina Abd Aziz[3]

[1]Department of Petroleum and Mining Engineering, Jashore University of Science and Technology, Jashore, Bangladesh; [2]Energy Conversion Laboratory, Department of Petroleum and Mining Engineering, Jashore University of Science and Technology, Jashore, Bangladesh; [3]Faculty of Civil Engineering Technology, Universiti Malaysia Pahang Al-Sultan Abdullah, Kuantan, Pahang, Malaysia

1. Introduction

Gas separation is required to remove contaminants, hazardous gases, and CO_2 in order to produce gas that meets consumer demands. Membrane gas separation is a typical gas separation method. Membrane gas separation processes, such as absorption, distillation, and adsorption, provide various benefits, including minimal cost, simplicity of handling, higher selectivity, and the potential to connect with other operations [1]. Presently, the primary emphasis is on developing technology and polymer manufacturing for improved selectivity and permeability in order to meet demand. Numerous studies are being conducted to separate CO_2/CH_4 and O_2/N_2 from air and natural gases [2]. Membrane performance is determined by its manufacture and material structure, as well as separation mechanism [3].

Benchold started the voyage of membrane gas separation in 1907 [4]. He was a pioneer in the development of nitrocellulose membranes with varying pore sizes. Collodion membranes with microparticle sizes first appeared on the market on a small scale in 1930s, when more refined membrane manufacturing procedures were devised [5,6]. Various polymers, such as cellulose acetate, were utilized to build membranes during the next 20 years, and its first significant use was filtering drinking water during the Second World War. Prior to 1945, membrane channels were used extensively to remove germs and other particles from fluids and to estimate macro size molecules. This approach is being utilized in the industry for water purification. The reverse osmosis method is a significant innovation in the history of membrane technology, and it is today employed in almost every industry.

Membrane can be used to segregate gases, according to researchers Moghiseh et al. [7]. Monsanto Prism successfully separated hydrogen gas in 1980, which was the first breakthrough in membrane gas separation technology. The first commercial membrane for air purification was prepared by Permea in the year of 1980. For separating hydrogen from methane, hollow fiber membrane was prepared [8,9].

Because of the tireless efforts of scientists, membrane gas separation technology has advanced significantly. Compared to a few years ago, fabrication and gas transport mechanism processes have advanced. It is now feasible to recover hydrogen, separate air, purify biogas, and remove CO_2 and pollutants utilizing membrane technology [10]. CO_2 can be converted to methane by hydrogenation of CO_2 and photocatalytic method [11]. Inorganic membranes outperform organic membranes in terms of gas separation capacity, thermal heat tolerance, and pressure tolerance. However, inorganic membranes are 100–1000 times more costly than organic membranes. Although membrane gas separation technique is widely used today, it has several drawbacks. The fundamental issue is that permeability diminishes as selectivity rises since they are inversely proportional. Furthermore, heat in the organic membrane deviates, and thus membrane is often destroyed. Impurities cause membrane aging in several circumstances. Some components have excellent selectivity; however, cellulose is the predominant element for natural gas processing [12]. In other circumstances, contaminants were shown to slow down the separation process. Many difficulties are solved by diol cross-linked polyimide and glassy members; however, there are still certain gaps that scientists are striving to fill.

2. Fundamentals of membrane gas separation for CO_2 removal

Membrane gas separation for separating CO_2 is a highly attractive moreover power-efficient method [13,14]. It is a kind of physical device that separates a few (one or more) gases from the gas mixture and discards others [15,16]. Membrane has shown a remarkable performance over conventional methods and it is used in many petroleum chemicals, water purifiers, and other industries today [15,16]. There are many advantages of membrane like its small size, diversity, maintenance, and simplicity. The separation of gas by membrane is used in several industries to enrich the quality of gas. In separating at least one gas from a feed combination and producing a pure gas-rich pervades, a membrane works as a "filter" that permits the special section of specific substances. Specifically, a membrane will separate if a few components from the combination can pass through the membrane more quickly than others [15,16]. Further, motion of the gas that passes quickly separates from the mixture.

There are mainly two types of membrane: porous membrane and nonporous membrane [15,16]. Mechanisms of the two types of membranes are different. Porous membrane separates gases through small pores based on molecular size. Diffusivity and solubility are nonporous membrane mechanisms of separating gases. Solution diffusion, surface diffusion, Knudsen diffusion, facilitated transport, molecular sieving, and capillary condensation are the mechanism for both porous and nonporous membrane gas separation [17–22]. For polymeric membrane, solution-diffusion is the most suitable procedure for removing CO_2 [22]. Gas transport depends on a solution dispersion system and results in particular separation of gases and, subsequently, their purification.

3. Membrane processes for efficient CO_2 removal

Membrane-based separation technology is widely used for gas purification, desalination of water, toxic metal removal, and recovery of valuables. The whole process depends on the membrane nature and manufactures from several materials among them polymeric, ceramic, and zeolites are common. Each membrane has a separate filtering property. Filtering property mainly depends upon the change of surface, different pore sizes, the morphology of membrane, and hydrophobicity characteristics [23].

Organic membrane partial pressure gradient is one method of gas separation. It is mostly the result of mole function and total pressure. Several gas transportation techniques have been suggested. It is mostly determined by the membrane's material qualities [24]. Poiseuille (viscous) flow, Knudsen diffusion, molecular sieving, capillary condensation, and solution–diffusion mechanisms are the common mechanisms that are proposed for gas transportation [25]. Table 18.1 shows gas transport regimes and their selectivity based on the pore size, and Fig. 18.1 is about the several gas separation mechanisms.

3.1 Hagen–Poiseuille (viscous flow in wide pores)

Hagen–Poiseuille technique is applicable when the pore width is enormous contrasted with the mean freeway of gas particles (λ) and transport is by mass liquid course through the huge pores [26].

Permeability and gas viscosity are proportional inversely in the Poiseuille regime [25] shown on Eq. (18.1).

$$P_e = \frac{\varepsilon \eta r^2}{8\mu RT} P_{av} (\text{mol m}^{-1} \text{ s}^{-1} \text{ Pa}^{-1}) \tag{18.1}$$

Here, ε is the porosity
η is the structural factor
μ is the viscosity
p_{av} is the pressure, Pa

3.2 Knudsen diffusion in narrow pores

Kundsen diffusion is a gas transportation technique in which pore size is larger than the size of the molecule but smaller than the gas path molecules. Instead of friction between gas particles, collision occurs between the pore divider and gas atoms in this system [27]. Although the bearing of bounce back is uneven,

Table 18.1 Main gas separation regimes and their selectivity based on the pore size in the membrane [25].

Separation mechanism	Turbulent flow	Laminar flow	Surface diffusion	Knudsen diffusion	Solution diffusion	Capillary condensation	Molecular sieving
Diameter of the pore	>50	>20	1–3	2–50	0	2–10	0–2
Selectivity	×	×	√	$\sqrt{m2/m1}, m2 > m1$	√√√	√√	√√

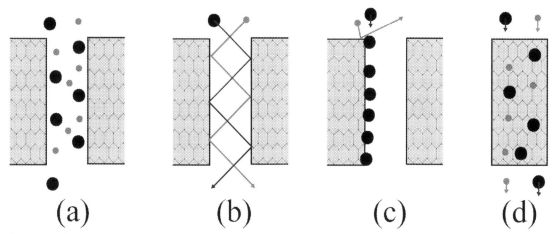

Figure 18.1 Several gas separation mechanisms: (A) Viscous flow, (B) Knudsen flow, (C) surface flow, and (D) solution diffusion [25].

collisions are diverse, with no proclivity for the atoms to attach to the surface [25,28].

Because the cooperation between diffusing atoms and pore divider is negligible, gas transport via Knudsen dissemination proceeds in the vaporous state without involvement of adsorption. Separation factor α of several typical gas mixtures by Knudsen diffusion is shown in Figs. 18.2 and 18.3.

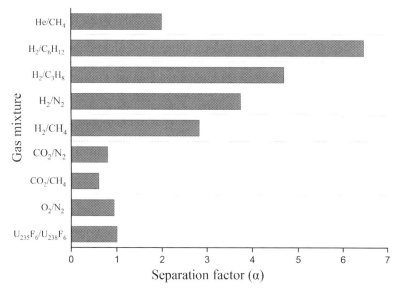

Figure 18.2 Separation factor α of some typical gas mixtures by Knudsen diffusion [25].

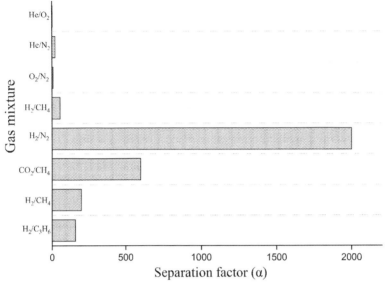

Figure 18.3 Separation factor of several gases [25].

Permeability of Knudsen for gas is [25] shown on Eq. (18.2).

$$P_e = \frac{2\varepsilon\eta r v}{3RT} \left(\text{mol m}^{-1}\ \text{s}^{-1}\ \text{Pa}^{-1}\right) \tag{18.2}$$

Here v is the velocity of the molecule (ms^{-1}). The equation shown on Eq. (18.3).

$$V = \sqrt{\frac{8RT}{\pi M}} \tag{18.3}$$

For the separation of gaseous Uranium U-238 and U-235, the Knudsen diffusion is used [25] shown on Eq. (18.4). It is used for nuclear fuel.

$$\alpha_{A/B} = \sqrt{\frac{MW_B}{MW_A}} \tag{18.4}$$

Here, α is the separation factor.
MW_B and MW_A are the weights of the molecule.

3.3 Capillary condensation

Capillary condensation is a process by which multifacet adsorption from a fume into a permeable media continues unabated, resulting in pore spaces being filled up with consolidated fume fluid [29].

3.4 Molecular sieving

By utilizing molecular sieving, pore membranes are typically segregated by the same gas particle size as the real molecule. In general, typical pore size for submolecular sieving is less than 2 mm [30].

4. Current application and cases

To enrich the quality of gas membrane-based gas separation technology is very important. Membrane separation strategies are contemporary cycles that might decrease capital speculation, working costs, and collaboration security. Separation of blended gas mixes is a basic unit activity in numerous organizations to recuperate key gases.

Conventional development, like cryogenic modification and adsorption on solid surface and dissolvable maintenance, is currently embracing such divides. Membranes are progressively being utilized for vaporous separations and have shown colossal commitment [31]. Membranes are types of gas purification techniques that are widely used in different modern areas.

4.1 Hydrogen recovery

Hydrogen is the most found component on earth. There are several components like water, biomass, and hydrocarbons from which hydrogen can be generated. Hydrogen can be separated from the gas mixture by using a membrane. Polymeric membrane is used for hydrogen separation based on its material [32]. 39%–65% of hydrogen contained by coke stove gas (CSG) and delivered at a 280–450 m^3 for each large load of coal in coke broiler batteries [33,34].

Recovery of hydrogen from NH_3 is another application of membrane gas separation. NH_3 is naturally composed by N_2 with H_2 at high temperature and tension. Membranes are typically supplemented in that reactor to rinse flow to storage of H_2 from washed gas and placed back to series for further use. Membranes have been similarly applied in the recovery of hydrogen in treatment ability hydrotreaters (H_2/CH_4) and to maintain the molar scopes of syngas (H_2/CO). Several methods are applied for producing of syngas, in light of that the H_2: CO (carbon monoxide) amount shifts somewhere in the variety of 1 and 5 [35–37]. This amount must be modified to explicit applications since of the great porosity of H_2 in the membrane. Polymeric layers with

H_2/CH_4 selectivity of 20−25 and high porosity are applied for this reason [38].

4.2 Air purification

Air purification is vital in intensifying industry for generating of pure O_2 and N_2. N_2 is frequently utilized for hardware incorporation, removal of lines and compartments besides equiponderant infiltration as an inefficient expansion to a multiple processor [39,40]. Air is essential and can be used in various ways. Pure oxygen is fundamental in various sectors: Fuel cell, boosting furnace combustion, in the medical sector, hazardous waste containment, aviation fuel, submarine control, and biotechnology. Manufacturing of pure oxygen and nitrogen is in great demand [41,42]. Air is ordinarily a mix of nitrogen, oxygen, argon, carbon dioxide, and a humble amount of water rage. About 79% of the ecological air is made from N_2, and 21% of it is O_2. Accordingly, encompassing air is utilized as a natural element for the combination of O_2 and N_2 of changing immaculateness [43−45]. Fig. 18.4 shows the manufacturing process of nitrogen and oxygen.

Several methods besides accurate parameters needed for balancing and optimized for designing the membrane structure that comprises gas flow rate, compressor expenses, cleanliness condition, and so many [47,48]. Most of the work was done in N_2 enrichment, while the concentration on increasing the selectivity of membrane because with a lower selectivity recovery

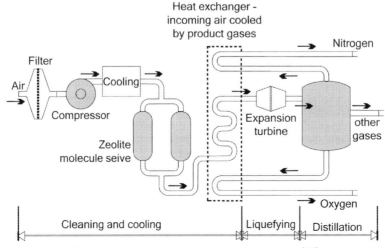

Figure 18.4 Manufacture of nitrogen and oxygen [46].

rate decreases with a high compressor cost [49]. First membrane for air separation was developed in the 1980s by Generon and Permea, but the usage of the membrane was limited because of the low selectivity [49]. Toward the mid-1990s, research into novel layer materials and configuration brought about the improvement of different new hollow fiber films, which incorporate the tetrahalogenated bisphenol-based polycarbonates by Generon [50], polyimides by Praxair and Medal [49], and polyamide by Medal [49]. During similar time, numerous asymmetric membranes had likewise been created for nitrogen-improvement applications.

Moreover, nitrogen-enhancement membranes likewise observe a wide possibility to be utilized in procurement oxygen-enriched air (OEA), that when utilized rather than surrounding air, can essentially further develop the energy productivity of burning cycles and decrease the expense of CO_2 capture from flue gases all through assembling industries. OEA likewise finds applications in modern cycles, for example, recovery of impetus in liquid reactant cracking, purification of wastewater, fractional oxidation of sulfur in Claus plants, and further developing indoor air purity [51].

Moreover, there have been crucial advances in manufacturing industrial membranes with the goal of air separation and purification by different mechanisms (nitrogen, oxygen, argon) [52]. Primary dehydration membrane was produced by Permea in 1987, which supplanted the desiccants in refrigeration dehydrators [53]. It was likewise utilized for generating dry air in different military applications, in electronics and correspondence frameworks [53]. Recently, because of the relative effortlessness and dependability of the layer dehydrators, they are generally utilized in the US, China, and Europe, for drying packed air [54]. Presently, air products markets dehydration films under the trademark CACTUS for use in drying out applications under high tension [55].

4.3 Purification of natural gas from acid gases

Gas treatment from a gas mixture is critical for retaining the useful qualities of gas. Numerous improvements have been made with the help of membranes [56]. Acid gases, namely CO_2 and H_2S, are often found in unrefined petroleum gas at varying fixation levels ranging from parts per million (ppm) to a few rates, making the gas to be caustic and dangerous [57].

Membrane is used for gas detachment solicitations (AG (acid gases) evacuation comprehensive) are of the nonpermeable (thick) type; membrane pore size is practically identical to that of the gas atomic width (Å) [58]. Versatility of a gas (for example carbon dioxide, hydrogen disulfide, or methane) across these membranes is depicted by the arrangement dissemination model, which can be improved and introduced by the Fick's dispersion law [55] shown in Eq. (18.5).

$$FluX_{AG} = -D_{AG}\left(\frac{\Delta C_{AG}}{t}\right) = -D_{AG}\, S_{AG}\left(\frac{\Delta P_{AG}}{t}\right) \quad (18.5)$$

Here,
ΔP_{AG} = Acid gas partial pressure difference
D_{AG} = Acid gas diffusivity
S_{AG} = Acid gas solubility
t = Membrane thickness

Dynamic breadth of a gas is inextricably linked to its diffusivity, since it regulates the rate of gas development in between polymeric membrane chain's free space [59]. Significant atomic weight gases often have a massive subatomic structure and, as a result, have high dissemination obstruction, such as poor diffusivity. Overall, due to their nature of condensing, such as a fast rate of sorption, these gases often exhibited high solvency in the polymeric membrane material. Gas's saturation rate is determined by its equilibrium nature, which consists primarily of two components: diffusivity and solvency [58]. The percentage of tried gases pervasion motion, while the gases are attempted on an unadulterated premise, and under indistinguishable strain proportions, is referred to as inborn membrane selectivity. In the AG evacuation application, (Eq. 18.6) is

$$\alpha_{\frac{AG}{CH4}} = \frac{Flux_{AG}}{Flux_{CH4}} = \left(\frac{D_{Ag}}{D_{CH4}}\right) \times \left(\frac{S_{AG}}{S_{Ag}}\right) \quad (18.6)$$

Where,
$\alpha_{AG/CH4}$ = Kink selectivity of acid gases relevant to methane
D_{AG} = Diffusivity of acid gas in the membrane structure
D_{CH4} = Diffusivity of methane in the membrane structure
S_{AG} = Solubility of acid gas in the membrane material
S_{CH4} = Solubility of methane in the membrane material

Membrane selectivity characteristics do not indicate the proportion of various gases pervasion rates when set in genuine activity; this is due to the blended idea of natural gas, which causes every one of the controlled gases (for example, CO_2) to apply diverse halfway strain across the membrane, for example, unique penetration potential. Characteristic selectivity is utilized as a standardized metric to evaluate and analyze membrane partition capabilities based on application-specific nuances [56].

Development of polymeric membranes that are extremely porous at a certain selectivity and highly selective at a specific penetrability is regarded as a breakthrough in the membrane sector. Some researchers have argued that no major improvements have been made in the typical polymeric membrane [60].

4.4 Biogas purification

Biogas is neat, environmentally friendly fuel generated by the anaerobic (no oxygen) fermentation of natural materials. Biogas is composed of around 55%−65% methane (CH_4), 30%−45% carbon dioxide (CO_2), hints of hydrogen sulfide (H_2S), and small amounts of water fume [61−63]. At 15.5°C and 1 atm, pure methane has a calorific value of 9100 kcal/m^3; biogas has a calorific value ranging from 4800 to 6900 kcal/m^3. To achieve the specified biogas production and calorific value of 5500 kcal/m^3, treatment processes such as retention or membrane division should be used [64−67]. Biogas has a significant role to play as an alternative energy source. Purified biogas is in high demand. Biogas may be used as a domestic fuel, transportation fuel, and to create energy. If biogas could be used as a viable fuel like automobile fuel by compressing and filling the cylinder after cleaning and drying, it would be a great energy source [68]. Different types of materials are used for purification of biogas.

Biogas can be cleaned by separating CO_2, H_2S, and other contaminants from the primary component of biogas utilizing membrane contractors and permeation membranes [69−71]. Gas pervasion membrane updates biogas at petroleum gas network pressure using typical gas detachment cycles and almost no extra ingredients. The interaction is ineffective for single-stage activities [69,72]. It is expensive because of the multistage application. CO_2 may be retained from biogas in layers with the addition of an alkali mixture or by applying vacuum recovery and vacuum membrane distillation (VMD) [72]. From a study, Park et al. [73] figured out that 98% of the methane adsorption can be possible by using membrane technology. Materials used for purification of biogas are shown in Table 18.2.

4.5 CO_2 removal

CO_2 is very harmful both for the environment and human body. Several technologies are developed to remove CO_2 from the gases. Among them, carbon membrane technology is such one which helps to separate CO_2 from the gases. With the strategy of controlled carbonization, carbon membranes are structured

Table 18.2 Assessment of materials used for purification of biogas [74].

Minimal-cost material	Clay	Wood ash	Activated carbon	Zeolite	Fly ash
Contaminant removal	Good for removal of CO_2 by adsorption At high p^H good for H_2S	High removal rate of both CO_2 and H_2S	CO_2 removal is high but low for H_2S removal	CO_2 removal is minimal High removal of H_2S and removes N_2	CO_2 removes by adsorption with carbonation Minimal efficiency of removal of H_2S by adsorption
Advantages	Simply accessible, minimal functional and installment cost, used for construction after application	Easy system, minimal functional and installment cost, favorable for rural setup, can be reused after application as soil boosting	Huge raw material for bracing, traditional activation method, ease of availability in the market of commercial products	Employ with membranes, CO_2 removal is high in PSA, used for reduce NH_3 in the digester	Easy system, minimal functional cost, used as admixture for the purpose of construction
Disadvantages	Required physical and chemical activation	Burning causes enhancing the CFC and leads to deforestation	Activation cost is high, uptake of H_2S is poor	CO_2 removal efficiency low, high pressure needed in PSA	Heavy metals use, fabrication liberates greenhouse gas
Remarks	In activation with caustic soda clay gives high adsorption of CO_2 while for H_2S removal it does with Fe and $CuCl_2$	Flaming tenets in release of CFC production and leads to deforestation	Caustic potash or soda uses for activation, infuse with the compounds of iron or basic leads high removal of H_2S	Generally high adsorption rate for H_2S with activation of zeolite, uptake of CO_2 is rich in system of high pressure	Carbonation systems take place, excellent CO_2 capturing system, mono ethyl amine or di ethyl amine in addition with activate fly ash gives excellent capacity of CO_2 adsorption

with an unbending design, which applied to antecedents of polymer at elevated temperature range 500–900°C, micro size structure of CMS membranes is an ordinary bimodal pore model, which comprises of micropores (∼7–20Å) and ultramicropores (<7Å) [75]. Various strategies utilized to portray the constructions and characteristics of carbon membranes [76–79].

Carbon membranes provide a huge green development while executed with potentially high thermal conditions over high pressure as an outcome of capturing CO_2 with low sustainable power sources (e.g., removal of H_2 from syngas, overhauling of biogas and petroleum gas improving) [80–83]. The partition factors of selectivity and permeability are commonly employed to illustrate membrane detachment activities, which impact the usefulness and separation effectiveness of a membrane detachment measure. Generally, gas porousness (P_i) is depicted as the result of dissemination coefficient (D_i) and solvency coefficient (S_i): $P_i = D_i \times S_i$. As needs be, gas pair selectivity is communicated by $\alpha_{i/j} = (D_i/D_j)(S_i/S_j)$. Dissemination coefficient is a dynamic quantity that describes the ability of gas particles to pass across the membrane. It relies upon the size and state of the entered gas particles and the basic ultramicropores measurements [84]. For a higher adsorbing gas like CO_2, a thermodynamically rectified fixation free dispersion coefficient is utilized, alludes to Maxwell–Stefan diffusivity. Kinetic dispersion coefficient is increased by following the Arrhenius law with rising temperature, while the co-efficient of thermodynamic sorption is restricted appropriately. Apart from the material's own features, module design and cycle functioning boundaries will have a substantial influence on the interaction execution of carbon membrane frameworks (fluid porousness and ideal selectivity), which are normally demonstrated by testing of solitary gas saturation [85–88]. Green process from materials to separation is shown in Fig. 18.5. Materials of membrane gas separation (MGS) used for CO_2 removal shown in Fig. 18.6.

5. Conclusion and future outlooks

Membrane technology is a well-known method of gas separation. This technology is one of the most promising for CO_2 separation. However, this technology has a number of issues when it is in use. Rather, absorption is a technique that is widely employed in the industry [90]. As a result, new developments in membrane technology are required as quickly as feasible in order to overcome the issues and most promisingly utilize membranes in the

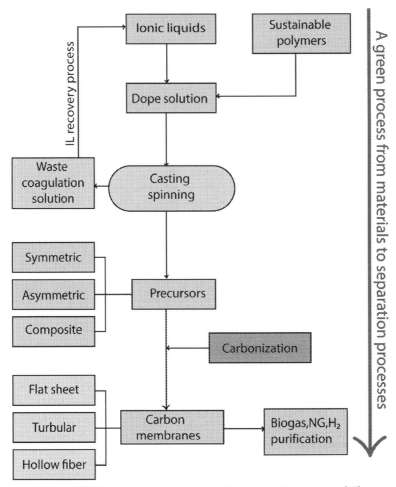

Figure 18.5 Green process from materials to separation process [77].

energy industry. There are several limitations to integrating the new age of membrane technology in the research industry, such as the accuracy of tests, true values of the tests, and antiaging stability. Based on the source of emission the percentages of nitrogen oxides (NOX), CO, and sulfur oxides (SOX) are low in flume gas [91]. These little amounts are equivalent to CO_2 separation. The components harmed the membrane and diminished its performance. As a result, real-world tested values are critical. It is vital to have a robust membrane that can endure the flume gas's high temperature and pressure while still working properly [92]. However, the developer's primary attention is on permeability and selectivity. More components should be identified to improve

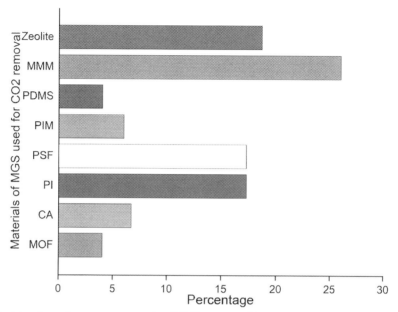

Figure 18.6 Materials of membrane gas separation (MGS) used for CO_2 removal (based on publications indexed by Scopus (TITLE (terms); January 9th, 2019). *CA*, cellulose acetate; *MMM*, mixed-matrix membrane; *MOF*, metal-organic framework; *PI*, polyimide; *PIM*, polymers of intrinsic microporosity; *PSF*, polysulfone [89].

membrane performance. Furthermore, the performance of CO_2 separation should be exact.

Undoubtedly, there have been many improvements in membrane materials research, which has opened the door to membrane technology. Sustainable membrane materials such as polysulfone, cellulose acetate, aramids, and polyimides show a very low separation capacity compared to newly developed membranes. Researchers have developed novel materials with enhanced gas transport properties. These materials include gas temperature rearranged membranes, mixed matrix membranes, polymeric room temperature membranes, perfluoro TR, and PIM, which is a highly selective polymer. However, further research is needed to fully characterize the gas transport properties of these materials. Perfluoropolymers show very satisfied values of N_2/CH_4 permeability and selectivity values in the separation process as compared to RTIL. There are many advantages of mixed matrix membrane with easy to handle, high selectivity, and minimum cost. Now, hydrogen, air separation, and natural gas purification are done in the industries. However, membranes are used in a broad area. It has been used for separating olefin/paraffin, dehydration of ethanol, and carbon separating purpose.

This is only possible by improvement of materials and the separation process. Several issues arise during gas separation, including physical aging, permeability, and selectivity reverse relationship, and the plasticization impact. Toxic CO_2 to CH_4 conversion improvement also needed to reduce the amount of CO_2 and enhance the value of quality fuel. Future patterns are vastly coordinated with planning and execution in order to tune the mixture of materials by selecting appropriate membrane materials, which can latterly deals with the rate of gas stream changes with respect to time. Broad examination should be done in creating nano-permeable and nano-composite based membrane materials in energy sectors, as they are highly beneficial with excellent penetrability and modest nature of selectivity.

Abbreviations and symbols

AG	acid gases
CA	cellulose acetate
CO	carbon monoxide
CSG	coke stove gas
MGS	membrane gas separation
MMM	mixed-matrix membrane
MOF	metal–organic framework
NOX	nitrogen oxides
OEA	oxygen-enriched air
PI	polyimide
PIM	polymers of intrinsic microporosity
PPM	parts per million
PSF	polysulfone
SOX	sulfur oxides
VMD	vacuum membrane distillation

References

[1] Kayvani Fard A, McKay G, Buekenhoudt A, Al Sulaiti H, Motmans F, Khraisheh M, et al. Inorganic membranes: preparation and application for water treatment and desalination. Materials 2018;11(1):74. https://doi.org/10.3390/ma11010074.

[2] Sridhar S, Bee S, Bhargava S. Membrane-based gas separation: principle, applications and future potential. Chemical Engineering Digest; 2014.

[3] Tien-Binh N, Vinh-Thang H, Chen XY, Rodrigue D, Kaliaguine S. Crosslinked MOF-polymer to enhance gas separation of mixed matrix membranes. Journal of Membrane Science 2016;520:941–50. https://doi.org/10.1016/j.memsci.2016.08.045.

[4] Valappil RSK, Ghasem N, Al-Marzouqi MJJOI, Chemistry E. Current and future trends in polymer membrane-based gas separation technology. A Comprehensive Review 2021;98:103–29.

[5] Wu X, Tian Z, Wang S, Peng D, Yang L, Wu Y, et al. Mixed matrix membranes comprising polymers of intrinsic microporosity and covalent organic framework for gas separation. Journal of Membrane Science 2017;528:273–83. https://doi.org/10.1016/j.memsci.2017.01.042.

[6] Yu S, Li S, Liu Y, Cui S, Shen X. High-performance microporous polymer membranes prepared by interfacial polymerization for gas separation. Journal of Membrane Science 2019;573:425−38. https://doi.org/10.1016/j.memsci.2018.12.029.
[7] Moghiseh M, Safarpour M, Barzin J. Cellulose acetate membranes fabricated by a combined vapor-induced/wet phase separation method: morphology and performance evaluation. Iranian Polymer Journal (English Edition) 2020;29(11):943−56. https://doi.org/10.1007/s13726-020-00847-z.
[8] Bernardo P, Drioli E. Membrane gas separation progresses for process intensification strategy in the petrochemical industry. Petroleum Chemistry 2010;50(4):271−82. https://doi.org/10.1134/S0965544110040043.
[9] Henis JMS, Tripodi MK. A novel approach to gas separations using composite hollow fiber membranes. Separation Science and Technology 1980;15:1059−68.
[10] Hasan MY, Monir MU, Ahmed MT, Aziz AA, Shovon SM, Ahamed Akash F, et al. Sustainable energy sources in Bangladesh: a review on present and future prospect. Renewable and Sustainable Energy Reviews 2021;155:111870. https://doi.org/10.1016/j.rser.2021.111870.
[11] Khatun F, Abd Aziz A, Sim LC, Monir MU. Plasmonic enhanced Au decorated TiO2 nanotube arrays as a visible light active catalyst towards photocatalytic CO2 conversion to CH4. Journal of Environmental Chemical Engineering 2019;7(6):103233. https://doi.org/10.1016/j.jece.2019.103233.
[12] Monir MU, Aziz AA, Kristanti RA, Yousuf A. Co-gasification of empty fruit bunch in a downdraft reactor: a pilot scale approach. Bioresource Technology Reports 2018;1:39−49. https://doi.org/10.1016/j.biteb.2018.02.001.
[13] Abetz V, Brinkmann T, Dijkstra M, Ebert K, Fritsch D, Ohlrogge K, et al. Developments in membrane research: from material via process design to industrial application. Advanced Engineering Materials 2006;8.
[14] Basu A, Akhtar J, Rahman MH, Islam MR. A review of separation of gases using membrane systems. Petroleum Science and Technology 2004;22:1343−68.
[15] Basile A, Iulianelli A, Gallucci F, Morrone P. 7 - advanced membrane separation processes and technology for carbon dioxide (CO2) capture in power plants. In: Maroto-Valer MM, editor. Developments and innovation in carbon dioxide (CO2) capture and storage technology, vol 1. Woodhead Publishing; 2010. p. 203−42.
[16] Basile A, Iulianelli A, Gallucci F, Morrone P. Advanced membrane separation processes and technology for carbon dioxide (CO2) capture in power plants. In: Developments and innovation in carbon dioxide (CO2) capture and storage technology. Elsevier; 2010. p. 203−42.
[17] Fritzsche A, Kurz JE, Porter MC. In: Park Ridge NJ, editor. The separation of gases by membranes. USA: Noyes Publications; 1990. p. 559−93.
[18] Gallucci F, Basile A. Pd-based membranes synthesis and their application in membrane reactors. In: Handbook of membrane research: properties, performance and applications. Nova Science; 2010. p. 1−65.
[19] Paul DR, Yampol'skii YP, editors. Polymeric gas separation membranes. 1st ed. CRC Press; 1994. https://doi.org/10.1201/9781351075886.
[20] Powell CE, Qiao GG. Polymeric CO2/N2 gas separation membranes for the capture of carbon dioxide from power plant flue gases. Journal of Membrane Science 2006;279(1−2):1−49.
[21] Spillman R. Economics of gas separation membranes. Chemical Engineering Progress 1989;85:41−62.

[22] Yampolskii Y, Pinnau I, Freeman BD. Materials science of membranes for gas and vapor separation. Wiley Online Library; 2006.
[23] Lalia BS, Kochkodan V, Hashaikeh R, Hilal NJD. A review on membrane fabrication: structure, properties and performance relationship. Desalination 2013;326:77−95.
[24] Xia Y, Wang Z, Chen L-Y, Xiong S-W, Zhang P, Fu P-G, et al. Nanoscale polyelectrolyte/metal ion hydrogel modified RO membrane with dual anti-fouling mechanism and superhigh transport property. Desalination 2020; 488:114510. https://doi.org/10.1016/j.desal.2020.114510.
[25] De Meis D. Gas transport through porous membranes. 2017.
[26] Loudon C, McCulloh K. Application of the Hagen—Poiseuille equation to fluid feeding through short tubes. Annals of the Entomological Society of America 1999;92(1):153−8. https://doi.org/10.1093/aesa/92.1.153.
[27] Liu C, Liu Z, Zhang Y. A multi-scale framework for modelling effective gas diffusivity in dry cement paste: combined effects of surface, Knudsen and molecular diffusion. Cement and Concrete Research 2020;131:106035. https://doi.org/10.1016/j.cemconres.2020.106035.
[28] Sidhikku Kandath Valappil R, Ghasem N, Al-Marzouqi M. Current and future trends in polymer membrane-based gas separation technology: a comprehensive review. Journal of Industrial and Engineering Chemistry 2021;98:103−29. https://doi.org/10.1016/j.jiec.2021.03.030.
[29] Petukhov DI, Berekchiian MV, Eliseev AA. Meniscus curvature effect on the asymmetric mass transport through nanochannels in capillary condensation regime. Journal of Physical Chemistry C 2018;122(51): 29537−48. https://doi.org/10.1021/acs.jpcc.8b08289.
[30] Ding L, Wei Y, Li L, Zhang T, Wang H, Xue J, et al. MXene molecular sieving membranes for highly efficient gas separation. Nature Communications 2018;9(1):155. https://doi.org/10.1038/s41467-017-02529-6.
[31] Zou X, Zhu G. Microporous organic materials for membrane-based gas separation. Advances in Materials 2018;30(3):1700750. https://doi.org/10.1002/adma.201700750.
[32] Spillman R. Chapter 13 economics of gas separation membrane processes. In: Noble RD, Stern SA, editors. Membrane separations technology - principles and applications, vol. 2. Elsevier; 1995. p. 589−667.
[33] Ramírez-Santos ÁA, Castel C, Favre E. A review of gas separation technologies within emission reduction programs in the iron and steel sector: current application and development perspectives. Separation and Purification Technology 2018;194:425−42. https://doi.org/10.1016/j.seppur.2017.11.063.
[34] Remus R, Monsonet MA, Roudier S, Sancho LD. Best available techniques (BAT) reference document for iron and steel production. Luxembourg: Publications Office of the European Union; 2013. p. 621.
[35] Liu KB, Song C, Subramani V. Hydrogen and syngas production and purification technologies. 2010.
[36] Monir MU, Abd Aziz A, Kristanti RA, Yousuf A. Syngas production from co-gasification of forest residue and charcoal in a pilot scale downdraft reactor. Waste and Biomass Valorization 2020;11(2):635−51. https://doi.org/10.1007/s12649-018-0513-5.
[37] Monir MU, Azrina AA, Kristanti RA, Yousuf A. Gasification of lignocellulosic biomass to produce syngas in a 50 kW downdraft reactor. Biomass and Bioenergy 2018;119:335−45. https://doi.org/10.1016/j.biombioe.2018.10.006.
[38] Stookey D. Gas-separation membrane applications. Membrane Technology: Chemical Industry 2006:119−50.

[39] Castro-Domínguez B, Leelachaikul P, Takagaki A, Sugawara T, Kikuchi R, Oyama ST. Perfluorocarbon-based supported liquid membranes for O2/N2 separation. Separation and Purification Technology 2013;116:19—24. https://doi.org/10.1016/j.seppur.2013.05.023.

[40] Murali RS, Sankarshana T, Sridhar S. Air separation by polymer-based membrane technology. Separation and Purification Reviews 2013;42(2):130—86. https://doi.org/10.1080/15422119.2012.686000.

[41] Abdollahi M, Khoshbin M, Biazar H, Khanbabaei G. Preparation, morphology and gas permeation properties of carbon dioxide-selective vinyl acetate-based polymer/poly(ethylene oxide-b-amide 6) blend membranes. Polymer 2017;121:274—85. https://doi.org/10.1016/j.polymer.2017.06.033.

[42] Khoshbin M, Abdollahi M, Abdollahi M, Khanbabaei G. Relationship of permeation and diffusion of carbon dioxide and methane gases with fractional free volume in the blend membranes of poly(ether-b-amide) and vinyl acetate- based copolymer. Journal of Petroleum Research 2017;27(96−2):122—33. https://doi.org/10.22078/pr.2017.757.

[43] Meshkat S, Kaliaguine S, Rodrigue D. Enhancing CO2 separation performance of Pebax MH-1657 with aromatic carboxylic acids. Separation and Purification Technology 2019;212:901—12. https://doi.org/10.1016/j.seppur.2018.12.008.

[44] Sanaeepur H, Ahmadi R, Ebadi Amooghin A, Ghanbari D. A novel ternary mixed matrix membrane containing glycerol-modified poly(ether-block-amide) (Pebax 1657)/copper nanoparticles for CO2 separation. Journal of Membrane Science 2019;573:234—46. https://doi.org/10.1016/j.memsci.2018.12.012.

[45] Shin JE, Lee SK, Cho YH, Park HB. Effect of PEG-MEA and graphene oxide additives on the performance of Pebax®1657 mixed matrix membranes for CO2 separation. Journal of Membrane Science 2019;572:300—8. https://doi.org/10.1016/j.memsci.2018.11.025.

[46] Chemical TE. Manufacture of nitrogen and oxygen by cryogenic separation of air. 2016. https://www.essentialchemicalindustry.org/chemicals/oxygen.html.

[47] Koros W, Pinnau I. Polymeric gas separation membranes. In: Polymeric gas separation membranes. Boca Raton: CRC Press; 1994. p. 202.

[48] Smith AR, Klosek J. A review of air separation technologies and their integration with energy conversion processes. Fuel Processing Technology 2001;70:115—34.

[49] Baker RW. Future directions of membrane gas separation technology. Industrial & Engineering Chemistry Research 2002;41(6):1393—411.

[50] Sanders Jr ES, Clark DO, Jensvold JA, Beck HN, Lipscomb IGG, Coan FL. Process for preparing POWADIR membranes from tetrahalobisphenol apolycarbonates. Google Patents; 1988.

[51] Lin H, Zhou M, Ly J, Vu J, Wijmans JG, Merkel TC, et al. Membrane-based oxygen-enriched combustion. Industrial & Engineering Chemistry Research 2013;52(31):10820—34.

[52] Bernardo P, Drioli E, Golemme GJI. Membrane gas separation: a review/state of the art. Industrial & Engineering Chemistry Research 2009;48(10):4638—63.

[53] Theis T, Titus S. The development of permeable membrane air dehydrators for the US Navy. Naval Engineers Journal 1996;108(3):243—65.

[54] Saidur R, Rahim NA, Hasanuzzaman MJR. A review on compressed-air energy use and energy savings. Renewable and Sustainable Energy Reviews 2010;14(4):1135−53.
[55] Sanders DF, Smith ZP, Guo R, Robeson LM, McGrath JE, Paul DR, et al. Energy-efficient polymeric gas separation membranes for a sustainable future: a review. Polymer 2013;54(18):4729−61. https://doi.org/10.1016/j.polymer.2013.05.075.
[56] Alcheikhhamdon Y, Hoorfar M. Natural gas purification from acid gases using membranes: a review of the history, features, techno-commercial challenges, and process intensification of commercial membranes. Chemical Engineering and Processing - Process Intensification 2017;120:105−13. https://doi.org/10.1016/j.cep.2017.07.009.
[57] Rufford TE, Smart S, Watson GCY, Graham BF, Boxall J, Diniz da Costa JC, et al. The removal of CO2 and N2 from natural gas: a review of conventional and emerging process technologies. Journal of Petroleum Science and Engineering 2012;94−95:123−54. https://doi.org/10.1016/j.petrol.2012.06.016.
[58] Hua B, Xiong H, Wang Z, Hill J, Buckler S, Gao S, et al. Physico-chemical processes. Water Environment Research 2015;87(10):912−45. https://doi.org/10.2175/106143015X14338845155228.
[59] Baker RW. Membrane technology and applications. John Wiley & Sons; 2012.
[60] Okamoto Y, Chiang HC, Fang M, Galizia M, Merkel T, Yavari M, et al. Perfluorodioxolane polymers for gas separation membrane applications. Membranes 2020;10(12). https://doi.org/10.3390/membranes10120394.
[61] Li M, Zhu Z, Zhou M, Jie X, Wang L, Kang G, et al. Removal of CO2 from biogas by membrane contactor using PTFE hollow fibers with smaller diameter. Journal of Membrane Science 2021;627:119232. https://doi.org/10.1016/j.memsci.2021.119232.
[62] Santos-Clotas E, Cabrera-Codony A, Comas J, Martín MJ. Biogas purification through membrane bioreactors: experimental study on siloxane separation and biodegradation. Separation and Purification Technology 2020;238:116440. https://doi.org/10.1016/j.seppur.2019.116440.
[63] Simcik M, Ruzicka MC, Karaszova M, Sedlakova Z, Vejrazka J, Vesely M, et al. Polyamide thin-film composite membranes for potential raw biogas purification: experiments and modeling. Separation and Purification Technology 2016;167:163−73. https://doi.org/10.1016/j.seppur.2016.05.008.
[64] Fuksa P, Hakl J, Míchal P, Hrevušová Z, Šantruček J, Tlustoš P. Effect of silage maize plant density and plant parts on biogas production and composition. Biomass and Bioenergy 2020;142:105770. https://doi.org/10.1016/j.biombioe.2020.105770.
[65] Hasan ASMM, Ammenberg J. Biogas potential from municipal and agricultural residual biomass for power generation in Hazaribagh, Bangladesh − a strategy to improve the energy system. Renewable Energy Focus 2019;29:14−23. https://doi.org/10.1016/j.ref.2019.02.001.
[66] Kapoor R, Ghosh P, Tyagi B, Vijay VK, Vijay V, Thakur IS, et al. Advances in biogas valorization and utilization systems: a comprehensive review. Journal of Cleaner Production 2020;273:123052. https://doi.org/10.1016/j.jclepro.2020.123052.
[67] Patinvoh RJ, Taherzadeh MJ. Challenges of biogas implementation in developing countries. Current Opinion in Environmental Science & Health 2019;12:30−7. https://doi.org/10.1016/j.coesh.2019.09.006.

[68] Kapdi SS, Vijay VK, Rajesh SK, Prasad R. Biogas scrubbing, compression and storage: perspective and prospectus in Indian context. Renewable Energy 2005;30(8):1195−202. https://doi.org/10.1016/j.renene.2004.09.012.

[69] Chmielewski AG, Urbaniak A, Wawryniuk K. Membrane enrichment of biogas from two-stage pilot plant using agricultural waste as a substrate. Biomass and Bioenergy 2013;58:219−28. https://doi.org/10.1016/j.biombioe.2013.08.010.

[70] Harasimowicz M, Orluk P, Zakrzewska-Trznadel G, Chmielewski AG. Application of polyimide membranes for biogas purification and enrichment. Journal of Hazardous Materials 2007;144(3):698−702. https://doi.org/10.1016/j.jhazmat.2007.01.098.

[71] Scholz M, Melin T, Wessling M. Transforming biogas into biomethane using membrane technology. Renewable and Sustainable Energy Reviews 2013;17:199−212. https://doi.org/10.1016/j.rser.2012.08.009.

[72] He Q, Yu G, Yan S, Dumée LF, Zhang Y, Strezov V, et al. Renewable CO2 absorbent for carbon capture and biogas upgrading by membrane contactor. Separation and Purification Technology 2018;194:207−15. https://doi.org/10.1016/j.seppur.2017.11.043.

[73] Park A, Kim YM, Kim JF, Lee PS, Cho YH, Park HS, et al. Biogas upgrading using membrane contactor process: pressure-cascaded stripping configuration. Separation and Purification Technology 2017;183:358−65. https://doi.org/10.1016/j.seppur.2017.03.006.

[74] Mulu E, M'Arimi MM, Ramkat RC. A review of recent developments in application of low cost natural materials in purification and upgrade of biogas. Renewable and Sustainable Energy Reviews 2021;145:111081. https://doi.org/10.1016/j.rser.2021.111081.

[75] Sanyal O, Zhang C, Wenz GB, Fu S, Bhuwania N, Xu L, et al. Next generation membranes —using tailored carbon. Carbon 2018;127:688−98. https://doi.org/10.1016/j.carbon.2017.11.031.

[76] Hamm JBS, Ambrosi A, Griebeler JG, Marcilio NR, Tessaro IC, Pollo LD. Recent advances in the development of supported carbon membranes for gas separation. International Journal of Hydrogen Energy 2017;42(39):24830−45. https://doi.org/10.1016/j.ijhydene.2017.08.071.

[77] Lei L, Bai L, Lindbråthen A, Pan F, Zhang X, He X. Carbon membranes for CO2 removal: status and perspectives from materials to processes. Chemical Engineering Journal 2020;401:126084. https://doi.org/10.1016/j.cej.2020.126084.

[78] Liu H, Cooper VR, Dai S, Jiang D-e. Windowed carbon nanotubes for efficient CO2 removal from natural gas. The Journal of Physical Chemistry Letters 2012;3(22):3343−7. https://doi.org/10.1021/jz301576s.

[79] Xiao Y, Low BT, Hosseini SS, Chung TS, Paul DR. The strategies of molecular architecture and modification of polyimide-based membranes for CO2 removal from natural gas—a review. Progress in Polymer Science 2009;34(6):561−80. https://doi.org/10.1016/j.progpolymsci.2008.12.004.

[80] Gamali PA, Kazemi A, Zadmard R, Anjareghi MJ, Rezakhani A, Rahighi R, et al. Distinguished discriminatory separation of CO 2 from its methane-containing gas mixture via PEBAX mixed matrix membrane. Chinese Journal of Chemical Engineering 2018;26(1):73−80. https://doi.org/10.1016/j.cjche.2017.04.002.

[81] Harrigan DJ, Lawrence JA, Reid HW, Rivers JB, O'Brien JT, Sharber SA, et al. Tunable sour gas separations: simultaneous H2S and CO2 removal from natural gas via crosslinked telechelic poly(ethylene glycol) membranes.

Journal of Membrane Science 2020;602:117947. https://doi.org/10.1016/j.memsci.2020.117947.

[82] Xin Q, Ma F, Zhang L, Wang S, Li Y, Ye H, et al. Interface engineering of mixed matrix membrane via CO2-philic polymer brush functionalized graphene oxide nanosheets for efficient gas separation. Journal of Membrane Science 2019;586:23–33. https://doi.org/10.1016/j.memsci.2019.05.050.

[83] Yang Z, Guo W, Chen H, Kobayashi T, Suo X, Wang T, et al. Benchmark CO2 separation achieved by highly fluorinated nanoporous molecular sieve membranes from nonporous precursor via in situ cross-linking. Journal of Membrane Science 2021;638:119698. https://doi.org/10.1016/j.memsci.2021.119698.

[84] Fu S, Sanders ES, Kulkarni SS, Koros WJ. Carbon molecular sieve membrane structure–property relationships for four novel 6FDA based polyimide precursors. Journal of Membrane Science 2015;487:60–73. https://doi.org/10.1016/j.memsci.2015.03.079.

[85] Kai T, Kazama S, Fujioka Y. Development of cesium-incorporated carbon membranes for CO2 separation under humid conditions. Journal of Membrane Science 2009;342(1–2):14–21. https://doi.org/10.1016/j.memsci.2009.06.014.

[86] Li J-L, Chen B.-H. Review of CO2 absorption using chemical solvents in hollow fiber membrane contactors. Separation and Purification Technology 2005;41(2):109–22. https://doi.org/10.1016/j.seppur.2004.09.008.

[87] Russo F, Galiano F, Iulianelli A, Basile A, Figoli A. Biopolymers for sustainable membranes in CO2 separation: a review. Fuel Processing Technology 2021;213:106643. https://doi.org/10.1016/j.fuproc.2020.106643.

[88] Wang H, Zheng W, Yang X, Ning M, Li X, Xi Y, et al. Pebax-based mixed matrix membranes derived from microporous carbon nanospheres for permeable and selective CO2 separation. Separation and Purification Technology 2021;274:119015. https://doi.org/10.1016/j.seppur.2021.119015.

[89] Siagian UWR, Raksajati A, Himma NF, Khoiruddin K, Wenten IG. Membrane-based carbon capture technologies: membrane gas separation vs. membrane contactor. Journal of Natural Gas Science and Engineering 2019;67:172–95. https://doi.org/10.1016/j.jngse.2019.04.008.

[90] Maab H, Touheed A. Polyazole polymers membranes for high pressure gas separation technology. Journal of Membrane Science 2021;642:119980. https://doi.org/10.1016/j.memsci.2021.119980.

[91] Li G, Kujawski W, Válek R, Koter S. A review—the development of hollow fibre membranes for gas separation processes. International Journal of Greenhouse Gas Control 2021a;104:103195. https://doi.org/10.1016/j.ijggc.2020.103195.

[92] Lai W-H, Hong C-Y, Tseng H-H, Wey M-Y. Fabrication of waterproof gas separation membrane from plastic waste for CO2 separation. Environmental Research 2021;195:110760. https://doi.org/10.1016/j.envres.2021.110760.

Index

'*Note:* Page numbers followed by "f" indicate figures and "t" indicate tables.'

Absorption, 11–16, 34, 97–99
  advances in, 209–210
  amine, 11–15, 212–222
  aqua ammonia process, 15–16
  carbon dioxide (CO_2)
    plant, 224–227
    removal flowsheet, 223–224
    specific characterizations
      and properties, 234–235
  column, 98–99, 225
    operating parameters, 226
    current applications and
      cases, 227–234
  liquids, 98–99, 101–102
  natural gas (NG)
    purification and processing,
      210–212
    sweetening, 140–142
  novel methods and solvents,
    235–237
    hybrid solvents, 235
    ionic liquids, 235
    membrane-based
      absorption, 236–237
    nanoparticles (NPs),
      235–236
  process, 4, 64–65, 335–336
  scales up of ionic
    liquid–based
    technologies, 237–239
Abu Dhabi Gas Liquefaction
  Company (ADGAS),
  59–60
Acetate (Ace), 120–121
Acid gas (AG), 90–92, 97–98,
  193–194, 453–456, 475,
  495–496
  absorption, 102
    in amines and alcohols,
      116–118
  elimination of, 333–336

  absorption process,
    335–336
  adsorption process, 336
natural gas (NG)
  composition, 77–78
  origins, 77
  purification, 495–497
  removing acid gases, 78–79
physical and chemical
  properties, 79–84
  boiling point, 79
  density, 80
  heat of vaporization, 83
  ionization potential, 83
  melting point, 80
  pH, 81–82
  vapor density, 80
  vapor pressure, 81
  viscosity, 81
removal
  amine-based techniques for,
    92–97
  cryogenic processes for,
    189–199
  Engelhard titanosilicate
    zeolite for, 349–350
  natural zeolites for, 338–343
  unit, 435–437
  13X, 4A, and 5A zeolites for,
    343–345
  zeolite Na-Y interchanged
    by metals for, 346–347
  zeolite SP-115, 348–349
  zeolite X interchanged by Na
    and Ca for, 345–346
  zeolite Y interchanged by Ce
    and Cu for, 346
  zeolites for, 350–352
  ZSM-5 and zeolite-A for,
    347–348
separation, 90–91

Acidic potential (AP), 55–56
Acidification potential (AP),
  66–67
Activated carbon (AC), 265,
  272–273, 404–405
  adsorption process, 16–17
  fibers, 407
Activated carbon monoliths
  (ACMs), 293–294
Activated methyl
  diethanolamine (aMDEA),
  104–105
Activation energy, 424–425
Adsorbate, 268–269, 288–289
Adsorbed gas, 277–278
Adsorbents, 268–269, 274–275,
  288–289, 291, 336,
  360–361
  effects of, 271–275
    carbon molecular sieve
      (CMSs), 272–273
    metal-organic framework
      (MOF), 274–275
    zeolites, 272
  material selection,
    401–409
    carbonaceous adsorbents,
      404–407
    metal-organic frameworks
      (MOFs), 407–409
    zeolites, 401–403
  regeneration, 396
  selection, 401
  in temperature swing
    adsorption (TSA) process,
    279–282
Adsorption, 16–19, 209,
  262–264, 360–361
  capability, 347
  capacity, 399–401
  equilibrium data, 404–405

Adsorption (*Continued*)
  of heavy hydrocarbons, 283–284
  isotherms, 402–403
  of mercaptans, 284–285
  process, 68, 263–264, 285, 310, 336, 399–401
    for natural gas (NG) sweetening, 142–143, 396–401
    process description, 396
    types of, 397–401
  technology, 268–269
  temperature swing adsorption (TSA) technology, 282–285
  time, 296
  of water, 282–283
Adsorptive desulfurization (ADS), 362
Air emissions, 36–42
  hydrogen sulfide (H_2S) and carbon dioxide (CO_2) emission, 36–37
  regulations for
    fugitive emissions, 37–39
    greenhouse gas (GHG), 39–40
    venting and flaring, 40–42
Air purification, 494–495
Air reactor, 22
Alaninate (Ala), 120–121
Alcohols, acid gas absorption in amines and, 116–118
Alkanol amines, 218–219
Aluminosilicate zeolites, 378
Aluminum oxides, 282–283
Aluminum zeolites, 402–403
Amines, 34, 44
  absorption, 212–223, 231–232
    carbon dioxide (CO_2) separation processes, 217–222
    plant details, 213–217
  acid gas (AG) absorption in, 116–118
  based chemical absorption process, 144
  for natural gas (NG) sweetening
    amine-based absorption process, 97–102
    amine-based techniques for acid gas removal, 92–97
    current applications and cases, 102–108
  process, 64–65, 335
  scrubbers, 117
  scrubbing method, 156
  sweetening process, 44, 59–60
  washing, 335
2-Amino-2-methyl-1-propanol (AMP), 96, 141–142, 219–220
Aminopropyltrimethoxysilane (APS), 375–376
Ammonia (NH_3), 210, 404–405, 433
Ammonium molybdate, 348–349
Analytic hierarchy process (AHP), 105
Aqua ammonia process, 15–16
Aqueous amines, 154–155
Aqueous amino acid salts, use of, 119
Arrhenius plot, 424–425
Arsenic (As), 331
  As-captured carbon dioxide (CO_2), 232
4A zeolites for acid gas (AG) removal, 343–345
5A zeolites, 272
  acid gas removal, 343–345

Bench-scale experimental tests, 229–230
Benfield HiPure process, 59–60, 62
Benzene tricarboxylic acid results, 409
Best available techniques (BAT), 35
Biogas purification, 497
Bis(trifluoromethanesulfonyl) imide (Tf_2N), 120–121
Blended alkanolamine method, 157
Blended amines, 97
Boiling point, 79
Branched polyethylenimine (PEI-B), 106–107
Brand-new bench scale system, 293
Brunauer–Emmett–Teller (BET) surface area, 366–367, 409
Butanes (C_4H_{10}), 330, 359–360

Calcium carbonate, 119
Capillary condensation, 8–9, 492
  membrane separation mechanisms, 10f
Carbon, 480–481
  atoms, 261–262
  based nanomaterials, 407
  ion, 476–477
  membrane technology, 497–499
Carbon capture and sequestration/storage (CCS), 154, 207, 276, 329
Carbon dioxide (CO_2), 90, 135, 153, 185–186, 279–280, 298, 333–334, 359–360, 404–405, 453, 471, 479, 497
  absorption, 216
    absorption column operating parameters, 226
    column, 225
    gas feed composition, 226
    gas pretreatment, 225
    kinetics, 235
    plant, 224–227
    purification parameters, 227
    reaction byproducts, 235
    regeneration column operating parameters, 226
    reversibility, 234
    selectivity, 234
    solubility, 234
    solvent regeneration, 226
    solvent type, 226

Index   **511**

specific characterizations and properties of, 234–235
  temperature dependence, 234
advances in absorption of, 209–210
capture, 145–147, 401
  external cooling loop cryogenic for, 195–198
compression, 224, 226
emission, 36–37
from natural gas (NG), 186–187, 191–192
physical and chemical properties of, 84
purification, 217, 226
regeneration, 216–217
removal, 193, 223, 401–402, 497–499
  amine absorption, 223
  cryogenic distillation, 223
  through electrochemical membrane separator, 475–480
  flowsheet, 223–224
  fundamentals of membrane gas separation for, 488–489
  gas conditioning, 224
  gas dehydration, 224
  gas pretreatment, 223
  membrane processes for efficient, 489–493
  membrane separation, 223
separation processes, 217–222
  chemical absorbent, 218–222
  gas separation, 189–191, 218
storage or utilization, 217
thermodynamic principles, 187–189
Carbon disulfide (CS_2), 139–140, 331
Carbon molecular sieves (CMSs), 272–273, 406–407
Carbon monoxide (CO), 404, 407, 433, 493–494

Carbon nanotubes (CNTs), 107, 144–145, 440–442
Carbonaceous adsorbents, 404–407
  activated carbons, 404–405
  fibers and carbon-based nanomaterials, 407
  molecular sieves, 406–407
Carbonaceous materials, 274
Carbonate, 477–478
  solutions, 119
  use of, 119
  washing, 335
Carbonyl sulfide (COS), 139–140, 280, 331, 345–346
  formation during temperature swing adsorption (TSA) process, 285–286
Cash flow diagram, 57
Catalyst toxins, 217–218
Cathode, 478
Catholyte, 478–479
Cationic poly(ionic liquid) (PIL), 462
Caustic wash, 335
Cellulose acetate (CA), 428–429
Cellulose triacetate (CTA), 428–429
Centrifugal compressor, regulation for, 42–43
Chelated iron treatment, 168–169
Chemical absorbent, 218–222
  alkanol amines, 218–219
  sterically hindered amines, 219–222
Chemical absorption, 11, 209, 220
  solvents, 141–142
  technologies, 117
Chemical Engineering Plant Cost Index (CEPCI), 59
Chemical looping, 22–24
Chemical looping combustion (CLC), 22–24
Chemisorbents, 397–398
Chemisorption, 34
Chitosan (CS), 119

Chloride (Cl), 120–121
Choline chloride (ChCl), 106
Circulating fluidized-bed (CFB), 163
Clean Air Act, The, 36
Clean coal technology (CCT), 338
Clinoptilolite, 338, 378, 401–402
Coal, 33–34, 329
Cobalt sulfide (CoS_2), 481
Coke stove gas (CSG), 493
Combined cryogenic carbon dioxide separation flowsheet with external cooling loop (CCCECL), 197
Convective flow, 420–422
Conventional absorption columns, 100
Conventional amine process, 59–60
Conventional development, 493
Conventional solvent, 456–457
Conventional technologies, 200
Copper (Cu), for acid gas removal
  zeolite SP-115 changed by, 348–349
  zeolite Y interchanged by, 346
Corrosion, 218, 232–233
CryoCell process, 19–22, 199–200, 231–232
Cryogenic fractionation, 186–187
  acid gas removal, 189–199
  cryogenic liquid, 198
  cryogenic packed bed, 191–193
  cryogenic separation based on multicompression stages with intercoolers, 193–195
  external cooling loop cryogenic for carbon dioxide (CO_2) capture, 195–198
  heat exchangers, 199
  current applications and improvements, 199–202
  CryoCell process, 199–200

Cryogenic fractionation (*Continued*)
   cryogenic–membrane hybrid system case, 200–202
   thermodynamic principles, 187–189
Cryogenics, 19–22
   CryoCell, 19–22
   distillation process, 194, 201, 223
   heat transfer, 199
   processes, 68
   separation, 209–210, 221
   technologies, 193–194
Cupric acetate, 348–349
Cycle time, effects of, 266–267
Cyclic adsorption systems, 294

Damage index (DI), 66–67
Decision-making process, 274–275
Deep eutectic solvents (DESs), 15, 106
Density, 80
Desorption process, 309
Desulfurization of gases, 362, 370
Diethanolamine (DEA), 45, 62, 92, 116, 141–142, 157, 207–208, 335, 359–360
Diethylenetriamine (DETA), 163
Diglycolamine (DGA), 13, 62, 92, 116, 141–142, 218–219, 335
Diisopropanolamine (DIPA), 47, 92, 141–142, 219–221, 335–336
Diisopropylamine (DIPA), 62
Dilution loading, 172
Dimethyl ethers of polyethylene glycol (DEPG), 118
Dimethylethylarsine (Me2EtAs), 331
Dimethyl formamide (DMF), 409
Dimethyl sulfide (DMS), 345–346
Direct electro-thermal regeneration might, 286–287

Direct immersion, 458–459
Dissemination coefficient, 499
Disulfide, 478–479
"Double-train" system, 303

Economic analysis, methods of, 57–59
Economic assessments, 57
   methods of, 57–59
      cash flow diagram, 57
      cost of equipment and operation, 58–59
      time value of money, 57–58
   model, 56–57
   techno-economic analysis of natural gas (NG) sweetening process, 59–67
Economic potential (EP), 60–61
Electric field, 473
Electric swing adsorption (ESA), 18–19, 263–264, 286–294. *See also* Pressure swing adsorption (PSA); Temperature swing adsorption (TSA); Vacuum swing adsorption (VSA)
   electrification step, 290–291
   literature, 292–294
   parameters affect, 291–292
      adsorbent, 291
      electric power, 292
      purge gas, 292
      voltage, 291–292
   procedure, 288–290
Electrification process, 288–289
Electrochemical membranes. *See also* Ionic liquid membrane (ILM) preparation
   electrode, 480–481
   membrane, 481–482
   removal of
      carbon dioxide (CO_2), 475–480
      hydrogen sulfide (H_2S), 473–475

Electrochemical technology, 471–472
Electrode preparation, 480–481
Electrolytic process, 472
Elevated temperature-pressure swing adsorption (ET-PSA) process, 301–304
Emulsification strategy, 159
Emulsions, 159
Encapsulated aprotic heterocyclic anion-ionic liquid (AHA-ENIL), 163–165
Encapsulated liquid sorbents, 157–165
   natural gas (NG) sweetening using encapsulated liquids, 166–174
   strategy
      emulsification, 159
      in situ and interfacial polymerization, 160–165
      spray-drying, 159
Encapsulated sorbent process, 163
Encapsulation proficiency, 159–160
Energy, 290, 306–307
   savings, 67
Engelhard titanosilicate (ETS), 349–350, 378
   zeolite for acid gas (AG) removal, 349–350
Enhanced coal bed methane (ECBM), 194
Enhanced oil recovery (EOR), 194, 261
Environmental assessment of gas sweetening process, 66–67
Environmental challenges of natural gas (NG) sweetening technologies, 67–68
Environmental policies and regulations of natural gas (NG), 35–42
   air emissions, 36–42
   regulating history, 36

Environmental Protection
    Agency (EPA), 35, 37
Erionite, 338
Ethane (C_2H_6), 330, 359–360
Ethyl mercaptan (EM), 345–346
1-Ethyl-3-methylimidazoliu
    methylsulfate IL, 464
Ethyl sulfate ($EtSO_4$), 120–121
Ethylene diamine tetraacetic
    (EDTA), 168–169
European Union (EU), 35
Eutectic carbonate functional
    group, 17
Evacuation time, 295–296
Exfoliated graphene oxide
    (EGO), 148
External cooling loop cryogenic
    for carbon dioxide (CO_2)
    capture, 195–198

Fabrication method of polymeric
    membrane, 425–428
Fajr Jam Gas Refining Company,
    104–105
Federal government, 42
Feed concentration, 297
Feed flow rates, 296–297
Feed pressure, 297–298
Feed process, 288–289
Ferrierite, 338
Fick's dispersion law, 495–496
Fick's law, 460
Fixed capital investment (FCI),
    233–234
Fixed operating cost (FOC),
    58–59
Flaring, regulations for,
    40–42
Flow rate, effect of, 267–268
Flue gas separation, 425
Fluor process, 115–116
Fluor solvent, 336
Fossil fuels, 77, 89, 329
Fuel cell, 494
Fugitive emissions
    policies and regulations,
        38
    regulations for, 37–39

Gas, 218, 496
    conditioning, 224
    dehydration, 224
    feed composition, 226
    flux, 423–424
    membrane-based gas
        separation technology,
        493
    mixture, 419–420, 495
    molecules, 421–422
    porousness, 499
    pretreatment, 223, 225
    purification, 489
    saturation rate, 496
    separation, 420–421, 487
    sweetening
        modified membrane
            composites for, 118–119
        process, 33–34, 64–65
    transport, 489–491
        supported ionic liquid
            membrane (SILM),
            460
    viscosity, 490
Gas hourly space velocity
    (GHSV), 378
Gas–liquid membrane
    contactors (GLMCs),
    100–102
Gas–oil separator, 333
Gas-to-liquid (GTL), 329
Genetic algorithm (GA),
    399–401
Gibbs free energy, 479–480
Girbotol process, 436–437
Global warming potential
    (GWP), 55–56
Glycerine, 125–126
Glycerol/glycerine, 125–126
Glycinate (Gly), 120–121
Graphene oxide (GO), 440–442
Graphite, 480
    electrodes, 478–479
Green process, 499
Greenhouse gas (GHG), 35,
    55–56
    emissions, 395
    regulations for, 39–40, 39t

Hagen–Poiseuille technique, 490
Heat exchangers, 199
Heat of vaporization, 83
Heat recovery steam generator
    (HRSG), 301
Heavy hydrocarbons, 280
    adsorption of, 283–284
Heavy mercaptans, 280
Helium (He), 135, 359–360
Henry's law, 460
    mode sorption, 424
Hexafluorophosphate (PF_6),
    120–121
Hexane (C_6H_{14}), 359–360
High-performance polyamines,
    106–107
High-temperature process gas
    combination, 473
Histidinate, 229
Hollow fiber, 427, 438–439
Hong Kong University of Science
    and Technology
    (HKUST-1), 362
Hybrid oxide ion, 476–477
Hybrid processes, 200
Hybrid solvents, 235, 237–238,
    336
Hybrid sorbents
    mechanism of solute take-up
        by, 126
    for natural gas (NG)
        sweetening, 116–126
Hybrid system, 65–66
Hydrocarbon dewpoint
    adjustment (HCDPA),
    232–233
Hydrocarbons, 277–278,
    283–284, 404, 407, 453,
    471
    separations, 437–438
Hydrogen, 475
    gas, 474
    recovery, 493–494
Hydrogen sulfide (H_2S), 90, 153,
    185–186, 331, 359–360,
    367, 375–376, 453, 471,
    473–474
    adsorption, 374

Hydrogen sulfide (H_2S)
(*Continued*)
  emission, 36–37
  physical and chemical properties of, 84
  removal
    adsorptive, 368–369
    electrochemical membrane separator, 473–475
    significance of eliminating, 336–337
Hydroxyethyl cellulose (HEC), 481

Ideal membrane, 428–430
Ifpexol process, 122
In situ polymerization strategies, 160–165
Industrial Emissions Directive (IED), 35
Industrial molecular sieve (IMS), 378
Infrared spectroscopy (IR), 407–409
Inorganic materials, 427–428
Inorganic membranes, 10, 488
Inorganic supports, 457–458
Integrated gasification combined cycle (IGCC), 23–24, 336–337
Intercoolers, cryogenic separation based on multicompression stages with, 193–195
Interfacial polymerization strategies, 160–165
Intergovernmental Panel on Climate Change (IPCC), 186, 395
International Energy Agency, 76, 186
Ionic liquid membrane (ILM), 454–465. *See also* Electrochemical membranes
  ionic liquid mixed-matrix membranes (ILMMMs), 462–463
  membrane contactors using ionic liquids (ILs), 463–465
  poly(ionic liquid) (PIL) membranes (PILMs), 462
  quasi-solidified ionic liquid membranes (QSILMs), 460–462
  supported ionic liquid membrane (SILM), 456–460
Ionic liquid mixed-matrix membranes (ILMMMs), 462–463
Ionic liquids (ILs), 15, 34, 116, 119–122, 141–142, 153, 235, 237, 454
  membrane contactors using, 463–465
Ionization energy. *See* Ionization potential
Ionization potential, 83
Iron chloride, 436
Iron oxide (Fe_2O_3), 46
Iron sponge process, 46
Isoreticular metal–organic frameworks (IRMOFs), 362
Isotherm data, 402–403

Kelvin's equation, 8–9
Khurmala field, 103
Kinetic dispersion coefficient, 499
Kinetics of carbon dioxide (CO_2) absorption, 235
Knock-out drum, 213–214
Knudsen diffusion, 6, 420–422
  in narrow pores, 490–492
Knudsen number, 421–422

Lactate (Lac), 120–121
Langmuir–Blodgett (LB) films, 427
Langmuir mode sorption, 424
Langmuir model, 406–407
Largest cavity diameter (LCD), 369–370
Layer-by-layer (LbL), 427
Lean solution cooler, 214–215
LieMather model, 63–64
Light product pressurization (LPP), 298
Liquefaction process, 90–91
Liquefied natural gas (LNG), 191–192, 210, 262
Liquid amine solution, 93
Lithium zirconate, 17
Long-term stability quasi-solidified ionic liquid membranes (QSILMs), 461–462
Long-term stability supported ionic liquid membrane (SILM), 459
Low-temperature desulfurization process, 374–375

Manganese acetate, 348–349
Mass transfer restrictions, 295–296
Mean free path, 421–422
Membranes, 4–5, 46–47, 420, 426–427, 456–457, 476–477, 488–489, 493, 495–496
  based absorption, 209, 236–238
  carbon dioxide (CO_2) removal, 497–499
    capillary condensation, 492
    Hagen–Poiseuille technique, 490
    Knudsen diffusion in narrow pores, 490–492
    molecular sieving, 493
    processes for efficient, 489–493
  configurations, 455–456
  contactors using ionic liquids (ILs), 463–465
  current application and cases, 493–499
    air purification, 494–495
    biogas purification, 497
    hydrogen recovery, 493–494

Index 515

purification of natural gas from acid gases, 495−497
fundamentals of membrane gas separation for carbon dioxide (CO_2) removal, 488−489
gas separation (MGS), 471, 487, 499
gas sweetening process, 46−47
performance, 487
preparation, 481−482
selectivity, 496
separation, 4−10, 201, 210, 221, 223, 437
  capillary condensation, 8−9
  Knudsen diffusion, 6
  molecular sieving, 7, 7f
  solution−diffusion separation, 7−8
  surface diffusion, 8
  types of membranes, 9−10
technology, 5, 437, 471, 487−488
types of, 9−10
Mercaptans, 284
  adsorption of, 284−285
Mercury (Hg), 135, 331
Mesoporous silica materials (MSMs), 375
Mesoporous silica structures, 375−378
Metal−organic frameworks (MOFs), 235−236, 271, 274−275, 360−370, 401, 407−409, 475
Metal oxides, 370−375
Metals, 369−370
  based sorbents, 362−375
    metal oxides, 370−375
    metal−organic frameworks (MOFs), 362−370
    zeolite Na-Y interchanged by metals for acid gas removal, 346−347
Methane (CH_4), 135, 268−269, 359−360, 399−401, 471, 497
  enrichment, 433−435

thermodynamic principles of, 187−189
Methanol, 122−125, 221, 283
  based procedure, 335
Methyl ethyl ketone (MEK), 293
Methyl sulfate ($MeSO_4$), 120−121
Methyldiethanolamine (MDEA), 13−14, 34, 92, 97, 116, 141−142, 157, 166, 207−208, 335−336, 359−360, 375−376
Methyldiethylarsine (MeEt2As), 331
Microcapsules, 163
Microporous membranes, 464−465
Microporous metal-organic frameworks, 407−409
Microwave, 286−287
Missan Oil Company/Buzurgan Oil Field of Natural Gas Processing Plant, 104
Mixed amine process, 64−65
Mixed matrix membrane (MMM), 427−428, 462−463
Mixed oxide-ion and carbon-ion conductor (MOCC), 476−477
Mixed swing adsorption processes, 300−313
  temperature−pressure swing adsorption (TPSA), 300−304
  vacuum pressure swing adsorption (VPSA), 310−313
  vacuum temperature swing adsorption (VTSA), 307−310
  vacuum-electric swing adsorption (VESA), 304−307
Mixed tertiary amines, 104
Modified membrane composites, for gas sweetening, 118−119
Molecular sieving, 7, 7f, 16, 406−407, 420−421, 493

Molten carbonate fuel cell (MCFC), 479
Monoethanolamine (MEA), 13, 34, 92, 116−117, 141−142, 156, 207−208, 335, 359−360, 399
Mordenite, 338
Multiple Lewis base functionalized proticionic liquids (MLB-PILs), 121−122

N-(2-aminoethyl)-3-aminopropyltrimethoxysilane, 375−376
Nano-encapsulation processes, 158
Nanocomposites, 373−374
Nanofluids, 144−145
Nanoparticles (NPs), 235−236, 238
  absorbents modified for natural gas (NG) sweetening, 144−148
Nanosorbents
  natural gas (NG) constitution, 330−331
  purification techniques, 332−352
Narrow pores, Knudsen diffusion in, 490−492
National Emission Standards for Hazardous Air Pollutants (NESHAP), 35
National Iranian South Oilfields Company (NISOC), 102
Natural gas (NG), 3, 33−35, 67−68, 75−76, 89, 115−116, 135, 137−138, 153, 185, 261−262, 276, 283, 329, 359−360, 395−396, 453−454, 471
  carbon dioxide (CO_2)-rich sweetening via adsorption processes
    adsorbent material selection, 401−409

Natural gas (NG) (*Continued*)
  adsorption processes in natural gas sweetening, 396–401
  composition, 77–78
  constitution, 330–331
  elimination of acid gases, 333–336
  origins, 77
  purification, 495–497
    and processing, 210–212
    techniques, 332–352
  removing acid gases from, 78–79
  selection of suitable sweetening technique, 24
    CO_2 removal with no H_2S present, 26t
    H_2S removal with no CO_2 present, 26t
    natural gas purification technologies, 25t
    selective H_2S removal with CO_2 present, 27t
    simultaneous H_2S and CO_2 removal, 27t
  significance of eliminating H_2S gas, 336–337
  streams, 282
  sweetening, 4–24, 5f, 92, 139–148, 277–278, 401, 406–407, 419–420, 435, 454
    absorbents modified with nanoparticles for, 144–148
    absorption process for, 11–16, 140–142
    acid gas absorption in amines and alcohols, 116–118
    adsorption process for, 16–19, 142–143, 396–401
    air emissions, 36–42
    chemical looping, 22–24
    cryogenics, 19–22
    using encapsulated liquids, 166–174

  Engelhard titanosilicate zeolite for acid gas removal, 349–350
  environmental policies and regulations, 35–42
  glycerol/glycerine, 125–126
  ionic liquids, 119–122
  mechanism of solute take-up by physical and hybrid sorbents, 126
  membrane separation, 4–10
  methanol, 122–125
  modified membrane composites for gas sweetening, 118–119
  natural zeolites for acid gas removal, 338–343
  operational and design standards for sweetening processes, 44–48
  physical and hybrid sorbents for, 116–126
  propylene carbonate (PC), 125
  regulating history, 36
  regulation for centrifugal compressor, 42–43
  regulation for pneumatic controller, 43
  regulation for pneumatic pump, 43–44
  regulations for controlling emissions from equipment, 42–44
  techno-economic analysis of, 59–67
  tetramethylene sulfone/2,3,4,5-tetrahydrothiophene-1,1-dioxide ((CH_2)$_4SO_2$), 126
  use of carbonate and aqueous amino acid salts, 119
  use of NMP, 125
  use of polyethylene glycol methyl isopropyl ethers, 125
  13X, 4A, and 5A zeolites for acid gas removal, 343–345

  zeolite Na-Y interchanged by metals for acid gas removal, 346–347
  zeolite SP-115 changed by Mo, Cu, and Mn for acid gas removal, 348–349
  zeolite X interchanged by Na and Ca for acid gas removal, 345–346
  zeolite Y interchanged by Ce and Cu for acid gas removal, 346
  via zeolite-based adsorbents, 337–352
  zeolites for acid gas removal, 350–352
  ZSM-5 and zeolite-A for acid gas removal, 347–348
Natural gas combined cycle (NGCC), 18
Natural gasoline, 78
Natural separation, 77
Natural zeolites, 272
  for acid gas removal, 338–343
Naturally occurring radioactive material (NORM), 331
(N,N-dimethylaminopropyl) trimethoxysilane (DMAPS), 375–376
Net present value (NPV), 232–233
Net profit (NP), 62
New Source Performance Standards (NSPS), 36
N-hexane, 284
Nickel (Ni), 481
Nickel oxides, 481
Nickel sulfides, 481
Nitrogen (N_2), 68, 135, 279, 359–360, 471, 475, 494
  enhancement membranes, 495
Nitrogen oxide (NOx), 279
N-methyl diethanolamine, 157
N-methyl-2-pyrrolidone (NMP), 335–336
  use of, 125

N-methylaminopro pyltrimethoxysilane (MAPS), 375–376
Nonporous fillers, 440–442
Nonsolvent induced phase separation (NIPS), 426–427
Nonwater sorbents, 154–155
Normal temperature-pressure swing adsorption (NT-PSA), 300–301
North Gas Company (NGC), 103
n-(3-trimethoxysilylpropyl) diethylenetriamine (TRI), 375–376
Nuclear magnetic resonance (NMR), 407–409

Occupational Safety and Health Administration, 37
Ohm's law, 290
One-dimensional zeolitic materials, 401–402
Operating temperature, 297
Organic membrane, 489–490
Organic supports, 457–458
Oxidative desulphurization process, 62
Oxygen, 494
Oxygen-enriched air (OEA), 495

Packed bed high-pressure cryogenic hybrid network, 21
Packed cryogenic beds, 192
Pentane (C_5H_{12}), 359–360
"Pentuple-train" design, 303
Perturbed chain-statistical associating fluid theory (PC-SAFT), 122
Phase-changing ionic liquids (PCILs), 154–155
Phenolic ionic liquid (ILs), 119–120
Phillipsite, 338
Physical absorption, 221
 of carbon dioxide (CO_2), 11
 processes, 217–218
 solvents, 142

Physical and chemical properties, 79–84
Physical solvent processes, 47
Physical sorbents, 157
 mechanism of solute take-up by, 126
 for natural gas (NG) sweetening, 116–126
Piperazine (PZ), 13–14, 34, 97
Pitch fibers, 404, 407
Plasticization, 439
Plasticizing pressure, 433–435
Pneumatic controller, regulation for, 43
Pneumatic pump, regulation for, 43–44
Polar oxidic adsorbents, 280
Polyamide blocks (PA), 429–430
Polyallylamine (PA), 375–376
Polydimethylsiloxane (PDMS), 428–429, 437–438, 460
Polyelectrolyte multilayers (PEMs), 427
Polyether (PE), 429–430
Polyether sulfone (PES), 428–429, 459
Polyethylene glycol (PEG), 115–116, 337, 436–437
 dialkyl ethers, 125
 dimethyl ether, 335
 methyl isopropyl ethers, 125
Poly(ethylene glycol) diacrylate, 460
Polyethylene oxide (PEO), 428–429
Polyethylenimine (PEI), 375–376, 404–405
Polyimide (PI), 428–429
 based membrane, 433–435
Poly(ionic liquid) (PIL), 462
Poly(ionic liquid) membranes (PILMs), 462
Poly(vinylidene fluoride-co-hexafluoropropylene), 460
Poly(vinyl trimethylsilane) (PVTMS), 119
Polymer electrode membrane (PEM), 345–346

Polymer of intrinsic microporosity (PIM), 419–420
Polymeric fibers, 404, 407
Polymeric layers, 493–494
Polymeric materials, 420, 439
Polymeric membranes, 9–10, 420, 428–433, 439, 489, 493, 497
 challenges, 438–442
 current application, 430–438
 gas separation fundamentals, 420–428
 mechanism, 420–425
 subclass and fabrication method of, 425–428
 hydrocarbon separations, 437–438
 ideal membrane, 428–430
Polymers, 461, 463, 487–488
 blends, 440
 chain, 425
 concentration, 422
 membranes, 456–457
 penetrant interaction, 425
Polyphenylene oxide (PPO), 428–429
Polyvinylidene fluoride (PVDF), 428–429, 459
Porous membrane separates gases, 489
Potassium, 119
Potassium carbonate (K_2CO_3), 160–162
Potassium lysinate (LysK), 227, 229–230
Potassium prolinate (ProK), 227, 229
Pressure, effect of, 268–269
Pressure swing adsorption (PSA), 17–18, 263–276, 360–361, 397. *See also* Electric swing adsorption (ESA); Pressure swing adsorption (PSA); Temperature swing adsorption (TSA); Vacuum swing adsorption (VSA)

Pressure swing adsorption (PSA) (*Continued*)
  literature, 275–276
  parameters affect, 266–275
    adsorbents, 271–275
    cycle time, 266–267
    flow rate, 267–268
    pressure, 268–269
    purge/feed (P/F) ratio, 269–271
  procedure, 265
Pressurization technique, 458–459
Pretreated gas stream, 225
Pretreatment, 216
Primary alkanolamine solvents, 141–142
Primary amines, 92
Primary dehydration membrane, 495
Process flow diagram (PFD), 169
Process industries, 59
Profit before tax (PBT), 66–67
Propane (C_3H_8), 330, 359–360
Propionate (Pro), 120–121
Purge gas, 292
Purge/feed (P/F) ratio, effect of, 269–271
Purification parameters, 227
Purisol procedure, 335–336
Purisol process, 118, 125
Pyridinium, 15

Quasi-solidified ionic liquid membranes (QSILMs), 455–456, 460–462
  long-term stability, 461–462
  preparation methods of, 461

Raw natural gas, 33–34, 90, 261, 453, 474
Reaction byproducts, 235
Rectisol process, 47–48, 122, 335
Reflux cooler, 45–46
Regeneration column operating parameters, 226
Regeneration method, 263–264

Regulations for controlling emissions from equipment, 42–44
Residence time, 266–267
Residue, 5
Retentate, 5
Reverse osmosis method, 487–488
Reversibility of CO_2 absorption, 234
Rich flash drum, 214
Rinse flow rates, 296–297
Rinse time, 294–295
Robeson upper limit, 476

SAPO-43, 380
Sarcosinate, 229
Scanning electron microscopy (SEM), 401, 409
Seawater, 211
Secondary alkanolamine solvents, 141–142
Secondary amines, 92
Selexol, 213
  procedure, 335
  process, 115–116
  stripper, 215
Separation factor, 491
Shell–Paques process, 64–65, 169–170
Silica, 235–236
  based sorbents, 375–382
    mesoporous silica structures, 375–378
    zeolites, 378–382
  gels, 277–278, 280–283
  alumina, 284
  membrane, 10
Simulation model process, 14–15
Single centrifugal compressor, 42–43
Single-stage Rectisol simulation study, 122
Sodium aluminate, 347
Sodium carbonate dilutions (Na_2CO_3), 160–161
Sodium hydroxide (NaOH), 210

Sodium swapped hydrogen zeolites, 350–351
Solubility of CO_2 absorption, 234
Solute take-up by physical and hybrid sorbents, mechanism of, 126
Solution diffusion, 420–423, 489
  model, 420
  separation, 7–8, 8f
Solvents, 463–464
  novel methods and, 235–237
  regeneration, 226
  type, 226
Sorbents, 475
  encapsulation, 158, 160, 166
Spray-drying strategy, 159
Sterically hindered amines, 219–222
Stirling cooler–based cryogenic CO_2 collecting system, 20–21
Stripper, 45–46
Substantial energy supply, 282
Sulfate, 473–474
Sulfide, 473–474, 477–478
  ion, 474–475
Sulfide stress cracking (SSC), 436
Sulfinol process, 47, 167–168
Sulfinol-M process, 64–65
Sulfite, 473–474
Sulfolane, 167–168
Sulfur, 453–454, 473–474
Sulfur oxides (SOx), 79
Supported ionic liquid membrane (SILM), 455–460
  gas transport properties in, 460
  long-term stability, 459
  preparation methods of, 457–459
Surface-area-to-volume ratio, 22
Surface diffusion, 8, 9f
Sweet natural gas, 454
Sweetening, 454, 462
  operational and design standards for, 44–48
  processes, 68, 261

Index   519

Swing adsorption processes, 264–313
  electric swing adsorption (ESA), 286–294
  mixed swing adsorption processes, 300–313
  pressure swing adsorption (PSA), 264–276
  temperature swing adsorption (TSA), 276–286
  vacuum swing adsorption (VSA), 294–299
Synthetic zeolites, 272, 378, 401–402

Taurinate, 229
Techno-economic analysis of natural gas sweetening process, 59–67
Techno-financial analysis, 163
Temperature approach, 44
Temperature dependence of CO_2 absorption, 234
Temperature swing adsorption (TSA), 263–264, 276–286, 360–361, 397. See also Electric swing adsorption (ESA); Pressure swing adsorption (PSA); Vacuum swing adsorption (VSA)
  adsorbents in, 279–282
  adsorption of components using, 282–285
  formation of carbonyl sulfide (COS) during, 285–286
  literature, 286
  procedure, 276–279
Temperature-concentration swing (TCS), 309
Temperature–pressure swing adsorption (TPSA), 300–304
  literature, 304
  procedure, 300–304
    elevated temperature-pressure swing adsorption (ET-PSA) process, 301–304
    normal temperature-pressure swing adsorption (NT-PSA), 300–301
Tertiary amines, 92
Tertiary butyl mercaptan (TBM), 345–346
Tetrafluoroborate (BF_4), 120–121
2,3,4,5-Tetrahydrothiophene-1,1-dioxide (($CH_2)_4SO_2$), 126
Tetramethylene sulfone/2,3,4,5-tetrahydrothiophene-1,1-dioxide (($CH_2)_4SO_2$), 126
Tetramethyl guanidiniumphenolate (TMGH)(PhO), 121–122
Tetramethyl hexanediamine (TMHDA), 375–376
Thermally induced phase separation (TIPS), 426–427
Thermally rearranged (TR) polymer, 430
Thermodynamic principles of CO_2 and CH_4 separation, 187–189
Thermogravimetric analysis (TGA), 409
Thin-film composite (TFC), 427
Thin polymeric layer, 427
Thiosulfate, 473–474
Three-dimensional zeolitic materials, 401–402
Titanosilicate zeolites, 378
Total capital investment (TCI), 62
Total capital requirement (TCR), 58–59
Total operating cost (TOC), 60–61
Triamine-grafted pore expanded MCM-41 (TRI-PE-MCM-41), 375–376
Triethanolamine (TEA), 92, 220–221, 335, 365–366
3-(Triethoxysilyl)propylamine, 375–376
Triethylarsine (Et3As), 331
Triethylenetetramine (TETA), 163
Trifluoroacetate (TfA), 120–121
Trifluoromethanesulfonate (TfO), 120–121
Trihexyl (tetradecyl) phosphonium 2-cyanopyrrolide, 163–165
Trimethylarsine (Me3As), 331
Tris(pentafluoroethyl) trifluorophosphate (eFAP), 120–121
Tubular Exchanger Manufacturers Association (TEMA), 44–45
Twister technology, 231–232
Two dimensional zeolitic materials, 401–402
Two-stage membrane gas sweetening process, 63–64
Two-stage Rectisol simulation study, 122

UK Climate Change Act (2008), 39
Ultrathin, 427
Unit operations and parameters, 224–227
Unmixed combustion, 22

Vacuum-electric swing adsorption (VESA), 304–307
  literature, 307
  procedure, 304–307
Vacuum membrane distillation (VMD), 497
Vacuum pressure, 297
Vacuum pressure swing adsorption (VPSA), 192, 310–313
  literature, 312–313
  procedure, 310–312

Vacuum swing adsorption (VSA), 263–264, 294–299, 360–361, 398–401. *See also* Electric swing adsorption (ESA); Pressure swing adsorption (PSA); Temperature swing adsorption (TSA)
  literature, 298–299
  parameters affect, 294–298
    adsorption time, 296
    evacuation time, 295–296
    feed and rinse flow rates, 296–297
    feed concentration, 297
    feed pressure, 297–298
    operating temperature, 297
    rinse time, 294–295
  procedure, 294
Vacuum temperature swing adsorption (VTSA), 307–310
  literature, 309–310
  procedure, 307–309
Vapor density, 80
Vapor pressure, 81
Vapor recompression process, 64–65
Vaporizable organic molecules, 345

Variable operating cost (VOC), 58–59
Venting, regulations for, 40–42
Volatile organic compounds (VOCs), 36, 264–265, 433
Volatile solvent loss, 120–121
Volumetric surface (VS), 369–370

Water, 211, 479
  adsorption, 282–283
    enthalpy, 282
  molecules, 345
  scrubbing, 211
  washing, 335
Water dew-point adjustment (WDPA), 232–233
Water vapor (H_2O), 135
Weak electrolytes, 217–218
Wobbe Index (WI), 311–312
Working capacity (WC), 305

X-ray diffraction (XRD), 343–345, 401, 407–409
13X zeolites, 272
  for acid gas removal, 343–345

Zeolites, 272, 279–280, 282, 286, 338, 378–382, 401–403
  for acid gas removal, 350–352
    Na-Y interchanged by metals, 346–347
    zeolite-A, 347–348
  membrane, 10
  sorbents
    natural gas (NG) constitution, 330–331
    natural gas purification techniques, 332–352
    SP-115 changed by Mo, Cu, and Mn for acid gas removal, 348–349
    X interchanged by Na and Ca for acid gas removal, 345–346
    Y interchanged by Ce and Cu for acid gas removal, 346
Zeolite Socony Mobil-5 (ZSM-5), 378
Zeolitic imidazolate frameworks (ZIFs), 362
  ZIF-8, 440–442
Zinc, 409
ZSM-5 for acid gas removal, 347–348
Zwitterion mechanism, 93–95

Printed in the United States
by Baker & Taylor Publisher Services